W9-BYV-842

Withdrawn
University of Waterloo

Terrestrial Slugs

Terrestrial Slugs

Biology, ecology and control

A. South

formerly
Department of Biological Sciences
City of London Polytechnic

Withdrawn
University of Waterloo

CHAPMAN & HALL
London · New York · Tokyo · Melbourne · Madras

Published by Chapman & Hall, 2–6 Boundary Row, London SE1 8HN

Chapman & Hall, 2–6 Boundary Row, London SE1 8HN, UK

Chapman & Hall, 29 West 35th Street, New York NY10001, USA

Chapman & Hall Japan, Thomson Publishing Japan, Hirakawacho Nemoto Building, 7F, 1-7-11 Hirakawa-cho, Chiyoda-ku, Tokyo 102, Japan

Chapman & Hall Australia, Thomas Nelson Australia, 102 Dodds Street, South Melbourne, Victoria 3205, Australia

Chapman & Hall India, R. Seshadri, 32 Second Main Road, CIT East, Madras 600 035, India

First edition 1992

© 1992   Chapman & Hall

Typeset in 10/12pt Palatino by Colset Pte Ltd, Singapore
Printed in Great Britain by St Edmundsbury Press Ltd, Bury St Edmunds, Suffolk

ISBN 0 412 36810 2

Apart from any fair dealing for the purposes of research or private study, or criticism or review, as permitted under the UK Copyright Designs and Patents Act, 1988, this publication may not be reproduced, stored, or transmitted, in any form or by any means, without the prior permission in writing of the publishers, or in the case of reprographic reproduction only in accordance with the terms of the licences issued by the Copyright Licensing Agency in the UK, or in accordance with the terms of licences issued by the appropriate Reproduction Rights Organization outside the UK. Enquiries concerning reproduction outside the terms stated here should be sent to the publishers at the London address printed on this page.
   The publisher makes no representation, express or implied, with regard to the accuracy of the information contained in this book and cannot accept any legal responsibility or liability for any errors or omissions that may be made.

A catalogue record for this book is available from the British Library

Library of Congress Cataloging-in-Publication data available

Contents

Preface

In recent years slugs have become increasingly important, partly because several species are agricultural and horticultural pests and partly because they have proved to be useful experimental animals, particularly in the field of neurophysiology. Most of the early works which included slugs were essentially taxonomic but the book by Taylor (1902–1907) contained a great deal of biological information about slugs, some of which is still relevant today. The publication of the book by Runham and Hunter (1970) represented a milestone in slug research, providing a comprehensive survey of current knowledge about slugs. The book by Godan (1983) on snails and slugs was mainly concerned with the economic importance of these animals.

The purpose of the present book is to present a review of current knowledge of the biology and ecology of slugs, together with their status and control as pests. Although relatively little is known about the biology and ecology of tropical slugs and most information is taken from work on European slugs, the European pest species have become widely distributed throughout temperate regions and this book should be of interest world wide. It is written as a source of information for people seeking to control slug pests and, also, for those wishing to use slugs for research or teaching purposes. The book is intended particularly to provide a starting point for those beginning research on slugs and an extensive bibliography has been provided.

The classification of slugs has been dealt with elsewhere and research of a purely systematic nature is not covered by this book unless it is of wider interest, for example the use of new techniques in taxonomy. Slug nomenclature in Britain has changed rapidly over the past few decades owing to the separation of closely related species, to the rediscovery of species that had been overlooked, or to the discovery of recently introduced slugs. I have adopted several conventions throughout this book. Slugs of the *Arion ater/Arion rufus* complex are referred to throughout as *Arion ater* following current usage in books such as Kerney and Cameron (1979) and Kerney *et al.* (1983). *Deroceras (Agriolimax) reticulatum* and *Deroceras (Agriolimax) agreste* were regarded as synonymous for many years in Britain and I have used *D. reticulatum* in place of *D. agreste* where the two species have obviously been confused (e.g. in Carrick, 1938). Where there is doubt I have retained the original nomenclature and I have not attempted to change these names where information is taken from non-British sources. Since individual species of the *Arion circumscriptus/Arion fasciatus agg.* and the *Arion hortensis agg.* have been recorded separately only recently by most authors, the names *A. fasciatus*, *A. circumscriptus* and *A. hortensis* refer to the species aggregate unless indicated by the use of the term *sensu stricto (s.s.)*. Slugs found in the Republic of Ireland are included with those found in the British Isles for the purposes of this book.

I should like to express my gratitude to Maureen South who has prepared the diagrams. Some diagrams are based on original material (slides, dissections, etc.) while others have been compiled from information drawn from numerous sources. Figure 3.3 is reproduced by permission of Harper Collins Publishers from the book *Terrestrial Slugs* by N.W. Runham and P.J. Hunter (1970). Figures 5.3–5.6 are reproduced by permission of the Royal Society of Edinburgh. Figures 8.1–8.4 are reproduced by permission of the Malacological Society of London. Figures 10.1 and 10.2 are reproduced by courtesy of the British Crop Protection Council.

I should also like to acknowledge the help and advice I have received over many years from colleagues in the former Department of Biological Sciences at the City of London Polytechnic and from members of the library staff at that Polytechnic. Dr Peter Mordan of the British Museum (Natural History) has also provided valuable advice on a number of occasions. My thanks are also due to Robert South who has provided assistance with the photography.

Finally, it would have been difficult to complete this book without the continued help and support of my wife, who provided much help in the preparation of the final manuscript.

1

Introduction

1.1 CLASSIFICATION OF SLUGS

The phylum Mollusca is probably the third most successful animal group after the arthropods and the vertebrates. Slugs and snails belong to the class Gastropoda, generally characterized by having a distinct head with tentacles and eyes, a broad flattened foot and a dorsal visceral mass at least partly covered by the mantle. The mantle extends downwards to enclose a mantle cavity, which contains gills in aquatic gastropods. The visceral mass exhibits torsion to a varying degree and is often coiled and typically covered by a calcareous shell.

Terrestrial snails, with a few exceptions, are pulmonates. These are gastropods in which the mantle cavity has been converted into an air-breathing organ which opens externally at the pneumostome. Several groups of pulmonate families show an evolutionary series, with a progressive reduction of the shell leading to the 'slug' form. For example, the arionid slugs have been derived from the shelled endodontid snails and the limacid slugs from the thin-shelled zonitid snails. Solem (1978) discussed the development of the slug form in relation to the evolution of land snails while Tillier (1984, 1989) gave detailed descriptions of 'limacization' in several families of snails. The shell, in the slug form, is usually very reduced and covered by the mantle, or is even absent. The organs of the visceral mass, which were originally coiled inside the shell, become incorporated into the cavity of the head and foot leaving a reduced mantle covering part of the dorsal surface of an otherwise naked body and enclosing the pallial complex. Thus land slugs are not a natural group of closely related gastropods but, rather, a polyphyletic group showing convergent evolution. Similar evolutionary trends are found in aquatic gastropod groups, particularly in the opisthobranchs. Terrestrial slugs have proved to be one of the most successful of all groups of molluscs. The reduction of the shell has lessened the need for calcium salts with the result that slugs can live in a wider range of habitats than most snails. Slugs are tolerant of water loss and the compressible and worm-like body allows the animal to squeeze through crevices in the soil in search of shelter.

The pulmonate families, which include slugs, are listed in Table 1.1, which is based on the classification of Vaught (1989). Because terrestrial slugs are a heterogeneous group and difficult to define, some arbitrary boundaries have

Table 1.1 Classification of pulmonate slugs (based on Vaught, 1989).

Class GASTROPODA
  Subclass Gymnomorpha
    Order SOLEOLIFERA
        Family Rathouisiidae
        Family Veronicellidae (= Vaginulidae)
  Subclass Pulmonata
    Order STYLOMMATOPHORA
      Suborder Orthurethra (no slugs)
      Suborder Mesurethra (no slugs)
      Suborder Elasmognatha
        Family Athoracophoridae
      Suborder Sigmurethra
        Family Testacellidae
        Family Urocyclidae
        Family Parmacellidae
        Family Milacidae
        Family Limacidae
        Family Boettgerillidae
        Family Trigonochlamydidae
        Family Arionidae
        Family Philomycidae

been drawn. Species completely lacking a shell or with a completely internal shell are included as slugs. Those genera with a reduced external shell have, with one exception, been excluded from this book even if the body cannot be fully retracted into the shell. Examples of such genera include *Semilimax* and *Daudebardia* and the shell of these snails still has recognizable whorls. The one exception is the genus *Testacella*, where the external shell is very reduced and has no clear whorls, since this mollusc has been traditionally regarded as a slug (e.g. Quick, 1960). Several groups have been excluded from this book because the literature is mainly concerned with their systematics. The succineid slugs have been excluded for this reason although the family includes at least one slug form, *Hyalimax*, while in others, e.g. *Omalonyx* *s.s.*, the reduced shell is not completely covered by the mantle folds. The evolution of the slug form in the Succineidae has been described by Tillier (1981, 1984). The large superfamily Helixarionoidea includes many species which have reduced shells, described as semislugs by Tillier (1984), who discussed the progressive limacization of this family. Tillier considered that a semislug evolved into a slug when the stomach sank into the body cavity, lower than the posterior edge of the foot cavity. The stages in this transition were illustrated very clearly by van Mol (1970). Since most literature on slugs of this family is concerned with systematics and comparative anatomy, these are considered to fall outside of the scope of this book. However, one helixarionoid family, the Urocyclidae, is included as this contains a number of slug forms. The taxonomy of slugs has been subject to frequent changes, particularly with the Limacidae. For example, in eastern Europe and the USSR, out of 125 species of slug belonging to 23 genera listed in the monograph by Likharev and Rammel'meier (1952), 68 species and three genera were regarded by Likharev and Wiktor (1980) as synonyms.

1.2 SURVEY OF SLUG FAMILIES

1.2.1 Rathouisiidae and Veronicellidae

These slugs have no shell and the mantle is enlarged and covers the entire dorsal surface of the animal (Fig. 1.1(a)). The curved dorsal surface of the mantle (the notum) is separated from the ventral side of the mantle (the hyponotum) by a ridge. The foot is separated from the hyponotum by a groove and is transversely ridged. The head has two pairs of tentacles, the upper pair with eyes are contractile but cannot be inverted as in most slugs while the lower pair are bilobed. In the Rathouisiidae, the anus, excretory duct and female ducts open on the right side of the body while the male duct opens on the right side of the head. In most other pulmonate slugs there is a common genital opening. In the Veronicellidae, the anus and the respiratory aperture (pneumostome) open behind the foot and below the hyponotum. Rathousid slugs have no jaw and feed on both animal and plant material, including snails. Little is known about their biology. The two genera *Rathouisia* and *Atopos* are found in south-east Asia, including China, Australia, the Philippines and New Guinea.

The jaw of veronicellid slugs is composed of a number of small parallel plates (polyplacognathous condition). The Veronicellidae are herbivorous slugs living in tropical and subtropical America, Asia and Africa. A number of veronicellid slugs are pest species and Godan (1983) lists several species of *Vaginulus* and *Veronicella* as pests of a wide range of crops, particularly in Central America and the West Indies. Pest species of other veronicellid genera in Central America, including an introduced species *Sarasinula plebeia* (Fischer), are also intermediate hosts of *Angiostrongylus costaricensis* (Morera and Cespedes), the cause of abdominal angiostrongyliasis in humans (Andrews, 1989). Dundee (1977) described how three species of *Veronicella*, together with *Vaginulus occidentalis* (Guilding) had entered the USA in the previous decade, become established and were extending their distribution through the movement of plant materials. Similar successful introductions of *S. plebeia* and *L. alte* into Queensland and the Northern Territory were reported by Smith and Dartnall (1976). Verdcourt (1981) described how *Veronicella sloanii* (Cuvier) had been accidentally imported into Britain in a cargo of bananas from Jamaica. Although veronicellid

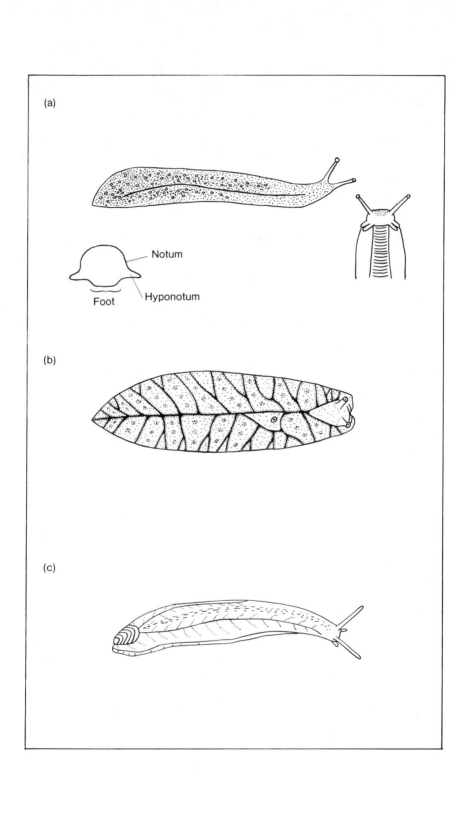

(a)

Notum

Hyponotum

Foot

(b)

(c)

biology is not well known, the slug *Laevicaulis alte* (Férussac) has been the subject of some research in India (e.g. Nagabhushanam and Kulkarni, 1971a, b) and reproduction in *V. sloanii* (Cuvier) and *Vaginulus borellianus* (Colosi) was investigated by Lanza and Quattrini (1964) and Quattrini and Lanza (1965, 1968).

1.2.2 Athoracophoridae

The athoracophorid slugs are flat and leaf-like (Fig. 1.1(b)) and unlike any other slugs. The mantle is reduced to a very small triangular area containing a rudimentary shell of calcareous granules. The head has a single pair of invertible tentacles with eyes. The slug has a pattern of dorsal grooves which are kept moist by a secretion and a tracheate type of respiratory system, which opens in the mantle area and is unlike that of any other mollusc. The jaw has an accessory plate which extends backwards along the roof of the mouth (elasmognathous condition). Unlike the previous families, the male and female ducts have a common opening on the right side of the head in common with other slugs. Athoracophorid slugs are found in Eastern Australia, the East Indies from the Bismarck Archipelago to New Caledonia and in New Zealand, where they are the only indigenous slugs (Martin, 1978). They are common in both bush and grassland and feed on fungi. There are several genera including *Aneitea* and *Athoracophorus*.

1.2.3 Testacellidae

These are carnivorous slugs with a small ear-shaped shell at the rear end of the body which covers the mantle and the associated organs (Fig. 1.1(c)). The jaw is lacking but the teeth on the rod-like radula are sharp and backwardly directed to impale the prey, usually earthworms. A prominent pair of lateral grooves runs along the side of the body between the front of the mantle and the back of the head. There is no caudal mucous gland. This is a small family with a single genus, *Testacella*, and three species in Britain and western Europe extending down to North Africa. *Testacella* has been introduced into North America where it has been recorded from greenhouses (Chichester and Getz, 1973) and into New Zealand (Barker, 1979).

1.2.4 Urocyclidae

The Urocyclidae are a group of slugs which retain a hump-backed appearance although the gut has moved into the cavity of the head and foot.

Fig. 1.1 (a) Veronicellid slug. The figure includes a transverse section of the slug to show the relationship of the foot to the hyponotum. (b) Athoracophorid slug. (c) Testacellid slug.

These slugs have a well-developed caudal pore which gives the tail a hooked appearance. Urocyclid slugs are found throughout the Ethiopian region, including Madagascar, and include the genera *Urocyclus*, *Trichotoxon* and *Atoxon*.

1.2.5 Parmacellidae, Milacidae, Limacidae and Boettgerillidae

These families have a number of features in common and Quick (1960) combined the first three families into a single family, the Limacidae. However, Wiktor and Likharev (1979) compared a number of external and internal characters for slugs in the last three families and concluded that they were distinct and that a fourth family, the Agriolimacidae, should be separated from the Limacidae, consisting of slugs belonging to the genus *Deroceras*. This division has been used by some recent authors (e.g. Likharev and Wiktor, 1980; Kerney *et al.*, 1983). However, the division used here (Table 1.1) follows the more recent classification used by Vaught (1989). Likharev and Wiktor (1980) emphasized the close relationship between the Parmacellidae and Milacidae by grouping them in the superfamily Zonitoidea and placing the Agriolimacidae, Boettgerillidae and Limacidae in the superfamily Limacoidea. In these families, the mantle covers only the anterior part of the back and is extended forwards as a loose lobe under which the head can be retracted (Fig. 1.2(a), (b) and (c)). The small calcareous shell is almost always enclosed by the mantle and the range of limacid shells was illustrated by Quick (1960) and by Wiktor and Likharev (1979). The pneumostome is located behind the midpoint of the right mantle margin. A dorsal keel is present and, unlike the Arionidae, there is no caudal mucous gland. The jaw is smooth and has a prominent beak (oxygnathous condition).

(a) Parmacellidae and Milacidae

The shell is either flattened, with a non-spiral nucleus in the midline near the posterior margin, and is completely enclosed by the mantle (*Milax* and *Tandonia*) or consists of a slightly curved plate with a very small spiral nucleus at one end and may not be completely enclosed by the mantle (*Parmacella*). The shell of *Parmacella* is illustrated in Germain (1930). The young animal at hatching is entirely enclosed in the small spiral shell which soon becomes too small to contain the slug. A calcareous plate is then secreted similar to the shell of other limacids and this becomes fused at the posterior border with the embryonic spiral shell (Germain, 1930). The mantle bears a horseshoe-shaped groove. There is a prominent keel in *Milax*, *Tandonia* and *Parmacella* running from the posterior end up to the edge of the mantle (Fig. 1.2(b)).

The genus *Parmacella* has a mainly Mediterranean distribution extending

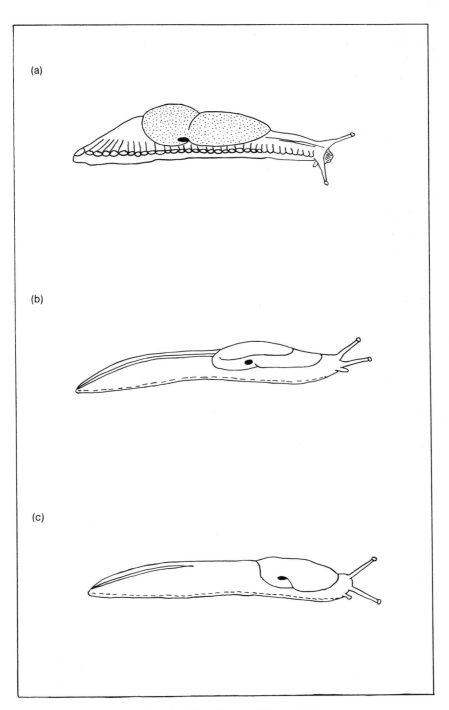

Fig. 1.2 (a) Parmacellid slug. (b) Milacid slug. (c) Limacid slug.

from the Iberian peninsula and the Canary Islands through to the southern USSR and Afghanistan, although it is missing from some countries, e.g. Italy (see Germain, 1930; Likharev and Rammel'meier, 1952). Two milacid genera, *Milax* and *Tandonia*, can be distinguished on morphological grounds (Norris, 1987). About 30 species of *Milax* and *Tandonia* have been described although some of these are probably synonyms. They occur mainly in the Mediterranean region through to the Balkan peninsula and in western Europe (Wiktor and Likharev, 1979) although they have also been introduced into other parts of the world, e.g. New Zealand, Australia and North America (Barker, 1979).

(b) Limacidae

The shell consists of a flattened plate with a terminal nucleus which lies to the left of the midline. The mantle is concentrically ridged and lacks a horseshoe-shaped groove. The keel is well developed but does not extend forwards to the edge of the mantle (Fig. 1.2(c)). This subfamily includes two major genera, *Limax* and *Deroceras* (formerly *Agriolimax*). Another genus, *Lytopelte*, is widespread in eastern Europe (Grossu and Lupu, 1961; Grossu, 1970). A number of authors, e.g. Quick (1960), have separated a genus *Lehmannia* from *Limax* on morphological grounds. Grossu and Tesio (1975) concluded that this division was not justified on biochemical grounds but recent authors have retained the use of *Lehmannia* (e.g. Kerney *et al.*, 1983; Vaught, 1989). The *Limax* species are medium to large slugs in which the keel slopes gently down to the tail. The centre of the pattern of concentric rings is in the midline of the mantle. The genus *Limax* has a western Palaearctic distribution, including the European USSR.

Species of *Limax* have been introduced into a number of other countries including North America, Australia and New Zealand. The genus *Limax* contains many more species than the previous genera, for example Likharev and Rammel'meier (1952) listed 23 species in addition to the European species, most of these coming from the Caucasus region. Thus the total number of species of *Limax* described may exceed 50, although it seems likely that some of those described will be shown to be synonymous with existing species.

The name *Deroceras* (Rafinesque-Schmaltz, 1820) has been used by most American and some European authors. However, British authors have generally used the name *Agriolimax* (Mörch, 1865) for this genus up until the publication of Waldén (1976). The *Deroceras* species are smaller slugs (up to 5 or 6 cm long) and the keel is truncated at the tail. The centre of the pattern of concentric rings is located to the right of the midline of the mantle. Wiktor and Likharev (1979) described the distribution of this genus as Holarctic but with the majority of species being essentially Palaearctic ones. Some species endemic to the Palaearctic, such as *D. reticulatum*, have been introduced into other regions such as North America and Australasia. Kerney and Cameron (1979) list six species of *Deroceras* from north-west Europe but, like the genus *Limax*, the genus *Deroceras* includes many additional species. Likharev and Rammel'meier (1952) listed 17 additional species from the USSR. A number of other *Deroceras* species have been described, particularly from eastern Europe. For example, about 20 additional species of *Deroceras* were recorded from the Balkan peninsula, including Romania, by Grossu (1969, 1972) and Dochita (1972). The number of species of *Deroceras* are probably similar to those of *Limax* although, like *Limax*, many of these species are probably synonyms. Likharev and Rammel'meier (1952) also included a limacid genus *Gigantomilax*, which superficially resembled *Milax*, and was distributed in the Caucasus, central Asia, northern Iran and Albania.

(c) Boettgerillidae

These are small to medium-sized, narrow worm-like slugs and have a reduced shell compared to other limacid slugs. There are very few species of *Boettgerilla* and the genus was restricted to south-eastern Europe (the Caucasus) until recently when it spread rapidly into north-western Europe, probably through its association with man (Kerney and Cameron, 1979; Wiktor and Likharev, 1979).

1.2.6 Trigonochlamydidae

The shell is very reduced or absent. The mantle is small and situated towards the rear of the animal (Fig. 1.3(a)) and sometimes has a horseshoe-shaped groove. The pneumostome lies behind the midpoint of the right mantle margin. The skin is thick and there is a keel along the back from the posterior margin of the mantle. The jaw is reduced while the radula is well developed with rows of claw-like teeth. In some features, this family resembles the Testacellidae although the shell is internal and the trigonochlamydid slugs also have a carnivorous diet. The Trigonochlamydidae are typically slugs of the eastern Mediterranean region and are found in the Caucasus region, extending down into Armenia and Iran (Likharev and Rammel'meier, 1952).

1.2.7 Arionidae

The mantle covers only the anterior part of the body with the pneumostome located in front of the middle of the right margin (Fig. 1.3(b)). The internal shell of arionid slugs is reduced to a flat calcareous plate (*Ariolimax*, *Geomalacus*) or to discrete calcareous granules which are sometimes absent (*Arion*). The jaw is crescent shaped and markedly ribbed (odontognathous).

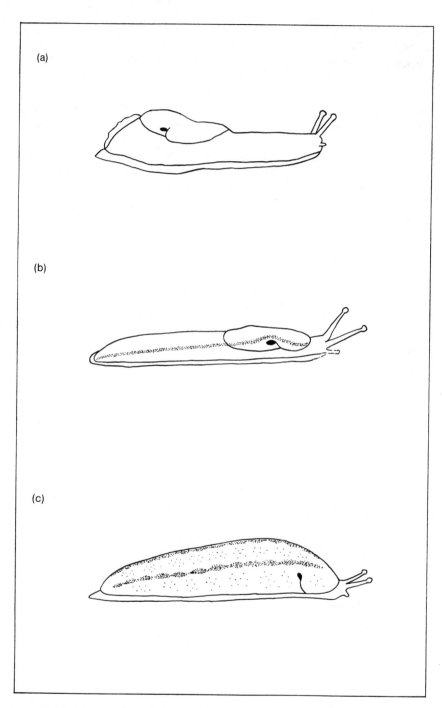

Fig. 1.3 (a) Trigonochlamydid slug. (b) Arionid slug. (c) Philomycid slug.

There are several subfamilies of arionid slugs which can be separated by anatomical features.

(a) Subfamily Ariolimacinae

The buccal and tentacular retractor muscles are inserted together at the posterior border of the floor of the mantle cavity. The penis and epiphallus are well developed and the second posterior loop of the intestine extends behind and beyond the first. According to Harper (1988), *Ariolimax* has a keel along the back. The caudal pore or tailpit, which has a well-developed mucus plug, is a characteristic feature of these slugs. The Ariolimacinae is restricted to western North America and includes some of the largest slugs. The genus *Ariolimax* is indigenous to the coastal forests of the west and the range of the banana slug *Ariolimax columbianus* (Gould) extends from Sitka, Alaska down to California although other species of *Ariolimax* are restricted to California. The large size of these slugs (up to 25 cm when extended) means that they are useful experimental animals and a great deal of research has been carried out using these animals. The biology of *A. columbianus* has been described in the book by Harper (1988). Banana slugs, so called because of their bright yellow coloration, are not generally regarded as pests although Godan (1983) included a reference to damage to a banana crop. However, this seems unlikely owing to their geographical distribution. Two other genera of smaller slugs, *Hesperarion* and *Zacoleus*, are included in this subfamily. These are also found in the western USA. Roth and Pressley (1983) recorded how *Hesperarion niger* (Cooper) was extending its range northwards across California.

(b) Subfamily Anadeninae

The buccal and tentacular retractor muscles are separated at their origin where they are inserted along the posterior border of the floor of the mantle cavity. The penis is either well developed and functional or much reduced while the second posterior loop of the intestine extends behind the first. The genus *Prophysaon* is widely distributed in the USA although it does not apparently occur in the north-eastern states (Chichester and Getz, 1968, 1973). A species of *Prophysaon* was recorded from the Aleutian Islands (Alaska) by Roth and Lindberg (1981). Roth and Pressley (1983) described how *Prophysaon dubium* Cockerell had extended its range south from Oregon to California. *Prophysaon andersoni* (Cooper) was listed as a garden pest in California by Godan (1983) and has been found in citrus fruits imported into Canada from California. This species was also considered to be endemic to Vancouver by Rollo and Wellington (1975). The genus *Anadenulus* contains a single species, *Anadenulus cockerelli* (Hemphill) from southern California, while a third genus, *Anadenus*, comprises several large species from the Himalayas and China, for example *Anadenus banerjeei* described by Rajagopal (1973).

(c) Subfamily Arioninae

The buccal and tentacular retractor muscles are widely separated at their origin from the posterior border of the mantle cavity floor. The first posterior loop of the intestine extends behind the second and there is no penis. The Arioninae have a caudal mucus gland just above the tip of the blunt tail, while the pneumostome is located in front of the midpoint of the mantle. Arionid slugs are bell-shaped or almost circular in cross section. There is generally no keel, although young slugs of the *A. fasciatus agg.* have a prominent row of tubercules running along the back, giving the appearance of a keel. The larger arionid species have a prominent foot fringe.

The Arioninae contains four genera, *Arion, Geomalacus, Letourneuxia* and *Ariunculus* (Platts and Speight, 1988). With the exception of *Arion*, these genera are very small and have a very restricted distribution. The most widespread species, *Geomalacus maculosus* Allman, occurs only in the south-west of Ireland and the Iberian peninsula (Platts and Speight, 1988). In contrast, the genus *Arion* comprises about 15 species most of which are distributed throughout the western Palaearctic region and have been introduced into America, Australia, New Zealand and South Africa by human activities. Chichester and Getz (1968), for example, have described how seven European species of *Arion* have become established and have spread in north-eastern North America. Slugs and snails, including *Arion hortensis* Férussac and *Arion ater* (Linnaeus) have also been introduced on to oceanic islands, such as St Helena (Crowley, 1978).

1.2.8 Philomycidae

The mantle of philomycid slugs covers the whole of the rounded dorsal surface except in *Pallifera* where the anterior edge of the mantle is free and the head is exposed (Fig. 1.3(c)). There is a shell sac but no shell. The buccal and tentacle retractor muscles have separate origins and are inserted towards the ventral side of the lateral body walls and not dorsally as in the Arionidae. The jaw is ribbed or striated. The reproductive opening is located on the right side of the head and the pneumostome is displaced anteriorly where it opens on the right side of the mantle near the head. The philomycid slugs are found in eastern Asia and North and Central America. Genera include *Philomycus*, which occurs in eastern North America from Canada to Florida and west across to Texas, and *Pallifera*, also occurring in eastern and midwestern North America except for one species in Arizona. *Philomycus* differs from other genera in having a vaginal, calcareous stimulating organ,

which is analogous with the 'dart' of some snails. *Philomycus carolinianus*
(Bosc) is a large (7–10 cm long), relatively inactive slug which, like the
smaller (2.5 cm) *Pallifera dorsalis* (Binney), is basically a woodland slug and
is fungivorous. The mantle of *P. carolinianus* is tan coloured with black or
brown blotches (Chichester and Getz, 1973). However, Chichester and Getz
considered that the more abundant *P. dorsalis* was more likely to be found in
ecotonal and semicultivated areas than *Philomycus*. While the philomycid
slugs are not generally agricultural pests, Godan (1983) recorded the Central
American species *Pallifera costaricensis* (Morch) as a pest on bromeliads and
orchids. The eastern Asia species belong mainly to the genera *Meghimatium*
and *Incillaria*. Likharev and Rammel'meier (1952) recorded a small species,
Meghimatium bilineatus (Benson), which was yellow or brown with three
longitudinal black stripes and which occurred in the extreme east of the
USSR as well as in China, Korea and Japan.

1.3 HISTORY OF THE BRITISH SLUG FAUNA

Slugs have a long history in the British Isles. So-called arionid shell granules
are frequently found in Pleistocene and Post-glacial deposits (Evans, 1972).
Limacid shells are also common but generally belong to the smaller
Deroceras spp. and Evans comments on the way in which a steady increase
of *D. reticulatum* shells in Post-glacial deposits reflects the beneficial
influence of agriculture on this species. *Milax* shells occur as Pleistocene
fossils but are virtually unknown from Post-glacial deposits. Kerney (1966),
in his review of the influence of man on slugs and snails during the Post-
glacial period, suggests that the present widespread distribution of milacid
slugs has been attained through the agency of man. Shells of *Parmacella* have
also been recorded from Pleistocene deposits although this genus has a more
southern distribution at the present day (Cambridge, 1981).

   In the British Isles, slugs are represented at the present time by the families
Testacellidae, Milacidae, Limacidae, Boettgerillidae and Arionidae. South
(1974) reviewed the changes that have taken place in the terrestrial mollusc
fauna over the past century using data from the censuses of non-marine
Mollusca published by the Conchological Society between 1885 and 1951
and from earlier sources. The history of recording in the British Isles was
reviewed by Kerney (1976) when the first *Atlas of the Non-marine Mollusca
of the British Isles* was published. This atlas was based on 10-kilometre
square records for terrestrial and freshwater snails and slugs. The number of
slug species recorded in Britain increased from 12 pre-1840 up to 23 in the
1951 census. It seems likely that some of the additional species were already
established before 1840, since Gray (1840) in the first critical checklist
included a number of species as either synonyms or varieties of existing
species. In addition, several species, such as the very local species *Limax*

Table 1.2 List of British slugs.

Arionidae
1. *Geomalacus maculosus* Allman, 1843
2. *Arion ater* (Linnaeus, 1758)
3. *Arion lusitanicus* Mabille, 1868
4. *Arion flagellus* Collinge
5. *Arion subfuscus* (Draparnaud, 1805)
6. *Arion circumscriptus* Johnston 1828
7. *Arion silvaticus* Lohmander 1937
8. *Arion fasciatus* (Nilsson, 1823)
9. *Arion hortensis* Férussac, 1819
10. *Arion distinctus* Mabille, 1868
11. *Arion owenii* Davies, 1979
12. *Arion intermedius* Normand, 1852

Milacidae
13. *Milax gagates* (Draparnaud, 1801)
14. *Milax nigricans* (Philippi, 1836)
15. *Tandonia sowerbyi* (Férussac, 1823)
16. *Tandonia budapestensis* (Hazay, 1881)
17. *Tandonia rustica* (Millet, 1843)

Boettgerillidae
18. *Boettgerilla pallens* Simroth, 1912

Limacidae
19. *Limax maximus* Linnaeus, 1758
20. *Limax cinereoniger* Wolf, 1803
21. *Limax flavus* Linnaeus, 1758
22. *Limax maculatus* (Kaleniczenko, 1851)
23. *Limax tenellus* Müller, 1774
24. **Lehmannia nyctelia* (Bourguignat, 1861)
25. *Lehmannia marginata* (Müller, 1774)
26. **Lehmannia valentiana* (Férussac, 1821)
27. *Deroceras laeve* (Müller, 1774)
28. *Deroceras agreste* (Linnaeus, 1758)
29. *Deroceras reticulatum* (Müller, 1774)
30. *Deroceras caruanae* (Pollonera, 1891)

Testacellidae
31. *Testacella maugei* Férussac, 1819
32. *Testacella haliotidea* Draparnaud, 1801
33. *Testacella scutulum* Sowerby, 1821

---

* Alien species found mainly in greenhouses in Britain.

tenellus Müller, were probably simply overlooked. A nomenclatural list of British land Molluscs by Waldén (1976) included 30 species of slug. Examination of the current species list (Table 1.2) shows that 33 species of slug have been recorded from the British Isles. The main reason for recent changes is that several established species have now been demonstrated to be aggregates of closely related species, for example *L. flavus* (Evans, 1982, 1985) and *A. hortensis* (Davies, 1979). Two recent additions to the British list are *B. pallens* by Colville *et al.* (1974) and *T. rustica* by Philip (1987). The monograph on British slugs by Quick (1960) provided detailed information on the anatomy of many of the British species while the diagnostic features of most British slugs were described and illustrated by Kerney and Cameron (1979). A recent key to the slugs of the British Isles (Cameron *et al.*, 1983) takes into account most of the recent changes to the slug fauna.

1.4 SLUGS OF THE BRITISH ISLES

1.4.1 Testacellidae

The species of *Testacella* are all medium to large-sized slugs, which may reach 10–12 cm in length when fully extended. At rest and when partly extended, the body is pear-shaped with the widest part at the back and associated with the external shell. These slugs are largely synanthropic species in Britain, being found mainly in gardens and cultivated land. They spend the day underground and hunt for earthworms on the surface of the soil at night. The Testacellidae occur mainly around the western Mediterranean and the Atlantic coasts and islands (Kerney and Cameron, 1979). In Britain, *T. maugei* has the most restricted distribution being confined to the south-west of England and to southern Ireland. *Testacella haliotidea* and *T. scutulum* are found in England and Wales and most of Ireland (Kerney, 1976). Ellis (1969) considered that the *Testacella* spp. were probably only native to the south-west of England and had been introduced elsewhere by man. However, there is no fossil record of this genus in the British Isles and *Testacella* spp. may be an introduction of very recent origin (Evans, 1972). *Testacella haliotidea* is also distributed over much of France. Ellis (1969) reported that *T. maugei* had been introduced into South Africa and New Zealand. *Testacella haliotidea* has been introduced into several regions of North America, where it has been recorded from greenhouses, and into Cuba (Chichester and Getz, 1973). It was also recorded as a free living species in Pennsylvania (USA) by Branson (1976). *Testacella haliotidea* has also been recorded from New Zealand (Barker, 1979).

1.4.2 Milacidae

Two species of *Milax* and three species of *Tandonia* have been recorded from Britain although the status of two of these remains uncertain. *Milax nigricans* is a Mediterranean species for which there was a single record (as *M.* cf. *insularis*) from a Sussex garden in 1948 (Quick, 1960). *Tandonia rustica*, a central and southern European species, was recorded recently from a single locality in old woodland in Kent (Philip, 1987). Kerney (1987) concluded that, since woodland rather than cultivated ground was the normal habitat for this species in Europe, its presence in south-eastern England probably represented an extension of its natural range rather than a recent introduction, especially as this species is not synanthropic and is not usually found in cultivated places.

The remaining species, *M. gagates*, *T. sowerbyi* and *T. budapestensis*, were described by Kerney (1966) as synanthropic species, associated with man-made environments such as gardens and agricultural land. The first British record for *T. budapestensis* was as recent as 1930 (Phillips and Watson, 1930). It is difficult to establish whether *T. budapestensis* is an indigenous species of slug. It may be a recent introduction as there is little evidence to suggest that earlier records were confused with other *Milax* spp. although some of the examples of *M. gagates* illustrated by Taylor (1902-1907) resembled *T. budapestensis* more closely than *M. gagates*. There is little doubt, however, that *T. budapestensis* has spread largely through the agency of man over the past 50 years. Evans (1972) suggested that milacid slugs had spread into natural habitats only in the south and west of England while Quick (1960) described how *M. gagates* is more common in natural habitats near the sea in the south-west of England. *Tandonia sowerbyi* is a medium to large slug (6–7 cm) while the other two species are medium sized (5–6 cm). *Tandonia budapestensis* can be distinguished from other species because it assumes a C-shape by bending laterally when at rest, compared with the compressed helmet shape of other milacid slugs. The three species are widely distributed over the British Isles although they are less frequent in Scotland (Kerney, 1976) and have not been recorded as far north as the Faroe Islands (Solhøy, 1981). In southern Scotland, they occur mainly in urban and agricultural areas where they may be locally abundant (e.g. Dobson, 1963). There seems little doubt that *T. budapestensis* has spread largely through the agency of man, and Dobson (1963) showed that its distribution has expanded in the Glasgow area over the past 30 or 40 years. *Tandonia budapestensis* is more common than the other species and is an important slug pest. However, its distribution, even on arable ground, tends to be patchy (see Chapter 10).

In Europe, *T. sowerbyi* is distributed along the coastal regions of western and northern France (Kerney and Cameron, 1979) extending into Belgium (van den Bruel and Moens, 1958b) and the Netherlands (Bruijns *et al.*, 1959).

Its distribution extends as far south as southern Spain (Alonso, 1975). *Milax gagates* has a similar distribution although it is more widely distributed in France, being absent only from the south-east of that country. This species has also been introduced with ornamental plants into greenhouses in Finland (Valovirta, 1969). The European distribution of *T. budapestensis* is unlike that of the other two species. It occurs sporadically in Belgium, Germany and north Italy (Quick, 1960) and then eastwards to the west Balkans and Bulgaria (Osanova, 1970) although it does not appear to have been reported from the USSR (Likharev and Rammel'meier, 1952). An isolated record for this species from southern Spain (Alonso, 1975) suggests that this may be an introduction. *Milax gagates* has been introduced into a number of countries including North America (Chichester and Getz, 1973), from California (Roth, 1986) up to Nova Scotia (Fox, 1962), Australia (Altena and Smith, 1975) and New Zealand (Barker, 1982). The other two species have been introduced into New Zealand (Barker, 1982), where *T. budapestensis* is a relatively recent introduction, although these two species have not yet been recorded from Australia or North America.

1.4.3 Limacidae

Two out of the eight British species of limacid slugs have been found mainly in greenhouses. *Lehmannia valentiana*, a slug endemic to the Iberian peninsula, has been introduced into a number of other countries by man. In Britain it had been recorded only from greenhouses in a few localities until recently when it was recorded from several open sites (Kerney, 1987). This species is widely distributed in greenhouses in North America and in California it also occurs in natural habitats (Chichester and Getz, 1973). Howe and Findlay (1972) suggested that *L. valentiana* had been introduced into greenhouses in the Winnipeg area (Canada) with imported plant material. This slug has also been recorded from gardens in the Netherlands by Gittenberger and de Winter (1980) and from greenhouses in Sweden, where it has caused a great deal of damage (Waldén, 1960). *Lehmannia nyctelia*, which is endemic to North Africa and south-east Europe, has been recorded from greenhouses in Scotland. It has also been introduced into South Africa and North America (Quick, 1960), New Zealand, where it occurs in greenhouses and gardens (Barker, 1982) and Australia (Altena and Smith, 1975) where it is common in natural habitats.

   Limax maximus is a very large slug (10–20 cm extended) which is widely distributed throughout the British Isles. It is found in a wide range of habitats including woodlands, hedgerows and gardens and is essentially a ground-living species (Cook and Radford, 1988). *Limax maximus* is distributed throughout north-western Europe although it becomes scarcer and increasingly synanthropic in the north-east, being generally found only in greenhouses in Finland (Kerney and Cameron, 1979). However, Holyoak and Seddon (1983) recorded this species from an open site in Finland and Waldén (1960) regarded it as a naturalized species in southern Sweden. Von Proschwitz (1988b) concluded that *L. maximus* had been introduced into south-west Sweden for over half a century and had spread rapidly. It is also distributed throughout southern and central Europe, including the western USSR (Likharev and Rammel'meier, 1952), the west Balkans and Bulgaria (Osanova, 1970) and Czechoslovakia (Lisicky and Ponec, 1979). *Limax maximus* has been introduced into a number of other countries including North and South America (Chichester and Getz, 1973), South Africa (Quick, 1960), New Zealand (Barker and McGhie 1984) and Australia (Altena and Smith, 1975). In most of these countries it is mainly restricted to areas associated with man, such as gardens and parks.

The largest British slug, *L. cinereoniger*, has an extended length of between 10 and 20 cm although it may reach 30 cm on the continent. Both this species and the smaller *L. tenellus* (2.5–4 cm) have a very local distribution although they are generally distributed throughout Britain and central and northern Europe, including the USSR. However, they are absent from northern Scandinavia and Iceland and, in the case of *L. tenellus*, from Ireland. Boycott (1934) described these slugs as anthropophobic species and intolerant of human disturbance and they are largely restricted to ancient woodland, both deciduous and coniferous. *Limax cinereoniger* and *L. tenellus* do not appear to have been introduced into any other countries through human agencies and this probably reflects their scarcity.

Limax flavus is a medium to large slug (7.5–10 cm) which was described by Boycott (1934) as being 'domesticated to about the same degree as the house mouse or cricket'. It is generally distributed throughout England, Wales and Ireland although it is scarce in Scotland (Kerney, 1976). In Europe *L. flavus* extends northwards to Denmark and southern Scandinavia (Kerney and Cameron, 1979) and eastwards to the Crimea and the Caucasus and it also occurs in North Africa and the Near East (Likharev and Rammel'meier, 1952). It is synanthropic for most of its range although Likharev and Rammel'meier recorded how, in Armenia, *L. flavus* lived in the subalpine and steppe zones in moist screes and cracks in cliffs. *Limax flavus* has been introduced into a number of other countries including North and South America (Chichester and Getz, 1973), South Africa (Quick, 1960), New Zealand (Barker, 1982), Australia (Altena and Smith, 1975) and the island of St Helena (Gittenberger, 1980).

Chatfield (1976a) recorded *Limax grossui* Lupu from Ireland where it was considered to be a widespread species. This species had previously been reported from Ireland as a variant form of *L. flavus*. According to Chatfield, the Irish individuals of *L. grossui* closely resembled *L. flavus* but, unlike that species, was not associated with man and occurred in woodlands and in open habitats. Evans (1978) investigated the variation in *L. flavus*, including a variant form taken from Ireland and from Merseyside, and concluded that this variant form and the *L. grossui* described by Chatfield (1976a), were synonymous. Evans (1978) described this form as a new species, *Limax pseudoflavus*, and made a critical comparison of this species with *L. grossui* and *L. flavus*. *Limax flavus* and *L. pseudoflavus* could be distinguished from one another and separated by differences in the external body coloration (Evans, 1982) and this separation was supported by differences in the pattern of variation of non-specific esterases (Evans, 1985). Wiktor and Norris (1982) concluded that *L. pseudoflavus* Evans was a synonym of *Limax maculatus* (Kaleniczenko), a forest species native to eastern Europe. This species was able to adapt to environments altered by man and occurred in greenhouses and cellars. Wiktor and Norris (1982) suggested that *L. maculatus* might have been brought to Ireland and to other countries as a synanthrobe. Evans (1986) reviewed the status of *L. maculatus*, *L. grossui* and *L. pseudoflavus* and concluded that the latter two species should not be regarded as junior synonyms of *L. maculatus*. Kerney (1986) suggested that the present widespread distribution of *L. maculatus* in Ireland and the lack of earlier records indicated that this species was a recent introduction. *Limax maculatus* has now been recorded from several localities in England, including Kent (Kerney, 1987), suggesting that this species is more widespread and had previously been confused with *L. flavus*.

Lehmannia marginata is a medium-sized slug (7–8 cm) and differs from other British limacids in having a gelatinous and translucent appearance in wet weather. It is a fairly common slug in woodlands where in damp weather it readily ascends tree trunks. It is also found on damp rocks and old stone walls in areas of high rainfall. *Lehmannia marginata* is widely distributed in Britain where, according to Quick (1960), it is more common in the west and north. It is widespread in north-western Europe except for northern Scandinavia and coastal areas of Iceland, while Solhøy (1981) found it to be very common in the Faroe Islands. *Lehmannia marginata* has been recorded to the south in Spain (Alonso and Ibanez, 1984) and eastwards to the Balkans and Bulgaria (Osanova, 1970). This slug has been introduced into Australia and New Zealand (Quick, 1960) and the USA (Dundee *et al.*, 1968).

Deroceras reticulatum is a medium-sized slug (3.5–5 cm) and is one of the most abundant British slugs and a major pest species. For many years this species was confused with, and even regarded as a variety of, the rarer *D. agreste*. Taylor (1902–1907), for example, included *D. reticulatum* as a synonym for *D. agreste*, although the first certain British record for *D. agreste* was from a Norfolk fen in 1941 (Quick, 1960). As a consequence, many records for the distribution of these species in Britain are uncertain. *Deroceras agreste* has now been recorded from localities that are widely separated from one another and it seems likely that it is a native species rather than a recent introduction. *Deroceras reticulatum* is widely distributed throughout the British Isles while *D. agreste* is mainly a northern species, being locally common in central Scotland. In Europe, *D. reticulatum* has a more westerly distribution while *D. agreste* has a mainly northern and eastern distribution, although there is a considerable overlap of their ranges. Luther (1915) considered that *D. agreste* was native to Finland and that *D. reticulatum* had been introduced in historical times. The distribution of these two species is discussed in more detail in Section 10.3.3. *Deroceras reticulatum* is probably indigenous to the western Palaearctic region but has now become introduced into most parts of the world (Quick, 1960) and has frequently become a serious pest after its introduction. This species has even been accidentally introduced on to sub-Antarctic islands (Dell, 1964). The distribution of *D. agreste* is less clear but Likharev and Rammel'meier (1952) describe the distribution of this slug as northern and central Europe, including the entire European USSR.

Deroceras laeve is a small (1.5–2.5 cm), translucent slug, which crawls very rapidly. It has been described as a hygrophilic species and in Europe is usually found in marshy habitats including wet woodlands. *Deroceras laeve* is widely distributed in Britain and Europe, including the northern and central regions of the European USSR, and it extends southwards into Italy (Quick, 1960) and Spain (Alonso, 1975). This species has a Holarctic distribution and is a native of North America, where it has been described as an ecologically ubiquitous slug in the north-east (Chichester and Getz, 1973). Its distribution extends as far north as the Aleutian Islands off Alaska (Roth and Lindberg, 1981) and south into Texas (Neck, 1984). *Deroceras laeve* has been introduced into New Zealand (Quick, 1960).

The remaining species of *Deroceras*, *D. caruanae*, is a small to medium-sized slug (2.5–3.5 cm) and can be distinguished from other *Deroceras*, except *D. laeve*, by the elongated head and neck extended in front of the mantle when active. This species is more active than *D. laeve* and is aggressive, lashing its tail when irritated. *Deroceras caruanae* has probably been confused with other species of *Deroceras* and was not recorded in Britain until 1930 (Quick, 1960). It is now clear that *D. caruanae* is generally distributed throughout the British Isles (Kerney, 1982). *Deroceras caruanae* was originally thought to be a largely synanthropic species, but Ellis (1964) found it to be ubiquitous in west Cornwall. It is found mainly in gardens, including greenhouses, parks and on waste ground although it also occurs in fields and hedges. There is evidence of a fossil record for this slug (Hayward, 1954), and Ellis (1951) suggested that *D. caruanae* was probably native in the south-west and introduced elsewhere in Britain, where it was especially common in gardens and on arable land. Kerney (1968) suggested that there was evidence for a rapid spread of this species as it was unlikely to have been overlooked for so long, while McMillan (1969) considered that it was probably a recent introduction into Ireland, where it was first recorded by Makings (1959). Outside the British Isles, *Deroceras caruanae* has also been recorded from the coastal regions of France (Chevallier, 1973a). Chevallier considered that this species had been introduced to those areas of France where it occurred away from the coast. *Deroceras caruanae* is also found in south-western Europe including Malta, Sicily, southern Italy and the Canary Islands, while it had been introduced into California, South Africa, the Netherlands, Denmark and Sweden (Chevallier, 1973a). In Sweden, however, it is only found in greenhouses (Waldén, 1960). Rollo and Wellington (1975) recorded *D. caruanae* from the Vancouver area of Canada and it is one of the most common and widespread of introduced slugs in Australia (Altena and Smith, 1975).

1.4.4 Boettgerillidae

Boettgerilla pallens, a species endemic to south-eastern Europe, was first recorded in the Caucasus in 1905 (Colville *et al.*, 1974). This species is a small to medium-sized slug (3–4 cm) and is worm-like when extended. It is able to burrow deep in the soil in a similar way to the earthworm, although it also occurs at the surface under wood and stones. This slug has spread rapidly in recent years throughout north-western Europe from Finland (in greenhouses) and southern Sweden, west to the Channel Islands and Ireland (Kerney and Cameron, 1979). In Germany, it is now regarded as one of the commonest slugs (Colville *et al.*, 1974). The first British record for this species was from the Lake District in 1972 by Colville *et al.* and in 1973 was also found in Northern Ireland (Anderson and Norris, 1976). Both areas were associated with tourists and picnickers, providing opportunities for dispersal. The Conchological Society recorder's reports in the *Journal of Conchology* have shown that this slug has spread rapidly and is now widely distributed throughout southern Ireland and Britain, including Wales and Scotland. North and Bailey (1989) described how this species had spread rapidly in the Greater Manchester region and was now widely distributed there.

1.4.5 Arionidae

The distribution of *Geomalacus maculosus* has already been described (Section 1.2.7(c)). Several authors, including Taylor (1902–1907), have noted the close association between *G. maculosus* and lichen-covered Old Red Sandstone rocks in the extreme south-west of Ireland and also, its dislike of limestone. Taylor also described how this species could elongate and flatten its body in order to pass through very small gaps and how it would roll up when irritated like a woodlouse. Platts and Speight (1988) have produced a comprehensive review of the taxonomy and biology of this species.

Eleven species of *Arion* have been recorded from the British Isles. The largest species, *A. ater*, may be 14 cm or more in length when extended. This species was divided by Quick (1960) into two subspecies, *A. ater ater*, which was black or brown with a grey sole and foot fringe, and *A. ater rufus*, usually yellow to reddish-brown with a creamy yellow sole and bright orange or red foot fringe. These subspecies have been regarded by some authors as distinct species and further research may confirm this. Evans (1977) compared these two forms, using morphological and biochemical characters, and concluded that they were conspecific. There is wide colour variation in *A. ater* and some examples of colour forms, including bicoloured varieties, were illustrated by Taylor (1902–1907). *Arion ater* has been fully described by Quick (1960) and is illustrated by both Kerney and Cameron (1979) and Cameron *et al.* (1983). A characteristic feature of *A. ater* is that, when disturbed or irritated, it assumes a hemispherical shape and may rock from side to side. This species is widely distributed throughout the British Isles and Europe, except Scandinavia where it occurs only on the coast of Norway and in the south of Sweden. Ellis (1969) suggested that *A. ater rufus* might have been a fairly recent introduction into the British Isles and that on the continent its range was more southerly than that of *A. ater ater*. *Arion ater* has been introduced into North America and New Zealand.

Arion lusitanicus was first reported in Britain in 1893 but this record was subsequently overlooked until its rediscovery in Durham (Quick, 1952). It is a medium to large slug, extending up to 10 cm in length and has been confused with *A. ater* although Conroy (1980) found that populations of these two species on the Wirral peninsula were mutually exclusive, with *A. lusitanicus* requiring a milder microclimate than *A. ater*. Ellis (1969) suggested that there were two distinct races of *A. lusitanicus* and Davies (1987) showed that two distinct species, *A. lusitanicus* and *A. flagellus* had been confused under the name *A. lusitanicus*. Davies (1987) described the characteristic features of these species and compared them with *A. ater* and *A. subfuscus*. Davies concluded that *A. flagellus* was more closely allied to *A. subfuscus* while *A. lusitanicus* was more similar to *A. ater*. The redis-covery of *A. lusitanicus* at Durham by Quick (1952) has now been shown to be a rediscovery of *A. flagellus* (Davies, 1987) and it seems likely that the *A. lusitanicus* reported by Conroy (1980) was in fact *A. flagellus*. Kerney (1989) described how this large *Arion* was locally common in Britain, especially in the west, while *A. lusitanicus* was much scarcer.

While problems of identification and nomenclature have obscured and confused records of the two species, Davies (1987) considered that both species had probably undergone an expansion of their distribution in Britain since about 1960 as colonizers of gardens and waste ground. Davies concluded that *A. flagellus* appeared to be confined to the British Isles, while *A. lusitanicus* also occurred in France, Switzerland and south-west Germany. It is locally frequent in natural habitats near western coasts while elsewhere it has been mainly spread by human agencies (Kerney and Cameron, 1979). Chevallier (1969, 1976) described the distribution of *A. lusitanicus* in France as mainly south of the Loire and, although it occurred in the Paris region it had probably been introduced there. *Arion lusitanicus* is also found in Portugal (Seixas, 1976) and in Austria where it is regarded as a serious pest, having spread rapidly from gardens to farms in 1982 and 1983 (Davies, 1987). Von Proschwitz (1988a) considered that this species was a native of south-west and west Europe which had extended its distribution rapidly during the last two decades by the aid of man to a range of countries including Algeria, Sweden, Jugoslavia and Bulgaria. The efficiency with which some slugs are dispersed by human activities was reflected in the occurrence of a slug, which was probably *A. lusitanicus*, in the Falkland Islands (von Proschwitz, 1988a).

Arion subfuscus is a medium-sized slug (5–7 cm in length) which may sometimes be confused with young forms of the previous three species. Some authors, e.g. Cameron *et al.* (1983), have considered that *A. subfuscus* may represent a complex of closely related species but there is no evidence to support this suggestion. It occupies a wide range of habitats although it is not common on arable ground. Like *A. ater*, this slug is widely distributed throughout the British Isles, although it may be less common in parts of eastern England and Scotland (Quick, 1960; Kerney, 1976). It is also common in Europe from north-west Finland (Fosshagen *et al.*, 1972) to southern Spain and Portugal (Alonso, 1975; Seixas, 1976), the Balkan states including Bulgaria (Osanova, 1970) and the European part of the USSR (Likharev and Rammel'meier, 1952). *Arion subfuscus* has been introduced into Venezuela, Iceland, New Zealand and much of central and north-eastern North America, where it is common, particularly in woodland (Chichester and Getz, 1973; Blanchard and Getz, 1979). Rollo and Wellington (1975) also recorded this species from western Canada.

Lohmander (1937) described three forms of *Arion* which he believed were closely related, but could be distinguished from one another on the basis of their distal genitalia. He gave a detailed illustrated account of these differences. He was uncertain whether these forms represented good species or were simply forms of *A. circumscriptus*. Thus, for many years three different species were known collectively as either *A. circumscriptus* (see Quick, 1949) or *A. fasciatus* (see Quick, 1960). Ellis (1969) recognized that three distinct species were involved and listed them as *A. fasciatus*, *A. circumscriptus* and *A. silvaticus* and this separation was also recognized by Waldén (1976) in his list of the land mollusca of the British Isles. These three species were illustrated by Kerney and Cameron (1979) and their diagnostic features were described. *Arion fasciatus s.s.* is a medium-sized slug (4–5 cm) while the other two species are slightly smaller (3–4 cm). All three species have an opaque white foot and are bell-shaped in cross section. A line of pale tubercles along the dorsal surface of the body forms a keel in the young animal but this usually disappears as the slug grows older. Because of the confusion between the three species, earlier records of individual species are unreliable.

Slugs of the *A. circumscriptus/A. fasciatus agg.* are widely distributed in the British Isles and Europe, except in Scandinavia where, like *A. ater*, it occurs on the coast of Norway and in the south of Sweden. However, Holyoak and Seddon (1983) showed that the ranges of *A. circumscriptus s.s.* and *A. silvaticus* extended further north in Scandinavia than previously recorded. The three species occupy a wide range of habitats although Kerney and Cameron (1979) considered that *A. fasciatus s.s.* was more often associated with gardens and waste ground than the other two species. Other differences in the ecological distributions of these species are described in Section 10.3. All three species have been introduced into north-eastern North America but *A. fasciatus s.s.* is by far the most widely distributed while *A. circumscriptus s.s.* and *A. silvaticus* are relatively uncommon (Chichester and Getz, 1973). Rollo and Wellington (1975) found that *A. circumscriptus s.s.* was abundant in the Vancouver region while *A. silvaticus* was present and *A. fasciatus s.s.* was not found there. *Arion silvaticus* has been recorded as far south as California in the USA (Roth, 1986), in northern Italy (Bishop, 1976) and is the only member of the group to occur in the Faroe Islands (Solhøy, 1981).

In a similar situation to that of *A. fasciatus*, three different species were known collectively as *A. hortensis*, until Davies (1977, 1979) identified these as *A. hortensis s.s.*, *A. distinctus* and a new species, *A. owenii* and listed their diagnostic features. Backeljau and Marquet (1985) applied multivariate statistical techniques to morphological characters to demonstrate that both *A. hortensis s.s.* and *A. distinctus* were present in Belgian populations. They concluded that the most reliable features to distinguish these species appeared to be the structure of the outlet of the epiphallus in the atrium and the presence or absence of an eversible portion in the oviduct. A comparison of isozyme patterns for *A. distinctus*, *A. owenii* and *A. hortensis s.s.* supported the separation of these species on morphological grounds (Dolan and Fleming, 1988) and also confirmed the conclusion of Davies (1977) that *A. intermedius* was not closely related to these species.

According to Kerney and Cameron (1979) the *A. hortensis agg.* is widely distributed throughout the British Isles and Europe. Although this species was widespread, it became rarer and more restricted to sites associated with man in the north and east although it reached Iceland and southern Scandinavia. The *Arion hortensis agg.* also occurs in the western and north-western regions of the European USSR (Likharev and Rammel'meier, 1952). It has been introduced into New Zealand (Barker, 1979) and into North America, where it is widely distributed, South Africa and Australia (Chichester and Getz, 1973). The distribution of the individual species is less well known as they have only recently been identified. *Arion owenii* has not been recorded outside of the British Isles and it has been recorded only from rather widely separated parts of England, Wales, Scotland and also from Ireland, where it is probably more widespread (Davies, 1977, 1979).

Backeljau *et al.* (1988) demonstrated an albumen gland polymorphism in *A. owenii* and suggested that the geographically isolated populations of this species might represent distinct forms or genetic strains. While *A. hortensis s.s.* has been recorded from England, Wales, parts of France and from the Netherlands, *A. distinctus* is probably more widely distributed and has been recognized in much of Europe and in North America while it is the only member of the group reported from Sweden (Davies, 1979). Holyoak and Seddon (1983) found this species much further north in Norway than previous records for the *A. hortensis agg.* It is also the only species of the *A. hortensis agg.* reported from the Faroe Islands. De Wilde (1983, 1986) showed that, while both *A. hortensis s.s.* and *A. distinctus* were common and widespread in Belgium, the latter was more widely distributed and that there was little ecological difference between both species. Davies (1979) considered that *A. hortensis s.s.* was more closely associated with gardens than *A. distinctus.* Both *A. hortensis s.s.* and *A. distinctus* have been introduced into the San Francisco Bay area of California (Roth, 1986).

Arion intermedius is the smallest British slug, reaching a length of about 2 cm when extended. This species can be distinguished from other *Arion* species by the prominent tubercles which become obvious when the animal is contracted, giving this slug the popular name of 'hedgehog slug'. It occurs over a wide range of habitats including woodland, grassland and wasteland but is less common on arable ground and in gardens. This slug is widely distributed throughout Europe, including the southern tip of the Italian peninsula (Backeljau, 1985a). It is at the northern edge of its range in southern Sweden (Kerney and Cameron, 1979) although it is the most common slug on the Faroe Islands (Solhøy, 1981). In New Zealand, *A. intermedius* is one of the most widely distributed of the introduced slugs (Barker, 1979) although in north-eastern North America its distribution is patchy but it may be locally abundant (Chichester and Getz, 1973). It has been introduced into Canada (Vancouver) (Rollo and Wellington, 1975) and into several other countries (Chichester and Getz, 1973).

2

Structure

2.1 EXTERNAL FEATURES

The outline of a typical slug (Fig. 2.1) is elongated and streamlined, with the visceral mass incorporated completely within the cavity of the head–foot. Tillier (1984) considered this character to be a distinguishing feature of the slug and suggested that a 'semislug' evolved into a slug when the stomach sank into the foot cavity, lower than the posterior edge of the foot cavity. The sequence of stages in limacization of urocyclids .has been clearly illustrated by van Mol (1970). Some urocyclid slug-like forms still retain a prominent dorsal visceral hump. The mantle covers about a third to a quarter of the anterior dorsal region of the body, giving some protection to the pallial complex and enclosing the vestigial shell when this is present. The anterior border is usually free, forming a flap which may cover the head and neck of the slug when these are contacted. The pneumostome, which opens into the lung-like mantle cavity, is located on the right side of the mantle edge and its position varies with the different slug families (Chapter 1). The pneumostome is opened and closed during respiration. The lips of the pneumostome are thickened and contain additional mucous gland cells. The structure of the pneumostome of *L. flavus* was described by Davies (1974) while Cook and Shirbhate (1983) showed that, in section, the pneumostome is surrounded by an annulus of large cells which may function as an elastic annulus.

The skin over the general surface of the body may be tough and bear prominent tubercles (e.g. *A. ater*). These tubercles stand up in little conical eminences in *A. intermedius*, giving this species the distinctive appearance of the 'hedgehog slug'. In other slugs, e.g. *D. laeve*, the skin is thin and translucent and has a smooth appearance. A row of prominent tubercles along the dorsal midline of the slug may form a keel from the posterior border of the mantle to the foot (Milacidae) or simply at the posterior end of the body (Limacidae). The mantle surface may be coarsely granulated (Arionidae), or may bear fine concentric ridges (Limacidae) or may have a horseshoe-shaped groove (Milacidae). The epithelial cells of the skin are generally non-ciliated, except around the pneumostome where they are responsible for rejection currents in the mucus. Cilia are also found on areas of the foot and foot fringe where they often form ciliated tracts.

The head usually bears two pairs of tentacles. There is an eye at the tip of

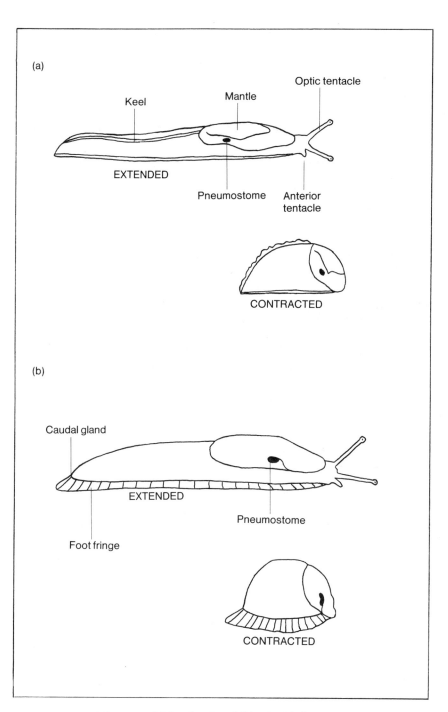

Fig. 2.1 External features of (a) milacid and (b) arionid slugs.

each of the posterior or optic tentacles while the shorter anterior tentacles are similar but have no eyes. The Athoracophoridae possess only a single pair of tentacles, the optic tentacles. The tentacles of most slugs can be completely retracted by inversion (see Section 6.2.1). Differences in the form of the tentacles of some primitive slug families have been described in Chapter 1. The mouth opens just below the anterior tentacles and is surrounded by several lips, including two mouth lobes which may represent a third pair of tentacles.

The sole of the foot is about the same width as the body, except in veronicellid slugs where it is very narrow and surrounded by the overhanging mantle, although remaining separated from the mantle by a deep groove. The sole is also relatively narrow in *T. budapestensis* which may reflect the predominantly subterranean habit of this species. The foot sole may be transversely ridged (Veronicellidae), undivided (Philomycidae, Arioninae) or longitudinally tripartite (Limacidae). In most slugs, including the Arionidae and Limacidae, the surface of the foot extends up the sides of the body (aulacopod condition) and is delimited by a pedal groove and a suprapedal groove. This region may be expanded to form a well-defined foot fringe, e.g. in the larger species of *Arion* and *Ariolimax*, or it may remain narrow, e.g. in *Limax* and *Milax*.

The caudal mucous gland, often marked by a mucus plug, is located on the dorsal surface of the foot just above the tail in arionid and ariolimacid slugs. This large gland produces a very thick mucus which may have a defensive role against predators or play a part in reproduction (Richter, 1980a). The large slit-like opening of the pedal gland is situated above the front of the foot and beneath the mouth. Slugs, in common with most gastropods, are primarily adapted for locomotion over relatively hard substrates and the foot is a muscular organ with a relatively large flat sole on which the animal crawls. During locomotion a continuous trail of pedal mucus is laid down at the anterior end of the animal and the sole passes over this trail using waves of muscular contraction. Slugs remain attached to the substrate using the adhesive properties of the mucus combined with the large area for attachment afforded by the flat sole.

2.2 INTERNAL ANATOMY

2.2.1 The shell

The internal shell lies immediately beneath the mantle in a shell sac (Figs 2.2 and 2.3). The shell sac has developed as the result of limacization by the growth of the mantle over the shell and in *Limax* and *Deroceras* a small mantle pore remains open between the shell sac and the exterior at the posterior end of the mantle (Fournié, 1979a). The embryonic origin of the

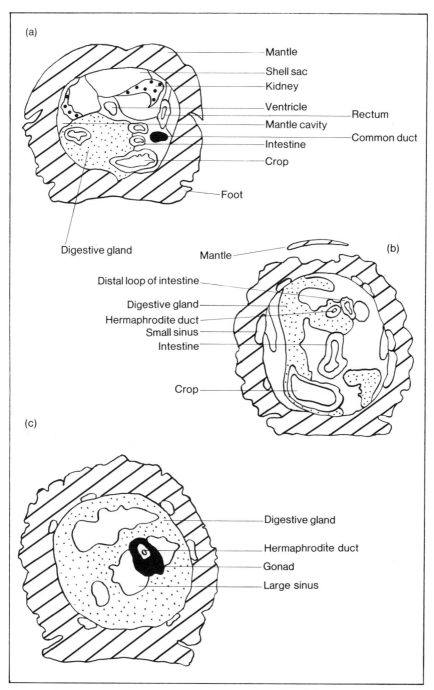

Fig. 2.2 Diagrammatic transverse sections through body of *Arion*: (a) anterior mantle region; (b) posterior to mantle; (c) posterior body region.

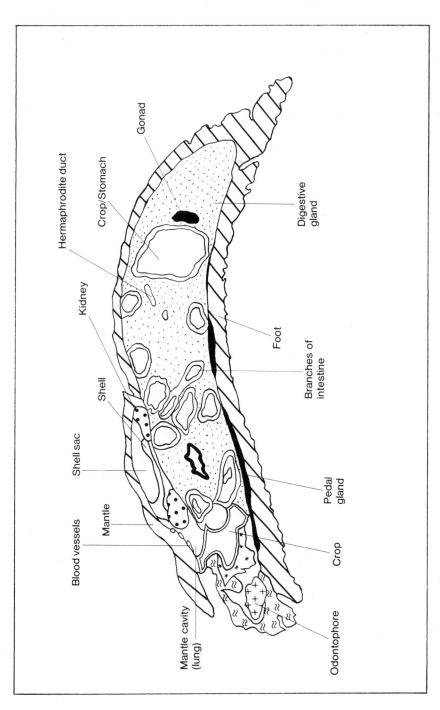

Fig. 2.3 Diagrammatic longitudinal section through body of *Limax* to right-hand side of midline.

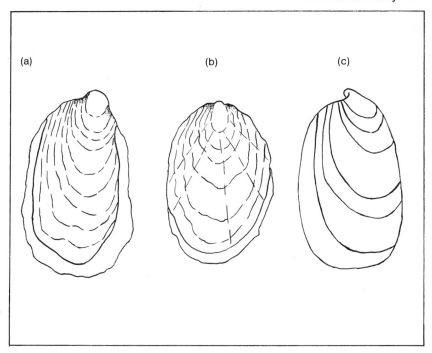

(a)   (b)   (c)

Fig. 2.4 Shells: (a) limacid, (b) milacid (c) testacellid slugs.

shell sac was questioned by Carrick (1938). He suggested that it was formed by local proliferation of ectodermal cells and that a lumen subsequently appeared in the centre of these cells. Granules of calcium carbonate were then secreted into this lumen. However, Kniprath (1980, 1981) showed that in *D. agreste* and *L. flavus* the embryonic shell field was formed by invagination and subsequently covered by the fusion of folds of ectodermal tissue. The shell in Limacidae is an oval flattened plate, which tends to be convex above and concave below. It has a terminal nucleus and generally shows concentric zones of growth (Fig. 2.4). The shell of *Boettgerilla* shows more reduction than that of other limacid and milacid slugs (Wiktor and Likharev, 1979). The shell of most arionid slugs has been reduced to discrete calcareous granules which may aggregate into an irregular mass, although an intact shell is present in *Geomalacus* and *Ariolimax*. The external shell of *Testacella* spp. ranges from flat to convex and ear-shaped (Fig. 2.4). It is situated at the posterior end of the slug where it covers the mantle and pallial complex of the animal. Reuse (1983) studied the shells of 12 species of limacid and milacid slugs and concluded that they were of limited value in taxonomy because of their variability.

2.2.2 Pallial complex

The pallial complex lies under the mantle and immediately beneath the shell sac and is protected by both the mantle and shell where this is present. It lies above the body cavity (Figs 2.2 and 2.3). There is considerable variation in the morphology of the pallial complex in slugs and it is an important systematic character (Wiktor and Likharev, 1980). Figures 2.5 and 2.6 show a generalized diagrammatic representation of this region. Full details of the variation in the arrangement of the pallial organs in most British slugs were given by Quick (1960). The lung-like mantle cavity is located towards the front of the mantle region (Figs 2.2 and 2.3) and part of the wall is folded and well supplied with blood vessels to form a respiratory surface. Although the pulmonate lung in snails is relatively large, changes in the pallial region with

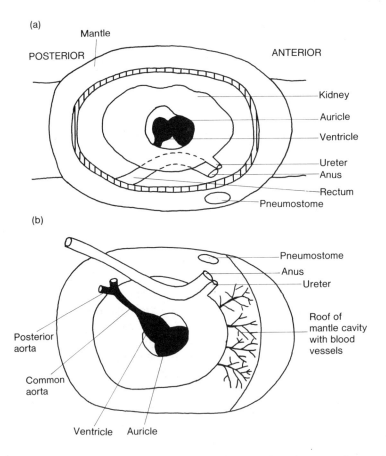

Fig. 2.5 Diagram of pallial complex in arionid slug: (a) dorsal view with top of mantle removed; (b) ventral view.

(a)

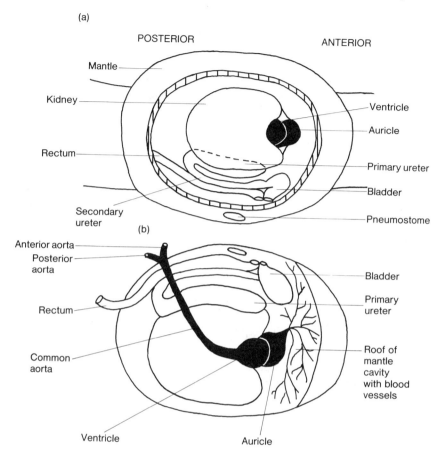

Fig. 2.6 Diagram of pallial complex in limacid slug: (a) dorsal view with top of mantle removed; (b) ventral view.

progressive limacization result in a reduction of this primary respiratory surface. However, Maina (1989) concluded that the lung in *Trichotoxon copleyi* continued to function as a true lung. Tillier (1983) discussed the function of possible secondary respiratory structures associated with the pallial complex including air sacs, alveoli in the upper wall of the mantle cavity, venous and air lacunae in the mantle shied, the shell sac and, in *Parmacella*, caecae of the primary ureter which are associated with the alveolar structure of the roof of the mantle cavity.

The rectum runs forwards along the right side of the pallial complex, opening into a groove associated with the pneumostome. The kidney is a large sac-like structure with internal folds which lies posterior to the mantle cavity or lung. In arionid slugs, the kidney is ring-like and surrounds the heart while in limacid slugs it is often bean-shaped with the heart lying in a notch at the anterior end, usually towards the left side of the kidney. The primary ureter runs over the surface of the kidney and then continues as the secondary ureter running alongside the rectum. In Limacidae this ureter discharges into a bladder which then opens to the exterior. A flap valve prevents back flow from the bladder into the ureter (Garner, 1974). The heart consists of a thin-walled auricle and muscular ventricle covered by a thin epithelium, the epicardium. The fine structure of the epicardial epithelium was described by Angulo and Moya (1989c). The heart is enclosed in a pericardium which is usually located towards the front of the kidney and on its ventral side. The common aorta arises from the ventricle and runs posteriorly before entering the body cavity and passing under a loop of the intestine before dividing into anterior and posterior aortae. The common aorta tends to be shorter in arionid slugs and in *A. ater* it divides near the ventricle, while it is longer in limacid slugs (Wiktor and Likharev, 1980).

2.2.3 Body wall

In section (Figs 2.2 and 2.3) it can be seen that the internal organs are contained within a well-developed body wall which is continuous with the foot ventrally and the mantle dorsally. Duval (1970b) described differences in the structure of the body wall of different species. *Deroceras reticulatum* had a softer body wall while *A. hortensis* and *T. budapestensis* had a rougher, fibrous wall, related to the more subterranean habit of these species. The detailed structure of the body wall was described by Lainé (1971), while Dyson (1964) described the structure of the mantle edge. The body wall consists of an outer single-layered epithelium composed of epithelial cells and mucous gland cells overlying connective tissue. The outer region of the connective tissue is pigmented, the middle region has a spongy appearance due to ramifications of the haemocoele and contains a lattice of smooth muscle fibres while the inner region is more dense and contains a mat of muscle fibres. The detailed structure of the body wall is described in Section 2.3 below.

The structure of the foot of *D. reticulatum* was described by Jones (1973). A layer of compact longitudinal muscle (about 0.1 mm) is located immediately below the pedal gland. This muscle layer also contains some transverse fibres. The rest of the thickness of the foot (about 0.4 mm) consists of connective tissue with an open network of muscle fibres running obliquely from the longitudinal muscle, where they are continuous with the muscle fibres, to the pedal epithelium. Two types of oblique fibres were described by Jones (1973b). The posterior oblique fibres ran forwards and downwards from the longitudinal muscle and exerted a posteriorly directed upward force on the epithelium. The anterior oblique fibres ran backwards and downwards from the longitudinal muscle and exerted an anteriorly directed upward force on the epithelium. The pedal gland in *T. sowerbyi* lies above the foot musculature (Barr, 1926) and that of *D. reticulatum* occupies a similar position (Jones, 1973b). In *A. ater* the gland lies on top of, but attached to, the pedal musculature while the gland in *Limax* is embedded in the musculature of the foot (Barr, 1928).

2.2.4 Digestive system

An outline of the digestive system of a typical slug is shown in Fig. 2.7. The mouth opens into the cavity of the buccal mass. On the roof of the buccal cavity is a small transverse chitinous plate, the jaw, except in *Testacella*. This works in conjuction with the radula which is a flexible chitinous strip with numerous rows of minute recurved teeth used for rasping plant tissue in most slugs or holding animal prey in the case of *Testacella*. The radula is supported by the odontophore which projects upwards from the floor of the

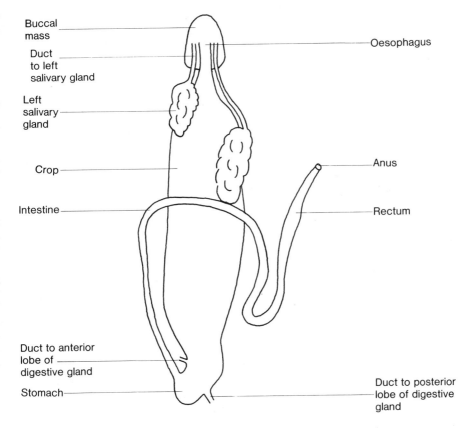

Fig. 2.7 Simplified diagram of slug digestive tract. The digestive gland is not shown. The intestine normally makes twisting loops between the lobules of the anterior lobe of the digestive gland.

buccal cavity and is formed in the radula sac at the base of the buccal mass. Further details of the jaw, radula and the odontophore are given in Chapter 3. The buccal cavity leads via a short oesophagus into a wide, thin-walled crop and thence to the stomach which is small and simple. The paired salivary glands are attached each side of the crop and empty their secretion by two ducts running forwards into the buccal cavity. The stomach lies partly embedded in the paired digestive glands and two small ducts run from the stomach into these digestive glands. The coiled intestine runs from the posterior end of the stomach to the thin-walled rectum which opens at the anus near the pneumostome. In some limacid slugs a small caecum is located at the junction of the intestine and the rectum.

2.2.5 Reproductive system

The reproductive system of the slug is a complex structure because most slugs are simultaneous hermaphrodites. The gonad lies embedded in the folds of the digestive gland towards the posterior end of the slug (Figs 2.2 and 2.3). It consists of numerous acini, containing both spermatocytes and oocytes. The narrow hermaphrodite duct runs from the gonad to join the reproductive tract and at this junction an albumen gland is also present. The reproductive tract is complex and in most slugs there is a common duct consisting of two tubes which are in open continuity along most of their length. They separate distally into separate male and female tracts, each with their associated structures, although there is a common external genital opening, located on the right side of the head below and just behind the posterior tentacle. However, the male and female openings are separated in the Soleolifera. The morphology of the reproductive tract, particularly the distal genitalia, is extensively used as a diagnostic character in slug systematics (e.g. Quick, 1960; Kerney and Cameron, 1979). Many of the structures are associated with the complex courtship behaviour that is associated with these animals. Further details of the reproductive system are given in Chapter 5.

2.2.6 Nervous system

The basic unit of the nervous system in slugs, in common with other pulmonates, is a nerve ring situated around the oesophagus and consisting of paired cerebral, pleural, parietal and pedal ganglia together with a single visceral ganglion. These ganglia are generally extensively fused so that it is difficult to distinguish between them. In addition, a pair of small buccal ganglia, lying on the dorsal surface of the buccal mass, are joined by connectives to the cerebral ganglia. The various organ systems in the body are served by a network of nerves and, particularly in the foot, digestive and reproductive systems, these form extensive plexuses. The nervous system is described more fully in Chapter 6.

2.3 INTEGUMENT

The skin of slugs is soft (Fig. 2.8), with no evidence of keratinization or cuticle production (Dyson, 1964; Machin, 1972). Szabó (1935b) described how in *D. agreste* the soft body wall became harder and darker in colour as the result of senescence, while South (1982) found that the skin of *D. reticulatum* tended to become translucent shortly before the death of the slug. The skin is particularly vulnerable to excessive gain or loss of water and to mechanical or chemical damage. Mucous secretions serve to lubricate the skin in order to reduce damage by friction and to protect the skin. Since the skin is not adapted to minimize water loss, slugs must live in moist environments. Barr (1928) examined the mucous and skin glands of *A. ater* and described three types of cutaneous gland: unicellular mucous glands, calcium glands (containing calcareous material) and pigment glands (containing black granules). Barr also noted red granules scattered among the mucous glands where they penetrated the connective tissue. The mucous glands of *T. sowerbyi* and *L. maximus* and their distribution over the surface of the body were described by Foster (1954). Campion (1957) described the development, histology and distribution of the types of gland cell present in the skin of *H. aspersa* and compared the results with other molluscs

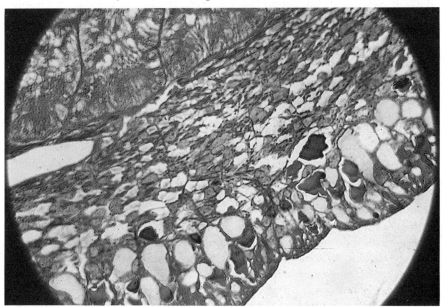

Fig. 2.8 Section through skin (*Arion intermedius*) (Mallory triple stain). The section shows the outer epithelium with mucous glands extending into the subepithelial connective tissue and an inner layer containing muscle fibres. A blood space is situated on the inner side of the muscle layer while the digestive gland can be seen beneath the skin.

including the slugs, *T. haliotidea* and *T. budapestensis*. In *H. aspersa*, the normal mucus secreted is a colourless or pale yellow, viscous fluid but when the animal is irritated the mucus became whiter in appearance due to the release of granules of calcium carbonate. Copious quantities of a less viscous, yellow material were also produced. This consisted mainly of the secretion from protein glands containing a flavone which coloured the secretion. Campion (1957) concluded that in *Helix*, mucus is mainly responsible for lubrication and, in conjunction with calcium granules, for adhesion while these granules also act in a defensive way, enhanced by the release of the protein component.

Dyson (1964) examined the structure of the anterior free edge of the mantle as a preliminary to a study of wound healing in *A. hortensis*. She described it as a glandular epithelial fold within which there was a zone of connective tissue comprising a dorsal pigmented region, a median vascular region with a lattice work of smooth muscle fibres and a ventral region containing a mat of muscle fibres. The single-layered dorsal epithelium consisted of epithelial cells (about 60% of cells) and unicellular gland cells which extended deeply into the subepithelial connective tissue and opened at the surface between the non-glandular cells. The dorsal epithelium contained three types of glands: mucous glands, protein glands and calcium glands. The ventral epithelium of the fold was similar but contained only mucous and calcium glands. Dyson (1964) found that four types of mucous glands were present in the mantle edge of *A. hortensis* although one type was restricted to the ventral epithelium. The secretion of all four types was an acid mucopolysaccharide. The secretion of the protein glands contained tyrosine and either cysteine, cystine or methionine while granules of calcium salts were present in the calcium glands. The dorsal-pigmented region of connective tissue contained black pigment granules of melanin but no uroporphyrins. The vascular region had a spongy appearance due to the ramifications of the blood spaces (haemocoele) and Dyson recognized three types of blood cell. While the smaller blood spaces were lined by a distinct endothelium, it seemed doubtful if the larger spaces were lined in this way. The smooth muscle fibres formed a ventral muscle mat of fibres arranged at right angles to one another, with a three-dimensional lattice of fibres above it and forming a network in the connective tissue. The size of the blood spaces was determined by the state of contraction of these muscle fibres.

Kennedy (1959) showed that both melanins and uroporphyrin I were present in the integument of *A. ater* and found that, in coloured forms of the slug, the amount of uroporphyrin, a photosensitive pigment, was directly proportional to the degree of masking by melanin. This contrasts with the finding of Dyson (1964) that melanin was present in *A. hortensis* although no uroporphyrin was found. Uroporphyrin I is the major porphyrin present in the shells of many molluscs and is present in the integument of gastropods (Florkin and Scheer, 1972). The red bilichrome-like pigment, first referred to as rufine, is also present in the epidermis of *A. ater rufus* (Fox, 1983). Pigment cells were not found in *D. reticulatum* by Lainé (1971), who suggested that there was unlikely to be much pigment present in the body of this light-coloured species although melanin was deposited under the epidermis in the grooves of the skin. This pigment extends on to the tubercles between the grooves in dark-coloured individuals. The darker grooves give this species its characteristic reticulate pattern. Lainé (1971) suggested that melanin granules may have been formed in melanocytes which had aged and died, depositing the melanin on the side of the epidermal cells away from the body surface.

Chétail and Binot (1967) investigated the structure and histochemistry of the pedal gland and the foot sole of *A. ater*. The well-developed pedal gland was divided into anterior and posterior regions with mucocytes of two histochemical types. Those from the anterior region secreted complex acid mucopolysaccharides associated with complex lipids while those from the posterior region secreted slightly acid, simple mucopolysaccharides. A similar result was obtained by Dalal and Pandya (1976) who described the structure of the pedal gland of *L. alte* and showed that the mucus produced by the gland contained acid mucopolysaccharides, neutral mucopolysaccharides and sulphomucin. Chétail and Binot (1967) found that the mucocytes of the foot sole were of four distinct histochemical types: type I containing slightly acid mucopolysaccharides; type II containing both simple and complex mucopolysaccharides; type III containing acid polysaccharides and lipoproteins; type IV containing mucoprotides, the mucous part of which was slightly acid. They concluded that *A. ater* showed a greater variety of secretions than *L. valentiana*, where only two types of mucocytes were described (Arcadi, 1963). Although Campion (1957) had described four secretions for *H. aspersa* skin, these were not directly comparable with those of *A. ater*. The mantle of *A. ater* contained two types of mucocytes together with a third type in the lateral groove between the mantle and the foot and these mucocytes originated from mesenchymatous cells rather than ectodermal cells (Binot and Chétail, 1968a). In a study of the development of the mucocytes of the pedal gland and the sole of *A. ater*, Binot and Chétail (1968b) demonstrated that the epithelium of the sole and duct of the pedal gland had an ectodermal origin while the mucocytes in these two tissues were derived from the mesenchyme. In *D. reticulatum*, the pedal gland was ectodermal in origin (Carrick, 1938). Van Mol *et al.* (1970) described the structure of the caudal gland of *A. ater*. They found that this region was similar to other regions of the skin with five different types of gland cells. The caudal gland supported a mucus plug formed by the accumulation of mucus and suggested that this might serve as a sexual signal for mature *A. ater*.

Lainé (1971) described the distribution of cilia on the exposed surface of the skin epithelium of *D. reticulatum*. Cilia were found on the median portion of the foot sole, on the foot fringe, around the pneumostome and on the ventral surface of the pedal gland duct, with the longest cilia being located in the most protected part, i.e. the pedal gland duct, and the shortest on the area most abraded, the sole of the foot. Mucous glands were found near the surface of the skin on all the exposed parts of the body, although they were not so evident on the underside of the mantle, and the structure of these glands was described in detail by Lainé. Two types of mucous gland were described from skin on the dorso-lateral surface of the body behind the mantle. The contents of both types of gland were histochemically similar with a pH of about 5.0 and consisting of acid mucopolysaccharides while their contents gave negative reactions to tests for protein, fats and calcium. The mucous glands on the sole of the foot had small cell bodies embedded deep in the foot muscle, with the secretion being transported to the surface of the sole by long winding ducts. Several types of granular gland of unknown function were described together with a protein gland type and calcium concretions. Lainé (1971) also discussed the development of these glands and concluded that they had their origin in the connective tissue, a conclusion similar to that of Binot and Chétail (1968a, b). The structure of three types of mucous cell from the skin of *D. reticulatum* was described by Triebskorn and Ebert (1989). These consisted of (i) pedal mucous gland cells, located in the sole of the foot, (ii) mantle gland cells, found in the mantle, the dorsal skin and the foot and (iii) club-shaped peripodial cells found along the foot margin. Mature cells contained enlarged mucous vacuoles which had displaced the nucleus and cytoplasm to the cell periphery. The fine structure of the pedal gland cells of *A. ater* was described by Angulo and Moya (1987).

Cook and Shirbhate (1983) suggested that, since different areas of the epidermis have different functions, the nature of their mucous secretions might also vary. In an investigation of the mucous glands of *L. maculatus*, Shirbhate (1987) and Cook and Shirbhate (1983) found that the body mucus could be divided into three types. The dorsal body surface was covered with a more viscous mucus composed of a sulphated acid mucopolysaccharide mixed with protein. The pedal mucus was more fluid and was formed by the mixture of the products of the pedal gland (a neutral mucopolysaccharide) and the glands at the leading edge of the foot (neutral mucopolysaccharides and a mucoprotein). The dorsal and pedal sheets of mucus were separated by a weakly acid mucus secreted by the peripodal groove, situated between the foot and the dorsal body surface. The dorsal and pedal mucus formed sheets, with the dorsal sheet remaining stationary relative to the slug's body while the pedal sheet remained stationary relative to the ground as the foot passed over it. In *L. maculatus* the pedal gland duct, the epidermis round the pneumostome, the ventral surface of the peripodial grove and the centre of the foot sole were all ciliated. The role of mucus in slug locomotion is discussed in Section 2.4. The mucus and the mucus-producing glands of the slug *Veronicella floridana* Leidy were compared with those of *L. maculatus* by Cook (1987). The dorsal mucus was similar in physical appearance although histochemically different from that of *L. maculatus*, being a carboxylated mucopolysaccharide. In *V. floridana*, the mucus was produced in deeply embedded gland cells which discharged into inflated common ducts where it was held in reserve. Little mucus was normally present on the dorsal surface of the slug giving it a dry leathery appearance and resulting in a lower rate of water loss than in *L. maculatus*, where mucus is freely distributed over the dorsal surface. The pedal mucus of *V. floridana* was of comparable function and composition to that of *L. maculatus* although it was produced in a different way. Most pedal mucus was secreted by gland cells associated with regular transverse ridges across the foot, while the pedal gland was small although it contributed some of the pedal mucus.

The presence of a brush border of tightly packed microvilli on the external surface of the epithelial cells of the optic tentacles and body wall of *H. aspersa* and *A. hortensis* was demonstrated by Lane (1963), who suggested that their function was to hold protective mucous secretions on to the body surface. Wondrak (1968, 1969a, b, c) described the fine structure of epithelial, gland and pigment cells from the skin of *A. ater*, including several types of intercellular junctions and the presence of small vesicles in some epithelial cells. The fine structure of the epithelia of the dorsal body surface and of the foot was compared for *A. hortensis* and *D. reticulatum* by Newell (1973, 1977). The dorsal epithelia were very similar, the columnar cells having their apical surface covered in microvilli and were well supplied apically with mitochondria. The epithelial cells were joined apically by septate junctions but Newell found no evidence for the basal macula occludens described by Wondrak (1968). There were no wide intercellular spaces and the lateral plasma membranes were sinuous with each cell interdigitating laterally with its neighbour. The ventral foot epithelium was composed of ciliated and non-ciliated cells, often arranged in tracts although the foot of *A. hortensis* was less ciliated than that of *D. reticulatum*. The epithelial cells of the foot also had microvilli and were well supplied with mitochondria. Newell (1973) concluded that cells from both the dorsal epithelia and the foot epithelia had more features in common with cells from absorptive epithelia, such as those from the molluscan digestive gland, than with skin cells. Newell (1977) also examined the enzyme histochemistry of epithelia in *D. reticulatum* and suggested from the distribution of phosphatases that the apical plasma membrane might be concerned with energy-dependent processes. Regional differences in the epidermal tissues of the skin of *I. fruhstorferi* were described by Chang and Lim (1989).

Ryder and Bowen (1977a) demonstrated, by the use of ferritin and horseradish peroxidase markers, that the foot epithelium of *D. reticulatum* was capable of functioning as a digestive epithelium, endocytosing material from the substratum. The slug possessed a vacuolar system in which both heterophagic and autophagic materials were hydrolysed. The larger marker ferritin was confined to fairly large vacuoles, while peroxidase was present in smaller pinocytotic vesicles which then amalgamated to form larger vacuoles. In addition to this endocytotic activity, evidence for autophagy was found in the epithelial cells and this process was enhanced during starvation. Small markers, such as peroxidase and ionic lanthanum, also appeared to pass through an epithelial paracellular pathway between adjacent cells and were able to enter through the septate desmosomes, while the larger marker ferritin was excluded from these sites (Ryder and Bowen, 1977b). However, the amount of peroxidase pinocytosed at the lateral membranes of the cell was insignificant compared with that taken up at the apical surface. Paracellular movement through the epithelium of the foot of *L. maximus* has also been demonstrated by Prior and Uglem (1984b) using ^{14}C-inulin. The presence of copper in the foot of slugs exposed to the molluscicide copper sulphate was demonstrated by Ryder and Bowen (1977c). They showed that the copper was absorbed both by the paracellular route and by pinocytosis.

Deyrup-Olsen and Martin (1982b) and Martin and Deyrup-Olsen (1986) described how controlled and often voluminous secretions containing water, ions, mucus and other substances were elicited by the mechanical, chemical and electronic stimulation of slug skin. The secretion included large molecules such as dextran and even slug haemocyanin, with a molecular mass up to 9×10^6 Da. The function of these secretions was to flood the skin surface and remove or neutralize toxins and to lubricate and moisten the surface. They were produced by cells of the epithelium, by mucous cells and by the channel cells, described by Luchtel *et al.* (1984), who also showed that when carbon particles were injected into the haemocoel of *A. columbianus*, they accumulated in large specialized cells and passed from these to the surface of the skin. These channel cells were large (up to 500 μm in length) and extended from the surface of the skin to deep within the connective tissue. The cells contained a large central fluid-filled channel which filled with ink when this was injected into the body cavity. A study of the fine structure of channel cells showed tubules of the smooth endoplasmic reticulatum crossing the cytoplasmic layer of the cell. Martin and Deyrup-Olsen (1986) suggested that the function of these channel cells could be explained by the ultrafiltration of blood components into the central channel of the cell and the modification of this ultrafiltrate in the cell, prior to its expulsion on to the body surface. They demonstrated that cellular transport of univalent ions took place from the central channel without change in osmotic potential. Sodium and chloride ions were conserved while potassium and bicarbonate ions were transferred into the fluid in the channel cell. There was no evidence for a paracellular route from the inside of the body to the surface of the skin although paracellular movement in the opposite direction had been demonstrated by Ryder and Bowen (1977a, b) and Prior and Uglem (1984b). Prior (1989) concluded that the paracellular pathway had a molecular mass cut-off of approximately 10^4 Da. The neurohormone arginine vasotocin stimulated fluid and particle movement through the channel cell and acetylcholine and 5-hydroxytryptamine also stimulated body wall secretion while norepinephrine and atropine inhibited fluid output (Luchtel et al., 1984; Martin and Deyrup-Olsen, 1986). Fluid production was also considerably reduced in slugs suffering from dehydration (Deyrup-Olsen and Martin, 1984). Martin and Deyrup-Olsen (1986) concluded that the production of fluid by channel cells was under neurohormonal control and a more detailed description of this control was given by Deyrup-Olsen et al. (1989).

2.4 LOCOMOTION

When a slug crawling across a sheet of glass is observed from below, alternating light and dark bands, representing waves of muscular contraction, are seen moving along the sole of the foot. In slugs the pedal waves move forwards in the same direction as the animal and are known as direct waves, while waves moving in the opposite direction are termed retrograde. These pedal waves are absent when slugs are stationary. In most slugs there is a single system of waves, involving only the central zone of the foot, and these are termed monotaxic, while two sets of waves, on alternate sides of the foot, are termed ditaxic. In slugs with a tripartite sole there is a clear line of demarcation between the central and lateral zones. Locomotor waves pass down the centre third of the sole only and the lateral zones play no part in muscular locomotion. However, Prior and Gelperin (1974) showed that these lateral bands were firmly in contact with the substratum in L. maximus. The number of waves passing along the sole at any one time varies and Crozier and Pilz (1924) recorded between 11 and 19 waves for L. maximus while Jones (1973b) recorded six or seven waves for D. reticulatum. Crozier and Pilz (1924) and Crozier and Federighi (1925a) investigated the relationship of step-length (the distance that any point on the foot sole moves as a single wave passes over it), wave frequency and wave speed to the rate of locomotion in L. maximus. They found a direct linear relationship between the frequency of pedal waves and the speed at which the slug moved, while wave frequency also increased with temperature ($Q_{10} = 2.1$ for 11 to 21°C). There was a similar relationship between step-length and speed of the slug although step-length decreased with temperature. The latter could be explained by a decrease in the viscosity of the mucus as temperature increases so that the foot would adhere less firmly to the substratum. Jones (1973b) found that D. reticulatum brought into a room at 20°C during winter were unable to adhere to vertical surfaces although they adhered readily at room temperatures during the summer months. However, Jones showed that slugs collected during the winter were able to adhere efficiently after being kept for some hours at room temperature and suggested that slugs were able to control the viscosity of the mucus according to the ambient temperature.

A number of authors have attempted to explain the locomotor waves of slugs and snails in terms of the pedal musculature, including the suggestion that these waves were convexities of the sole rather than concavities, and this literature was critically reviewed by Jones (1975). Lissmann (1945, 1946) described the mechanics of locomotion in the snails *H. aspersa* and *H. pomatia* and demonstrated conclusively that the waves were concavities and that any point on the sole was stationary between waves and moved forwards during waves. Thus each wave was a region of longitudinal compression, the sole remaining stationary on the substratum when it was relatively elongated and lifting off the substratum and moving forwards when it was relatively compressed. Jones (1973b) showed that as crawling commenced in *D. reticulatum*, the first pedal wave to appear did so near the anterior end of the sole. This wave moved anteriorly and other waves appeared which also moved forwards until the central strip of the sole had six or seven waves passing along it and new waves were subsequently propagated only at the posterior of the foot. Prior and Gelperin (1974) also found that at the onset of movement the initial pedal waves appeared at the anterior end of the foot of *L. maximus*.

Jones (1973b) demonstrated that the oblique muscle fibres in the foot (see Section 2.2.3) were the only elements capable of accounting for the locomotory waves and described the mechanism of the pedal waves. When the anterior oblique fibres contracted, they exerted an anteriorly directed upwards force on the epithelium, lifting it upwards and forwards and initiating a forward-moving wave. As the region of contraction of these fibres moved forwards, the fibres then relaxed and the epithelium was re-applied to the substratum where it became stationary, adhering to the sheet of pedal mucus. The posterior oblique fibres then contracted and, since the epithelium at the end of each fibre adhered to the substratum, the layer of longitudinal muscle to which the fibres were also attached moved forwards together with the body of the slug. Thus the force of the pedal waves was coupled to the substratum by the thin but viscous layer of pedal mucus. The replacement of the sole on the substratum was probably due to a pressure difference between the haemocoel and the space under the foot. Jones (1973b) concluded that this mechanism probably applied to most slugs, since similar oblique muscle fibres had been found in the sole of *L. cinereoniger* and *A. ater*. This type of adhesive crawling requires that the frictional resistance to forward movement be minimized by the foot being lifted during the passage of a pedal wave. As the result of a study of adhesive locomotion in *A. columbianus*, Denny (1981) concluded that this lifting of the foot could not be reconciled with the properties of the pedal mucus and epithelium and that *A. columbianus* did not lift its foot when crawling. Denny proposed a new mechanism to explain locomotion which did not involve the lifting of the foot.

Denny (1981) and Denny and Gosline (1980) showed how the pedal mucus of *A. columbianus* was well suited to its function as a consequence of its unusual physical properties. At small deformations the mucus was a viscoelastic solid with a shear modulus of 100–300 Pa. The mucus showed a sharp yield point at a strain of 5–6, the yield stress increasing with increasing strain rate. At strains greater than 6 the mucus became a viscous liquid although it recovered its solidity if it was allowed to heal for a time. Denny and Gosline (1980) also found that resting slugs attached to a vertical surface could remain in position for many hours although the pedal mucus might be expected to flow over long periods of time and the slugs slide down under the influence of gravity. They found that the mucus under these resting slugs contained a dense network of fine, fairly long (up to 0.5 mm) fibres and they showed that they could induce fibre formation in pedal mucus by immersing it in a salt solution. Denny and Gosline (1980) suggested that when these fibres formed in the pedal mucous film under a slug, the fibre reinforced material would creep more slowly than one that was not reinforced, and this would explain the ability of slugs to remain attached to vertical surfaces without slipping.

The pedal mucus plays an essential role in slug locomotion. Barr (1926) showed that if the pedal gland of *Milax* was cauterized, slugs were unable to crawl normally although pedal waves continued to pass along the body. Barr (1926) found that the less viscous mucus secreted by the mucous glands of the pedal epithelium was distributed laterally over the surface of the sole by cilliary action while the more viscous mucus secreted by the pedal gland passed backwards along the sole of the foot. This mucus is flattened as the animal passes over it to leave the familiar slug trail. Jones (1975) suggested that as the anterior oblique muscle fibres contracted the contents of the sole mucous glands might be squeezed out and that, as the sole was replaced, some of this mucus might be resorbed into the glands. Although this would act against the internal haemostatic pressure responsible for replacing the sole on the substratum, Jones (1975) considered that unless this happened, this more fluid mucus would be carried forwards by the pedal waves and accumulate at the anterior end of the foot. The mechanism by which mucus is secreted by the pedal gland is not clear. The ciliated duct assists in expelling mucus but other mechanisms are also necessary. Barr (1926) suggested that, as the result of the high viscosity of the pedal mucus, there would be a tendency for this to be drawn out of the gland like a thread. Where the gland is embedded in the musculature of the foot, e.g. in *Limax*, muscle contractions could assist in the ejection of mucus from the gland, but this is not possible for species of *Milax* and *Deroceras* where the gland lies above the longitudinal muscle of the foot.

A number of earlier writers, including Taylor (1902–1907), noted that *L. maximus* mated while suspended in mid air from a stout thread of mucus. While this habit is not shared by many other species, several species use mucous threads to descend from overhanging situations, including *D. reticulatum*, *L. maximus*, *A. ater* and *A. columbianus*. While an animal is descending pedal waves still move along the foot sole and the mucus thread down which the slug crawls consists of the more viscous pedal mucus which has not been flattened by the weight of the slug (Barr, 1926). This mucus is supplemented by mucus from the caudal mucus gland in *Arion* (Barr, 1928) and in *Ariolimax* (Richter, 1980a).

2.5 SHELL

The range of form of the slug shell has been described in Section 2.2.1. Despite their fragile appearance most shells are sufficiently robust to survive in their subfossil state as intact shells (Limacidae) or calcareous granules (Arionidae) (Evans, 1972). Although the shell of *D. reticulatum* was internal and vestigial, Fournié (1979a, b) demonstrated that it showed the basic features of the molluscan shell. An upper organic periostracum secreted by the mantle edge covered an ostracal layer composed of calcite crystals. This periostracum was extended laterally, separating a ventral extra-pallial compartment from the rest of the shell sac. The inner layer of the shell (hypostracum) consisted of a discontinuous layer of organic and calcareous material formed by the growth and fusion of numerous spherules. Amoebocytes at the surface of the hypostracum produced microgranules of organocalcareous material which acted as centres of calcification for the formation of these spherical particles. These amoebocytes originated in the connective tissue under the mantle and their presence was associated with shell repair and the mobilization of calcium reserves from the hypostracum for reproduction (Fournié and Chétail, 1982b). Meenakshi and Scheer (1970) showed that the organic matrix of the internal shell of *A. columbianus* was composed mainly of protein, with small amounts of neutral and acidic mucopolysaccharides and lipids. Hydrolytic products of the matrix included fructose, galactose and glucosamine. Seventeen amino acids were present in hydrolysates of this matrix. Poulicek and Voss-Foucart (1980) showed that the internal shell of *D. reticulatum* contained chitin, much of which was free and not bound to proteins. They demonstrated cyclic fluctuations in the amounts of calcium carbonate and chitin in the shell at different times of the year. There was a decrease during the hibernation of the slug (October to March) and the amounts of calcium carbonate and chitin then rose during the period of activity of the slug (March to October), reaching a maximum just before hibernation. It seems unlikely that these slugs hibernated from October to March (South, 1989a) and these changes were more likely related to the mobilization of calcium reserves for reproduction described by Fournié and Chétail (1982b).

Poulicek and Jaspar-Versali (1982) described seasonal variations in the structure and calcium content of the shell of *T. rustica*, *L. maximus*, *L. cinereoniger* and *L. marginata*. They confirmed that the shell structure in these species was similar to that described for *D. reticulatum* by Fournié (1979a, b) and described how an additional, very thin discontinuous layer of organic material partially covered the ventral surface of the hypostracum in *L. maximus* under optimum culture conditions. Poulicek and Jaspar-Versali (1982) described how the seasonal mobilization of calcium from the ostracum and hypostracum was accompanied by the hydrolysis of the chitin-based organic matrix of the shell with a consequent decrease in the chitin content of the shell. The fine structure, development and regeneration of the periostracum of *D. reticulatum* was described by Fournié and Zylberberg (1987) who showed that it consisted of two layers, an inner dense layer and an outer fibrous layer. The structure and development of the inner dense layer supported the hypothesis that the reduced shell of *D. reticulatum* could be considered as a typical molluscan shell. However, the outer fibrous layer appeared to be a peculiar characteristic of this species. The development of the shell of *L. maximus* was studied by Furbish and Furbish (1984) who described three stages in the growth of the shell. Primary growth occurred radially from the umbonal region. Secondary growth was represented by shell thickening while tertiary growth was characterized by a lateral component, where the shell extended beyond the primary growth boundaries, and a ventral component, in which the shell continued to grow in thickness.

3

Feeding, digestion and metabolism

3.1 FEEDING AND DIGESTION

3.1.1 Buccal mass

The digestive tract of *A. ater* has been described and illustrated by Roach (1968) and that of *D. reticulatum* by Walker (1969) and by Runham and Hunter (1970). The mouth opens into the buccal cavity and, in *D. reticulatum*, it is surrounded by a number of small lips, including three anterior lips, five pairs of lateral lips and one single posterior lip (Walker, 1969). These are flanked by a pair of larger mouth lobes (oral lappets). The two most posterior pairs of lateral lips fuse inside in the mouth to form lateral folds, which extend back through the buccal cavity and are continuous with the lateral folds in the oesophagus (Runham, 1975). Runham also described how the surface of these lips, the mouth and the buccal cavity were lined by a cuticle which was thickened inside the buccal cavity to form a jaw and probably also the radula. The jaw consisted of chitin and protein together with mineral salts. The form of the jaw varies with different families (see Chapter 1) and it is lacking in the family Testacellidae. The range of jaw types was described by Taylor (1902–1907) and the jaws of most slugs may be divided into two groups. They are either crescent-shaped and markedly ribbed (odontognathous; Fig. 3.1(a)) or relatively smooth and usually with a median anterior projection or beak (oxygnathous; Fig. 3.1(b) and (c)). The former group includes the arionid slugs while the latter group includes the limacid and milacid slugs.

Runham and Thornton (1967) examined the ventral surface (i.e. the surface exposed to the buccal cavity) of the jaw of *D. reticulatum*. They found numerous poorly defined fine groovings approximately parallel to one another but at an angle of 20° to the longitudinal axis of the jaw and these were absent where the centre of the jaw was produced forwards as a beak. These grooves were caused by the action of the radula teeth against the jaw and the shape of the jaw was determined by the abrading of its surface by the radula. South (1980) found that the jaw of *T. budapestensis* showed, in addition to these poorly defined oblique groovings, deep distinct grooves covering the centre of the jaw and running parallel to the longitudinal axis of the jaw. These were also observed on jaws of other milacid and limacid slugs but were absent from the jaw of *D. reticulatum*. The number and form of the

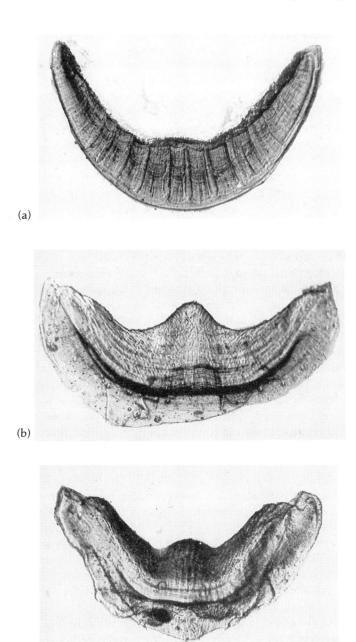

(a)

(b)

(c)

Fig. 3.1 Slug jaws: (a) odontognathous (*Arion circumscriptus*); (b) oxygnathous (*Deroceras reticulatum*); (c) oxygnathous (*Tandonia budapestensis*).

ribs on the odontognathous jaw varies with different genera and species (Quick, 1960). However, Taylor (1902–1907) described how the number of ridges and strength of the jaw could be modified by the type of food, with the jaw becoming stronger if the food was tough.

Most of the buccal cavity is taken up by the odontophore (Figs 2.3 and 3.3). The radula covers much of the anterior and dorsal surface of the odontophore. The odontophore is supported internally by the odontophore cartilage, which is associated with a complex musculature. Runham (1975) described the structure of this cartilage which contained many muscle fibres, mainly directed in a dorso-ventral direction, with vesicular connective tissue cells between the muscle fibres. A connective tissue sheath was present over most of the surface of the cartilage and the latter consisted of two rods, which were fused anteriorly where the cartilage was mainly muscular. Runham (1975) suggested that the fluid-filled vesicular cells within the limiting sheath might function as a hydrostatic skeleton. The odontophore is dorsally grooved and posteriorly this groove leads into the radular gland, located in a radular sac which protrudes externally from the ventral posterior surface of the buccal mass. The radular gland consists of an epithelium above the radula (the supraradular epithelium), an epithelium beneath the radula (the subradular epithelium) and, at the base of the gland, the odontoblasts which secrete much of the radula (Runham, 1963). The radula lies along the dorsal groove and, at its anterior end, it is spread over the surface of the odontophore. A rod of connective tissue, the collostyle, lies above the dorsal groove. The epithelium of the collostyle was continuous with that of the oesophagus dorsally and the radular gland ventrally and it secreted the cuticle of the collostyle hood (Runham, 1975). Curtis and Cowden (1977) examined the fine structure of the odontophore cartilage and the collostyle of *L. maximus*. They showed that, while the tissue superficially resembled vertebrate cartilage, it was histologically and biochemically different. The cells of the odontophore cartilage contained glycogen-filled cores as well as bundles of peripherally located filaments resembling myofilaments and were innervated like muscle cells. The extracellular matrix of the odontophore was sparse and contained glycogen and fibrillar material, but no acid mucosubstances were found. The collostyle consisted of a gelatinous type of tissue with an abundant extracellular matrix containing cross-banded filaments and other material, including neutral and weakly acidic mucosubstances. The cell component of the collostyle included solitary muscle cells and fibrocytes containing large quantities of glycogen.

The gastropod radula is a flexible chitinous membrane to which are attached transverse rows of hard teeth. The form of the teeth in a transverse row varies and it is possible to distinguish a central tooth, flanked by lateral teeth which are flanked in turn by marginal teeth. Each tooth consists of one or more cusps attached to a large basal plate (Fig. 3.2(a)). While the number and size of the teeth vary with age within a species, there are specific

(b)

(a)

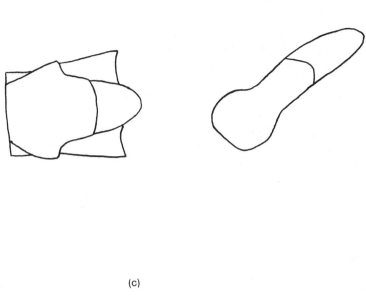

(c)

(d)

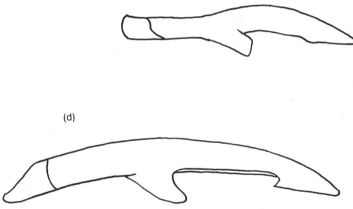

Fig. 3.2 Examples of radula teeth: (a) central and (b) lateral teeth from *Limax* radula; (c) lateral and (d) central teeth from *Testacella* radula.

differences in the arrangement of the radula teeth and a radular formula is used to describe this arrangement. For example, the adult radular formula for the small slug, *A. intermedius*, is C.16.16 and this means that there are 16 lateral teeth and 16 marginal teeth on each side of a central tooth (C). The radular formula for the larger slug, *L. cinereoniger*, C.19.50, showing a much larger number of marginal teeth. The form of the teeth and the radular formulae for most British species of slug have been described by Quick

(1960). Moens and Rassel (1985) made a comparative study of the radula of three slugs (*A. hortensis, D. reticulatum* and *T. budapestensis*) and three zonitid snails and showed fundamental differences in the functioning of the three different types of teeth and in the type of radula in relation to diet. Jungbluth *et al.* (1980) assessed the value of the slug radula in taxonomy. The radula of the slug *Testacella* is modified for a carnivorous diet (Fig. 3.2(b)). The central tooth is vestigial or absent and there are only 16 or 18 teeth on each side of the radula. These elongated teeth are sharp and barbed and increase in size up to the penultimate tooth in each V-shaped transverse row (Quick, 1960). Similar modifications are shown by the radula of the Rathouisiidae which are semicarnivorous slugs (Grassé, 1968).

The teeth on the most anterior part of the radula become worn down and break off. Runham and Thornton (1967) found that wear in the unmineralized teeth of *D. reticulatum* was initially restricted to a shortening and rounding of the cusps, but that in the oldest rows the whole tooth was worn down to a small stub. Runham (1963) demonstrated that the radula in the snails *Helix pomatia* Linnaeus and *Lymnaea stagnalis* (Linnaeus) was continually replaced and moved forwards into the buccal cavity where it was shed. Continuous replacement was also found to occur in all 15 prosobranchs and pulmonates examined by Isarankura and Runham (1968), including *D. caruanae, D. reticulatum* and *A. ater.* A radula replacement theory was put forward by Runham (1963). This described how the radula was secreted continually in the radular gland by the permanent odontoblasts and continually moved forwards. The cells of the supraradular epithelium, produced by the division of cells near the odontoblasts, moved forwards at the same rate as the radula. These cells produced materials which hardened the newly formed radula and then died. Runham (1975) described how the youngest teeth were very soft while further forward they became hardened. The radula consisted of chitin and protein and the hardening was due to chemical changes similar to those associated with quinone tanning. Runham (1963) suggested that the subradular epithelium, which was also produced by the division of cells at the base of the radular gland, secreted the subradular membrane which attached the radula to the epithelium. This attachment was essential as, during feeding, movement of the radula over the odontophore was the result of the action of muscles that were attached to the basement membrane of the subradular epithelium. At the anterior end of the radula, the subradular epithelium secreted a cuticle continuous with the buccal cuticle and, as a result, the subradular membrane and radula became detached and broke up. The process of radular transport towards the mouth was investigated further by Kerth (1976).

The rates of radula replacement vary with different species and with age. Isarankura and Runham (1968) showed that the rate of replacement was very rapid in newly hatched slugs and that this rate steadily decreased with age. The rate of replacement was also dependent on temperature. Isarankura and

Runham (1968) measured the rate of replacement by marking the newly formed teeth in various ways, including cold shock, cauterization and the injection of magnesium chloride and colchicine. The estimated rates of replacement (in rows of teeth per day) for mature slugs were: *D. reticulatum* 5.1, *D. caruanae* 5.6 and *A. ater* 3.9 at 20°C. X-rays were used to mark the radula teeth of *L. flavus* by Kerth and Krause (1969) and the estimated rates of replacement for the radula were 3.1 rows of teeth per day in young slugs, reducing to 1.4 rows per day in adult slugs. Kerth (1973) made further investigations into the effect of X-rays on the radular gland of *L. flavus*. In the supraradular epithelial area, many cells died and cell proliferation was reduced for a week after irradiation, returning to a normal level after nine weeks. Few cells died and mitotic activity in the subradular epithelium was less reduced while proliferation was almost normal in the fifth week after exposure. No odontoblasts died after irradiation although from the third week onwards after exposure they were deformed but had reassumed their normal shape in the tenth week. Kerth (1979) found that in *H. pomatia* and *L. flavus* the teeth secreted in the odontoblast region were already in their final form.

The odontophore is a muscular structure and Roach (in Runham and Hunter, 1970) showed that the musculature of the buccal mass of *A. ater* was similar to that described for *L. stagnalis* by Carriker (1946). Carriker described this musculature as one of the most intricate structures in the invertebrates, having 28 different muscles, some single, some paired, some within the buccal mass and some connecting the buccal mass to other structures such as the body wall. Gelperin *et al.* (1978) has described the functional morphology of the buccal mass in *L. maximus*. The important role of the musculature involved in feeding is reflected in the rich blood supply to this region (Duval and Runham, 1981). The behavioural sequence in *L. maximus* leading up to feeding includes (i) cessation of locomotion in the presence of food, (ii) partial retraction of the superior tentacles, (iii) eversion of the lips, (iv) initiation of rhythmic lip movements and (v) cyclical protraction and retraction of the radula against the food (Gelperin *et al.*, 1978; Schagene *et al.*, 1989). The information available concerning the process of feeding (e.g. Runham and Thornton, 1967; Runham, 1969) suggests that the sequence is similar to that described for *L. stagnalis* by Carriker (1946) and Hubendick (1957). The cycle of feeding had four main phases. In the resting phase, the mouth was closed and the odontophore was directed almost vertically upwards. During the first feeding phase, the mouth was opened by dilator muscles and applied to the surface of the food. The buccal mass was rotated forwards through nearly 90° while the odontophore rotated independently through another 45–90°, resulting in a total rotation of the odontophore through between 125 and 170°. During this movement, contraction of the odontophore cartilage muscles acting against the vesicular cells straightened the cartilage making it rigid and

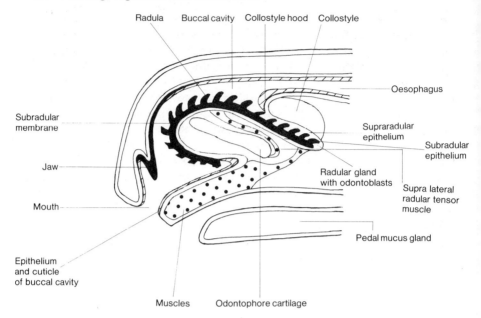

Fig. 3.3 Sagittal section through the head of a slug (after Runham and Hunter, 1970).

stretching the radula over the surface of the odontophore. The odontophore, which was now held vertically downwards, was moved down through the open mouth to meet the substrate at the beginning of the second feeding phase. Since the lateral portions of the cartilage extended further out through the mouth than the central part, the tip of the odontophore in contact with the substrate was U-shaped and the lateral portions of the radula contacted the substrate before the central part. During the second and third phases, the tip of the odontophore moved forwards in contact with the substrate and, at the end of this feeding stroke, first the lateral edges and then the central region of the odontophore were lifted from the substratum. The odontophore was withdrawn into the mouth and returned to its resting position during the final phase of feeding and the mouth was closed. The actions of the various muscles involved in these movements were fully described and illustrated by Hubendick (1957).

Runham and Thornton (1967) and Runham (1969) showed that the radula teeth of *D. reticulatum* moved forwards over the cartilage during the feeding stroke and, as they passed over the bending plane of the cartilage, they rotated from an anteriorly directed position on the outside of the odontophore to a posteriorly directed position in the odontophore groove. Runham and Thornton (1967) concluded from the grooving of the jaw that this must be closely applied to the odontophore while the radula is passing forwards and backwards over the cartilage. The angle of the grooves indicated that the tooth cusps were diverging during their forwards movement, showing that the jaw was assisting in the spreading of the radula while it passed over the bending plane of the odontophore. The pattern of wear over the whole surface of the teeth was consistent with their being dragged through a soft substratum. Runham (1975) described how, after a meal of lettuce, the contents of the crop of *D. reticulatum* contained pieces of lettuce rather than the fine particles that might be expected if the food had been rasped. The passage of teeth over the edge of the odontophore made this into a cutting edge, and when this met the jaw at the end of a forward feeding movement a piece of lettuce was cut off. This kind of feeding was also possible with thicker food such as carrot. Thus *D. reticulatum* could feed by biting off pieces of food as well as by rasping. In carnivorous slugs, such as *Testacella*, the region of the buccal mass behind the opening of the oesophagus is elongated backwards to form an evaginable snout or buccal bulb (Tillier, 1984). Runham (1975) described feeding in the carnivorous slug, *Testacella*. When an earthworm is attacked, the odontophore is rapidly shot out of the mouth and the erected radular teeth penetrate its skin. As the radula passes back over the odontophore cartilage, the worm is held in the groove by the backwardly directed teeth. The prey is then dragged into the mouth when the odontophore is retracted although the process of swallowing a large worm may take some time. The detailed mechanism by which this specialized odontophore can be rapidly everted and then retracted while grasping the prey was described for *T. maugei* by Crampton (1975).

3.1.2 Salivary glands

The paired, flattened salivary glands are situated on either side of the crop, to which they are joined by connective tissue. Each gland consists of numerous glandular acini connected to small collecting ducts which join to form the salivary duct. The salivary glands are well supplied with blood from the salivary arteries (Duval and Runham, 1981). The paired salivary ducts enter the buccal mass on the dorso-lateral surface on either side of the junction with the oesophagus. The structure of the salivary gland has been described for *D. reticulatum* (Walker, 1970) and for *L. maximus* (Beltz and Gelperin, 1979). Walker described ten different cell types in the salivary gland, including two types of mucous cell, which produced acid mucopolysaccharides, together with granular, grain and pseudochromosome cells, all three of which produced proteins together with neutral polysaccharides. Beltz and Gelperin (1979) described four major types of secretory cells from the salivary gland of *L. maximus* while Goldring *et al.* (1983) distinguished three different secretory cell types from the salivary glands of *Ariolimax* on the basis of their electrical activity. These were described as large granule, medium granule and small granule cell types. Kulkarni (1972) found only two types of cell in the salivary glands of *L. alte*.

Only the intralobular ducts were ciliated in *D. reticulatum* and, although the initial collection of salivary secretions resulted from ciliary action, subsequent movement of saliva through the interlobular ducts and salivary ducts was by peristaltic action. A muscular layer was described in the walls of these ducts in *D. reticulatum* by Walker (1970). Beltz and Gelperin (1979) also showed that the salivary duct of *L. maximus* was composed of several muscular layers surrounding the epithelium which lined the duct lumen. The morphology of the duct epithelium in both species indicated that it might have a function in ionic regulation or water balance. Extracts of salivary glands have been shown to contain amylase in *A. ater* (Evans and Jones, 1962a) and in *D. reticulatum* both amylase and a trypsin-like protease (Walker, 1969, 1970). Evans and Jones found that the salivary gland extract hydrolysed a narrower range of carbohydrate substrates than extracts of the crop juices and digestive gland.

3.1.3 Oesophagus and crop

The oesophagus, which is short and narrow, runs backwards from the mid-dorsal region of the buccal mass to merge with the crop after passing through the circumoesophageal nerve ring. The lining of the oesophagus is arranged into six longitudinal folds and includes a dorsal food groove which is continuous with the dorsal food groove of the buccal cavity. Food particles are also transported through the oesophagus by ciliary currents from areas of ciliated epithelium which also contain mucous cells (Walker, 1969). One type of mucous cell was shown by Triebskorn and Ebert (1989) to be predominant in the oesophagus, stomach and intestine of *D. reticulatum*. They described the fine structure of this type of cell, which contained mucous vacuoles, and demonstrated that the vacuoles were released into the lumen of the digestive tract. Rapid transport of food though the oesophagus could be achieved through fast peristaltic waves (three contractions per minute) running along layers of muscle underlying the epithelium of the oesophagus (Roach, 1968). This prevented food from accumulating in the collecting area of the buccal mass. Roach suggested that the most posterior part of the oesophagus in *A. ater* might act as a sphincter to prevent regurgitation of food from the crop into the buccal cavity.

Babula and Skowronska-Wendland (1988) showed that the lining of the alimentary tract of the slug *D. reticulatum* was formed from a single layer of epithelium resting on a layer of connective tissue. The wall of the tract also had an outer circular layer and an inner longitudinal layer of muscle cells (Walker, 1969). The crop is a large, thin-walled sac, which is much longer than the oesophagus and it may represent a specialized region of that organ. It functions as an area for the storage of food and also for extracellular digestion and some absorption (Babula and Skowronska-Wendland, 1988; Babula and Wielinska, 1988). Walker (1972) showed that the crop was a site for the absorption of soluble food substances such as glucose, galactose, glycine and palmitic acid. Babula and Skowronska-Wendland (1988) described the epithelium of the crop as smooth although Walker (1969) found scattered patches of ciliated cells near the junction with the oesophagus. The epithelial cells of the crop wall contained many apical granules and both Walker (1969) and Babula and Skowronska-Wendland (1988) showed that these contained non-specific esterase, an enzyme associated with lysosomes. However, acid phosphatase, an enzyme most frequently used as a marker for lysosomes, was absent from the crop wall of *D. reticulatum* although Bowen (1970) found it was associated with the crop wall of *A. ater*. The detailed structure of the crop wall was described by Bourne *et al.* (1991). Using the electron-opaque markers horseradish peroxidase and lanthanum nitrate, they also demonstrated that both pinocytotic and paracellular uptake of endogenous material was possible in the crop.

Roach (1968) described how, although there was no structural basis for distinguishing between different regions of the crop, it could be divided into two regions, forecrop and hindcrop, based on the patterns of peristaltic activity. Contractions in the forecrop were stronger than those in the oesophagus and lasted 1.2 minutes. They were of a characteristic type with fast relaxation and contraction between successive waves but with a period of maintained contraction between. The hindcrop showed large slow changes in muscular tone taking three minutes each. These movements in the crop ensured that the contents were thoroughly mixed.

The empty crop in slugs and snails is usually distended with a reddish-brown liquid known as crop juice. This is capable of digesting a wide range of materials, including carbohydrates, proteins and lipids. Because of its broad range of enzyme activity, including β-glucuronidase and cellulase, snail crop juice is available commercially although slug crop juice does not appear to have been used for this purpose. Runham (1975) listed 24 enzymes which had been recorded from the crop juice of pulmonates and most of these have been found in slugs, mainly by Stone and Morton (1958), Evans and Jones (1962a, b) and Walker (1969). Evans and Jones (1962a) listed the carbohydrases present in *A. ater* crop juice and digestive gland extracts together with their substrates. The higher levels of proteases in arionid and limacid slugs than in other terrestrial mollusc families was considered by Marcuzzi and Lafisca (1975) to reflect the omnivorous diet of these slugs. However, Stone and Morton (1958) and Walker (1969) reported low protease activity in the crop juice of *D. reticulatum* although Evans and Jones (1962b) found strong protease activity in *A. ater*. Walker (1969) found that a crop wall extract from *D. reticulatum* showed only weak activity for amylase, invertase, lipase and protease and concluded that most of the crop juice was produced elsewhere. The major source of the extracellular digestive enzymes in the crop juice seems to be the digestive gland (Stone and Morton, 1958; Evans and Jones, 1962a, b). Roach (1968) described how the duct of the anterior lobe of the digestive gland opened directly into the posterior end of the hindcrop in *A. ater*. This arrangement appears to be unusual since Tillier (1984) defined the stylommatophoran stomach as that part of the digestive tract which received the two ducts of the digestive gland. In *D. reticulatum*, however, both digestive gland ducts opened into the stomach (Walker, 1969). According to Quick (1960) the ducts of the digestive gland in *Geomalacus* open further back at the junction of the stomach and intestine.

Jeuniaux (1954) showed that chitinase was present in the gut of eight pulmonates, including *A. ater*, *L. cinereoniger* and *D. reticulatum*, and showed that large numbers of chitinolytic bacteria were also present. Stone and Morton (1958) reported that cellulase had been found in the digestive gland and crop fluid of *A. ater* and *L. flavus* and they demonstrated the presence of cellulases and carbohydrases in the crop and digestive gland of *D. reticulatum*. The presence of cellulase and chitinase in the gut of slugs raises the problem of the origin of these enzymes. While it is generally considered that these are produced by bacteria in the gut (Evans and Jones, 1962a), little is known about the bacterial gut flora of slugs. Watkins and Simkiss (1990) have shown that the bacterial flora of the alimentary tract of the snail, *H. aspersa*, were similar to those found in the soil, while Charrier (1990) concluded that the snail, *H. aspersa*, did not possess an endogenous bacterial flora but that bacteria were ingested with plant food. However, it has been shown that cellulase and chitinase can also be produced by the digestive gland of some snails, such as *H. pomatia*, as well as by bacteria (Runham, 1975). During a study of the contribution of soil invertebrates to the heterotrophic decomposition process, Hartenstein (1982) demonstrated catalase and cellulase activity in *A. hortensis*, *D. reticulatum* and *P. carolinianus* and also peroxidase and aldehyde oxidase activity in the first two species. The specific activity of catalase and cellulase was considerably higher in the molluscs and oligochaetes than in other soil invertebrates and Hartenstein suggested that these groups were probably more effective in degrading cellulosic material in the litter than other animals.

3.1.4 Stomach

The crop passes directly into the small stomach with no sphincter between them. Roach (1968) described a slight constriction in this region but also suggested that the stomach would remain full because of the maintained pressure of the distended crop. Walker (1972) showed how, when radio-opaque material was fed to *D. reticulatum*, it appeared almost immediately in the crop and then first showed in the stomach after 20–40 minutes. Several hours were needed for all of the material to pass into the stomach. Movement of food into the stomach from the crop was also assisted by very slow peristaltic waves which moved over the crop surface, taking 15–20 minutes to pass from forecrop to stomach. The stomach was essentially a sorting area and food remained there for only a short time. In *D. reticulatum* the two digestive gland ducts open separately into the stomach (Walker, 1969). From these openings a pair of large folds (typhlosoles) run back along the dorsal wall of the stomach to converge posteriorly to form the walls of the intestinal groove. This is considered to be homologous to the style sac in primitive gastropods (Walker, 1972). An anterior triangular accessory fold is situated between the converging typhlosoles, delimiting a gastric groove between the typhlosole on each side and the accessory fold. The pair of gastric grooves merge posteriorly into the intestinal groove. The epithelium of the stomach wall includes a number of ciliated folds particularly around the openings of the digestive glands. The ciliary tracts of the stomach and their function have been described and illustrated by Walker (1972). Walker found that the stomachs of other slugs examined, including *A. subfuscus*, *T. budapestensis*, *L. maximus* and *L. flavus*, possessed similar stomach folds to *D. reticulatum* although *A. ater* was an exception to this arrangement.

Food material entering the stomach moved across the ciliated folds surrounding the openings of the digestive glands. Fine particles and soluble material passed into the ducts of the digestive glands. In experiments, particles of 0.1–0.4μm in size passed into the digestive gland ducts while materials with a particle size of 1–3μm were rejected. The larger rejected particles were transported by ciliary currents back along the intestinal groove. Because ciliary currents in the posterior part of the intestinal groove transported particles forwards along the groove, these particles were compacted into faeces where the opposing currents met (Walker, 1972). Pallant (1970) described how *D. reticulatum* produced two types of faecal matter. The bulk of the material (75–90% of wet weight) consisted of pellets, of regular shape and size, consisting of comminuted, relatively undigested fragments enclosed in mucus and represented the rejected material which had passed directly from the stomach to the intestine. The other form of faeces consisted of a membranous sac of irregular shape containing a brown fluid followed by small particles similar to those found in the digestive gland. This is derived from strings of material which emerge at intervals from the digestive gland openings into the gastric grooves and are transported back to join the other waste material in the intestinal groove. The movement of material to and from the digestive gland is rapid and is achieved by the combined action of muscular movements in the stomach and by peristaltic contractions in the digestive glands (Roach, 1968; Walker, 1969).

3.1.5 Digestive gland

The digestive gland occupies a large part of the body cavity (Figs 2.2 and 2.3) and in most slugs it is paired, with an anterior gland, which may be reduced in size, and a larger posterior gland. Both glands have a separate duct leading to the stomach in *D. reticulatum* (Walker, 1969) or to the crop (anterior gland) and stomach (posterior gland) in *A. ater* (Roach, 1968). Each gland is subdivided into several lobes and, according to Runham and Hunter (1970), the number and arrangement of these lobes is constant within each species. For example, in *D. reticulatum* there are six lobes in the posterior and nine in the anterior digestive glands. Each lobe consists of many acini at the end of branched tubules which merge to form ductules and lobular ducts and then, finally, the digestive gland duct. The ducts have convoluted linings with an epithelium which is highly ciliated. The ciliary currents in the ducts are all directed towards the stomach (Walker, 1972), confirming that the movement of material into the digestive gland must be the result of muscular action. The connective tissue beneath the epithelium contains muscle fibres. There is general agreement on the types of epithelial cells present in the ducts and acini. Both Walker (1971) and Babula and Skowronska-Wendland (1988) described three cell types in *D. reticulatum*: digestive cells, excretory cells and calcium cells. Walker also described a thin cell type which was an undifferentiated precursor of the other cell types. Walker (1971) suggested that digestive cells might eventually become excretory cells, which could be distinguished by a large central vacuole containing excretory granules, histochemically similar to the yellow granules of the digestive cells. However, Sumner (1966) concluded that degenerating calcium cells in *Testacella* became excretory cells. Morton (1979) showed that the excretory cell was a form of digestive cell which appeared during the normal sequence of cytological changes that occurred every 24 hours. The digestive gland tubules therefore contain three cell types: the digestive cell, the calcium cell and the thin cell.

The calcium cells were pyramid-shaped and the apical cytoplasm contained small granules and protein granules. The remainder of the cytoplasm contained numerous large calcium-containing spherules formed from concentric layers of material (Walker, 1969; Babula and Skowronska-Wendland, 1988). The calcium cells appeared to be storage cells for calcium and Walker (1972) found that ten hours after a ^{45}Ca meal, labelled calcium had been incorporated into the spherules and could still be found there after five days. The digestive cells were columnar cells containing apical granules, green granules and larger yellow granules. Walker (1972) demonstrated that the digestive cells took up Thorotrast (colloidal thorium dioxide) particles by active endocytosis within two hours of a meal. Walker concluded that green granules observed in the digestive cells were vacuoles containing material which had entered the cell apically by endocytosis and that intracellular digestion occurred as these vacuoles migrated towards the base of the cell where they released their undigested material, in the form of yellow granules containing lipofuscin, into the supranuclear vacuole. Sumner (1969) showed a weak diffuse acid phosphatase reaction in the apical cytoplasm of digestive cells of *D. reticulatum, A. intermedius* and from apical granules in *T. maugei,* evidence that there were primary lysosomes present. Walker (1969) also found acid phosphatase containing granules apical to the yellow granules and concluded that lysosomal enzymes were not added to the green granules until they had migrated deep into the cell and indicated that lipofuscin granules might contain enzymes of lysosomal origin. Bowen and Davies (1971) recorded acid phosphatase activity within the lysosomal vacuoles of actively phagocytosing digestive cells and suggested that a release of enzyme into the cytoplasm which occurred in some cells of the digestive epithelium might be linked with cellular lysis and be a prelude to cell replacement. Bowen and Davies (1971) also found acid phosphatase activity associated with the brush border of the excretory cells in the digestive gland.

Thorium dioxide particles were accumulated in the supranuclear vacuole in large numbers initially but they were almost completely eliminated from the cells within five days of the Thorotrast meal (Walker, 1971). Although some particles probably passed out of the cell basally, since amoebocytes containing these particles were found in the underlying connective tissue of the tubules, most labelled particles passed out apically into the lumen of the digestive gland. However, no Thorotrast was found in the excretory cells of the digestive gland. Several soluble ^{14}C-labelled compounds including glucose, galactose, glycine and palmitic acid were absorbed readily by the cells of the crop, intestine and all cell types in the digestive gland (Walker, 1972).

Morton (1979) showed that *D. caruanae* possessed an endogenous cycle of activity entrained by the normal 24-hour day and night cycle and that most food was consumed at night (see Section 7.4.2). Structural and functional changes in the digestive system, especially in the digestive gland, were also recorded and these were correlated with this 24-hour cycle. Morton showed that the endocytosis and intracellular digestion of food particles took place in the digestive cells during the hours of daylight. These cells showed a considerable increase in size during the day with a resulting decrease in tubule lumen diameter. Numerous excretory vacuoles formed in many of the cells and, later in the day, the cells became packed with apical granules. By early evening food was absent from the lumina of the tubules, which were now much narrower. The digestive cells had reached a maximum size and were packed with apical granules and excretory vacuoles. In the late evening a process of breakdown occurred in the distal part of the digestive cells, releasing fragmentation spherules and excretory concretions into the stomach. By early morning the process of breakdown was almost complete and at the end of the 24-hour cycle, and the beginning of the next period of daylight, the digestive cells had returned to their original size with few cell inclusions and the tubule lumina were wide. Morton (1979) suggested that the fragmentation spherules, by releasing unused intracellular enzymes, might assist in the extracellular digestion of food in the crop during the next feeding cycle. Calcium cells in the digestive gland were unaffected by this cycle although they might have undergone a phase of calcium absorption during the early evening.

Czarna *et al.* (1985) investigated the fine structure of the digestive gland cells of *A. ater* and recorded the presence of tubular structures with a characteristic six-ray rosette appearance surrounding a central seventh annulus. Czarna *et al.* concluded that they might originate from the membranes of the smooth endoplasmic reticulum but were unable to suggest any function for these structures. David and Götze (1963) also found similar tubular structures in cells of the digestive gland of *A. hortensis* and the snail *H. pomatia*. Moya and Rallo (1975) described intracisternal polycylinders from the digestive epithelium of *A. ater*. They consisted of cytoplasmic cylinders arranged in sheaves within cisterns of the endoplasmic reticulatum. Moya and Rallo suggested that these structures might be involved in exchange, transport and oxidation processes contributing to the excretory function at a subcellular level. Szabó (1935a, b) described marked changes that occurred in the digestive glands of *D. agreste* with the onset of senescence. Ageing was accompanied by atrophy of the tubules and an increase in connective tissue between the tubules so that eventually the amount of connective tissue interfered with the functioning of the tubules and death resulted. In *L. flavus* the connective tissue increased with age but to a lesser extent.

3.1.6 Intestine and rectum

In most slugs the stomach is the part of the digestive tract which extends furthest from the mouth and forms a bend from which the intestine runs forward. The intestine opens from the right side of the stomach and then goes forwards along the left side of the crop to turn around the anterior aorta. From the aorta the intestine bends backwards once before going forwards again to the anus. In some slugs, e.g. *Arion* and *Limax*, the backward loop of the intestine is short and the stomach remains the most posterior part of the digestive tract. In other slugs, e.g. milacid slugs and *Deroceras*, the backward loop of the intestine extends well beyond the stomach. The form and arrangement of the intestine is constant within a species and has been used in systematic investigations (e.g. Wiktor and Likharev, 1979; Tillier, 1984). In *Limax* there is a second forwardly directed loop before the intestine goes forwards again to the anus. A small rectal caecum is present in *D. reticulatum* and *D. agreste* but not in other British *Deroceras* spp., while in

L. flavus and *L. marginata* a long diverticulum extends from the rectum backwards along the body (Quick, 1960). The position of the rectum in relation to the pallial complex has already been described in Section 2.2.2. In *Testacella* the intestine makes one forwardly directed loop before the long rectum runs backwards instead of forwards to the anus which is located by the pneumostome at the rear of the body (Quick, 1960). The anus in the Veronicellidae also opens below the hyponotum behind the foot in the vicinity of the pneumostome (Grassé, 1968). Modifications in the form of the intestine and other parts of the digestive tract in urocyclid and athoracophorid slugs were described by Tillier (1984).

The intestine is a thin-walled tube of uniform diameter and merges imperceptibly with the narrower rectum (Roach, 1968; Walker, 1969). The structure of the wall is similar to that of the crop and stomach and the epithelium shows adaptations for absorption and secretion (Babula and Skowronska-Wendland, 1988; Babula and Wielinska, 1988). However, Orive *et al.* (1979) found that amino acids were not actively transported across an isolated intestinal sac preparation from *A. ater*. There is evidence that intestinal tissues can metabolize materials and Orive *et al.* (1980) showed that L-glutamic acid was absorbed and broken down by intestinal tissue from *A. ater*. Bowen (1970) found that intestinal cells contained vacuoles which contributed to intracellular digestion of food materials and might represent lysosomes. Most of the intestine has folded walls with thin layers of circular and longitudinal muscle and a lining of ciliated and goblet mucous cells (Walker, 1969). Roach (1968) distinguished two regions of the intestine of *A. ater* on the basis of their muscular activity. The initial part of the intestine, the duodenum, showed a pattern of three or more small contractions per minute superimposed on large contractions occurring once every one or two minutes. The remainder of the intestine showed a similar pattern of three small contractions per minute superimposed on large contractions occurring every 2.5 minutes. The intestine of *D. reticulatum* was divided by Walker (1969) into the pro-, mid- and post-intestinal regions based on histological differences.

Although the walls of the intestine and rectum are ciliated, movement of faeces along the intestine was by peristalsis (Roach, 1968). Roach also described antiperistaltic waves in the intestine of *A. ater* which would result in a thorough mixing of gut contents and would delay food transport, thus increasing the digestive and absorptive time in the intestine. Antiperistalsis was more pronounced in *L. maximus* and *T. budapestensis* where the crop and stomach were smaller (Roach, 1968). However, Roach concluded that since most digestion was completed anterior to the intestine and most absorption took place in the digestive gland, the main functions of the intestinal contractions were the production and transport of faeces although rectal reabsorption of water might be important to *A. ater* as it was vulnerable to water loss. Walker (1972) found that the passage of food along the intestine was slow (about seven hours), allowing time for other processes to take place and he demonstrated the uptake of soluble materials from the lumen of the intestine. Runham (1975) described how further digestion could take place during the extended stay of faeces in the intestine. Enzymes from the crop juice would be present among the faecal material and bacterial action was also possible. Further digestion by cellulases and chitinases, which are slow in their action, would also be possible here. Estimates of the time taken for food to pass through gut and into the faeces vary. Roach (1966), by feeding *A. ater* with different coloured foods such as carrot and cabbage or lettuce, found that these took between 11 and 15 hours or more to appear in faeces and that these foods were still present in faeces 25 hours after the meal. Lyth (1982) also used carrot to mark the faeces of several species of slug and found that it took rather longer, about two days, to appear in the faeces. Three days' starvation was needed to clear the gut of food. Garner (1974) showed that *D. reticulatum* produced a string of faeces once every 24 hours, usually in the early evening before emerging to feed and Morton (1979) concluded that *D. reticulatum* and *D. caruanae* did not generally defaecate while feeding. In contrast to these timings, Walker (1972) found that faecal pellets of *D. reticulatum* containing a barium sulphate marker were deposited approximately eight hours after a meal.

3.1.7 Meal size

Ridgway (1971) described how chromic oxide incorporated into food could be used to provide a reliable estimate of food consumption. The amount of chromic oxide recovered from faeces could be related to the quantity of food consumed. The consumption by *D. reticulatum* and *T. budapestensis* of a wholemeal flour-based food labelled with chromic oxide was measured by Wright and Wlliams (1980) using this technique. The chromic oxide content of each slug, together with faeces, was estimated by atomic absorption spectrophotometry after exposure to labelled food for 24 hours. The mean consumption of food by slugs in 24 hours, following 48 hours without food, was 35.6 mg (5°C) and 33.0 mg (10°C) for *D. reticulatum* and 24.8 mg (5°C) and 35.4 mg (15°C) for *T. budapestensis*. There was no clear correlation between slug body weight and the weight of food eaten. A similar value of 29.7 mg (10°C) for *D. reticulatum* has been recorded by Bourne *et al.* (1988). Meal size for *D. reticulatum* has also been determined by visually scoring the amount eaten from agar discs made up with a water-soluble extract of wheat bran and then converting these scores to weight losses (Frain, 1981). Frain and Newell (1982), using this technique, estimated that slugs could eat about 50 mg of food in 24 hours. However, adding attractive materials such as pressed yeast extract to the agar-based food, produced a fivefold increase in food consumption to over 250 mg slug^{-1} day^{-1}. This high rate of food consumption was not sustained when the same food was presented on successive days.

Deroceras reticulatum did not feed continuously but rather fed for a few minutes several times each night (Frain and Newell, 1982). Dobson and Bailey (1982) found that *D. reticulatum* had an average of three feeding periods or meals during the dark period of a L:D 16 : 8 photoperiod. They showed that slugs tended to feed for a set time over the whole night (about 2.5 hours) and also during separate meals. The first meal lasted longer than the second or third meals while the intervals between meals were of variable length. There was a tendency for meal length to become shorter with higher consumption rates, and Senseman (1978) found a similar relationship for *A. californicus*. Senseman concluded that feeding was terminated by two processes: sensory adaptation and post-ingestional feedback (see Section 7.1.6). Dobson and Bailey (1982) provided evidence that crop fullness in *D. reticulatum* could be a suitable homoiostat for terminating feeding. Senseman (1977) described how the perception of an important nutrient such as starch, through contact chemoreceptors, served to initiate and sustain the feeding response in *A. californicus*. Dobson and Bailey (1982) showed that the crop was not always full when feeding stopped, nor always empty when it resumed. They suggested that other factors were involved in the termination of feeding, such as the adaptation of the external chemoreceptors which initiate feeding, thus supporting the conclusion of Senseman (1978). Dobson and Bailey (1982) recorded mean meal lengths for *D. reticulatum* of 59 ± 28 min for the first meal and 40 ± 27 min and 41 ± 28 min for the second and third meals. A mean meal length of 23 ± 15 min for *A. hortensis* was reported by Wedgwood and Bailey (1986) with a mean rasp cycle length of three seconds.

A study of food consumption in *D. reticulatum* by Rollo (1988a) showed that this species normally ate two to four meals per night but that the number, duration and size of meals and the interval between them varied widely between individuals, with the type of food and the duration of starvation. Although the number and frequency of meals increased with increasing starvation, starved adult *D. reticulatum* showed no compensation following starvation, in terms of either consumption or growth. Rollo found similar results for the larger and longer-lived species, *L. maximus*. However, immature *D. reticulatum* showed strong compensatory growth following starvation. One response to starvation was degrowth and Rollo (1988a) described how, after 40 days of starvation, immature *D. reticulatum* were 37% of their original body weight although they were completely normal in appearance. Rollo suggested that one phase of the life cycle was devoted to the accumulation of resources, with compensatory control, while the other phase was associated with output and compensation was lacking. Thus slugs were capable of long-term regulation during the growth phase of their life cycle but during reproduction adults lost weight even when fed in the absence of any compensatory mechanism. *Deroceras reticulatum*, when starved, survived for a shorter length of time (23.0 ± 1.9 days) than *A. hortensis* (54.2 ± 4.2 days) (Kemp, 1987).

3.1.8 Synthetic diets

Diets for slugs in laboratory cultures have tended to consist mainly of natural or semi-natural foods. Galtsoff *et al.* (1937) used bread and a variety of vegetables together with milk as a diet for *L. flavus*. Stephenson (1962) recommended a varied diet although he found that carrot slices were more suitable than potato, lettuce, apple and bran because these foods tended to decay quickly and contaminate the culture. Alfalfa and clover were used by Byers and Bierlein (1982) to rear *Arion* and *Deroceras* while South (1982) used germinating wheat grains to feed *D. reticulatum* and *A. intermedius*. 'Readybrek' breakfast cereal, supplemented by leaf litter and fungi, was used for cultures of *Limax* spp. by Cook and Radford (1988). Herrera (in Rueda, 1989) found that veronicellid slugs tended to grow faster and with lower mortality when they were reared on a polyphagous diet rather than a monophagous diet. The nutritional value of natural foods is variable and they are not suitable for nutritional studies.

Ridgway (1971) described the development of a synthetic diet for *A. ater*. Diets bound in a calcium alginate gel, and used previously for invertebrates, were rejected as unsuitable as they varied in texture and contracted, slowly losing water and soluble nutrients. Ridgway developed an agar-based diet in which the water content remained constant for at least 15 days and showed that it was acceptable to slugs. This diet supported a rapid rate of growth and development to maturity in *A. ater* but reproductive performance was poor (Wright, 1973). Wright described how the fertility of slugs reared in captivity tended to be generally low, although the reason was not clear, and concluded that the diet was satisfactory for use in nutritional studies. The effects of B vitamin deficiencies when *A. ater* was reared on this diet were described by Ridgway and Wright (1975). Pantothenic acid deprivation rapidly reduced the growth rate and survival was poor. The effects of deprivation of thiamin, riboflavin, niacin, pyridoxine and folacin were not detected until the second phase of growth, when the growth rate was reduced and food conversion was impaired. In trials comparing the value of various commercially available animal diets for rearing the veronicellid slug, *Sarasinula plebeia*, Rueda (1989) demonstrated that the commercially available feeds for guinea pigs, rabbits and fish resulted in faster growth rates and were more suitable for rearing slugs than the traditional laboratory slug diets described above. However, fungi grew over these feeds reducing their suitability to only two or three days.

3.2   ENERGY METABOLISM

3.2.1   Assimilation

Stern (1970) made a study of assimilation in the slug *A. ater* at a temperature of 15°C, relative humidity of 100% and photoperiod (L : D) of 12 : 12. The slugs were fed on lettuce leaves. The maximum body size was reached in 52 weeks and the following conclusions were based on mean values. During this period, 119 kJ of energy were consumed and 74% of this energy (assimilation efficiency, A/C) was assimilated to produce body weight equivalent to 24 kJ. The energy lost through respiration (i.e. metabolic energy) during this period was 64 kJ. The energy cost of synthesizing fresh tissue was, therefore, 4.96 kJ for each kilojoule of organic matter produced. An estimated 20% of the energy consumed (gross growth efficiency, P/C) and 27% of the energy assimilated (net growth efficiency, P/A) was used in the production of living material. The metabolic rate is known to depend on environmental factors, particularly temperature, and Stern (1975) investigated the effect of temperature on growth and consumption in *D. reticulatum* fed on lettuce leaves. Stern found that at 13°C the relationship was positive – for example, a slug weighing 100 mg would consume 43.3 J day^{-1} with a resulting production of 4.6 J day^{-1}. However, at 18°C, the relationship was negative with the slug consuming energy while losing weight, and this tendency was accentuated at 23°C with an energy loss of 2.1 J day^{-1} although consumption reached a maximum rate of 50.4 J day^{-1}. Stern (1975) concluded that survival of the slug at high temperatures was achieved at the expense of greater food consumption and the breakdown of body tissue.

In a study of feeding by *D. reticulatum* in woodland, Pallant (1970) estimated that a slug of average live weight 254.6 mg consumed 60.9 mg (24 h)$^{-1}$ of fresh *Ranunculus repens* Linnaeus leaf and that 47.7 mg (78%) of this was assimilated. Pallant (1974) suggested that *D. reticulatum* of average live weight 349 mg consumed 62.8 mg (24 h)$^{-1}$ of fresh leaves of the grass *Dactylis glomerata* Linnaeus and that 48.3 mg of this was assimilated, giving an assimilation efficiency of 77%. Pallant (1974) estimated the mean assimilation rate for *D. reticulatum* fed in the laboratory on carrot or grass leaves to be 134.9 J (24 h)$^{-1}$ per 100 mg dry wt at 10°C. This was slightly lower than the corresponding rates of assimilation estimated for slugs feeding in woodland, 141.5 J (24 h)$^{-1}$ per 100 mg dry wt, and slugs feeding on grassland, 161.3 J (24 h)$^{-1}$ per 100 mg dry wt and Pallant concluded that the difference reflected other activity in the field such as seeking shelter and reproduction. The mean net growth efficiency for *D. reticulatum* obtained from the laboratory experiments was 58 ± 28.7%, which was considerably higher than the 27% estimated for *A. ater* by Stern (1970) although the standard deviation of the estimate indicates considerable variation in individual estimates. Pakhorukova (1976) found that the assimilation efficiency of *D. agreste* and *D. reticulatum* was between 70 and 80%.

Jennings and Barkham (1976) compared the rates of feeding by mature and immature *A. ater* at 10°C using several different plant foods. The mean ingestion rate by immature slugs (15.53 ± 1.22 mg dry wt per gram of live wt per day) was higher than the corresponding rate for mature individuals (4.65 ± 0.32 mg dry wt per gram of live wt per day) although the assimilation efficiencies were similar, 72 and 70% respectively. These values were of the same order as those obtained by Pallant (1970) and Stern (1979). Jennings and Barkham (1976) suggested that the relatively high assimilation efficiency of slugs might be due to the large complement of enzymes present in the digestive system, which include cellulase, xylanase and chitinase, the latter being particularly important when slugs fed on fungal material. In a comparison of the assimilation efficiencies of *L. flavus* and *A. hortensis* fed on carrot root and potato tuber at a range of temperatures between 5 and 20°C, Davidson (1976) recorded mean assimilation efficiencies of 77% for *L. flavus* feeding on carrot and potato while the equivalent values for *A. hortensis* were 89%. Assimilation efficiencies of 76 and 72% were also recorded for *D. reticulatum* feeding on carrot and potato respectively. These values were similar to those described previously although the value for *A. hortensis* was higher than other recorded values. Davidson also found that immature *L. flavus* had a higher ingestion rate than adult slugs but, unlike Jennings and Barkham (1976), the assimilation efficiency was also higher for the immature slugs.

The relationship between daily food consumption and body weight of seven terrestrial slugs, two snails and two aquatic snails was examined by Rollo (1988b) using original and published data. Intraspecific regression analyses of food consumption on body weight showed considerable variation with respect to the regression coefficient, described by Rollo as the 'mass exponent', with a range of -0.617 to 1.394. Multiple regression performed for the intraspecific analysis of the data showed that 77% of the variation in the pooled data could be accounted for by the variation associated with temperature and food hydration. Food consumption was found to be almost directly proportional to body weight, with a mass exponent of 0.919, while there was a marked negative relationship between consumption and the hydration of the food. Rollo (1988b) also calculated a Q_{10} value of 1.715 for consumption by terrestrial molluscs over a temperature range of 5 to 23.5°C.

Stern (1979) recorded energy budgets for the juvenile and adult stages in the life cycle of *D. laeve*. The assimilation efficiency was highest (82%) during the most rapid stage of growth, the first juvenile stage, and the efficiency was then reduced to 75 and 72% during the second juvenile and adult stages. Stern found a strong correlation between assimilation and ingestion at each stage. The gross growth efficiency fell gradually from 16% in the first juvenile stage to 14% in the adult stage. Calorific values for the stages in the life cycle of *D. laeve* were given by Stern (1979) as: egg 17.6 kJ g^{-1} dry wt; juvenile stages 20.2 kJ g^{-1} dry wt and adult 19.6 kJ g^{-1} dry wt. Several other authors have recorded calorific values for slug tissue. Stern (1970) found a mean calorific value of 20.7 ± 0.19 kJ g^{-1} dry wt for *A. ater*. Values for the calorific value of *D. reticulatum*, estimated by two different methods by Pallant (1974), were 21.2 ± 0.98 kJ g^{-1} dry wt and 23.2 ± 0.44 kJ g^{-1} dry wt. Wallwork (1975) determined the calorific values for a range of soil invertebrates and included values of 20.28 ± 0.28 kJ g^{-1} dry wt for *A. hortensis* and 21.97 ± 0.57 kJ g^{-1} dry wt for *L. maximus*. The value for *A. hortensis* was significantly lower than the value for *L. maximus* and Wallwork suggested that this might indicate a relationship between calorific value and mobility since *L. maximus* was the more mobile of the two species. Lutman (1978) recorded a calorific value for slugs generally of 20.08 kJ g^{-1} dry wt.

Ecological efficiencies for different species of slug are compared in Table 3.1. The assimilation efficiency (A/C) in slugs is relatively high for herbivorous animals (between 70 and 90%), being higher than for snails (40–75) (Lamotte and Stern, 1987). These relatively high assimilation efficiencies reflect the efficient cellulose degradation which occurs during digestion. Lamotte and Stern also described a marked difference in net growth efficiency (P/A) between the slugs *A. ater* (39%) and *D. reticulatum* (26%) and the snail *H. aspersa* (17%). This could be partly attributed to the energy required for shell growth. The measurement of mucous production had been neglected in mollusc energy budgets and Lamotte and Stern suggested that this might represent more than 10% of production in some instances.

3.2.2 Respiration

(a) Breathing

Breathing involves the uptake of oxygen and the release of the products of cell respiration, mainly carbon dioxide and water. This is achieved by the process of diffusion through a biological membrane separating the internal environment of the slug from its external environment. Any thin epithelial layer separating two media in which a gas is present at different pressures can function as a respiratory surface. Respiratory exchange occurs in slugs through the lung-like mantle cavity, which is associated with the pallial complex described in Section 2.2.2, and also through the skin. The evidence for skin respiration was reviewed in Section 2.2.4 and Schuurmans-Stekhoven (1920) showed that over 50% of the normal oxygen uptake still took place through the skin of *D. agreste* if the mantle cavity was filled with paraffin. The mantle cavity was shown by Prior *et al.* (1983) to be responsible for about 20% of the total oxygen uptake in resting *L. maximus*

Table 3.1 Ecological efficiencies of slugs.

Species	Assimilation efficiency, A/C (%)	Gross growth efficiency, P/C (%)	Net growth efficiency, P/A (%)	Source
A. ater				
	74	20	27	Stern (1970)
Immature	72	–	–	Jennings &
Mature	70	–	–	Barkham (1976)
	–	–	39	Lamotte & Stern (1987)
A. hortensis				
	89	–	–	Davidson (1976)
D. reticulatum				
	78	–	–	Pallant (1970)
	–	–	58	Pallant (1974)
	70–80	–	–	Pakhorukova (1976)
	76, 72	–	–	Davidson (1976)
	–	–	26	Lamotte & Stern (1987)
D. laeve				
	70–80	–	–	Pakhorukova (1976)
1st juvenile	82	16	–	Stern (1979)
2nd juvenile	75	–	–	Stern (1979)
Adult	72	14	–	Stern (1979)
L. flavus				
	77	–	–	Davidson (1976)

and this figure increased to 50% in active slugs. The primary respiratory surface is the vascular roof of the mantle cavity but it may extend over the walls and floor of the cavity. The reduction in the size of the respiratory surface which has accompanied the evolution of the slug form and the varied form of the mantle cavity in different groups of slugs has been discussed by Wiktor and Likharev (1980) and by Tillier (1983). Tillier described a number of secondary respiratory surfaces associated with the pallial complex (Section 2.2.2). Maina (1989) concluded, from a study of the structure of the lung in *Trichotoxon copleyi*, that this functioned as a true lung in all respects.

In the snail *H. pomatia* the respiratory surface in the mantle cavity occupies $8.3\,cm^2\,g^{-1}$ body wt but in *Arion* it is reduced to only $0.7\,cm^2\,g^{-1}$ body wt, reflecting the relative importance of skin respiration in slugs (Wells, 1980). The volume of the mantle cavity in slugs is also relatively small, being about 0.3 ml in an *Arion* weighing 10 g (Ghiretti and Ghiretti-Magaldi, 1975). The tracheate type of lung in athoracophorid slugs is unlike that of other slugs. The dorsal pneumostome opens into a short, narrow duct which ends in a small mantle cavity whose walls are produced into thin-walled tubules which ramify through the dorsal blood sinus, recalling the tracheal system of the insects. The athoracophorid lung was described in detail by Grassé (1968). Ventilation of the mantle cavity in pulmonate slugs and snails may be achieved by a rhythmic opening and closing of the pneumostome (Section 2.1) and by contraction of the muscular floor of the mantle cavity. However, the nature of gaseous exchange in the lung remains uncertain and Ghiretti and Ghiretti-Mazgaldi (1975) reviewed the evidence for and against the existence of a ventilation lung in pulmonates. Dahr (in Ghiretti and Ghiretti-Mazgaldi, 1975) considered that pneumostome opening was normally controlled by the partial pressure of carbon dioxide in the mantle cavity. Dahr found that, in *Arion*, when the oxygen partial pressure fell to 15 mmHg the pneumostome closed and that the pressure inside the cavity was then increased to 20–30 mmHg by muscular contractions. The closing of the pneumostome may simply serve to reduce evaporation and Dickinson *et al.* (1988) found that, in fully hydrated *L. maximus*, the pneumostome was held open most of the time, closing an average of 0.17 ± 0.19 times per minute.

(b) Oxygen consumption

Newell (1967) investigated the effect of the level of activity on the respiration rates of *D. reticulatum*, *A. hortensis* and *T. budapestensis*. The rate of consumption of oxygen by an active slug, the active metabolism, was considerably higher than the standard or quiescent metabolism. For example, the mean oxygen consumption per 100 mg dry weight of quiescent *D. reticulatum* at 15°C was $67.5\,\mu l\,O_2\,h^{-1}$ compared with $412.3\,\mu l\,O_2\,h^{-1}$ for an active slug. Corresponding figures for *A. hortensis* were $26.7\,\mu l\,O_2\,h^{-1}$ (quiescent) and $152.6\,\mu l\,O_2\,h^{-1}$ (active) and for *T. budapestensis* were $41.1\,\mu l\,O_2\,h^{-1}$ (quiescent) and $272.9\,\mu l\,O_2\,h^{-1}$ (active). Newell (1966b) showed that *D. reticulatum* was active for only about six hours in every 24 hours. Thus it was necessary, when calculating the estimated oxygen requirement for a slug over 24 hours, to take into account the duration of quiescent and active phases (Newell, 1967). Newell concluded that the oxygen requirement for *D. reticulatum* at 10°C estimated from a constant respiration figure (mean of active and resting rates) was considerably higher ($2923.2\,\mu l\,O_2$ per 100 mg dry wt) than an estimate which took into account the different durations of the active and resting phases (2043.0 μl O_2 per 100 mg dry wt). Estimates obtained by Opalinski (1981) for the respiratory rate of a *Limax* sp. ranged from 167.3 to 234.8 μl O_2 h^{-1} per 100 mg dry body wt at 20°C, with a respiratory quotient (RQ) of between 0.72 and 0.79, indicating that energy was being mainly produced by fat metabolism. Janssen (1985) recorded an RQ value of 0.76 for *A. ater* and stressed the importance of fat metabolism, especially during periods of starvation. However, Horne (1979) suggested that RQ values of 0.97 and 0.86 for feeding and fasting *L. flavus* respectively represented a shift from the normal utilization of carbohydrate towards protein utilization during starvation. Horne (1979) recorded a 30% decrease in the rate of respiration by the 16th day of fasting. This was considerably less than that recorded for two species of snail. He concluded that *L. flavus* was less well adapted to starvation, probably because the slug was active for most of the year unlike the snails which became dormant in response to starvation. Reddy *et al.* (1981) demonstrated that the total ninhydrin positive substances in the blood of *L. alte* increased during starvation due to protein degradation.

The effect of temperature on the rate of respiration was also measured by Newell (1967) for the same three species of slug. The rate of oxygen consumption by quiescent slugs was relatively unaffected by temperature changes over the normal range, values of $Q_{10}^{0°C-10°C}$ and $Q_{10}^{5°C-15°C}$ for each species being approximately 1. However, the rate of oxygen consumption by active slugs was strongly temperature dependent with corresponding values of $Q_{10}^{0°C-10°C}$ ranging from 1.705 to 1.765 for the three species. The effect of temperature on the rate of respiration of an animal may be modified by its previous history with respect to temperature. This modification – which may be in response to a range of environmental factors – takes place naturally when animals are exposed to seasonal climatic changes, and is known as acclimatization. Where these changes are experimentally induced under laboratory conditions the process is known as acclimation. When *A. circumscriptus* were acclimated at a wide range of temperatures between 5 and 25°C, Roy (1963) found a decrease of 1–1.5% in metabolic rate, measured by oxygen consumption or by heat produced, for every degree of increase in the acclimation temperature. Roy (1963) also found that the effect of thermal acclimation on the metabolic rate was relatively rapid, becoming apparent after only three hours' exposure to the acclimation temperature. The temperature quotient (Q_{10}) was also found to vary inversely with the weight of slug and experimental temperature, and in an irregular manner with the acclimation temperature. Roy (1969) described the effect of these three factors on the variation in oxygen consumption in terms of a multiple regression equation. Rising and Armitage (1969) found similar tendencies for acclimation in *L. maximus* to those described by Roy (1963) for *A. circumscriptus*, except that the relationship between body weight and oxygen consumption was less clear. The metabolic responses of *A.*

circumscriptus and *L. maximus* to temperature were also similar to those described for *L. flavus* by Segal (1961). However, Rising and Armitage (1969) showed that the pattern of acclimation was different in *P. carolinianus* and related this difference to the greater commitment that this species shows to hibernation. The temperature preference of *L. maximus* was also altered as a result of temperature acclimation while the responses of *P. carolinianus* showed no such alteration.

3.3 CARBOHYDRATE METABOLISM

Relatively little work has been carried out on carbohydrate metabolism specifically in slugs. According to Meenakshi and Scheer (1968), the major carbohydrate in the blood of *A. columbianus* was glucose, at levels of about 28 mg $(100 \text{ ml})^{-1}$, and representing 84% of the total carbohydrates present. Glycoproteins were also present in appreciable amounts, representing 11% of the total carbohydrate content. Circadian fluctuations in carbohydrate levels were recorded in the slug *L. alte* by Kumar *et al.* (1981). The total carbohydrate level of several tissues was highest during the inactive light phase and lowest during the active dark phase while the changes were greatest in the foot, indicating a correlation with locomotor activity. When the snail, *H. pomatia*, feeds, the level of glucose in the blood rises and then returns to normal level, suggesting that there must be a regulatory process (Runham and Hunter, 1970). Meenakshi and Scheer (1968) showed, by the use of ^{14}C glucose, that this was the precursor of the oligosaccharides maltose, maltotriose and maltotetrose which were present in the digestive gland. Three polysaccharides were present in appreciable amounts in the slug. Glycogen was present in large amounts in the digestive gland and foot and in small quantities in other parts of the body, including the albumen gland (Meenakshi and Scheer, 1968). Deyrup-Olsen *et al.* (1986) found that the autotomy section in the tail of *Prophysaon foliolatum* (Gould) was packed with glycogen cells. If food was lacking, the stored glucogen was utilized by the slug. Deyrup-Olsen *et al.* also recorded an insulin-like substance present in the blood which might have a role in glucogen metabolism. Galactogen was present in large amounts in the albumen gland, comprising 96% of the total carbohydrates present in the gland. ^{14}C glucose injected into slugs was incorporated into galactogen, demonstrating that glucose was a precursor for this polysaccharide (Meenakshi and Scheer, 1968).

Since *D. reticulatum* was more susceptible to starvation than *A. hortensis*, Kemp and Newell (1989) compared the total lipid and glycogen contents of these species to determine whether *A. hortensis* possessed greater energy reserves than *D. reticulatum*. They found that in a natural population the glycogen content of *D. reticulatum* was consistently lower than that of *A. hortensis* with no distinct seasonal variation. In the laboratory the highest mean glycogen contents recorded were 9.3% for *A. hortensis* and 4.8% for *D. reticulatum*. These values were more than twice the value found in slugs from the field, and Kemp and Newell suggested that conditions in the field did not allow the slugs to build up their reserves to maximum capacity. The glycogen content of *A. hortensis* showed seasonal changes, becoming lower in spring and autumn with the onset of reproduction. Kemp and Newell (1989) suggested that the fall in glycogen content in *A. hortensis* represented the conversion of glycogen to galactogen. In *A. columbianus*, the galactogen content of the albumen gland showed a seasonal variation with highest levels during the breeding season and lowest during the period following egg laying (Meenakshi and Scheer, 1969). Galactogen has been shown to be the reserve material for the developing embryo in the egg of *D. reticulatum* (Bayne, 1966). The third polysaccharide, a sulphated acid mucopolysaccharide, was present in the glandular region of the common duct of the reproductive tract (Meenakshi and Scheer, 1968). A similar polysaccharide was recorded by Bayne (1967a) in the oviducal gland and eggshell of *D. reticulatum*, suggesting that the egg received its shell covering from the glandular part of the common duct. The synthesis of the complex polysaccharides is by way of uridine diphosphate compounds and the presence of UDP-glucose, involved in the biosynthesis of glycogen and of UDP-galactose a precursor of galactogen, has been demonstrated in *L. maximus* (Goudsmit, 1972). Dutton (1966) showed that phenolic glycosides were formed by glycosyl transference from UDP-glucose in the digestive tract and the digestive gland of *A. ater*. Kumar *et al.* (1981) described how the total carbohydrate and protein content of selected organs of the slug, *L. alte*, showed circadian fluctuations. The higher carbohydrate levels were recorded during the day while high protein levels were found at night.

3.4 LIPID METABOLISM

The lipid content of a large *A. ater* was estimated by Thompson and Hanaham (1963) to be 0.8% of fresh body weight (8.0% of dry weight) and that of young *A. ater* was 2.3% of body weight. The corresponding value for *A. columbianus* was 1.12% of fresh body weight. Lipid droplets were recorded in the basal region of calcium, digestive and excretory cells of the digestive gland of slugs (Walker, 1969; Janssen, 1985). Janssen described how after three weeks of starvation in *A. ater*, areas of high lipid content coexisted with those of low content while the droplets had become both smaller and fewer. All the large droplets had disappeared after eight weeks' starvation and the distribution of lipids had become more restricted. However, Janssen (1985) found that the total body lipid content of normal and starved *A. ater* was similar and about 16% of dry body weight, although starved animals lost weight in proportion to the time they were starved. Janssen concluded that, although the lipid content of the digestive gland was reduced and the lipid content of the body wall increased during starvation, a redistribution of lipids in the body was unlikely but, rather, that body wall materials were transformed into lipids before being metabolized. The low RQ (below 0.8) of starved slugs supported this emphasis on lipid metabolism. A study of lipid levels in the digestive gland, reproductive system and foot muscle during the development of *A. ater* showed considerable variation in lipid levels and these variations were related both to the life cycle and to environmental conditions (Catalan *et al.*, 1977). In a comparison of total lipid contents of *D. reticulatum* and *A. hortensis*, Kemp and Newell (1989) found that the total lipid contents were similar for both species and showed little seasonal change.

Several groups of lipids have been investigated in slugs. Voogt (1972) reported that cholesterol was the main sterol occurring in *A. ater*, representing 78% of the sterol content, while at least six other sterols were present. Voogt found no ergosterol present but Bock and Wetter (in Voogt, 1972) found that ergosterol represented 25% of the sterols present in *A. ater*. Bock and Wetter also concluded that *A. ater* was able to synthesize ergosterol because the presence of this sterol was independent of the diet of the slug. Voogt (1967) investigated lipid biosynthesis in *A. ater* by the incorporation of sodium acetate-1-^{14}C into lipids and showed that the slug could synthesize sterols from this acetate. Cholesterol was used by Gottfried and Dorfman (1970c) as the initial substrate to follow *in vitro* steroid biosynthesis in young male phase gonad tissue in *A. californicus*. The phospholipids in *A. ater* and *A. columbianus* were studied by Thompson and Hanahan (1963), with special reference to α-glyceryl ethers. The lipids of *A. ater* contained about 40% phospholipids, with a relatively high glyceryl ether content. The remainder of the lipid content of the slug comprised approximately 20% triglycerides and 40% steroids. Further investigations into the biosynthesis of ether-containing phospholipids in *A. ater* have been described by Thompson and Lees (1965) and Thompson (1966, 1968). Liang and Rosenberg (1968) described another phosphorus-containing lipid, 2-aminoethylphosphonate, from the slugs *L. valentiana* and *L. flavus* and suggested that the precursor of this lipid was probably an intermediate compound formed during glycolysis. Czeczuga (1984) recorded the occurrence of the carotenoid, α-doradexanthin, in the slugs *A. ater, A. subfuscus* and *D. agreste*.

3.5 CALCIUM METABOLISM

Calcium storage compartments in the slug *D. reticulatum* include the internal shell and the calcium cells in the digestive gland, the integument and in the general connective tissue, where they are associated particularly with the vascular system and the nervous system (Fournié and Chétail, 1982a, b). The distribution of connective tissue calcium shows interspecific differences. For example, the arteries of some slugs are transparent but in *L. maculatus* they are semi-transparent and in *A. ater* they are white and opaque, due to the presence of calcium cells in the connective tissue of the artery wall (Duval and Runham, 1981). Calcium cells have also been recorded from the connective tissue of the arteries of *L. maximus* (Fournié and Chétail, 1982a). Calcium measurements in the different compartments indicated that the shell, the integument and the digestive gland contained 47, 15 and 18% of the total calcium content in young slugs just before the egg-laying period (Fournié and Chétail, 1982b).

Fournié and Chétail investigated the mobilization of calcium reserves during the reproductive cycle in *D. reticulatum*. During the male phase of the cycle, calcium was accumulated in all the storage compartments of the slug. Calcium levels then decreased progressively from the beginning of oogenesis, reaching their lowest level when the first eggs were laid. During the interval between the first and second egg-laying periods (about eight days), calcium levels in the storage compartments increased and this was followed by a decrease just before the second period of egg laying. The stored calcium content increased rapidly after the last egg laying. Just before the death of the slug, shortly after egg laying was completed, the total calcium content was higher than it had ever been before. The internal shell was well calcified with numerous hypostracal spherules while the digestive gland and integument were also heavily loaded with calcareous spherules. Fournié and Chétail (1982b) suggested that this calcium overload might be a contributory cause of death in the slug (Section 8.3). Calcium accumulated in the common duct just before each egg-laying period and this was followed by a fall in the calcium level after eggs had been laid. Similar changes in calcium levels were recorded in the albumen gland but earlier than those in the oviduct. The calcium level remained low in the common duct after the last egg-laying period until death although calcium continued to accumulate in the albumen gland. The total calcium loss as the result of egg laying represented 32% of the total metabolized calcium or 44% if calcium stored after the last eggs were laid was ignored (Fournié and Chétail, 1982b).

Both Burton (1972) and Janssen (1985) found evidence that the spherules in the calcium cells contained calcium phosphate as well as carbonate. Burton also showed that magnesium was also present in the spherules of *L. maximus* and that the concentration of magnesium was positively correlated with that of calcium. Metal binding has been frequently associated with calcium salts, especially carbonates, in molluscs and the relationship between zinc and calcium metabolism was described by Recio *et al.* (1988a, b). Sac preparations of isolated body wall of slugs were found to respond by fluid output to external stimulation, both tactile and osmotic, and to the internal administration of neurotropic agents. Within the physiological range, these responses varied with the concentration of calcium ions, both inside the sac and in fluids in contact with the external surface of the body. Stimulated body wall was able to increase the concentration of calcium above that in the blood or fluid filling the sac, suggesting the presence of a calcium secreting process. Calcium secretion could play a part in mucus formation and the internal concentrations of calcium ions might serve as cues in the regulation of fluid exchange by the body wall (Deyrup-Olsen and Martin, 1982b).

3.6 NITROGEN METABOLISM

Although several publications, e.g. Delaunay (1931), give values for the rates of excretion of the nitrogenous products of protein catabolism by slugs, the first details of nitrogenous excretion in these molluscs were given by Jezewska (1969). Potts (1967), in a general review of excretion in molluscs, summarized the values found previously for rates of excretion of the various nitrogenous compounds. Potts concluded that ammonia excretion was small or negligible in terrestrial molluscs and that the ornithine cycle was absent in *Helix*. However, the synthesis of urea might take place by other routes, such as the breakdown of uric acid and also from arginine in the diet. Horne and Barnes (1970) showed that the ornithine cycle functioned in several pulmonate snails which they examined, including *H. aspersa*, although urea did not accumulate in helicid snails, even during aestivation. The evidence for the ornithine cycle in slugs was less clear. Horne and Boonkoom (1970) showed that many of the ornithine cycle enzymes were present in the digestive gland of several pulmonate snails that were examined but levels in the slug *L. maximus* were below the level of detection, except for ornithine transcarbamylase and arginase, although high blood urea levels were found in *L. maximus*. They were also unable to detect urease activity in this slug. Horne (1977a, b) recorded tissue levels of urea ranging from 13.8 to 25.0 μmol g^{-1} wet wt and ammonia levels of 6.0–10.0 μmol g^{-1} wet wt in *L. flavus*, and suggested that the ammonia was efficiently incorporated into purines and urea. Delaunay (1931) described how 60–80% of the total non-protein nitrogen (NPN) in *Arion* was in the form of urea after 'auto-digestion' of the excreta and this urea might have been derived from uric acid. Young *L. maximus* ingested about twice as much of the total nitrogen and protein nitrogen and excreted the same amount of protein nitrogen as adult slugs (Jezewska, 1971). Jezewska concluded that the young slugs were retaining about 50% of the protein nitrogen from food ingested. Spitzer (in Potts, 1967) found that *Arion* excreted only 3% of NPN as urea. Potts (1967) concluded that slugs appeared to be only slightly uricotelic and indicated that the study of the purines generally had not received much attention.

Jezewska (1969) investigated nephridial excretion in snails and in the slugs

L. maximus and *D. agreste* and concluded that slugs, like snails, were purinotelic. The dry nephridial excreta contained 72–91% of purines, with the lower values being found in slugs. Uric acid accounted for about a half of the total purines in the limacid slugs. The guanine content was also high in these slugs and equalled that of uric acid in *L. maximus*. Xanthine was a minor component of the excreta of the slugs examined. The total purine content of dry excreta was estimated at 72–77 mg per 100 mg of dry material for the two slug species in the autumn at the time eggs were laid. Purine excretion in arionid and limacid slugs was compared by Jezewska (1972). The arionid slugs *A. ater* and *A. circumscriptus* were not seen to excrete purines and Jezewska concluded that these species were probably not purinotelic. All eight limacid slugs examined, except *L. flavus*, excreted sufficient quantities of purines to be regarded as purinotelic animals. The proportions of the three purines present – uric acid, xanthine and guanine – varied depending on the species, although they resembled those reported by Jezewska (1969), with guanine and uric acid levels tending to be higher than xanthine levels. Garner (1974) showed that *D. reticulatum* was purinotelic, excreting uric acid and xanthine but, unlike previous authors, found little evidence for guanine as an excretory product. Less than 1.5% of the nitrogen excreted by *D. reticulatum* was in the form of gaseous ammonia from the mantle cavity. Loest (1979) found no evidence for gaseous ammonia production by four species of slugs (*P. carolinianus*, *V. floridana*, *D. laeve* and *L. maximus*) but showed that they absorbed gaseous ammonia from the air via the epidermal mucous cover and suggested that this might act as a salvage mechanism for endogenous nitrogen.

A study of ureotelism in *L. flavus* by Horne (1977a) showed that urea and purines accounted for up to 59 and 41% of the excretory nitrogen, respectively, with little or no ammonia excreted. Uric acid (64%) was the predominant purine excreted with guanine (22%) and xanthine (14%) also being produced, but other purines, such as adenine and hypoxanthine, were not detected. The purines were excreted irregularly as semi-solid, yellowish nubs while urea was excreted in liquid urine. Horne (1977b) studied the regulation of urea biosynthesis in active and starved *L. flavus*. When starved, the slugs utilized larger amounts of tissue protein and smaller quantities of carbohydrates. Since amino acid carbon was being channelled towards Krebs cycle oxidation, ammonia became readily available via the deamination of these amino acids. With the increased availability of ammonia, urea biosynthesis increased twofold or threefold above the rates in normal feeding slugs. The enzyme activities of glutamine synthetase and the urea cycle enzymes were not altered by starvation. Horne (1977b) concluded that since *L. flavus* produced ample quantities of urea during both active feeding and starvation and did not have an active urease, slugs differed considerably from many other gastropods, such as *Helix* and *Otala*, in their nitrogen metabolism. The effect of starvation on purine biosynthesis in *L. flavus* was investigated by Horne and Beck (1979). The total body purines decreased by 42% during starvation and rates of purine synthesis were also reduced by about 56%. Starvation appeared to result in a shift from purine synthesis to urea synthesis. Horne and Beck (1979) also found hypoxanthine and adenine in small quantities in the excreta of *L. flavus*. It may be concluded that *L. flavus* and probably other slugs have a combined ureotelic–uricotelic pattern of nitrogenous excretion. During starvation, ureotelism is enhanced by increased urea biosynthesis and decreased uric acid production. Since there is evidence that some slugs aestivate under very dry conditions (Section 4.2.7), it would be interesting to discover whether this pattern of excretion is also adopted during aestivation.

4

Vascular system, water relations and nitrogenous excretion

4.1 VASCULAR SYSTEM

The location of the heart and the common aorta was described in Section 2.2.2. The anatomy of the heart of *L. maximus* has been described by MacKay and Gelperin (1972). Injections into the heart have demonstrated that there is an effective auriculo-ventricular valve preventing back flow of blood from the ventricle into the auricle (Duval and Runham, 1981). Duval and Runham also showed that there were no valves between the aorta and ventricle and between the main veins and the auricle. In most slugs the arteries were transparent although in *A. ater* they were white and opaque due to the presence of calcareous deposits on the outside of the vessel walls (Runham and Hunter, 1970). Curtis and Cowden (1979) examined the structure of the wall of the aorta of *L. maximus*. Both the anterior and posterior aortae were similar in structure with a loosely organized endothelial layer surrounded by two layers of innervated smooth muscle. The outer (adventitial) layer of the aorta consisted of large, glycogen filled-cells with characteristic arrays of pores in their plasma membranes. These resemble the Leydig cells, described by Runham and Hunter (1970), which are associated with the walls of the arteries and may store glycogen and calcium. Curtis and Cowden (1979) also described blood cells in the form of amoebocytes containing large glycogen deposits which were occasionally found in the walls of vessels. Furuta *et al.* (1988) have described blood cells in the slug *Incilaria*, which probably have a defensive role. There is an arterial gland in *D. reticulatum* which consists of irregularly shaped masses of opaque whitish tissue arranged in patches along the distal portion of the cephalic artery and its branches (Laryea, 1969). The role of this gland appeared to be secretory and it appeared that the secretion was released into the haemocoele. Runham and Hunter (1970) described how the arterial gland began to disintegrate following the end of egg laying towards the end of life. Laryea (1969) showed that this gland was also present in *D. caruanae* and *L. flavus*.

The vascular system of *Limax*, including arteries, veins and some sinuses, was illustrated by Grassé (1968). The arterial systems of six species of slugs were described by Duval and Runham (1981). They showed that, while there was some variation between the families studied, the differences within families were relatively small and the distribution of vessels to the head region was similar for all species. The anterior aorta supplied blood to most of the body including the foot while the posterior aorta supplied mainly the digestive glands, intestine and gonad. In the foot, the large pedal arteries branched to give a large number of small arteries which penetrated the thick muscle layers of the foot. The smallest arteries emptied into small blood spaces or lacunae in the connective tissue of the foot. This blood was eventually returned back to the heart through spaces in the body wall probably assisted by contractions of the foot musculature. Duval and Runham (1981) suggested that the spaces in the body wall, which lack any limiting membrane (see Dyson, 1964), might serve for a venous return of blood and also provide a pathway for skin respiration. Garner (1974) described how most of the blood from the body wall eventually entered two pallial sinuses, running through the connective tissue of the dorsal body wall from the posterior end of the animal to the posterior edge of the kidney. A third central sinus was often present in larger individuals.

The heart and kidney are isolated from the sinuses of the body cavity and have no direct arterial blood supply but instead receive blood returning from the rest of the body. Blood enters the auricle from vessels in the roof of the lung-like mantle cavity and so the heart receives blood which has been well oxygenated and has also passed through the kidney. Garner (1974) showed that the epithelia of the kidney and mantle cavity were closely opposed so that venous blood came simultaneously into contact with both epithelia. A large volume of venous blood also came into contact with the epithelium of the ureter. The roof of the mantle cavity was supplied by an afferent vessel, the *circulus venosus*, which circled the edge of the cavity in *Limax* (Grassé, 1968). This was described as the *vena circularis* by Tillier (1983). Garner (1974) found that the *circulus venosus* was absent from *D. reticulatum*. In a diagram of the vascular system of *Limax*, Grassé (1968) showed that blood from sinuses in the body walls flowed into the *circulus venosus*, which also received blood returning from other parts of the body including the foot. However, Tillier (1983) suggested that blood from the lacunae of the foot was not returned directly to this vessel but first circulated through blood spaces in the mantle wall, providing an opportunity for it to become oxygenated through skin respiration before entering the afferent vessel and passing to the roof of the mantle cavity. Duval (1982) also concluded that the main function of the spaces in the integument was to return blood to the heart and that they had a role in skin respiration. Duval and Runham (1981) described how the haemocoele, the main body cavity in slugs, was divided into compartments or large sinuses around the organs. These included an anterior or cephalic sinus, a middle or visceral sinus and a posterior or gonadal sinus.

Little has been written about the fluid mechanics of the slug circulatory system although this subject has been investigated for *Helix* and other snails and the literature on the fluid mechanics of circulation in gastropods and bivalves was reviewed by Jones (1973b). Duval (1983) measured the blood pressure of *D. reticulatum* and *L. maculatus*. In *D. reticulatum*, mean diastolic and systolic auricular pressures were 1197 and 2158 N m^{-2} respectively, while the corresponding values for diastolic and systolic ventricular pressures were 2060 and 3443 N m^{-2}. Within the haemocoele of resting slugs the blood pressure was equivalent to the ambient pressure. Higher pressures were recorded for *L. maculatus*. Mean diastolic and systolic pressures within the auricle were 2874 and 5297 N m^{-2} respectively, and within the ventricle 5091 and 7309 N m^{-2}. These blood pressures were higher than those recorded for other pulmonates and Duval (1983) discussed the implications of this difference. Duval (1983) also investigated heart beat in *D. reticulatum*, *L. maculatus*, *T. budapestensis* and *A. ater*. There was an inverse relationship between heart beat and body weight. A characteristic specific heart rate was found which was related to body weight, with a value of 97 beats per min for the small species *D. reticulatum* and a heart rate of 55 per min for the larger *A. ater*. These values were similar to those recorded for these species by Schwartzkopff (1956), i.e. 91 and 49 per min respectively. Carrick (1938) recorded a heart beat of 48 per min for *D. reticulatum* just prior to emergence from the egg. The innervation of the heart and neural control of heart beat together with the effect of neurotransmitters on the heart are discussed in Chapter 6.

4.2 WATER RELATIONS

4.2.1 Water content of body

Terrestrial slugs and snails lose water by evaporation from exposed surfaces if the ambient humidity falls below blood equilibrium humidity, i.e. about 99.5% r.h. at 20°C (Machin, 1975). This will occur even in inactive slugs because they have no protective shell, while in active slugs the water loss increases rapidly due to accelerated mucus production. Dainton (1954a) found that the rate of evaporation from several species of slug at 45% r.h. was 3–5% of the original body weight per hour. However, the increased mucus production by *D. reticulatum* during locomotion in a saturated atmosphere resulted in a loss of 17% of original body weight over a period of 40 minutes (Dainton, 1954a). Prior *et al.* (1983) obtained a similar result using *L. maximus*. Howes and Wells (1934a) found that the weight of *H. pomatia* showed continuous weight fluctuations, corresponding to variations in water content, even under fairly constant environmental conditions. Weights of the snail varied by up to 50% in an irregular cycle over a few days. Similar fluctuations were demonstrated in the slugs *A. ater* and *L. flavus* (Howes and Wells, 1934b). The slugs were kept in the laboratory in vertical cylinders of perforated zinc, closed with glass plates and water was supplied in a dish. The temperature of the laboratory varied between 15 and 19°C while the relative humidity ranged from 40 to 80%. It might be expected that slugs would lose weight under these conditions through evaporation and mucus production and, in addition, weight changes would result from feeding and defaecation and these could explain the irregular fluctuations. Foster (1954) repeated these experiments and obtained similar results using *A. ater*, *D. reticulatum*, *T. sowerbyi*, *L. maximus* and *A. hortensis*. However, when droplets of water were freely available on the floor of the cage, these fluctuations were considerably reduced and Foster concluded that the variations in slug weights had been exaggerated by the experimental conditions. Foster (1954) also demonstrated that free water was required and that if this was not provided, slugs rapidly lost weight and died within three or four days. Both Howes and Wells (1934b) and Foster (1954) concluded that the major intake of water was by ingestion into the crop. Foster (1954) suggested, as the result of a series of experiments using *A. ater*, that the slug was unable to absorb water through the skin. However, Dainton (1954a) demonstrated that *L. maximus* was able to compensate for water loss by absorbing water through the skin surface.

Lyth (1982) found that when 11 species of slug were reared under more favourable conditions, i.e. at a lower temperature (11 ± 1°C) and at a high humidity, they showed no marked fluctuations in hydration but exhibited relatively stable water contents within a species, indicating that slugs possess some method for the control of body water content. This stability was confirmed by Lyth (1983), who measured the daily weights of individuals of *A. ater* kept over a four-week period. Although weights were rarely constant, daily fluctuations were almost invariably small, changes over 6% of body weight being uncommon. The mean water contents (% wet weight) of adult slugs recorded by Lyth (1982) ranged from 80.2 and 80.8% in *A. fasciatus* and *A. hortensis*, to 88.3, 88.9 and 89.8% in *A. ater*, *D. laeve* and *A. intermedius* respectively. The water contents of these two groups were significantly different from those of other species examined. Other mean water contents were: *T. budapestensis* 82.5%, *T. sowerbyi* 85.1%, *D. reticulatum* 85.1%, *A. subfuscus* 86.1%, *L. flavus* 86.2% and *D. caruanae* 87.7%. Within a species, the water contents became less with increasing body weight; for example, young *L. flavus* (mean weight 1.36 g) had a significantly higher water content (87.9%) than adults (5.4 g and 86.2%). This relationship did not apply between species as the larger species did not have the lowest water contents. Water contents of these species, with the exception of *L. flavus*, were estimated for slugs taken from a range of habitats between April and October (Lyth, 1983). Mean water contents ranged between 81.2% (*A. fasciatus*) and 90.6% (*D. laeve*), with values for *A. fasciatus* and *A. hortensis* again being the lowest and values for *A. ater*,

D. laeve and *A. intermedius* being among the highest, although the water contents for *D. reticulatum* were also high for slugs taken from the field. Water contents within species were again relatively stable. Other estimates of the water contents of slugs fall within this range, e.g. 87.5% for *A. ater* (Pusswald, 1948) and 86.3% for *D. laeve* (Forsyth *et al.*, 1983). Thus, although the water contents are not characteristic for a species, because they vary with environmental conditions, the variation within a species under optimum conditions is narrow and some species, such as *A. hortensis* and *A. fasciatus*, appear to have consistently lower water contents than others, such as *A. intermedius* and *D. laeve*. Lyth (1983) concluded that the marked fluctuations in water content observed under laboratory conditions by Howes and Wells (1934b) and, to a lesser extent by Foster (1954), did not occur under natural conditions because slug water content was maintained by water uptake, mainly through the integument, together with appropriate cryptozoic behaviour patterns. A 24-hour cycle in body weight and hydration in fasting *L. maximus*, with maxima at midnight and minima at noon, was described by Prior *et al.* (1983). There was also a direct relationship between blood osmolality and loss in body weight.

The water present in slugs is distributed between intra- and extracellular compartments, with the haemocoele forming the main extracellular storage compartment. Pusswald (1948) concluded that the blood formed the largest water storage compartment in slugs. Martin *et al.* (1958) found that extracellular water made up nearly a half of the total body water content of *A. ater*, although the blood volume in *A. ater* could vary from 25 to 47% of the fresh weight of the slug. Since there is no separation of blood from the interstitial tissue fluid, the skin also contains large quantities of water and is an indicator of the state of hydration of the slug (Pusswald, 1948). In addition to the intra- and extracellular water, water is present in the crop and in mucus distributed over the surface of the body. Pusswald (1948) suggested that the gut in *A. ater* and *L. maximus* might serve as a temporary reservoir for water, while Deyrup-Olsen and Martin (1987) found that the rectum of *A. columbianus*, which was often distended with fluid at an osmolality below that of the blood, served as a small but readily available source of water for use under conditions of water deficiency. Lyth (1982) reported the discharge of stored pallial water, a process originally described from snails by Blinn (1964), from the pneumostome of slugs. However, Martin and Deyrup-Olsen (1982) showed that relatively large volumes of blood were vented through the pneumostome of several species of slug in response to osmotic, electrical and mechanical stimuli and this might explain the discharge of 'pallial water'.

The regulation of body volume in response to mechanical and chemical stimuli was investigated by Martin *et al.* (1990) using *A. columbianus*. They used an *in vitro* preparation of the posterior chamber of the body wall, i.e. that part of the body normally occupied by the viscera and blood.

This preparation was separated from the ganglia of the head and so was only under peripheral nervous control. The volume contained by the posterior chamber preparation was determined, firstly, by the contraction or relaxation of muscle cells causing exchange of fluid between the chamber and the manometer controlling pressure within the sac, and secondly, by secretion of epidermal channel cells (Section 2.3). These cells transported fluid from inside the posterior chamber to the surface of the skin. Mechanical stimulation of the epithelium, when the posterior chamber was maintained at a constant internal pressure, resulted in a short burst of secretion. Prolonged stretching of the body wall caused intermittent bursts of secretion and muscle contraction while, at very high pressures, muscle contraction was suppressed and large volumes of fluid were secreted so that the volume of the posterior chamber gradually returned to its initial value. When the posterior chamber preparation was immersed in fluid containing garlic extract, the effect was similar to that of mechanical stimulation, initiating a burst of secretion. However, other chemicals known to be aversive to slugs did not initiate this secretory response.

4.2.2 Blood

Roach (1963) found considerable variation in the concentration of the blood of A. ater, the osmolality of which ranged from 97 to 231 mOsm kg^{-1} water. Similar values for this species of 69–220 mOsm kg^{-1} water were recorded by Rouschal (in Machin, 1975). The osmolality of D. reticulatum blood was shown by Hughes and Kerkut (1956) to vary from about 161 mOsm kg^{-1} of water in active slugs to 430 mOsm kg^{-1} water in desiccated slugs. Prior et al. (1983) showed that during the progressive dehydration of L. maximus, the blood osmolality increased exponentially, from 140 mOsm kg^{-1} water at 100% initial body weight to 200–220 mOsm kg^{-1} water at 65% initial body weight. This indicated that the water lost during the initial stages of dehydration originated from the blood. Bailey (1971a) recorded osmolalities of 349 and 410 mOsm kg^{-1} water, respectively, for D. reticulatum taken from the wild and for laboratory-reared animals. Garner (1974) demonstrated a diurnal rhythm in the blood concentration of D. reticulatum, with the lowest concentration being found between 21.00 and 05.00 hours. An analysis of the blood of A. ater by Roach (1963) showed that, of the cations measured, potassium and calcium were the most variable while sodium was the most constant. About 20% of the blood calcium existed in an organically bound form. Chloride was the most concentrated of the anions measured, and the least variable. Sulphate was more variable while the phosphate concentration was low.

The pH of the blood of A. ater was high, 8.83, and varied only within narrow limits (Roach, 1963). Bailey (1971a) recorded a lower value of 7.72 for the pH of D. reticulatum blood measured in situ in the haemocoele between the viscera using a microcapillary electrode. However, the pH of a pooled sample was similar to that of *A. ater* at 8.86. Parivar (1980) found that the pH of *A. ater* blood was 7.8 when measured *in situ*. Since Burton (1969) showed that most of the buffering capacity of the blood was due to bicarbonate in the snail, *H. pomatia*, it seems likely that when blood was exposed to the air it would rapidly lose carbon dioxide by breakdown of the bicarbonate and the alkalinity would increase. Thus pH measurements not made *in vivo*, such as those of Roach (1963), are probably too high. Roach (1963) concluded that the greater part of the osmotic fluctuation that occurs in the blood was due to changes in concentration of non-electrolytes rather than electrolytes, which were gained or lost with the water. The activity of slugs appeared to be regulated by the degree of hydration of the slug (Howes and Wells, 1934b; Dainton, 1954a). Since the gain and loss of water could result in a change of concentration of Ca^{2+} and K^+, ions which affect nervous and muscular irritability, Roach suggested that this might be the mechanism causing changes in activity in the slug, in addition to the direct osmotic effect. Two physiological salines for use with slugs were also described by Roach (1963). Similar concentrations of sodium, potassium and magnesium were found in the blood of *A. ater* by Burton (1968) although the calcium levels were about twice as high as those reported by Roach (1963). However, it has already been shown in Section 3.5 that calcium levels in slugs vary widely, depending on their physiological state.

Roach (1963) estimated the mean protein content of *A. ater* blood at 10.55 mg ml^{-1} although the protein content was very variable (coefficient of variation 28.6). According to Runham and Hunter (1970), this protein was largely haemocyanin and Burton (1964) also found that haemocyanin was the most abundant solute in the blood of the snail *H. pomatia*, with a concentration of between 9 and 77 g kg^{-1} water. Bailey (in Runham and Hunter, 1970) found that after haemocyanin was removed by centrifugation, some protein still remained in solution. In *D. reticulatum* this residual protein comprised about 4% of the total blood protein content and consisted of up to ten constituents, although its function was not known. Murthy and Ramamurthi (1978) found that the most abundant free amino acids in the blood of *L. alte* were serine + glycine, glutamic acid and methionine. The blood composition differed from that of two snails examined, asparagine being absent from the snails but present in the slug while glutamine was not detected in the blood of the slug. Although arginine, which is associated with the urea (ornithine) cycle, was not present in the blood, it was present in several tissues such as the digestive gland and the foot muscle. The highest tissue concentration of free amino acids was found in the digestive gland, comprising 54% of the total free amino acid pool, while the foot contributed 34% and the blood only 6% to the pool.

The respiratory pigment haemocyanin is a copper protein which is rarely found in cells but is dissolved in plasma as a high molecular weight polymer (up to 9×10^6 Da) which dissociates into smaller units upon oxygenation. Haemocyanin is blue when oxygenated and colourless when deoxygenated. Although a considerable amount of research has been carried out on gastropod haemocyanins, including those of *Helix*, relatively little is known about slug haemocyanins specifically. A comprehensive review of the literature on molluscan haemocyanins has been prepared by Bonaventura and Bonaventura (1983), including information on the properties of haemocyanins from five species of slug. In *Deroceras*, the oxygen affinity of the blood shows an increase as the blood becomes more alkaline (Runham and Hunter, 1970). For example, at pH 8.78, the blood was 50% saturated when in equilibrium with oxygen at a partial pressure of 3 mmHg, while to produce 50% saturation at pH 7.60, an oxygen partial pressure of 32 mmHg was required. Thus, in the vessels of the mantle cavity, the alkaline blood absorbs oxygen, while in the tissues, where carbon dioxide is being produced, the blood becomes more acid and oxygen is given up. Haemocyanin formation in *Limax* sp. has been described by Reger (1973). Reger described how the connective tissue of tentacular and subdermal regions of the body wall contained interstitial cells, which were morphologically recognizable as secretory cells. The fine structure of these cells was examined and they were shown to contain haemocyanin particles. Reger (1973) showed that haemocyanin was produced in cisternae of the endoplasmic reticulum and then released from the cell by exocytosis. Skelding and Newell (1974) have suggested that pore cells (globular cells) in the connective tissue of *A. hortensis* were involved in the synthesis of haemocyanin.

4.2.3 Water balance

Water is gained by drinking, from the food and by absorption through the skin. It is lost in the faeces, through excretion and by evaporation from exposed surfaces such as the skin and through increased mucus production during locomotion. The importance of drinking as a source of water to slugs is uncertain. Künkel (1916) suggested that movements of the mouth and buccal mass as slugs crawled along a surface containing water droplets, represented drinking movements. Howes and Wells (1934b) and Foster (1954) concluded that drinking was the major source of water for slugs, although Foster's experiments with free water did not exclude the possibility of water uptake through the skin. Dainton (1954a) never observed slugs drinking water and suggested that, although some water was ingested with the food, the bulk of water absorption was probably through the body surface. Machin (1975) stressed the difficulties that were experienced with experimental technique when investigating the problem of drinking in slugs. Prior (1985) showed, by the use of dyes and ^{14}C-labelled dextran, that oral ingestion was not involved in the process of contact-rehydration in slugs.

Fresh plant material has a high water content and the results of a study by Pallant (1970) of feeding by *D. reticulatum* indicate that between two-thirds and three-quarters of water ingested is retained. Thus there is a net gain of water to the slug during feeding. The water gained through feeding is, however, probably small in comparison with the total water requirement of the slug. *Arion ater* and *T. sowerbyi* both lost water when kept at 40–80% r.h., even when fed with lettuce and provided with a dish of drinking water (Foster, 1954). Hunter (1968c) found that slugs in captivity lost weight, even in saturated air, although they continued to feed. Purine excretion in slugs takes the form of a semi-solid paste excreted with the faeces. However, some species, such as *L. flavus*, are also ureotelic and urea is excreted in liquid urine (Horne, 1977a). Urine was produced by unstressed, well-hydrated *A. columbianus* at a mean rate of $4.7 \, \mu l \, g^{-1} \, min^{-1}$ (range 2.9–$9.1 \, \mu l \, g^{-1} \, min^{-1}$). This represented about 22% of the fluid lost, less than 1% of body weight per hour, while the remaining loss was in the form of surface fluids (Deyrup-Olsen and Martin, 1982a). Thus although water is lost from slugs in the faeces and by excretion, the greatest source of water loss is by evaporation from the moist skin of these animals and through the copious secretion of mucus during locomotion. It also seems likely that the intake of water by drinking and feeding is relatively small compared with the uptake of water through the skin.

4.2.4 Evaporation of water from the skin

Relatively little is known about evaporation from the skin of slugs, but an extensive study of the evaporation of water from the snail *H. aspersa* was made by Machin and many of his conclusions are applicable to slugs. Machin (1964a) showed that evaporative water loss did not take place from the surface of the skin directly but from a superficial layer of secreted mucus. If this superficial mucus was allowed to dry out, the skin became very dry and brittle. The continuous production of fresh mucus is, therefore, essential to prevent damage to the skin. Machin also demonstrated that the mucous coating of the skin was maintained by haemocoelic pressure caused by muscle contraction and that, under conditions of rapid water loss, more intense muscular undulations served to spread the mucus via grooves between the skin ridges to more exposed areas. The mean water content of normal mucus was high (88.9% of fresh body weight) while that of mucus produced under stimulation was even higher (95.6%) and this water was readily evaporated in dry air. Machin (1964a) concluded that, for practical purposes, the rate of evasion of water molecules from mucus could be considered as identical to that of a free water surface. The effect of air flow on the removal of water vapour from the evaporating surface by diffusion was investigated by Machin (1964b). The distribution of the aerodynamic boundary layer over the snail depended on the orientation of the animal in relation to wind direction and this resulted in changes in the evaporation rate. Thus a snail at right angles to the direction of wind flow lost only 83% of the water that was lost by evaporation when facing into the wind, and the corresponding loss for a snail facing downwind was only 68%. These differences might explain the response of *D. reticulatum* and *L. maximus* when they were exposed to air currents, as they increased their speed of locomotion when facing downwind and, when facing any other direction, they turned to face downwind (Dainton, 1954b). Thus the slugs were able to reduce evaporative water loss by this kind of behaviour. Machin (1964c) described the development of a mathematical model which could be applied to evaporation from the surface of a moist-skinned animal.

When an inactive *H. aspersa* withdrew into its shell, the only part of the body left exposed was the mantle collar, which was a potential source of water loss for the inactive snail. However, Machin (1966) found that the rate of evaporation from the exposed surface of the mantle was reduced during the inactive state, implying that the snail was able to regulate water loss from the mantle. Machin (1966) suggested the existence of an active or passive barrier to water which was near or at the integumental surface although it did not lie in the mucus itself. Machin (1974) found that the reduction in water loss in the dormant snail, *Otala lactea* Müller, was very marked and that mantle permeability was of the same order as insect cuticle. Machin also demonstrated osmotic gradients within mantle epithelial cells, indicating that a barrier existed in the apical layers of these cells. Newell and Machin (1976) showed that the relative concentrations of ions on a transect through the epithelium of inactive and active *O. lactea* were similar except in the region of the apical microvilli of the epithelial cells of inactive snails. The relative concentrations of several ions in this region exceeded those in other parts of the cell by a factor of at least five. This gradient was subsequently shown to be a passive gradient caused by the dehydration of the apex of the cells (Newell and Appleton, 1979). Newell and Machin (1976) also found that the apical cytoplasm of these cells developed a mass of lamellate vesicles in the inactive snail and concluded that these vesicles constituted a barrier to water movement within the epithelial cells, with the parts of the cell distal to the vesicles becoming desiccated and the proximal parts remaining hydrated and maintaining continuity with the haemocoele. The temporary dense layer of lamellate vesicles was lost within 30 minutes of mechanical stimulation of inactive snails.

In addition to water lost during the normal secretion of mucus at the surface of the skin, appreciable quantities of water derived directly from the blood may be lost through the voluminous secretions elicited by mechanical, chemical and electrical stimulation of the skin (Section 2.3). These include the surface exudation described by Deyrup-Olsen and Martin (1982b) and blood venting through the pneumostome described by Martin and Deyrup-Olsen (1982). Martin and Deyrup-Olsen suggested that the function of this fluid loss might be either for the rapid adjustment of body volume to restricted environmental spaces or the discarding of hypotonic body fluid.

Künkel (1916) found that an inactive *L. flavus* lost 2.4% of its initial body weight per hour when exposed to dry air. Howes and Wells (1934b) showed that, when deprived of water, *A. ater* lost about one-third of its weight per day, mainly through evaporation. The rate of water loss by evaporation from the skin of a slug exposed to a relative humidity of 45% varied between 3 and 5% of the initial body weight per hour, depending on the size of the slug (Dainton, 1954a). These rates were measured for *A. ater, A. subfuscus, L. maximus, T. sowerbyi* and *D. reticulatum* in moving air. The increased rates of water loss for active compared with inactive slugs, due to mucus production, recorded by Künkel (1916) and Dainton (1954a), were confirmed by Lyth (1972). Lyth also showed that smaller *A. ater* lost weight more rapidly than larger individuals, as the result of the higher surface area/volume ratio. For example, a slug weighing 0.42 g lost over 40% of its initial body weight in four hours compared with a loss of about 25% of initial body weight over a similar period by an individual weighing 1.57 g at a temperature of 11.5°C and 5 ± 1% r.h. Dainton (1954a) found that the rate of water loss was similar for all species if size was taken into consideration. However, Lyth (1972) found significant differences between the rates of water loss for different species of slug after the rates were corrected for differences in body weight. Lyth suggested that these differences could partly be attributed to the different shapes assumed by inactive slugs. The rate of evaporation from *A. ater* was found to be significantly less than that for other species. For active slugs, *L. flavus* lost water more rapidly than most other species. These observations support the statement by Howes and Wells (1934b) that *A. ater*, when deprived of water, lost under one-third of its weight in a day compared with a loss of two-thirds by *L. flavus*. Lyth (1972) also showed that rates of water loss were similar for different species of narcotized slugs and concluded that this might indicate some form of regulation of water loss in living slugs.

4.2.5 Effects of desiccation

Slugs are unable to reduce water loss through the skin by retiring within a shell and so behavioural mechanisms, such as homing, 'huddling' and the modification of locomotor activity appear to play an important role in water conservation. Homing and huddling behaviour are discussed in Section 7.2.2. Dainton (1954a) found that when several species of slug were dehydrated to about 83% of initial body weight, this resulted in a decrease in locomotor activity following mechanical stimulation. She concluded that the level of body hydration influenced the ability of slugs to respond to this stimulation. The rate of water loss at lower humidities was shown by

Pusswald (1948) to be reduced when slugs were grouped together. Cook (1981a) showed that when two slugs were in close contact, the evaporation rate was reduced by 34%, while Prior *et al.* (1983) demonstrated that huddling behaviour in *L. maximus* was an effective means of reducing evaporative loss from individual slugs. Small slugs have a greater specific surface area to volume ratio than larger slugs and so their rate of evaporative water loss is likely to be greater. Waite (1987) found that small individuals of *D. reticulatum* tended to conserve water by spending more time in a contracted or resting posture and less time moving when a moist microclimate was available. When no moist microhabitat was available, however, small slugs lost a greater percentage of their initial body weight and were more active than larger slugs. Waite concluded that smaller slugs were under such severe constraints, with their surface area to volume ratios being so large, that any water-conserving tactic other than finding a moist microclimate would have been inadequate and so they were compelled to remain active and to search for shelter.

A sequence of different behaviours in *L. maximus* and *L. maculatus*, aimed at minimizing further water loss by evaporation and corresponding to levels of body hydration, was described by Prior (1981) and Prior *et al.* (1983). When the slug had lost 15–20% of its initial body weight the frequency of pneumostome closures, less than 0.5 closures per minute in a fully hydrated slug, began to increase. The frequency increased with further dehydration with the result that the proportion of time during which the pneumostome was closed increased. Prior *et al.* (1983) found that the duration of closures increased with further dehydration although Dickinson *et al.* (1988) concluded that the mean duration of each closure remained constant. The diameter of the pneumostome when open was also reduced. These changes in the pneumostome resulted in a reduction of water loss by evaporation from the mantle cavity. Locomotor activity declined and finally ceased and this served to reduce water loss from increased mucus production during locomotion. The final response to excessive dehydration (30–40% loss of initial body weight) was a general contraction of the body, resulting in a reduction in the area of exposed integument. Hess and Prior (1985) compared the circadian locomotor rhythms of *L. maximus* under moist and drying conditions. They showed that exposure to drying conditions increased the intensity and duration of locomotor activity and slugs remained active well beyond the time they would normally have stopped. The normal circadian rhythm of activity was rapidly re-established when slugs were returned to moist conditions. Prior (1985) suggested that progressive dehydration in slugs initiated an escape or water-seeking response in addition to the initial short-term reduction in activity referred to above.

Water orientation behaviour in response to dehydration in *L. maximus* consisted of an increase in locomotor activity and a preference for the moist arm of a Y-maze (Banta *et al.*, 1990). Welsford *et al.* (1990) described how the dehydration-induced increase in locomotor activity was a function of body size, with small and medium-sized slugs (0.3–2.9 g) showing significantly more activity than large individuals (3.5–6.0 g). Since the rate of dehydration affected the dehydration-induced behavioural responses of small slugs, they suggested that the variation in response between individuals of different sizes might be due to the different rates of dehydration they experienced. The modifications in behaviour, which result from desiccation, are partly controlled by the effect of changes in blood osmolality on the nervous system and on the release of hormones, and are discussed in Chapter 6.

Dainton (1954a) found that *A. ater* and *L. flavus* were able to survive a water loss of 50% of initial body weight and that they fully recovered after some hours on damp filter paper. Hughes and Kerkut (1956) showed that *D. reticulatum* that had lost 35% of initial body weight, regained their initial body weight within an hour when allowed to come into direct contact with water. Lyth (1972) concluded for eight species of slug (*T. budapestensis, T. sowerbyi, D. reticulatum, D. laeve, A. fasciatus, A. hortensis, A. ater* and *L. flavus*) that weight losses of less than about 40% of initial body weight rarely caused death, while losses greater than about 58% killed all slugs. *Arion fasciatus* was an exception to this with no survivals at weight losses over 40% of initial body weight. This sensitivity to desiccation by *A. fasciatus* has been observed in the field by Getz (1959), South (1965) and Lyth (1972). The tolerance of other species to desiccation did not show any interspecific differences (Lyth, 1972). Foster (1954) suggested that the speed of desiccation was as important to slugs as the degree of desiccation but Lyth (1972) found that this conclusion was not supported by his results. The lethal limit of desiccation for *L. maximus* was estimated to be 40–50% of initial body weight by Prior *et al.* (1983).

4.2.6 Water uptake through the skin

Pusswald (1948) suggested that water was taken up through the skin of slugs, while Howes and Wells (1934b) and Foster (1954) found that a saturated atmosphere was not sufficient to prevent water loss in slugs and that a source of free water was required to maintain a fully hydrated body. Dainton (1954a) suggested that the bulk of water absorption was probably through the skin surface. A specific pattern of behaviour which allowed dehydrated slugs to rehydrate by absorbing water through the foot was described by Prior (1982). Dehydrated slugs moved directly on to a moist surface, expanded the margin of the foot and then remained stationary. When they had absorbed enough water through the sole of the foot to rehydrate themselves they moved away. This contact-rehydration behaviour was initiated when slugs had been dehydrated to a threshold level. For

L. maximus this was between 60 and 70% of initial body weight. Contact-rehydration behaviour was terminated once slugs had rehydrated to their 'rehydration set-point' which for *L. maximus* was about 94% of initial body weight (Prior *et al.*, 1983). The existence of a dehydration threshold and a rehydration threshold indicated that contact-rehydration behaviour might be regulated by a dual threshold control system. Prior (1982, 1989) showed that contact-rehydration behaviour could be initiated or terminated by, respectively, increasing or decreasing the osmolality of the blood, suggesting that variation in the osmolality of the blood was involved in the control of this sequence of behaviour. The neural and hormonal control of water uptake is discussed in Chapter 6.

The mechanism for contact-rehydration through the foot, with the bulk flow of water passing via an epithelial paracellular pathway in the integument of the foot, has been described in Section 2.3. Prior and Uglem (1984a, b) and Uglem *et al.* (1985) showed that the intercellular spaces involved were reduced in dehydrated slugs but were rapidly enlarged during contact-rehydration. Prior (1983a) described hydration-induced changes in feeding in *L. maculatus* and *L. valentiana*. Slugs dehydrated at 25–35% r.h. to 60–70% of initial body weight ceased to feed although feeding responsiveness was rapidly restored when slugs were allowed to rehydrate in contact with a moist surface. Prior (1983a) demonstrated that blood osmolality was involved in this modification of the feeding response. Contact-rehydration behaviour was initiated when the osmolality of the blood reached about 200–210 mOsm kg^{-1} water and was terminated when it was diluted to 117.3 ± 32.0 mOsm kg^{-1} water (Prior, 1989). This rehydration process is rapid and the time between the initiation and termination of contact-rehydration in *L. maximus* was only 10–20 minutes (Prior *et al.*, 1983). Oral ingestion of water was not involved in this sequence of behaviour. Prior (1989) showed that the osmolality of the external substrate could modify the flow of water. Increase in the osmolality of the external medium reduced the level of integumental absorption, and at 150–200 mOsm kg^{-1} water no net absorption of water occurred although the pattern of contact-rehydration behaviour was maintained. Prior suggested that the paracellular pathway was not opened when the external osmolality was higher than 150 mOsm kg^{-1} water.

An excess of water in the environment appears to result in further, probably involuntary, dilution of the blood. Slugs are exposed to a diluting environment as they move across the ground and over vegetation after rain or heavy dew. Arvanitaki and Cardot (1932) observed a decrease in the osmolality of blood from 253 to 108 mOsm kg^{-1} water following rain. The osmolality of *L. maximus* examined after a heavy rainstorm was found to be 97–117 mOsm kg^{-1} water compared with about 140 mOsm kg^{-1} water in normally hydrated slugs (Prior, 1985). Experiments by Hess (in Prior, 1985) showed that when *L. maximus* was placed in continuous contact with distilled water there was a gradual reduction in blood osmolality to 130 mOsm kg^{-1} water after two hours and 107 mOsm kg^{-1} water after four hours. Prior (1985) concluded that this result was consistent with the idea that the kidneys of slugs and snails were capable of maintaining body volume during excessive exposure to free water.

4.2.7 Aestivation

In snails a common response to dry conditions is aestivation, when the snail withdraws into its shell and secretes an epiphragm to cover the shell aperture. This leads to a considerable reduction in water loss, enabling snails to survive even dry desert conditions (Schmidt-Nielsen et al., 1971). In the absence of a protective external shell, aestivation appears to be a less appropriate strategy for slugs to avoid dry conditions. However, aestivation is a well-developed adaptation to adverse conditions in earthworms and in some species this even takes the form of an obligatory diapause (Edwards and Lofty, 1977). At the beginning of diapause, earthworms stop feeding and construct a small round cell lined with mucus, which serves to reduce water loss to a minimum. There is evidence that some slugs avoid very dry conditions by aestivation. Woodward (1913) stated that slugs burrowed into the soil and formed small chambers lined with mucus. Hora (1928) described how the arionid slug *Anadenus dalhousiensis* Bhatia aestivated in a contacted form enclosed in a mucous cell attached to the underside of a stone rather than in loose soil. The slug became active a short time after it rained. Lyth (1972) described how apparently healthy slugs were sometimes found inactive and with their tentacles retracted and completely covered by a dry mucous film. Poulton (in Stokes, 1958) recorded a *T. scutulum* resting in a hardened mucous secretion, while Stokes (1958) described how an individual of this species in a culture, was found with particles of soil adhering to its mucous covering. The slug was in a humped position and did not respond to handling. A similar observation was made by Oldham (1915) for this species in the field. Stokes (1958) found that the slug subsequently became active and throughout the summer months alternately rested in this way and then fed actively, with the resting periods lasting from 1 to 13 days. South (1989a, b) described a quiescent stage in the life cycle of *A. intermedius* on grassland in late spring which was a response to dry conditions and reported that individuals of *A. intermedius* had been found in mucus-lined cells in loose soil in the field. There was no evidence that *D. reticulatum* aestivated in response to dry conditions although slugs were killed directly by drought. It seems likely that some species of slugs do aestivate in response to dry conditions.

4.3 NITROGENOUS EXCRETION

4.3.1 Kidney

The location of the kidney and the primary and secondary ureters in the pallial complex has been described in Section 2.2.2 and Figs 2.2, 2.3 and 2.5. The kidney is a large thin-walled sac with many folds, which bulges into the posterior wall of the mantle cavity. The walls are folded and are lined by a simple columnar epithelium. The kidney is closely associated with the pericardium, their cavities are connected by a renopericardial canal and these represent the coelom in gastropods. The primary and secondary ureters differ in their structure and function (Garner, 1974) and in their origin (Martin, 1983), the primary ureter being derived from nephridial sac mesoderm and the secondary ureter from ectoderm. The detailed structure and function of the kidney of *D. reticulatum* has been described by Garner (1974). The inner surface of the kidney sac is lined by a continuous layer of columnar cells (nephrocytes). Beneath this layer lies a narrow band of connective tissue which is in direct contact with the haemocoele. The nephrocytes have either a number of small apical vacuoles or a single large apical vacuole, which often contains crystalline material. The fine structure of the nephrocytes and the basement membrane of the kidney sac was described by Angulo and Moya (1985, 1986).

The walls of the primary ureter are also much folded and lined with two types of cell, columnar cells being the most common while pyramid cells occur mainly at the apices of the folds in the walls. The cell membrane of the columnar cells is deeply folded giving a greatly increased surface area to the cell and these cell processes interdigitate extensively with neighbouring cells. The pyramid cells are ciliated on their inner surface. The structure of the wall underneath the epithelium is similar to that of the kidney sac. Garner (1974) suggested that the folded plasmalemma of the columnar cells (β-cytomembrane systems) indicated that this was a site for the resorption of water and ions. The wall of the secondary ureter has a well-developed muscle layer and more connective tissue than the kidney or primary ureter. There is a simple epithelium and most cells are unciliated, with well-developed intercellular spaces between many cells owing to their irregular shape. The wall of the secondary ureter is less folded and so has a smaller surface area than the walls of the kidney and primary ureter. Garner (1974) described a diurnal rhythm in excretion in *D. reticulatum* with excreta generally being voided during the late afternoon or early evening although there was considerable variation in timing. Peak purine accumulation took place in the kidney between 12 and 17 hours after voiding an excretory mass. This accumulated material was released rapidly over a short period between 14 and 19 hours after voiding and almost immediately after the maximum kidney purine content was reached. A gradual maturation of vacuoles occurred within the nephrocytes during the cycle. Small vacuoles accumulated in each cell and these gradually coalesced to form a single, large vacuole. After the formation of a single vacuole, the contents crystallized to form a purine concretion which was then released into the lumen of the kidney. The cytoplasmic structures involved in the formation of these purine concretions were described by Angulo and Moya (1989a). Excretory material was stored in the bladder before being voided at intervals to ensure the conservation of water. There was some evidence that the site of ultrafiltration of the blood was the pericardium where purines were concentrated before being passed to the lumen of the kidney via the renopericardial canal. Purines also appeared to pass from the kidney lumen into the nephrocyte vacuoles (Garner, 1974). Deyrup-Olsen and Martin (1982a) found that carbohydrates ranging in size from inulin up to dextrans of molecular weight up to 5×10^6 were present in the urine within 1 hour after injection into the haemocoele. The penetrability of the barrier in the kidney to the passage of these substances appeared to drop off significantly between dextrans in the molecular weight range 90 000 Da and below, and dextrans in the molecular weight range 100 000 Da and above. Very large dextrans appeared in the urine at concentrations relative to blood concentrations which were lower than those seen with the smaller dextrans.

4.3.2 Rate of excretion

The rates of excretion of purine nitrogen were measured by Jezewska (1969). The rate of excretion for *D. agreste* was found to be 8.4 mg $(100\,\text{g})^{-1}$ fresh body weight per day (approximately 23 μg g^{-1} dry weight per hour). The rates for *L. maximus* ranged from 11.0 to 13.7 mg $(100\,\text{g})^{-1}$ fresh body weight per day (approximately 31–38 μg g^{-1} dry weight per hour). The total purine content of dry excreta was found to be 72% for *L. maximus* and 77% for *D. agreste* during September and October. These values were lower than the corresponding values measured by Jezewska for several species of snail. The mean purine content of dry excreta from *D. reticulatum* was found by Garner (1974) to be 99%, including 83% of uric acid while the total purine nitrogen excreted by the kidney was 58 μg g^{-1} dry body weight per hour.

5

Reproduction

5.1 REPRODUCTIVE TRACT

5.1.1 Introduction

The reproductive system of the slug, in common with most pulmonates, is a complex structure because terrestrial slugs are hermaphrodites. The many variations in the form and organization of the pulmonate reproductive system were reviewed by Duncan (1975). Webb (1961) discussed the changes that had taken place in the reproductive system in the phylogeny of the Arionidae. Illustrations and descriptions of the morphology of the reproductive systems of many testacellid, arionid, limacid and milacid slugs were included by Germain (1930), Quick (1960) and Likharev and Rammel'meier (1952) in their systematic accounts of slugs. The distal genitalia have been widely used as diagnostic characters in the taxonomy of slugs. Webb (1961) included illustrated descriptions of the reproductive systems of some nearctic slugs, including *Ariolimax*, *Prophysaon* and *Philomycus*, while the reproductive systems of veronicellid and athoracophorid slugs have been described by Lanza and Quattrini (1964) and Burton (1978) respectively. The organization of the reproductive systems of some families of slugs are compared in Fig. 5.1.

5.1.2 Early development of the gonad

The term gonad is used here to describe the structure where the gametes are produced. Other terms used for this structure in slugs have included the hermaphrodite gland and the ovotestis. The gonad is located among the lobes of the digestive gland towards the posterior part of the body (Figs 2.2 and 2.3). It consists of numerous acini (alveoli of some authors, e.g. Smith 1966a). The number of acini may exceed several hundred in large species such as *L. maximus*. Each acinus is bounded by a basal lamina covered externally by a pigmented layer of connective tissue and there is a small amount of loose connective tissue between the acini, which are usually grouped together into lobes. Lusis (1961) and Parivar (1978) described how the gonad of *A. ater* consisted of two major lobes separated by the genital artery. Sabelli *et al.* (1978) showed that each acinus narrowed proximally into a canaliculus (ductule of Runham and Hogg, 1979) and the canaliculi

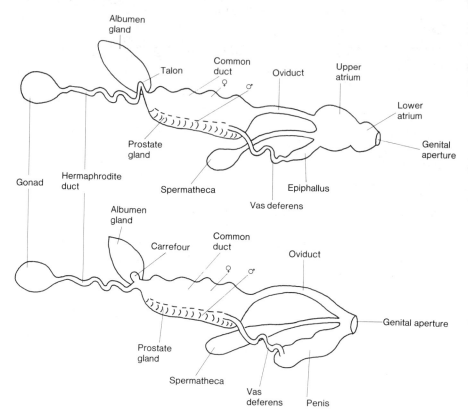

Fig. 5.1 Diagram comparing reproductive systems of (a) arionid and (b) limacid slugs.

from several acini merged to form larger canals or intercalated ducts in the gonad of *D. reticulatum*. These ducts, which were ciliated, eventually joined to form the hermaphrodite duct, which transported ripe gametes produced in the acini to the reproductive tract.

Lusis (1961) described how, in newly hatched *A. ater*, small groups of cells in the wall of the undifferentiated genital artery developed into the first acini of the gonad, although he concluded that they were not initially associated with the hermaphrodite duct. However, Luchtel (1972b) showed that in *D. reticulatum* and *A. ater* the primary acini originated as branches of the hermaphrodite duct with additional acini then budded off from these original acini. Runham and Hogg (1979) obtained a similar result, showing that, at hatching, the reproductive system of *D. reticulatum* consisted of a simple blind ending duct opening to the exterior at the genital pore. The closed tip of the duct formed a terminal acinus and other tubular outgrowths close to this acinus also developed acini at their tips. In this way the closed end of the duct gradually developed into the gonad with 70 to 180 acini. Thus the gonad developed from the hermaphrodite duct and was closely associated with it from the earliest stages of development. Runham and Hogg (1979) concluded that, since the duct was continuous with the skin of the slug, the reproductive system was ectodermal in origin, including the gonad. Luchtel (1972a) similarly concluded that the gonad of *A. circumscriptus* was ectodermal in origin while Kulkarni (1971) came to a similar conclusion for *L. alte*. There are two conflicting opinions as to the origin of the gametes within the pulmonate acinus. Ancel (1902) concluded that, in the snail *H. pomatia*, any of the cells in the epithelium lining the acinus could develop into spermatogonia, oogonia or sertoli cells, indicating that there was no specialization of cells within the acinus. However, Joosse and Reitz (1969) found that in the aquatic snail *L. stagnalis*, there was a zoning of different types of cell within the acinus, with a localized germinal epithelial ring and a maturity gradient of gametes from the proximal to the distal part of the acinus and some separation of male and female gametes.

No germinal epithelial ring was reported by Lusis (1961) or Smith (1966a) from their examination of the gonad of *A. ater*. Kugler (1965) described how, in *P. carolinianus*, the oocyte developed from cells of the flattened germinal epithelium. Sabelli *et al.* (1978) examined the distribution of germ cells in the acini of *D. reticulatum* and concluded that there was no localized germinal epithelial ring and no maturity gradient. Instead, in *D. reticulatum*, a wide germinal epithelium lined the acinus almost to the duct. This epithelium could differentiate and mature all the kinds of cell found in the acinus at any point along its surface. This was similar to the structure described by Ancel (1902) for *H. pomatia*. Sabelli *et al.* (1978) suggested that the acinar organization described by Joosse and Reitz (1969) for *L. stagnalis* represented the condition in more primitive pulmonates, including the veronicellid slugs while in the stylommatophoran slugs there was a uniform germinal epithelium lining the whole of the acinus. Luchtel (1972a) rejected the concept of a germinal epithelium in which germ cells differentiated from indeterminate germ cells throughout the reproductive life of the slug. He suggested instead that the gonad was composed of two cell lines – germinal and non-germinal – which were established in the embryo. The germinal line was first represented by primordial germ cells which differentiated into spermatogonia and oogonia about the time of hatching. The non-germinal line was represented by auxiliary cells which differentiated into sertoli cells, associated with the male germ cells, and follicular cells, associated with individual oocytes.

Luchtel (1972a) concluded that all subsequent germ cells differentiated from the initial populations of spermatogonia and oogonia rather than from a germinal epithelium. He described how, at about the time of hatching, some of the primordial germ cells became localized at the periphery of the gonad, where they became separated from the rest of the gonad by a layer of auxiliary cells. Other primordial germ cells remained in the lumen of the gonad. Thus two compartments were established in the gonad with oogonia differentiating at the periphery of the acinus and spermatogonia differentiating in the acinar lumen, with a barrier of auxillary cells between them. Luchtel (1972a) proposed the hypothesis that, as the sexually non-determined primordial germ cells were distributed into two physiologically isolated compartments, spermatogonia and oogonia could then differentiate under the influence of two different microenvironments. Luchtel (1972b) obtained similar results for both *D. reticulatum* and *A. ater*. He described how the sertoli cell developed a cytoplasmic extension which penetrated into the acinar lumen between the spermatogonia and, subsequently, between the spermatocytes, spermatids and spermatozoa. The latter were attached to the cytoplasmic extension by means of long pseudopod-like structures which extended into the sertoli cell. However, the function of the sertoli cells remained unclear although some previous authors had suggested that they had a nutritive function. Parivar (1980) described two different stages, active and atrophic, in the development of the sertoli cells of *A. ater*. The cytoplasm of the active stage showed evidence of high metabolic activity and Parivar suggested that their role was that of an intermediary on a nutrient pathway between the haemolymph on one side and the developing spermatozoa on the other. They also provided support for the latter. Parivar (1980) also suggested that these cells might have a role in the production of steroid hormones. The atrophic sertoli cells developed from active sertoli cells after the sperm had been shed and attached themselves to the post-reproductive epithelial cells, which by that time had completely lined the inner part of wall of the acinus. The cytoplasmic processes of the atrophic sertoli cells projected into the acinar lumen and the cells were larger than active sertoli cells.

In a study of the early development of the gonad in *D. reticulatum* by Runham and Hogg (1979), the acini were found to be lined by an epithelium continuous with that in the remainder of the hermaphrodite duct. Runham and Hogg identified the cells which lined the acinus as gonadal stem cells as they subsequently differentiated into all of the types of cell found in the mature gonad. In a process similar to that described by Luchtel (1972a), some cells accumulated in the lumen of the acinus while similar cells appeared beneath the gonadal stem cell layer, accompanied by a few very small cells. The gonadal stem cells persisted in the ductules and as an epithelial layer around the neck of the acinus and Runham and Hogg (1979) identified the latter cells as a germinal epithelium (germinal epithelial ring). The remaining gonadal stem cells lining the acinus differentiated into sertoli cells. The cells in the lumen of the acinus developed into spermatogonia while those beneath the sertoli cell layer differentiated into oocytes and follicle cells. Unlike Luchtel (1972a), however, Runham and Hogg concluded that further follicle cells and oocytes differentiated from the edge of the germinal ring and they demonstrated a maturity gradient in the acinus, with the smallest oocytes nearest the germinal epithelial ring and the largest oocytes located at the base of the acinus opposite the opening of the duct. This arrangement is similar in some respects to that described by Joosse and Reitz (1969) for *L. stagnalis*. There was no evidence that secondary production of spermatogonia took place from the germinal epithelial ring or anywhere else in the acini of *D. reticulatum* (Bailey, 1970).

5.1.3 Maturation of the gonad

The later development and maturation of the gonad has been subdivided into several stages, distinguished by the relative proportions of different cell types present. Four phases in the development of the gonad of *A. californicus* were described by Gottfried and Dorfman (1970a,b). The following sequence is based on the stages described by Lusis (1961), and Smith (1966a) for *A. ater* and Runham and Laryea (1968) and Bailey (1970) for *D. reticulatum*:

(a) Undifferentiated stage

The gland consists of a solid mass of undifferentiated cells enclosed in a connective tissue sheath.

(b) Spermatogonia and spermatocyte stage

Acini begin to develop as outgrowths from the hermaphrodite duct but the gonad is still filled with a solid mass of cells. Differentiation of the gonadal stem cells has begun and several cell types can be recognized. These include spermatogonia, together with spermatocytes towards the centre of the acini, and oocytes, together with differentiating sertoli cells and follicle cells in the acinar wall.

(c) Spermatid stage

Secondary spermatocytes give rise to spermatids and these become arranged in clumps around the sertoli cells projecting into the developing lumen of the acinus. Oocytes begin to develop while ducts and blood vessels begin to appear between the acini in the gland. Smith (1966a) described a late spermatid stage with spermatids being the predominant constituent of the acini and a few spermatids showing small tails.

(d) Early spermatozoon stage

Spermatids still predominate and the earlier stages of spermatogenesis still line the acinar walls (Fig. 5.2). Some mature spermatozoa are now found

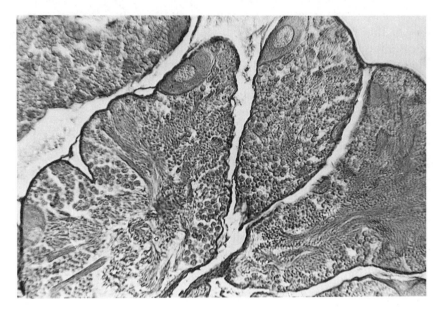

Fig. 5.2 Section through gonad (*Arion circumscriptus*) (Mallory triple stain). Spermatid/early spermatozoon stage.

attached to the sertoli cells, with their tails filling the lumen of the acini. By this stage the gonad has become very large with thin walls.

(e) Late spermatozoon stage

Mature free spermatozoa fill the lumen of the acini and some spermatozoa have already passed out into the hermaphrodite duct, giving it the opaque, white appearance described by Bett (1960) as characteristic of mature slugs. From this stage onwards the gonad becomes smaller as sperm is lost. The earlier stages of spermatogenesis have become much fewer. Many oocytes are present in the wall of the acinus.

(f) Early oocyte stage

Oocytes become enlarged and are often surrounded by a thin layer of cells forming a follicle. Additional small oocytes are produced by the germinal epithelial cells. Spermatozoa are still present.

(g) Late oocyte stage

The number of spermatozoa are considerably reduced while some oocytes are found free in the acinar lumen (Fig. 5.3). During the early and late oocyte

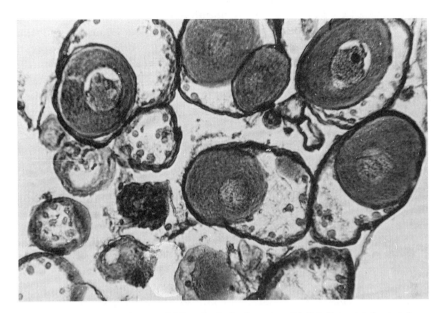

Fig. 5.3 Section through gonad (*Tandonia budapestensis*) (Mallory triple stain). Late oocyte stage.

stages the gonad becomes progressively smaller and begins to undergo structural changes which continue into the post-reproductive stage. Lusis (1961) and Parivar (1980) described how, at the completion of spermatogenesis, the active sertoli cells in *A. ater* underwent nuclear changes while their cytoplasmic processes became enlarged, extending further into the acinar lumen to become atrophic sertoli cells. Angulo and Moya (1989b) described how rough endoplasmic reticulum-rich cells were found among the sertoli cells at the time of egg laying. The latter part of the late oocyte stage was termed the atrophic stage by Parivar (1978) who found that most *A. ater* died at this stage in laboratory cultures.

(h) Post-reproductive (atrophy) stage

The nuclei of the atrophic sertoli cells finally disintegrated by the post-reproductive stage and the cells themselves disappeared. Runham and Laryea (1968) and Bailey (1970) found that, in *D. reticulatum*, the post-reproductive stage was characterized by the presence of a cuboidal epithelium which lined the acini. Bailey (1970) suggested that this cuboidal epithelium was produced by the ring of germinal epithelial cells and that the cuboidal cells then migrated around the wall of the acinus causing the remaining germ cells to be sloughed into the lumen of the acinus. Parivar (1980) described these cuboidal cells as post-reproductive epithelial cells and suggested that they developed from undifferentiated cells in the acinar wall. They first appeared as a single layer between the basal lamina of the acinar wall and the atrophic sertoli cells, gradually increased in number and, after the degeneration of the sertoli cells, they filled the lumen of some acini. Parivar suggested that these cells were syncytial.

Lusis (1961) described how, during the post-reproductive stage, proliferating interacinar connective tissue cells also invaded the acinar lumina, gradually obliterating the space to the extent that it was difficult to distinguish the outline of acini in the gonad of the ageing animal. It seems likely that these cells were the same as the post-reproductive cells recorded by Parivar (1980) rather than interacinar connective tissue cells. This proliferation of tissue accompanying the atrophy of the acini in the gonad of senescent *D. agreste* has also been described by Szabó and Szabó (1936). Similar changes in other parts of the body, which took place as the result of ageing, were discussed by Szabó (1935a, b) and these are reviewed in Section 8.3. These changes included the accumulation of a brownish-black pigment in different organs (Szabó and Szabó, 1934b) and both Lusis (1961), Smith (1966) and Bailey (1970) reported the accumulation of pigment in the connective tissue between the acini and around the acini as the gonad matured. Lusis (1962) described how the gonad was translucent in the early stages of development but during spermatogenesis it became opaque and grey in colour, turning dark brown at the oocyte stage and ultimately, black after oviposition. This increase in pigmentation with ageing was due partly to a gradual increase in the amount of pigment and partly to the concentration of existing pigment as the gland atrophied. Lusis (1962) showed that the pigment was lipofuscin and suggested that, because the pigment slowly accumulated with age, it was a waste product.

5.1.4 Hermaphrodite duct and carrefour

The ducts from the gonad merge to form the hermaphrodite duct which runs forwards from the gonad to the carrefour, a complex region formed by the junction of the albumen gland ducts and the common duct with the hermaphrodite duct. In the region of the carrefour, the male and female gametes, which pass together along the hermaphrodite duct, are directed along separate pathways in the common duct. The hermaphrodite duct in slugs is a simple tube lined with an epithelium resting on a basement membrane. The original epithelium, which is continuous with that of the acini, contains both ciliated and secretory cells (Runham, 1982). This inner layer is convered externally by thin layers of connective tissue and of muscle (Lusis, 1961). In *A. ater*, the hermaphrodite duct is straight initially when it emerges from the gonad but becomes convoluted distally and finally makes a loop by running backwards between the lobes of the albumen gland and then turning sharply forwards (Lusis, 1961). This loop, also termed a 'claw' or 'talon' is characteristic of the genera *Arion* and *Geomalacus* and a similar arrangement was described in *Philomycus* by Kugler (1965). The epithelium in the region of the loop is thicker and folded and the muscle and connective tissue layers are also thicker. Large mucous cells occurred sparingly embedded in the connective tissue of the loop in *Arion* and *Philomycus* (Lusis, 1961; Kugler, 1965). The sharp loop is absent from other genera but in some, e.g. *Testacella, Milax, Deroceras* and *Limax*, the hermaphrodite duct bears a small sac-like diverticulum at its junction with the albumen gland (Quick, 1960). This diverticulum has been variously described as a seminal vesicle, a fertilization pocket and a spermathecal sac although the functions of the diverticulum and of the arionid loop remain uncertain.

Focardi and Quattrini (1972) found that gametes were absent from the diverticulum in *M. gagates*. However, Nicholas (1984) showed that sperm were present in the diverticulum of *D. reticulatum* shortly after copulation although they were stored there only temporarily and, in both this species and *A. hortensis*, foreign sperm passed into the hermaphrodite duct. The convoluted region of the hermaphrodite duct acted as a seminal vesicle, becoming distended at the spermatozoan stage by the pressure of accumulating sperm. Nicholas (1984) described how the epithelial cells of the hermaphrodite duct were adapted for transporting, immobilizing, phago-cytosing and maintaining the sperm. Lusis (1961) suggested that the loop or claw in arionid slugs, which differed histologically from the rest of the duct, acted as a barrier to control the movement of spermatozoa from the herma-phrodite duct through the carrefour. The detailed structure of the carrefour in *D. reticulatum* was described by Nicholas (1984), who concluded that the carrefour played an important role in controlling the movement of the gametes and in the packaging of the fertilized egg. Nicholas also showed that the site of fertilization of the egg was not necessarily the diverticulum in *D. reticulatum* or the carrefour loop in *A. hortensis* as fertilization could be delayed, with sperm waiting within the oocyte cytoplasm, until both maturation divisions were complete and this could occur at other locations along the reproductive tract.

5.1.5 Albumen gland

Lusis (1961) described the formation of the albumen gland. It was formed initially from a small hollow diverticulum of the reproductive tract at the junction of the hermaphrodite duct and the common duct and was lined on the inside by a folded columnar epithelium. A compound tubular gland developed as the result of the proliferation of primary and secondary side branches, forming glandular units while the original tube remained as the central ciliated duct of the gland. This duct had a thick inner connective tissue layer and an outer muscular sheath (Smith, 1965). The albumen gland underwent rapid growth towards the end of the juvenile growth phase, when secretion began to appear in the collecting ducts and the gland represented up to 6% of body weight at its maximum size (Runham and Laryea, 1968). During later development of the glandular cells, they became filled with droplets or granules which merged into globules, the cells became distended and the apical region of the cell degenerated, releasing secretion into the collecting ducts (Lusis, 1961; Smith, 1966a). The albumen gland may grow much larger and Lusis (1961) described instances where newly mature A. ater were found dying with part of the gut or reproductive tract protruding through the body wall. This was caused by hypertrophy of the albumen gland to the extent that it filled almost the whole of the body cavity. Runham and Laryea (1968) and Smith (1966a) described the stages in the development of D. reticulatum and of A. ater respectively and Runham and Laryea showed that these stages were closely related to stages in the development of the gonad.

The function of the albumen gland is the production of albumen or perivitelline fluid for the egg (Bayne, 1967a,b). Galactogen has been shown to be the reserve carbohydrate for the developing embryo in the slug egg (Section 3.3). Large amounts of galactogen were shown to be present in the albumen glands of A. columbianus (Meenakshi and Scheer, 1968) and D. reticulatum (Bailey, 1970). Glycogen was present in the epithelium of the albumen gland canal of Philomycus (Kugler, 1965) and traces (about 4.5%) were present in the albumen gland of Ariolimax (Meenakshi and Scheer, 1968). Seasonal variations in the galactogen content of the albumen gland of Ariolimax have been described in Section 3.3. Amino groups and acid groups were identified by Smith (1965) in the albumen gland secretion of A. ater and, according to Duncan (1975), the egg albumen is a complex fluid containing proteins, glycoproteins, some free amino acids and calcium although it was not certain that all these substances were produced by the albumen gland. For example, it seems unlikely that calcium was derived from the albumen gland as histochemical reactions specific to calcium were found by Smith (1965) to give negative reactions with albumen gland secretion.

5.1.6 Common duct

In most slugs, the male and female gametes follow separate pathways from the carrefour along a common duct or spermoviduct, and separate male and female ducts then diverge at the distal end of the common duct. In the Veronicellidae the male and female ducts are separate along their whole length from their origin at the carrefour (Lanza and Quattrini, 1964). The stages in the development of the common duct have been described for A. ater (Smith, 1966a) and for D. reticulatum (Runham and Laryea, 1968). The basic structure of the common duct, which has been described for A. ater (Lusis, 1961; Smith, 1966a), for P. carolinianus (Kugler, 1965), for *M. gagates* (Focardi and Quattrini, 1972) and for *D. reticulatum* (Bayne, 1967b; Bailey, 1970), is similar for each species. The lumen of the duct is incompletely divided into two grooves by lateral folds. These groves are lined by ciliated epithelial cells and are unequal in size, with a larger female groove and a narrow male groove. According to Lusis (1961), spermatozoa were confined to the male groove by the dividing fold and by long cilia beating inwards although eggs could fill and expand both grooves of the duct. Kugler (1965) described how, in *P. carolinianus*, mucus produced by mucous cells in the male groove was used to bind the sperm into a sperm thread to restrict them to the male groove. The proximal end of the female groove or duct is very wide and has side pockets or haustra in *Philomycus* (Kugler, 1965) and *Arion* (Smith, 1965). The walls of the grooves are glandular with the oviducal gland opening into the female duct and the prostate gland opening into the male duct. The oviducal gland in *A. ater* was made up of large flask-shaped gland cells opening individually into the lumen of the female duct via short ducts passing between the epithelial cells, and in mature slugs these gland cells constituted about 95% of the tissue in the proximal part of the common duct (Smith, 1965). The basic structure of both oviducal and prostate glands of *D. reticulatum* was found to be very similar in the early stages of development although accumulated secretion obscured this similarity at maturity (Bailey, 1970). Runham and Laryea (1968) described at least three types of glandular cell from the diverticula of the prostate gland, while Smith (1966a) described a second gland, the male flask gland, opening into the male duct of *A. ater*. The ultrastructure of the prostate glands of *V. borellianus* and of *M. gagates* has been described by Quattrini (1966a, 1967).

Connective tissue covered the glandular regions of the common duct together with muscular tissue which was more extensively developed on the female side of the duct (Lusis, 1961). Bailey (1970) found evidence for nervous control over the release of prostate gland secretion in *D. reticulatum* and suggested that this was necessary for the rapid release of secretion just prior to copulation. Bayne (1967b) investigated the composition of extracts from the oviducal gland and concluded that after the fertilized egg had received perivitelline fluid from the albumen gland, the perivitelline membrane formed when the fluid came into contact with a jelly layer secreted in the upper region of the common duct by the oviducal gland. The shell was subsequently formed in the lower part of the duct. Bayne (1967b) suggested that the egg shell was essentially jelly material which had been modified by a later secretion. The changes which occur in calcium levels in the common duct during egg laying as the result of the mobilization of calcium reserves in the body have been described in Section 3.5. The function of the prostate gland secretion was less clear. Lusis (1961) suggested that it might be a source of nutritive material for the spermatozoa while both Lusis and Bayne (1967b) have suggested that the gland might have some role at the time of copulation. Smith (1965) showed that, in *A. ater*, the prostate was active in spermatophore formation. Lusis (1961) and Bayne (1967b) described how prostate gland secretion continued during egg formation and so might contribute to the formation of the egg. Bayne (1967b) also suggested that the prostate secretion might facilitate the passage of foreign sperm during its ascent via the male groove to the carrefour.

5.1.7 Vas deferens, penis and associated structures

The common duct divides distally into two distinct ducts, a short oviduct and a longer, narrower vas deferens which is often coiled and in most slugs runs forwards to join the penis. The penis is absent from the Arioninae although it is well developed in the related *Ariolimax* and *Prophysaon* (Webb, 1961). In *Arion* and *Geomalacus*, the vas deferens, which is a simple, ciliated duct, is modified and enlarges distally to form the epiphallus which enters the genital atrium directly. An epiphallus is also present in the Milacidae and Parmacellidae where the vas deferens runs first into an epiphallus which then passes into the penis (Quick, 1960). There is no epiphallus in the Limacidae (Quick, 1960) or in *Philomycus* (Kugler, 1965), where the penis is a long, V-shaped, cylindrical organ. Lusis (1961) showed that, in *A. ater*, the histological structure of the epiphallus was similar to that of the vas deferens but that the folds of the epithelium were taller and more numerous. Pabst (in Lusis, 1961) suggested that the spermatophore was formed in the epiphallus of *Arion* and this was confirmed by Smith (1965) for *A. ater*. Smith showed that secretions of the prostate and other male glands in the common duct were mixed with the sperm mass as it passed down the duct, and that this mass was converted in the epiphallus into a spermatophore by further secretions from a glandular layer in the epiphallus. Focardi and Quattrini (1972) described how the wall of the epiphallus of *M. gagates* had a thick compact muscle layer and suggested that the complicated infoldings of the epithelium were responsible for the elaborate design of the spermatophore that was formed in this region. Lusis (1961) also showed that, in *Arion*, the vas deferens and epiphallus were surrounded by circular and longitudinal muscles, with the muscle layers becoming much thicker just before the epiphallus entered the atrium. This thickening was considered by Webb (1961) to represent a true penis homologue. The penis in *D. reticulatum* is an eversible sac with a large erectile appendage, the sarcobellum or stimulator, attached to the wall of the sac (Bayne, 1966; Runham and Laryea, 1968). Runham and Laryea (1968) found that gland cells were scarce in the penis sac except in the sarcobellum where secretion began in the early spermatozoan stage. In *Limax* the penis sac contains one or more long prominent folds which may be expanded distally into crests or combs. These folds are everted during copulation (Quick, 1960). There is no stimulatory structure associated with the penis of milacid slugs although in M. *gagates*, M. *nigricans* and T. *sowerbyi* a sarcobellum or stimulator is found attached to the genital atrium (Quick, 1960; Kerney and Cameron, 1979). This is absent from *Boettgerilla*, where the epiphallus is separated from the penis by a narrow duct.

Quick (1960) described how the penis sac of *Deroceras* was equipped with an appendage near its apex which was completely everted during mating. This apical appendage has been used as a taxonomic character (e.g. Quick, 1960; Kerney and Cameron, 1979) and is a tifid appendix in D. *reticulatum*; it has four or five branches in D. *caruanae* and is a single appendix with a bifid tip in D. *laeve*. Runham and Laryea (1968) described how the appendage was hollow until the early spermatozoan stage but after this stage the lumen became progressively filled with cell debris from the epithelium lining the appendage. Some individuals of D. *laeve* have been found to be aphallic. These slugs have a vestigial penis and the vas deferens ends blindly although the remainder of the reproductive tract is normally developed, including the common duct and prostate gland, and sperm is usually present (Quick, 1960). Babor (1894) suggested that this species was protogynous, with young adults being always female and some older slugs reaching a purely male phase characterized by an atrophied albumen gland and spermatheca and with a hypertrophied penis. Quick (1960) concluded that aphallism in D. *laeve* was a form of sexual dimorphism as no individuals were found in which the albumen gland and spermatheca had atrophied, and in young individuals which would become euphallic the vas deferens always reached the developing penis. In addition, euphallic and aphallic forms occurred together in varying proportions at the same site in different seasons. Mordan (1973) described a seasonal variation in the proportion of aphallic and euphallic forms of D. *laeve*, with aphallic forms abundant only in the late spring and early summer, following a period of breeding activity. Lupu (1977) described aphallism in D. *laeve* as a form of polymorphism, consisting of protogynous and hermaphrodite forms.

The penis was exceptionally well developed in *Ariolimax* (Webb, 1961) and Harper (1988) described how, in this species, pairs sometimes became so entangled that they were unable to separate and so resorted to biting off the penis. This phenomenon was termed apophallation and Mead (1943) suggested that the severed penis might regenerate in this species. During courtship the penis sac and its associated structures are everted by haemostatic pressure (Runham and Hunter, 1970). A well-developed penis retractor muscle arises from the body wall and is inserted at the apex of the penis at the point where it joins the vas deferens in limacid slugs or the epiphallus in milacid slugs (Quick, 1960). The penis sac in *Deroceras* was retracted after copulation by the retractor muscle but the appendage remained in the lumen of the sac for several hours before being completely returned to its pre-coital state (Webb, 1961). In *Arion*, where the penis is absent, the genitalia are retracted by a retractor muscle with insertions on the free oviduct and the spermathecal duct.

5.1.8 Oviduct, spermatheca and atrium

The free oviduct is similar histologically to the vas deferens (Lusis, 1961) and it opens into the genital atrium. The base of the oviduct in *P. carolinianus* was termed a vagina by Kugler (1965). Kugler described how eggs collected temporarily in the highly distensible vagina during egg laying before passing into the atrium. The wall of the vagina contained muscle and elastic fibres and the epithelium was much folded in the relaxed state. Most slugs lack the dart sac found in some snails such as *Helix*. This has been retained in *Philomycus* and Kugler (1965) described it as a highly muscular evagination of the wall of the vagina. Dissection of freshly killed slugs revealed a dart composed of calcium carbonate secreted by the epithelium of the sac. The spermatheca (bursa copulatrix of some authors) is a simple sac-like structure lined with a columnar epithelium which Smith (1965) described as being laden with secretory material in the mature phase. A thin layer of connective tissue with some muscular tissue lay beneath the epithelium. The spermatheca has a narrow duct opening at the apex of the atrium near the junction with the oviduct and the penis or epiphallus. In some species the duct opens at the base of the male or female tracts. Smith (1966a) des-cribed how the spermatheca in *A. ater* received the partner's transferred spermatophore during copulation. Runham and Hunter (1970) suggested that when the spermatophore or sperm mass was transferred at copulation it passed to the spermatheca where the outer material was digested, releasing the sperm. Some of this foreign sperm then subsequently moved up the male tract into the carrefour region where fertilization takes place (see Section 5.1.4). Bayne (1970) found early traces of breakdown of sperm in sections of *D. reticulatum* spermatheca contents and suggested that cell debris from the reproductive tract collected in the spermatheca. It seems likely that excess sperm undergoes autolysis in the spermatheca and the evidence for this was reviewed by Duncan (1975).

The atrium is a simple structure in the Limacidae and in Milacidae but in some milacid slugs it is equipped with a stimulator (see Section 5.1.7) and, in most milacids, accessory or atrial glands open into the atrium. Quattrini and Focardi (1976) described how the atrial gland of *M. nigricans* produced small crystals of calcium carbonate in the form of calcite. This secretion was associated with the stimulator and Quattrini and Focardi suggested that these crystals were analogous to the dart of other gastropods. In mature *Arion* the atrium is divided into upper and lower atria, with the oviduct, spermatheca duct and epiphallus opening into the upper atrium. A stimulatory organ of variable form, the ligula, is present in many *Arion*. The position of this structure varies and in *A. ater* it is located at the opening of the oviduct into the upper atrium, while in *A. lusitanicus* it is situated in the free part of the oviduct. The ligula is absent from the small species *A. intermedius* and *A. fasciatus* but its functional equivalent is present in the lower part of the oviduct in *A. hortensis* (Quick, 1960). The atrium, particularly the upper atrium, was described by Smith (1965) as a very muscular structure. The atrium in most slugs opens close behind and below the right upper tentacle.

5.2 SPERMATOGENESIS, OOGENESIS AND FERTILIZATION

5.2.1 Spermatogenesis

Spermatogenesis in *A. ater* began soon after hatching and continued until the oocyte stage of gonad development was reached (Lusis, 1961). Lusis found that there was a continuous differentiation of spermatogonia from the acinar epithelium until the late phase of spermiogenesis. The stages in spermatogenesis were described for *A. subfuscus* by Watts (1952) and, briefly, for *D. reticulatum* by Bailey (1970). Tuzet and Galangau (1967) described the stages of meiosis in primary spermatocytes of *D. agreste* and also recorded sperm abnormalities from this species. Bayne (1970) found similar abnormalities and also a group of ten closely associated sperm tails which he suggested represented the failure of ten spermatids to separate. Rousset-Galangau (1972) described sperm dimorphism in *M. gagates* and *D. agreste* which originated at the spermatid stage. Morphological differences between the two types of sperm included differences in nuclear size and in the organization of the middle piece of the flagellum. A detailed description of the fine structure of the stages in spermatogenesis in *A. ater* was given by Parivar (1981). Parivar also described sperm dimorphism involving two different types of sperm, eupyrene and apyrene. Both eupyrene and apyrene spermatids differentiated side by side in the gonad. The most obvious difference between the two was the lack of an acrosome in the apyrene form although this was present in the eupyrene spermatozoan. Since the acrosome was considered important for the penetration of the spermatozoan into the ovum, Parivar (1981) concluded that only eupyrene sperm were involved in fertilization and suggested that the apyrene sperm might have some physiological or endocrinological role in the gonad. It was also suggested that the abnormalities reported by Tuzet and Galangau (1967) and Bayne (1970) could be explained on the basis of apyrene spermatozoa. This sperm dimorphism occurs in other gastropods. Maxwell (1983) described how some streptoneuran gastropods had developed two types of spermatozoa, eupyrene and apyrene (oligopyrene). The eupyrene spermatozoan contained genetic material and was capable of fertilization while the apyrene spermatozoan was much larger, lacked genetic material and was incapable of fertilization.

The fine structure of the mature spermatozoan of *D. reticulatum* was described by Bayne (1970). The acrosome was the most anterior part of the spermatozoan and in *D. reticulatum* this was a relatively simple structure. The anterior tip of the nucleus extended forwards as a thin, elongated appendage, the granular acrosomal substance was deposited around this and was bounded by the plasma membrane. The head of the sperm consisted of nuclear material present in a homogeneous state. There was no nuclear membrane but the entire head was covered by the same two-layered plasma membrane which enclosed the whole cell. The head was bounded anteriorly by the acrosome and posteriorly by a neck region at the head–tail junction. The sperm tail or flagellum core (axoneme) was composed of 9 + 2 axial fibres and was bounded by the periflagellar and peripheral mitochondrial sheaths enclosed within the cell plasma membrane. Takahashi *et al.* (1973) investigated the origin and fine structure of the flagellum of the spermatozoan in *L. flavus*. They confirmed the 9 + 2 pattern of axial fibres but showed that nine doublets differentiated from the nine coarse fibres. The mature sperm displayed a helical structure in the acrosome, head and middle piece. This helical pattern was confirmed by Maxwell (1977), who described the trihelical organization of the *A. hortensis* spermatozoan.

The fine structure of the mitochondrion, which extends the entire length of the spermatozoan in *Limax* sp., was described by Reger and Fitzgerald (1983). They showed that a paracrystalline complex was situated between the inner and outer mitochondrial membranes. The spermatozoan described by Bayne (1970) resembled the modified molluscan type described by Maxwell (1983) as characteristic of internal fertilization. Maxwell described the fine structure of the spermatozoa of *T. sowerbyi*, *D. reticulatum* and *A. hortensis* and confirmed that the axoneme had a typical 9 + 9 + 2 structure with peripheral doublets rather than the singlets originally described by Bayne (1970). Details of the events which took place during the development of the mature spermatozoan from the early spermatid (spermiogenesis) were described by Maxwell (1983), using *A. hortensis* as a model for the nuclear changes that occurred. Spermatozoan lengths for several species of slug were listed by Maxwell (1983) as: *D. reticulatum* 140 μm, *T. sowerbyi* 246 μm, *A. hortensis* 300 μm and *A. columbianus* 265 μm.

Since the sperm has become highly specialized for motility, much of its cytoplasm has been lost and little space has been left for the intracellular storage of the polysaccharides which could provide the energy for motility. Although exogenous nutrients possibly derived from the secretion of glands, such as the prostate gland, may be important, considerable interest has been shown in identifying intracellular reserves in the spermatozoan. The presence of glycogen in the mitochondia of the spermatozoan has been demonstrated for *T. haliotidea* (Personne and Andre, 1964), for *Arion* sp.

(Anderson and Personne, 1970) and for *A. hortensis* and *L. flavus* (Maxwell, 1983). Maxwell (1980) found a non-linear distribution of glycogen deposits within the mitochondrial derivative of mature spermatozoa in *A. hortensis*, with the distal part of the sperm tail containing significantly more glycogen than regions nearer the head. Maxwell (1983) described how glycogen deposits were not formed until the mitochondrial derivative was mature and how much of the glycogen was synthesized after the spermatozoan had left the gonad. Maxwell concluded that the mature spermatozoan possessed the necessary amylophosphorylase system which would control the metabolism of polysaccharides. The existence of such a system was demonstrated by Anderson and Personne (1970) and this operated in the direction of synthesis in free sperm incubated in a medium containing glucose-1-phosphate.

Kugler (1965) described how in *P. carolinianus* mucus was used to bind the sperm into a sperm thread. This thread was passed to the partner during copulation rather than a spermatophore. A similar form of sperm transfer was described for another philomycid slug, *Eumelus wetherbyi*, by Webb (in Kugler, 1965) although Ikeda (in Kugler, 1965) described how a spermatophore was used to transfer sperm between partners in the Asian philomycid *M. bilineatus*. No spermatophore is produced in slugs belonging to the family Limacidae and this is reflected in the absence of an epiphallus from the reproductive tract. In both *Deroceras* and *Limax* a mass of spermatozoa enclosed in mucus is transferred from one partner to the other during mating (Quick, 1960). Webb (1961) has described how during mating in *Deroceras*, this mass of spermatozoa became smeared over the area of contact between the everted penis, becoming entangled in the everted appendages so that it was engulfed as the penis sac was retracted. The spermatozoa are enclosed in a spermatophore in the Milacidae and the Arioninae, although spermatophores have not been recorded for *G. maculosus* (Platts and Speight, 1988). The formation of the spermatophore in the epiphallus of *Arion* from the secretions of the prostate and other glands has been described previously (see Section 5.1.7). Quick (1950) described how, in *Arion*, the spermatophore was probably formed shortly before or even in the early stages of mating. The spermatophore has a complex shape, reflecting the internal shape of the epiphallus. It is a curved, chitinous tube, up to 20 mm in length in larger species, and in *Arion* bears ridges which may be serrated while in *Milax* and *Tandonia* it bears prominent spines. Davies (1987) described how the strongly serrated ridge, which was well developed on the wider portion of the spermatophore of larger arionid slugs, provided great strength with controlled flexibility. Quick (1950) showed that the spermatophore of *T. sowerbyi* was capped by a conical structure and travelled from the penis with the cap end in advance. This cap disappeared rapidly after mating when the spermatophore reached the spermatheca. Similar caps have not been recorded from other species. The shape of the spermatophore is characteristic of the species and has been widely used in the taxonomy of slugs (Quick, 1960). They differ markedly even within a species complex and can be used, for example, to distinguish between *A. hortensis* s.s., *A. distinctus* and *A. owenii* (Davies, 1977). Davies also described how the spermatophore of the *A. hortensis agg.* was equipped with an adhesive collar, enabling it to adhere to thickened receiving areas near the entrance of the recipient's spermatheca duct.

5.2.2 Oogenesis

The reviews by Raven (1975), Adiyodi and Adiyodi (1983a) and Dohmen (1983) show that, although there is a considerable body of information about oogenesis, fertilization and the development of the egg in pulmonate molluscs, particularly relating to the aquatic snail *Lymnaea*, relatively little is known about these processes in slugs. Sabelli *et al.* (1978) concluded that oogenesis in *D. reticulatum* was similar to that described for other pulmonates. Guraya (1969) described the results of a histochemical examination of the components of the developing oocyte of *A. columbianus*. The synthesis and processing of rRNA in *D. reticulatum* was described by Hill *et al.* (1978). Oogenesis continues throughout the life of the slug (Runham and Hogg, 1979). Lusis (1961) was unable to observe oogonia in the gonad of *A. ater* and Sabelli *et al.* (1978) noted that oogonia were rarely seen in the gonad of *D. reticulatum*. In laboratory reared slugs, Lusis (1961) reported that oogenesis started later than spermatogenesis in *A. ater* although Parivar (1978) suggested that oocytes differentiated first before the spermatogonia in the same species and they were visible within a few days of hatching. Although oocytes differentiated first they reached maturity after the spermatozoa (Runham and Hunter, 1970). The fine structure of the oocyte of *D. reticulatum* was described by Bailey (1970) and by Hill and Bowen (1976) who showed that the development of the oocyte in the gonad could be divided into three stages. The structure of the amphinucleolus of the *Limax* oocyte was described by Yamamoto (1977). Lusis (1961) found that as the oocyte began to grow during the later spermatozoan stage, several adjacent cells extended around the oocyte to form a follicular membrane and this persisted until the oocyte was released into the lumen of the acinus by the rupture of the membrane. These follicular cells probably developed from the gonadal stem cells (Jong-Brink *et al.*, 1983). Quattrini and Lanza (1968) showed that while the inner part of the oocyte was completely covered by follicular cells, the basal part remained in contact with the basal membrane of the follicle wall in *Vaginulus*. A cavity appeared between the follicular membrane and the developing oocyte as this grew larger.

Development of the oocyte in the gonad proceeded no further than the primary oocyte stage and the maturation division was delayed until the oocyte reached the carrefour region (Kugler, 1965). Oocytes of various sizes and at different stages are present in the acinus at any time and Lusis (1961) found that many of the earlier oocytes disintegrated before reaching maturity. Lusis showed that oocytes reached their maximum number per gonad (964) and maximum mean size during the late oocyte stage in *A. ater*. Mature oocytes first appeared free in the acinar lumen during this stage (Runham and Hunter, 1968). Jong-Brink *et al.* (1983) discussed the possibility that the follicular cells in pulmonates were involved in vitellogenesis, i.e. the production of macromolecular yolk components. There was evidence that the follicular cells were involved in the synthesis of yolk components and that exogenous vitellogenesis occurred in cephalopods but the evidence was less certain for gastropods, although the follicular cells had the morphological characteristics of protein-producing cells in other molluscs. However, Bailey (1970) described organelles in the oocyte of *D. reticulatum* which were indicative of vitellogenic activity and Hill and Bowen (1976) described how the plasma membrane of the oocyte of *D. reticulatum* became specialized during its development with the appearance of microvilli and pinocytotic tubules. This would enable yolk proteins and other components to pass from the follicular cells to the oocyte. The structure and functions of the epithelial cells associated with the oocyte of *D. reticulatum* were examined by Hill (1977). While the oocyte initially remained in contact with the germinal epithelium, as oocyte vitellogenesis proceeded it became totally encapsulated by the layer of follicular cells, which were shown to be active in secretion. Hill also found that the follicular cells were active in digestion and Jong-Brink *et al.* (1983) concluded that follicular cells would have the capacity to phagocytose materials and so degenerating oocytes and cell debris remaining after the release of the oocyte could be broken down intracellularly by the follicular cells. The contents of the egg of *A. ater* were studied by Sathananthan (1970, 1972) using centrifugation. The general stratification of contents conformed to the molluscan pattern although an unusual band of inclusions, consisting mainly of phagosomes and associated lysosomes was detected.

5.2.3 Fertilization and hermaphroditism

The sperm probably enters the egg as it passes through the carrefour region of the reproductive tract although the actual fusion of the male and female nuclei may be delayed (Section 5.1.4). Carrick (1938) showed how at the moment of laying in *D. reticulatum*, the oocyte had not completed the maturation divisions although the sperm had already entered the egg. Raven (1966) described how, in *Limax*, the sperm entered the egg and moved through the cytoplasm. The tail then broke away and the head changed shape and rounded off and the sperm tail was resorbed into the egg cytoplasm after the second maturation division. Nicholas (1984) similarly found that fertilization in *D. reticulatum* and *A. hortensis* could be delayed after the sperm had entered the egg in the carrefour region, with the sperm waiting in the oocyte cytoplasm until both maturation divisions were completed. Raven (1966) described the formation of the second maturation spindle in *L. flavus* and *D. reticulatum*, showing that in slugs the sperm aster had become a more or less rudimentary structure. *Arion ater* has been described as a protandrous hermaphrodite and Runham and Hunter (1970) described how most of the sperm were lost before the oocytes became mature. However, Parivar (1978) suggested that this species was not a strict protandrous hermaphrodite as both male and female gametes grew and developed simultaneously. Runham and Laryea (1968) found that some egg laying could even precede the release of sperm in *D. reticulatum*. Nicholas (1984) described *D. reticulatum* and *A. hortensis* as simultaneous hermaphrodites and examined the possibility of self-fertilization in these species. Nicholas found that while both species were capable of self-fertilization, only *A. hortensis* did this regularly when reared in isolation. Although *D. reticulatum* did not normally resort to self-fertilization, egg production was not usually suppressed and consequently many slugs laid infertile batches of eggs, suggesting that ovulation was independent of mating. South (1982) also showed that individuals of *D. reticulatum* reared in isolation readily laid eggs, although these proved to be infertile. Nicholas (1984) also discussed the possible mechanisms involved in self-fertilization.

Self-fertilization in molluscs was discussed by Duncan (1975) who listed examples of self-fertilization in a number of slugs, including the family Arionidae, *Philomycus*, *L. cinereoniger*. *M. gagates*, *Ariolimax*, *V. borelli-anus* and *L. alte*. In the last three examples, successive generations have been maintained by self-fertilization. According to Runham and Hunter (1970), a few species of slug, such as *D. reticulatum*, only rarely produced self-fertilized eggs. However, self-fertilization has been reported from other species of *Deroceras*. Luther (1915) bred two successive generations of the related species *D. agreste* by self-fertilization and Maury and Reygrobellet (1963) recorded self-fertilization in *D. laeve*, *D. agreste* and *D. meridionale* (Reygrobellet). Chen *et al.* (1984) showed how five generations of *D. agreste* could be produced experimentally per year by self-fertilization. Fecundity by self-fertilization was estimated to be two to four times greater than that of normal reproduction, the growth rate of the young slugs was higher and the life span longer than for slugs produced by sexual reproduction. Williamson (in Duncan, 1975) showed that mixed clutches of eggs were produced by *Arion* using both autosperms and allosperms.

Electrophoretic studies of genetic variation in 14 species of slug in the eastern USA indicated that facultative or obligatory self-fertilization (automixis) was the normal breeding system in six of the species (McCracken and Selander, 1980). These included *A. fasciatus s.s.*, *A. circumscriptus s.s.*, *A. silvaticus* and *A. intermedius*. In addition, two out of three strains of *A. subfuscus* were monogenic and so probably self-fertile. Eight species showed high levels of heterozygosity, indicating that they were largely outcrossing species. There were: *A. hortensis* s.s., *P. carolinianus* and three other species of *Philomycus*, *L. maximus*, *L. valentiana* and *D. reticulatum*. In addition, one strain of *A. subfuscus* was also amphimictic. McCracken and Selander (1980) also found that *A. columbianus* consisted of several monogenic strains, although there was some outcrossing locally between them. The most complex pattern of genetic variation occurred in *D. laeve*, which was capable of both outcrossing (amphimixis) and self-fertilization. This survey was followed by surveys of genetic variation in arionid, limacid and milacid slugs in the British Isles and France by Foltz *et al.* (1982b, 1984).

Foltz *et al.* (1982a) found three categories of *Arion* species on the basis of their breeding systems. *Arion lusitanicus*, *A. hortensis* s.s., *A. distinctus* and *A. owenii* were highly polymorphic and heterozygous and reproduced predominantly by outcrossing. *Arion circumscriptus* s.s., *A. silvaticus* and *A. intermedius* consisted of one or more monogenic strains, probably resulting from self-fertilization while *A. ater* and *A. subfuscus* consisted of both a polymorphic form and a monogenic strain, between which hybridization occurred. Foltz *et al.* (1982a) concluded that the genetic structures and breeding systems in populations of *Arion* spp. introduced to North America were similar to those of British populations. Foltz *et al.* (1984) included three milacid species and seven limacid species in their survey. They found that outcrossing was the normal breeding system in most of these species, with *D. agreste* being the only self-fertilizing species of those examined. This conclusion was supported by the observations of Luther (1915) and Maury and Reygrobellet (1963) described above. Foltz *et al.* (1984) concluded that slug breeding systems had not been modified in the process of colonization of North America although the type of system was related to colonizing ability, since self-fertilizing European species appeared to be disproportionately successful in colonizing North America. Endemic North American species, which are restricted to undisturbed habitats and have not been successful colonizers, might therefore be expected to outcross rather than self-fertilize. Anderson and McCracken (1986) examined the breeding systems and genetic variation in four species of the family Philomycidae, which is endemic to North America. They concluded that outcrossing was the main breeding system for these slugs, supporting the suggestion that endemic species are less likely to use self-fertilization than colonizing species.

Foltz *et al.* (1982b) studied gametic disequilibria in populations of *D. laeve* and concluded that these were caused by random genetic drift and maintained by self-fertilization. Nicklas and Hoffmann (1981) described how the normal breeding pattern in *D. laeve* was by apomictic parthenogenesis rather than self-fertilization, although reciprocal outcrossing in complete hermaphrodite forms occurred occasionally. Both hermaphrodite and female (i.e. aphallic) forms reproduced parthenogenetically. Nicklas and Hoffmann provided evidence that apomixis was also the rule in natural populations and that most of these consisted of only a single clone. They also confirmed that *D. reticulatum* was an obligate outcrossing species. Hoffmann (1983) reviewed the experimental evidence supporting the conclusion that reproduction in *D. laeve* in the absence of a mate was by parthenogenesis rather than by self-fertilization and concluded that this could be explained by automictic rather than apomictic parthenogenesis. Foltz *et al.* (1984) considered that the conclusions of Nicklas and Hoffmann (1981) about apomictic parthenogenesis in *D. laeve* were incompatible with the low level of heterozygosity reported for this species in the literature and suggested that this subject required further investigation.

5.3 EGG

5.3.1 Structure of the egg

The egg laid by the slug tends to be oval in shape, ranging in size from 2.0 × 1.5 mm (*A. intermedius*) up to 5.0 × 5.5 mm (*L. maximus*) or even larger, e.g. 7.0 × 4.0 mm in *T. haliotidea* (Quick, 1960). The eggs of the large slug *A. columbianus* were described by Harper (1988) as nearly spherical and measuring about 7 mm in diameter, while those of *V. ameghini* were oval and measured 8.5 × 4.0 mm (Tompa, 1980). Harper (1988) described how the whitish outer coat could be easily removed to reveal a translucent egg inside. Smith (1965) showed that the completed egg of *A. ater* had a fairly thick outer layer containing crystals of calcium carbonate which were acquired during the passage of the egg through the free oviduct. Smith suggested that this calcium was derived from calcium stores in the digestive gland and Fournié and Chétail (1982b) described how calcium accumulated in the female duct just before egg laying. An account of the mobilization of calcium reserves during the reproductive period was given in Section 3.5. Most arionid eggs are provided with a calcareous coating which is white or yellow in colour when first laid but may turn brown later (Quick, 1960). A calcareous shell is absent from limacid and milacid slug eggs which are translucent, although in some species, e.g. *D. reticulatum*, the outer coat is speckled with calcareous particles. Carrick (1938) suggested that these were absorbed during development but Bayne (1966) showed that considerable amounts of calcium remained in the shell even after hatching. Bayne concluded that the embryo derived most of its calcium from the egg albumen rather than from the jelly layer or the shell. Tompa (1980) found that the surface of the egg of *V. ameghini* was lightly sprinkled with spherules of calcium carbonate and that these spherules dissolved and disappeared during development. Tompa also demonstrated that the calcium content of the egg albumen was insufficient to account for the amount of calcium found in the

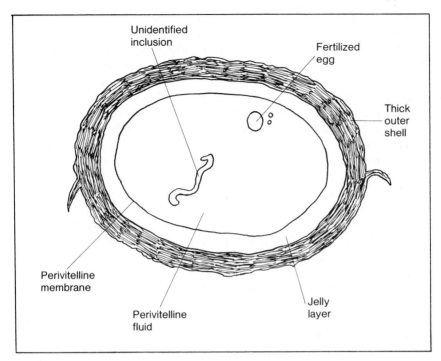

Fig. 5.4 Limacid egg.

newly hatched embryo. A third form of calcium available to the egg was a coating of calcium-rich soil and faeces applied to the surface of the shell of the eggs when they were laid.

The structure of the egg (Fig. 5.4) can be easily seen when the egg is immersed in water and viewed under transmitted light. The detailed structure of the egg of *D. reticulatum* has been described by Carrick (1938) and by Bayne (1966, 1988a), who investigated the biochemical composition of the layers sorrounding the embryo. The shell had a granular appearance due to the small calcareous granules embedded in it and at each end it was often produced as a short tapering thread. Batches of eggs were sometimes found joined by these threads to form a string. Lusis (1961) described how the last batches of egg laid by *A. ater* were always few in number, abnormally formed and laid in strings. He suggested that the reason for this was that the slug was left with insufficient material to produce fully developed eggs. The shell was composed of concentric layers and had a thickness of about 150 μm (Carrick, 1938). Free lipids were present on the surface of the egg of *A. ater* but not from the surface of *D. reticulatum* and *D. caruanae* eggs (Bayne, 1968c). This was probably added to the egg by a secretion of the spongy gland in the atrium described by Smith (1965).

Bayne (1966) found that the lipid layer did not effectively reduce desiccation from the egg. Bayne (1966) concluded that the shell acted as a mechanical support and partial barrier to the entry of harmful organisms. The possible role of lectins derived from the albumen gland in conferring immunobiological protection on the egg is discussed in Section 9.2.4. Bayne (1966) suggested that the calcium found in the shell and jelly layers might contribute towards the physiological buffer that was necessary between the embryo and the external environment, by chelating toxic ions and other substances diffusing into the egg. The jelly layer lying between the shell and the perivitelline membrane was a homogeneous layer. Bayne (1966) showed that if eggs were immersed in distilled water for more than about an hour a wrinkling of the perivitelline membrane was seen. He concluded that this was due to pressure being exerted upon the perivitelline sac by the swollen jelly that had taken up water. He concluded that the jelly layer provided a short-term buffer to changes in the water relations of the egg as the shell provided no appreciable resistance to water loss. The perivitelline membrane enclosed the perivitelline fluid or egg albumen and was formed by coagulation of the albumen surface at the albumen/jelly interface. This layer in *A. ater* was thicker and had histochemcal reactions similar to the shell membranes (Bayne, 1968a). The origin and composition of the egg albumen is described in Section 5.1.5. Bayne (1966) found that lipid compounds were absent from the perivitelline fluid of *D. reticulatum* and concluded that the lipid requirements of the developing slug were probably met by the metabolism of galactogen. Lipid compounds were also shown to be absent from the egg of *L. alte* by Nagabhushanam and Kulkarni (1971a). They described how the embryo grew at the expense of the albumen until just before hatching when it filled the entire perivitelline sac.

Carrick (1938) described a twisted membranous structure within the perivitelline sac of *D. reticulatum* and concluded that it represented the remains of the sperm body. Bayne (1966) recorded a similar unidentified inclusion in the egg of the same species but did not agree with Carrick's conclusions. Carrick (1938) also described a number of abnormalities from eggs of *D. reticulatum*. Polyembryonic eggs, containing more than one normal embryo, each fertilized by a separate sperm, were occasionally produced and the parent slug was found to possess a relatively small albumen gland. These eggs took longer to hatch than normal eggs and the young slugs which hatched were smaller than usual. A few large abnormal eggs with up to 30 unfertilized egg-cells were produced before normal eggs were laid by young slugs or by senescent slugs, and where these also contained one or two normal zygotes, the latter did not complete their development. However, Carrick (1938) found that abnormalities were generally uncommon and South (1989b) showed that, out of more than 1800 eggs of *D. reticulatum* taken from a population in the field over two years, less than 1% of these eggs were abnormal and most of these were polyembryonic. Mason and Copeland (1988) recorded a number of abnormalities from eggs of laboratory raised *L. valentiana*. These included viable conjoined twins, fused tentacles, supernumerary eyes and a healthy, two-headed, one-tailed individual among newly hatched slugs. The proportion of abnormalities ranged from 2.1% up to 9.4–14.3% from eggs laid by an individual which had been isolated from birth. They also recorded small numbers of abnormalities from *D. reticulatum* (0.1%) and *L. maximus* (0.7%). Mason and Copeland (1988) concluded that *L. valentiana* produced viable conjoined twins naturally at a higher rate (over 1%) than the other two species. Self-fertilizing individuals also produced conjoined twins. *Deroceras reticulatum* and *L. maximus* produced no conjoined twins during the investigation. Joined zygotes and ova, with cytoplasmic continuity, were found in the oviduct and fertilization chamber but not in the hermaphrodite duct. Mason and Copeland (1989) suggested that conjoined twins were formed either by the random fusion of two or more zygotes enclosed in a single egg capsule or, more probably, by the development of a central constriction around a primary oocyte.

5.3.2 Development of the egg

A general account of the early stages in the development of the pulmonate egg, including spiral cleavage and gastrulation, has been given by Raven (1966, 1975) and was also included in the review by Verdonk and van den Biggelaar (1983). Focardi and Quattrini (1972) briefly described the first cleavage stages in *M. gagates*, Carrick (1938) described the stages of cleavage and gastrulation in the egg of *D. reticulatum* and Kulkarni (1971) made a similar study for *L. alte*. Verdonk and van den Biggelaar (1983) described how a cleavage cavity appeared between the blastomeres during the intervals between successive cleavages in some land and freshwater pulmonates, including *Limax* and *Deroceras*. The cavity grew until its contents were expelled and the egg then resumed its normal size. A regular succession of formation and extrusion of cleavage cavity contents continued until gastrulation. Raven (1966) had suggested that this recurrent cleavage cavity served as a mechanism for osmotic regulation since the eggs were surrounded by a hypoosmotic medium. As the result of experiments on segmentation and organogenesis in *L. maximus*, Guerrier (1971) concluded that the relative position of the blastomeres did not play a significant role in controlling the early features of segmentation as these depended essentially on the properties of each blastomere, i.e. cleavage of the egg is determinate. The effect of temperature on the development of the egg is described in Section 8.1. Carrick (1938) found wide variation in the rate of development of eggs in the same batch soon after segmentation began although the eggs had all undergone the first segmentation division simultaneously. Embryonic development can be easily followed in the transparent limacid egg and Carrick (1938) described and illustrated a series of six recognizable stages in the development of *D. reticulatum*. Stages I and II were the blastula and advanced gastrula stages (Fig. 5.5). The rudiments of the body and mantle became differentiated during Stage III while the rudiments of the tentacles and the podocyst appeared in Stage IV (Fig. 5.6). Both the podocyst and the anterior sac reached their maximum development by Stage V (Fig. 5.7). In Stage VI (Fig. 5.8) the anterior sac was retracted, the podocyst atrophied and the young slug was formed.

The podocyst (posterior sac, foot vesicle) developed as an enlargement of the caudal portion of the embryonic foot. The podocyst showed powerful contractions and contained muscular elements. The anterior sac (visceral sac, hepatic lobe) contains the albumen sac or larval digestive gland and lies wholly outside of the foot, later becoming partly surrounded by it. This development recalls the gradual movement of the visceral mass into the foot during the evolution of the slugs (Tillier, 1984). Quattrini and Sacchi (1971) suggested, from a study of the structure of the podocyst of *M. gagates*, that its functions included gaseous exchange between the embryo and its environment, the absorption of trophic substances from the albumen and the

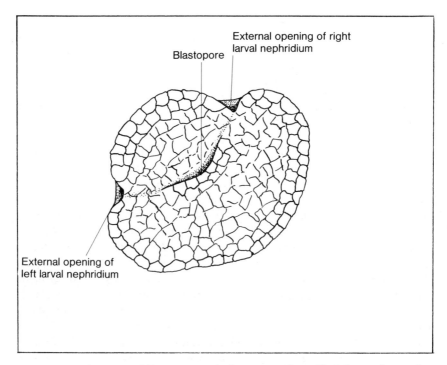

Fig. 5.5 Development of *Deroceras reticulatum* egg. Stage II. Advanced gastrula. (Reproduced by permission of the Royal Society of Edinburgh and R. Carrick from the *Trans. Roy. Soc. Edinburgh*, **59**, 577.)

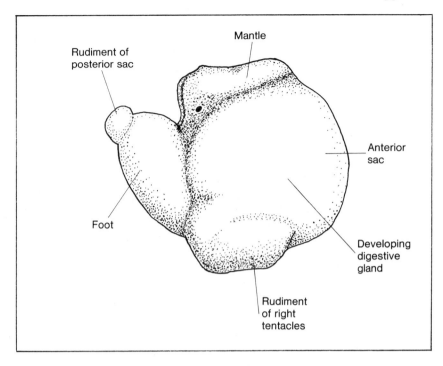

Fig. 5.6 Development of *Deroceras reticulatum* egg. Stage IV. Differentiation of tentacles and posterior sac. (Reproduced by permission of the Royal Society of Edinburgh and R. Carrick from the *Trans. Roy. Soc. Edinburgh, 59, 578.*)

removal of waste products, and that these activities were assisted by its contractile movements. Carrick (1938) also described how gaseous exchange occurred through the permeable membranes of both anterior sac and podocyst and how contractile movements of the podocyst helped to circulate oxygenated fluid through the embryonic tissues and also assisted with flow through the larval nephridia. Quattrini and Sacchi (1971) also showed that the embryo rotated within the albumen, possibly as the result of ciliary movement. Cather and Tompa (1972) found that embryos of two pulmonate snails could survive for a time outside their capsules if immersed in a mixture of capsular albumen and Thorotrast. They demonstrated that Thorotrast particles were incorporated within the ectodermal cells of the podocyst in less than one hour of exposure. Cather and Tompa (1972) also described the fine structure of the podocyst of several slugs and snails including *D. reticulatum* and *L. flavus* and concluded that this specialized organ had an albumenotrophic function.

In the early embryonic stages of limacid and arionid slugs the digestion of albumen takes place in the albumen sac (Weiss, 1968). Meisenheimer (in

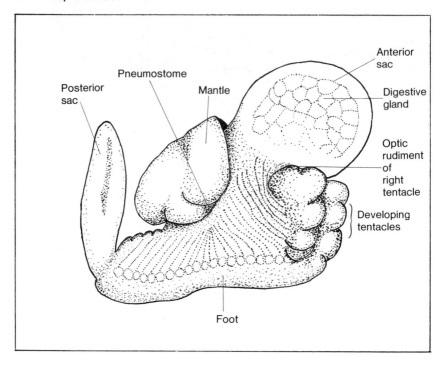

Fig. 5.7 Development of *Deroceras reticulatum* egg. Stage V. Maximum development of posterior and anterior sacs. (Reproduced by permission of the Royal Society of Edinburgh and R. Carrick from the *Trans. Roy. Soc. Edinburgh*, **59**, 579.)

Verdonk and van den Biggelaar, 1983) found that the uptake of albumen in *Limax* started as early as the 16-cell cleavage stage. Towards the end of embryonic development when the anterior sac, containing the albumen sac, was withdrawn into the body cavity, the albumen sac was transformed into a part of the digestive gland. Shortly before hatching, the remaining albumen in the egg capsule was ingested. In limacid slugs, the reserve albumen was stored mainly extracellularly in the greatly dilated cavities of the stomach and digestive gland, while in arionid slugs it accumulated intracellularly in the digestive gland (Weiss, 1968). This was confirmed by Arni (1973) who also found that, at hatching, the development of several organs, such as the digestive gland and the kidney, was still incomplete while the ganglia were relatively large compared with the proportions of the adult ganglia. Organogenesis in *D. reticulatum* has been fully described by Carrick (1938) although controversy still remains over the origins of some organs, such as the gonad, and these are discussed in other chapters (e.g., see Section 5.1.2). The process of hatching was described by Carrick (1938). The young slug was fully formed before hatching took place and, just prior to hatching,

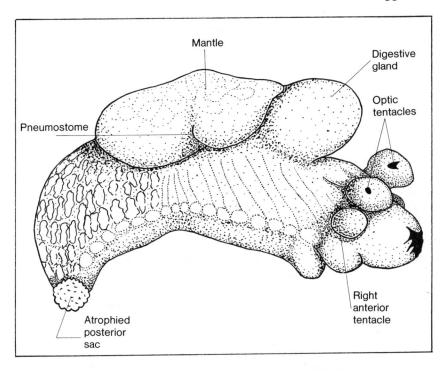

Fig. 5.8 Development of *Deroceras reticulatum* egg. Stage VI. Fully developed slug. (Reproduced by permission of the Royal Society of Edinburgh and R. Carrick from the *Trans. Roy. Soc. Edinburgh*, **59**, 580.)

movements of the embryo became more pronounced. The radula was used to penetrate the perivitelline membrane, the slug rasping indiscriminately at any part of this membrane (Miles, 1924). As a result, the membrane bulged at the point attacked and was eventually ruptured. The slug then pushed its way out and had no difficulty in penetrating the outer shell of the egg.

5.3.3 Desiccation of the egg

Slug eggs have no structural provision for the retention of water and desiccate rapidly unless maintained in a moist environment. Reports in the early literature of eggs being desiccated for weeks and then resuming normal development when exposed to moisture (e.g. Lovatt and Black, 1920) were rejected by Carrick (1942) and Bayne (1967a) as unlikely. Arias and Crowell (1963) allowed a large number of *D. reticulatum* eggs to desiccate and found that after five months, when the eggs were rehydrated, they rapidly regained their original form but no embryonic development took place. Slug eggs are exposed to drying conditions in the field (South, 1965) and eggs are killed directly by drought (South, 1989b). Carmichael and Rivers (1932) dehydrated eggs of *L. flavus* containing developing embryos. They found that eggs with very young embryos still hatched after they had lost between 80 and 85% of their weight. However, eggs with fully developed embryos failed to hatch after losing about 75–80% of their weight although they still hatched after losing 70–75% of their weight, equivalent to a loss of about 40% of the weight of the embryo. Thus older embryos were less tolerant of desiccation. Bayne (1967a) found that the reverse of this was true for eggs of *D. reticulatum*, the younger eggs being more susceptible to desiccation. Embryos of *D. reticulatum* were found to survive weight losses of 60–80%. Bayne also found that hatching was neither delayed nor prolonged by desiccation. In a comparison of the desiccation rates of eggs of eight gastropods, Bayne (1968a) found that differences in rate were due to differences in size of the eggs. Thus eggs of *A. ater* desiccated more slowly than those of *D. reticulatum* and *D. caruanae*, while eggs of *D. caruanae* showed the highest rate of desiccation. Egg sizes for these species are *A. ater* 5 × 4 mm, *D. reticulatum* 3 × 2.5 mm and *D. caruanae* 1.5 × 1.5 mm (Quick, 1960). Both Carrick (1942) and Arias and Crowell (1963) found that maximum numbers of *D. reticulatum* eggs were laid in cultures with soil which was 75% saturated with water. Slugs did not lay eggs when the soil was below 10% saturation (Carrick, 1942). Eggs laid at 100% and 25% saturation failed to hatch. However, Karlin and Naegele (1960) demonstrated that *D. reticulatum* eggs could withstand complete immersion in water for at least four days without normal development being affected.

5.4 MATING AND EGG LAYING

Copulation in slugs is usually preceded by elaborate courtship behaviour and often involves the reciprocal exchange of sperm. Detailed descriptions of mating behaviour in different species of slug include those given by Taylor (1902–1907), Gerhardt (1933, 1935), Quick (1960), Karlin and Bacon (1961) and Webb (1965). Mating usually takes place at night, frequently on the soil surface although Webb (1965) recorded a subterranean courtship for *D. reticulatum*. Barnes and Stokes (1951) described how copulation in *T. haliotidea* and *T. scutulum* took place under the surface of the soil. The most spectacular displays of courtship behaviour are shown by *L. maximus* and *L. cinereoniger* where mating takes place on vertical or overhanging surfaces or, in the case of *L. maximus*, the slugs hang suspended from stout threads of viscous pedal mucus. Taylor (1902–1907) described how mating by *L. maximus* began with the slugs following each other in a circle, caressing with their tentacles. This lasted for up to two hours or more but could be much shorter, with the circle gradually becoming smaller and the slugs becoming more and more excited. Eventually, the slugs launched themselves off the surface, becoming suspended by a mucous thread which gradually increased up to about 40 cm or more in length. The penis sacs were everted at this time and underwent a series of changes in form as the penes intertwined in a tight spiral and then expanded into an umbrella shape. The sperm mass was transferred at this point and the animals separated, either descending to the ground or climbing up the thread which was then sometimes consumed.

Nicholas (1984) concluded that the same basic elements – 'following', 'pairing' and 'circling' – were common to slug courtship behaviour. 'Following' was undertaken by slugs prepared to mate and involved trail-following behaviour (see Section 7.2.2). 'Circling' behaviour involved the pair of slugs circling slowly in a clockwise direction and gradually coming closer together until their genital openings were apposed. 'Circling' behaviour was accompanied by the production of copious mucus and by caressing and biting in some genera of slugs. In several species of *Arion*, courtship may be abandoned at this stage (Davies, 1977, 1987). 'Pairing' occurred after the two slugs took up their copulatory positions, with the head of one facing the tail of the other. This arrangement allowed the genital openings on the right-hand side of the head behind the tentacle to become apposed to one another. Copulation took place at this point and the sperm mass or spermatophore was transferred. Webb (1961) made a very detailed study of the changes which took place in the form and function of the distal genitalia during courtship and copulation in *L. marginata* and in *D. reticulatum*, *D. caruanae* and also included some observations on *D. laeve*. Webb described how the sarcobella or stimulators were used extensively for caressing in *Deroceras*. The atrial stimulator of milacid slugs is used in a similar way (Focardi and Quattrini, 1972). The use of a stimulatory structure is more variable in the genus *Arion*. In *A. hortensis s.s.* there is no ligula but the proximal part of the oviduct is everted into a club-shaped process, comparable to the ligula, and this acts as a stimulator (Quick, 1960; Davies, 1977). In contrast, in *A. fasciatus*, there is little or no eversion of the genitalia (Quick, 1960). Davies (1987) described how the ligula could be applied to the partner during copulation in *A. lusitanicus*, *A. subfuscus* and *A. flagellus* although it was also sometimes applied to the ground. Caressing in *Limax* is restricted to the use of the tentacles in the absence of a stimulator.

Webb (1965) described how *D. reticulatum* and *D. caruanae* bit one another during courtship. In the case of *D. caruanae* this biting was particularly vicious and tail lashing also occurred after a slug had been bitten by its partner. The stimulator was also sometimes bitten during courtship behaviour. Biting was also violent in *A. columbianus*, where Harper (1988) described how pieces of flesh could be bitten off during mating. Adult *A. ater* are sometimes found with white lesions on the skin surface which may have originated from the rasping of a partner. Davies (1977, 1987) described how some species of *Arion* nibble a potential partner early in courtship. *Arion hortensis*, *A. subfuscus* and some other species of *Arion* also lick the caudal mucus trail left by the partner during circling behaviour (Quick, 1960). The time taken for mating varies widely within species although it appears to be prolonged in milacid slugs and Quick (1960) described how in *T. budapestenis* copulation usually started in the evening and continued until the following midday or later. Slug eggs are usually laid in holes or crevices in the soil or at the surface of the soil under the cover of stones or pieces of wood. Tompa (1980) found that in *V. ameghini* the eggs were laid in a clockwise spiral held together by a mucous strand which was so strong that if one egg was picked up the others remained attached to the strand. Each mass contined 3–15 eggs. Eggs of *V. borellianus* and *L. alte* are also laid in strings or mucous-bound masses and faecal pellets are deposited next to the strings (Lanza and Quattrini, 1964; Raut and Panigarhi, 1988b). The numbers of eggs and of egg batches vary widely between and within species. Carrick (1938) estimated that *D. reticulatum* laid up to 500 eggs in a year, with an average of 22.0 eggs per batch (range: 9–49 eggs). During egg laying the tentacles were kept retracted and there was little movement on the part of the slug. Davies (1977) found that the usual numbers of eggs in a batch for species in the *A. hortensis agg.* ranged from 10 to 30 in captivity. Information about the size of egg batches is given in Quick (1960).

5.5 CONTROL OF REPRODUCTION

5.5.1 Control of reproduction

Carrick (1938) concluded that breeding in *D. reticulatum* took place whenever environmental conditions were suitable, and South (1989a) suggested that under suitable conditions this species might breed continuously throughout the year. However, this species appears to be the exception and most species of slug have clearly defined life cycles. For example, the species of *Arion* have an annual life cycle, some *Milax* and *Tandonia* spp. have a biennial life cycle, while most *Limax* spp. are pluriennial species (see Section 8.4). Other aspects of reproduction are also regulated. Intraspecific competition was shown to influence the rate of reproduction in *L. maximus* by Rollo (1983a). Rollo described how *L. maximus* could make adjustments with respect to egg weight, numbers/batch and the number of batches produced, and the amount of material allocated to any particular batch of eggs was very variable. As slugs aged they produced smaller batches of larger eggs, and at the end of the breeding season some very large eggs and large masses of unpackaged albumen were sometimes deposited. There is evidence, therefore, that reproduction in slugs is carefully controlled and there is substantial evidence for the endocrine control of reproduction in slugs.

5.5.2 Control of development of the reproductive tract

A close relationship between the development and maturation of the gonad and of the regions of the reproductive tract has been demonstrated for *A. ater* by Lusis (1961) and Smith (1966) and for *D. reticulatum* by Runham and Laryea (1968). This relationship was confirmed by the experiments of Abeloos (1943) and Laviolette (1950a, 1954). Laviolette showed that when the gonad in *L. maximus* was removed, the albumen gland and common duct underwent no further development although the penis sac was unaffected. However, McCrone and Sokolove (1979) demonstrated that removal of the gonad in *L. maximus* prevented penis development. Laviolette (1954) showed that, when an immature reproductive tract was transplanted into a mature slug, it showed a marked enlargement. This suggested that a hormone present in the blood controlled the maturation of the reproductive tract. Transplantation of gonads from mature slugs into castrated immature individuals also resulted in maturation of the reproductive tract. However, injection of gonad extracts into slugs did not affect the tract. Runham *et al.* (1973) found that, when immature common ducts were transplanted into the haemocoele of older *D. reticulatum*, the changes observed in the transplants reflected the stage in reproductive maturation of the host slug. However, no change occurred when common ducts were transplanted into slugs that had been castrated by removal of the gonad. Runham *et al.* suggested that at least two hormones were involved, one controlling differentiation and enlargement of the prostate gland and the other controlling development of the oviducal gland. They concluded that substances produced by the gonad caused the brain to produce these two hormones.

The dorsal bodies are groups of cells which extend over the surface of the cerebral ganglia and down to the pleural ganglia and discharge their secretion directly into the haemocoele (Runham and Hunter, 1970) (see Section 6.3.2(*b*)). Removal of the dorsal bodies in *D. reticulatum* had no effect on spermatogenesis or on the prostate gland and penial mass but oocyte maturation was slowed down resulting in increased numbers of small oocytes and few large oocytes (Wijdenes and Runham, 1976). The development of the oviducal and albumen glands was also retarded. Wijdenes and Runham suggested that the dorsal bodies produced a hormone that promoted maturation of the oocytes and differentiation and growth of the female accessory sex organs. A major change in the pattern of neurosecretion in *A. ater* was shown by Smith (1967) to be associated with the onset of maturation of the female accessory glands and with the beginning of copulation. This critical time occurred during the mid and late spermatozoan stages and was marked by the appearance of large amounts of secretory material in groups of cells in the pleuro-parieto-viscera ganglion mass. The secretion was only found in slugs in which the female glands had started to mature and secretion reached a maximum by the late spermatozoa stage when copulation occurred. A rapid decrease in secretion then followed until very little was present by the late oocyte stage. Smith (1967) noted the similarity between this pattern of neurosecretion and the brain hormone proposed by Pelluet and Lane (1961) which controlled oocyte production. The onset of neurosecretion was termed critical by Smith (1967) because experimental adverse conditions could retard maturation if applied before this stage but had much less effect if applied after the critical stage was reached.

5.5.3 Tentacle/brain axis

Pelluet and Lane (1961) and Pelluet (1964) investigated the hormonal control of differentiation in the germinal epithelium of *A. subfuscus*, *A. ater* and *Milax* sp., including the effect of removing the optic tentacles. They concluded that there were two distinct hormones controlling differentiation. One, from the optic tentacles, stimulated spermatogenesis while the other, from the brain, was concerned with oocyte production. The two hormones formed a balanced system, with the tentacle hormone appearing first. When Renzoni (1969) removed the tentacles from *V. borrellianus* at the age of ten days, however, no effect was observed on the gonad although Nagabhusahanam and Kulkarni (1971c) found that the number of oocytes was increased by a similar operation in the veronicellid *L. alte*. In contrast to the previous findings, Gottfried and Dorfman (1970a) found that removal of the optic tentacles from immature *A. californicus* resulted in precocious spermatogenesis. However, investigations into the effect of the removal of the optic tentacles of *A. subfuscus* at birth by Wattez and Durchon (1972) and Wattez (1973) showed a clear feminization of the gonad. When the tentacles were removed from juvenile castrated individuals this also resulted in an increase in the number of oocytes in the regenerating gonad. Wattez (1973) concluded that the optic tentacles exerted an inhibitory influence on oogenesis. Kulkarni *et al.* (1988) came to a similar conclusion of *L. alte* and also showed that the optic tentacles did not affect spermatogenesis. Tentacular extracts were injected into the body cavity of *A. subfuscus*, whose optic tentacles had been removed at hatching, and into juvenile castrated individuals without tentacles (Wattez, 1975). Results showed that optic tentacles from juveniles had an inhibitory influence on oogenesis while those from slugs in the female phase of development were ineffective.

Wattez (1976, 1978, 1979) used organ cultures of infantile gonads of *A. subfuscus* to examine the role of the optic tentacles in sexual differentiation. When cultured in isolation, there was a tendency for an infantile gonad to develop towards the female line. When gonads were cultured with brain–tentacle complexes, Wattez was able to confirm the results of previous experiments that the optic tentacles exerted an inhibitory influence on female differentiation during the infantile phase but were ineffective during the female phase. The results also suggested that the brain might stimulate the female line although the feminization tendency of the isolated gonad needed to be taken into account. The stimulating effect of cerebral ganglia extracts on oocyte development in adult *A. subfuscus* was confirmed by Wattez (1982). Takeda (1977) showed that brain hormone (homogenate of the central ganglia) stimulated egg laying in *D. reticulatum* and *L. flavus* while tentacular hormone (homogenate of the optic tentacles) inhibited the action of the brain hormone. Both hormones remained active after boiling or dialysis. Takeda (1977) suggested that the tentacular hormone inhibited the activity of neurosecretory cells in the central ganglia, resulting in the inhibition of egg laying. The optic tentacle was found to be involved in the control of growth of the albumen gland in *L. alte* (Kulkarni, 1978). The galactogen content of the albumen gland shows seasonal variations, being highest in the breeding season (Section 3.3). Meenakshi and Scheer (1969) found that the removal of the optic tentacles was followed within three weeks by an increase in the weight of the albumen gland and in the galactogen concentration in the gland. Both of these effects could be reversed by injection of optic tentacle extract. Homogenates of the brain were shown to increase galactogen synthesis in albumen gland explants by van Minnen *et al.* (1983). The major sources of the galactogen synthesis stimulating factor (GAL-SF), which was probably a polypeptide, were found to be the cerebral ganglia and their surrounding connective tissue. The dorsal body cells, present in the connective tissue around the circumoesophageal brain, were shown to be the source of GAL-SF in *L. maximus* by van Minnen and Sokolove (1984). They also showed that the dorsal body cells were small and released little secretory product in immature and early male phase slugs while release was higher and the cells were larger in the late female phase. Van Minnen and Sokolove (1984) also described a technique for the partial purification of GAL-SF.

5.5.4 Effect of light and other physical factors

The experimental exposure of *D. reticulatum* to low temperatures causes a general disorganization within the cells of the gonad resembling induced senescence (Bridgeford and Pelluet, 1952). Henderson and Pelluet (1960) showed that when slugs (*D. reticulatum*) were exposed to long periods of light or higher light intensities, changes took place in the gonad, including an increase in the rate of spermatogenesis. Lusis (1966) examined the effect of temperature and humidity on the development of the gonad in *A. ater*. High temperature had an injurious effect on the development of both male and female gametes although the effect was more marked on oocytes. Low temperatures resulted in a general delay of gametogenesis while lowered relative humidities disturbed the ratio of male to female gametes, favouring the development of oocytes. Smith (1966a) found that in *A. ater* there was a close correlation between maturation of the reproductive tract and the season in animals collected from the field. Smith also showed that although the reproductive cycle could be advanced or retarded by varying the environmental conditions, the relationship between the maturation stage of the gonad and of the remainder of the tract could not be altered. Takeda (1977) suggested that since the eye was located near the neurosecretory cells of the optic tentacle, environmental stimuli such as light and day length could influence the release of tentacular hormone.

Maturation of the reproductive tract of *L. maximus* was studied in a natural population by Sokolove and McCrone (1978). Male phase maturation took place mainly in June and July while female phase maturation occurred in September. Female phase maturation was not significantly affected by exposure to artificial photoperiods of long (L:D 16 : 8) or short (L:D 8 : 16) days. However, male phase maturation was induced by a short-day to long-day transition although no gonadal development was seen when slugs were maintained only on short-day photoperiods. Sokolove and McCrone (1978) concluded that the development of the reproductive tract in *L. maximus* was triggered mainly by the annual spring increase in day length while McCrone *et al.* (1981) described how approximately four weeks of long-day exposure was required to trigger irreversibly the development of juvenile slugs. However, once fully initiated, maturation of the reproductive system was no longer significantly affected by photoperiod. Sokolove and McCrone proposed a scheme for the control of reproductive maturation. Firstly, a lengthening photoperiod in spring could trigger spermatogenesis in the gonads. This might involve the stimulated release of a male phase brain gonadotrophin, a direct photoperiodic response of gonadal tissue, or a combination of both. Secondly, male phase differentiation in the gonad could be associated with the release from the gonad of a hormone that stimulated the production of a female brain hormone. Thirdly, female brain hormone could induce the development of the accessory female reproductive organs and might also control vitellogenesis.

This scheme was modified by McCrone and Sokolove (1979) who suggested that the brain produced a maturation hormone (MH) rather than a male phase hormone as it appeared to initiate the full maturation sequence in the reproductive system. However, they did not rule out the possibility of other brain hormones which might be individually targeted for male or female tissues. Their modified scheme for the initiation of male phase maturation in *L. maximus* now became (a) the direct photoperiodic activation of the brain by the long-day photoperiod, (b) the secretion by the brain of a maturation hormone (MH) which induced gonad development and (c) secretion by the developing gonad of a male phase sex hormone which promoted the development of the male accessory glands and structures. McCrone and Sokolove (1979) and McCrone *et al.* (1981) demonstrated that whole brains removed from maturing slugs and implanted into the haemocoele of juvenile slugs could induce maturation. However, no development followed the implantation of short-day brains. McCrone *et al.* (1981) also showed that the cerebral ganglia were as effective as whole brains in causing implantation-stimulated maturation although suboesophageal implants were ineffective. Although the dorsal body cells associated with the cerebral ganglia of *D. reticulatum* had been found to secrete a hormone which promoted oocyte development (Wijdenes and Runham, 1976), the dorsal body cells in *L. maximus* were scattered in the sheath around the suboesophageal ganglia as well as around the cerebral ganglia. For this reason McCrone *et al.* (1981) concluded that the probable location of MH secreting cells was within the cerebral ganglia rather than in the dorsal body cells. McCrone *et al.* also described how photoperiodic release of MH from the brain could occur if castrated slugs were exposed to long days, demonstrating that the gonad was not involved in the photoperiodic stimulation of MH release.

The synthesis and secretion of a male gonadotrophic factor released from the cerebral ganglia of male phase *L. maximus* as the result of prior exposure to long-day photoperiods was described by Melrose *et al.* (1983). This factor produced, directly or indirectly, an increase in the proliferation of spermatogonia in the gonad. Sokolove *et al.* (1983) demonstrated that implant-stimulated spermatogonial DNA synthesis in *L. maximus* depended upon implantation of supraoesophageal ganglia. Neither suboesophageal nor immature supraoesophageal ganglia implants had any effect. Spermatogonial DNA synthesis was stimulated by exposure of slugs to long-day light cycles (L:D 16 : 8) for three to four weeks. Exposure for this length of time was sufficient to trigger male phase development even when slugs were returned to short-day light cycles (L:D 8 : 16). Sokolove *et al.* concluded that three to four weeks' exposure was necessary to promote irreversibly the release from the slug cerebral ganglia of the male-phase gonadotrophic factor which caused spermatogonial proliferation either directly or indirectly. Using castration and gonad implantation, McCrone and Sokolove (1986) confirmed that a factor from the gonad was required for development of the female accessory organs in *L. maximus*. This gonadal factor promoted rapid growth of the gonad and sperm production and also gradual development of the penis. While inductive photoperiods failed to produce maturation of the reproductive tract in a castrated immature slug, the brains of such animals were able to promote maturation when an immature gonad was subsequently implanted into the slug. McCrone and Sokolove (1986) suggested that these observations provided evidence for the photoperiodic induction of MH secretion by the brain in the absence of a gonad. They also suggested that, while one or more gonad hormones were not required for the photoperiodic stimulation of MH release, they were necessary for the production of sperm and for the development of the female accessory organs.

McCrone and Sokolove (1979) showed that tentacular eyes were not required to distinguish between short- and long-day light cycles, indicating that *Limax* used an extraocular photoreceptor system to mediate its photoperiodic response. Beiswanger *et al.* (1981) also demonstrated that the eyes of *L. maximus* and *L. flavus* were not essential for light:dark entrainment (see Section 7.4.2). McCrone and Sokolove (1979) suggested that the extraocular receptor system might lie in the central nervous system since increased spontaneous activity had been observed upon illumination of the isolated brain. *Limax maximus* slugs that had been blinded by the removal of their optic tentacles still matured when exposed to a white light long-day photoperiod (L:D 16 : 8) but failed to mature in response to a long-day red light photoperiod (>600 nm) (Sokolove *et al.*, 1981). Intact slugs matured in either white or red light photoperiods. Sokolove *et al.* concluded that the extraocular receptors in *L. maximus* were relatively insensitive to wavelengths longer than 600 nm.

5.5.5 Hormones

Bridgeford and Pelluet (1952) showed that the injection of a mixture of the hormones FSH and LH from the vertebrate anterior hypophysis into *D. reticulatum* accelerated the growth and maturation of the gonad. The number of eggs per acinus was found to increase with the number of hormone injections. However, both Rose and Hamon (1939) and Laviolette (1954) had previously found that the injection of synthetic sex hormones produced no appreciable change in the gonads of *M. gagates* or *A. ater*. Gottfried and Lusis (1966) detected testosterone, 11-keto-testosterone and 17α-hydroxyprogesterone in the egg of *A. ater*, while the oestrogens, oestrone and oestradiol-17β, were found in the spermatheca of this slug by Gottfried *et al.* (1967). Gottfried and Dorfman (1970a) identified several steriods from the reproductive tract of the slug, *A. californicus*, including dehydroepiandrosterone and 11-keto-testosterone from the gonad, pregnenolone, oestradiol-17β and oestrone from the albumen gland and oestrone from the spermatheca. The *in vitro* biosynthesis of steroids by the subcellular fraction of the male phase gonad of *A. californicus* was investigated by Gottfried and Dorfman (1970b,c). They also showed that the metabolism of androstenedione to 3α-androstandiol was severely inhibited by an optic tentacle homogenate although the formation of 3β-androstandiol was relatively unaffected. The role of some of the steroid hormones identified by Gottfried and Dorfman in reproduction in the slugs, *D. reticulatum* and *L. flavus*, was investigated by Takeda (1979). These hormones included dehydroepiandrosterone, testosterone, oestradiol, oestrone and pregnenolone. Gonadotrophin and metopirone were also used. Egg laying was stimulated by oestrogens such as oestradiol and oestrone but not by the androgens, dehydroepiandrosterone and testosterone. However, the number of eggs which began to develop was relatively higher in the androgen group than in the oestrogen group. Takeda concluded that a balance of both groups of steroid hormones was required for successful egg laying. Takeda (1977) has demonstrated that the brain hormone stimulated egg laying in *D. reticulatum* and *L. flavus* and Takeda (1979) suggested that egg laying was directly controlled by the steroid hormones and that the brain hormone acted as a gonadotrophic hormone stimulating the gonad to secrete the steroid hormones. In this context, Angulo and Moya (1989) have described rough endoplasmic reticulatum-rich cells in the gonad of *A. ater* at the time of egg laying. The number of eggs and the proportion of eggs developing was increased by removing the tentacles, and Takeda suggested that the tentacular factor inhibited egg laying. Sokolove and McCrone (1978) suggested that, since Takeda (1977) had demonstrated an inhibitory role for the tentacular hormone on female brain hormone, the gonadal hormone might not act directly to cause release of female brain hormone but rather it might counteract production of the inhibitory tentacular hormone, leading indirectly to an increase in female brain hormone. Since Gottfried and Dorfman (1970a) had postulated that the tentacular factor of *A. californicus* was a gonadal inhibitor, whose release was controlled directly by a gonadal steroid, Takeda (1979) introduced the concept of a negative/positive feedback mechanism, where an anti-gonadotrophin inhibited the secretion of gonadal sex steroids, while the tentacular factor itself was controlled by the secreted gonadal steroid. Krajniak *et al.* (1989) suggested that the peptide SCP$_B$ (see Section 6.8) might be involved in the regulation of reproduction, particularly in the stimulation and relaxation of the penis.

5.5.6 Organ culture and regeneration

(a) Organ culture

Early studies on the control of maturation of the reproductive tract were carried out using castration by the removal of the gonad (e.g. Abeloos, 1943; Laviolette, 1950b, 1954). Laviolette (1954) included full details of the techniques involved. The role of the optic tentacles and the circumoesophageal ganglia in the control of reproduction has been extensively investigated by a large number of experiments, which were described in Section 5.5.3. An alternative method for the study of the control of maturation of the reproductive tract is the use of *in vitro* organ culture and a physiological saline suitable for use with *A. ater* was described by Roach (1963). A review of the use of molluscan organ culture was published by Bayne (1968b). A new organ culture technique was described by Bailey (1970, 1973) for use with *D. reticulatum*, including full details of the procedures used. Bailey found that, using this technique, isolated explants of gonad changed rapidly with the male and female germinal elements undergoing necrosis. However, when explants of the gonad together with the tentacle–brain complex were cultured together there was an improvement in the survival of the germinal elements, particularly the male germ cells. The effect was related to the state of maturity of the slug from which the tentacle–brain complex explant was removed. Gonad explants cultured with tentacle–brain complex from a large female stage slug showed an improvement in the appearance of the oocytes although the male line did not benefit. Explants of common duct were also cultured using this technique.

Wattez (1978) also described a technique for culturing infantile gonads, containing no oocytes, and juvenile gonads, with oocytes, *in vitro* alone or associated with tentacle–brain complexes taken from slugs at various phases of maturation and the results of these experiments were discussed in Section 5.5.3. Organ cultures were used to investigate cell differentiation in the gonad of *A. ater* by Parivar (1982), who used a modification of the technique described by Bailey (1973). The effect of the addition of the hormones oestradiol, oestriol, testosterone and FSH to cultures of the gonad was examined. Explants from young slugs showed mitotic divisions in spermatogonia and newly differentiated spermatogonia were seen in explants of gonads from young and old animals. In the presence of oestriol, new germinal cells differentiated from the neck region of the acini of mature slugs and new acini appeared to be forming. However, the male stages of spermatogenesis degenerated in the course of three-day cultures in the presence of oestriol while oocytes and sertoli cells survived longer, for up to ten days. Other hormones did not affect the explants. Primary tissue culture of cells from the slug *Incilaria bilineata* was maintained for two to three months by Manaka *et al.* (1980). When slug blood was added to the medium, it inhibited the migration of slug cells. However, blood heated at 56°C for 30 minutes neither inhibited this migration nor induced cell division.

(b) Regeneration

Luchtel (1972b) described how the primordial germ cells appeared to be distributed for some distance along the distal end of the hermaphrodite duct and gonad regeneration could take place from segments of this duct. The rate of regeneration of the gonad after castration varied with the age of the slug. In *D. reticulatum* there was rapid regeneration of the gonad in slugs castrated during the undifferentiated and spermatocyte stages but regeneration was slow during the spermatid stage and largely absent from older animals

(Runham and Hogg, 1979). Runham and Hogg described the stages in regeneration for this species. A short time after castration, muscular contractions of the wall resulted in sealing of the hermaphrodite duct. A mass of tissue was formed around the wound by the proliferation of connective tissue. The duct cells dedifferentiated and outgrowths of the duct penetrated the surrounding connective tissue. These outgrowths were at first tubular and then acini developed at their tips. The sequence of differentiation within these regenerated acini was similar to that within normal acini during the undifferentiated stage. Runham and Hogg (1979) suggested that, since the lining epithelium of the acini normally differentiated during the early spermatozoon stage, this might be the reason for the lack of regeneration in older animals. Hogg and Wijdenes (1979) also concluded that regeneration of the gonad after castration would occur from the hermaphrodite duct provided it was still in an immature state, i.e. the gonadal stem cells were present. A similar process of gonad regeneration after castration was described for *L. marginata* and *L. flavus* by Takeda and Sugiyama (1984). They also showed that the capacity for regeneration depended upon the age of the slugs.

The body wall of the limacid slugs, *D. agreste* and *L. maximus*, has been shown to possess a capacity for regeneration by Abeloos (in Reygrobellet, 1970). Dyson (1964) showed that wounds in the mantle of *A. hortensis* healed rapidly. Reygrobellet (1970) investigated the effect of the removal of the caudal extremity, which consisted simply of skin, connective tissue and muscle from the slugs *D. agreste*, *L. tenellus* and *L. maximus*. The epidermal and subepidermal tissues of the sole of the foot were found to be capable of actively regenerating the lost tissues while the tissues of the dorsal body wall remained inactive in this respect. This capacity of cells for regeneration was demonstrated in experiments on the primary tissue culture of cells from the slugs, *Incilaria bilineata* and *I. fruhstorferi*, by Manaka et al. (1980) and Furuta and Shimozawa (1983). Manaka et al. found that cells from explants of minced tissue survived for two or three months although no cell division was observed. When cells from the mantle and foot were cultured (Furuta and Shimozawa, 1983), fibroblast-like cells and macrophage cells migrated from the explants, with the former cells attaching themselves to the glass dish and building up a multilayer focus of cells. Plisetskaya and Deyrup-Olsen (1987) found that the tail region of the slug *P. foliolatum* could regenerate after autotomy had taken place and that an insulin-like substance was involved in the process of regeneration. When the tentacles of slugs and snails are removed surgically, they regenerate rapidly and Chétail (1963) has described regeneration of the tentacle in *A. ater* and *D. agreste* which was completed within 30–60 days and 10–20 days respectively. Renzoni (1969) described how the regenerated tentacle of *V. borellianus* lacked an eye.

5.6 GENETICS

5.6.1 Chromosomes

A general account of the chromosomes of pulmonate slugs and snails, including the chromosome cycle, has been given by Patterson and Burch (1978). They described the chromosome numbers for several slug families. In Philomycidae, a haploid number of 24 had been reported for two Japanese species. Haploid chromosome numbers in Arionidae ranged from 25 to 29, in Limacidae from 24 to 31 and in Milacidae from 33 to 34. Beeson (1960) recorded haploid numbers of 25 for *A. subfuscus*, 26 for *A. ater*, 28 for *A. hortensis* agg. and *A. intermedius* and 29 for *A. circumscriptus* agg. Patterson and Burch (1978) described how, in limacid slugs, each genus had a different but constant chromosome number (*Limax*, $n = 31$; *Lehmannia*, $n = 24$; *Deroceras*, $n = 30$). *Limax tenellus*, however, corresponds more closely with *Lehmannia* in having a haploid number of 24 (Perrot and Perrot, in Beeson, 1960). Beeson (1960) recorded a haploid number of 32 for *T. haliotidea*. Patterson and Burch (1978) recorded haploid numbers for Veronicellidae of 16 (*V. floridana*) and 17 (*L. alte*). Methods for pulmonate chromosome preparations were described by Babrakzai and Miller (1974) and Patterson and Burch (1978).

5.6.2 Pigmentation

Taylor (1902–1907) described colour varieties for many British species of slug and, in the case of *A. ater*, illustrated at least ten distinct forms. Taylor concluded that two distinct pigments, red and black, were involved in the variation of *A. ater* and that their development depended upon climatic conditions, with more extensive development of the black pigment being associated with colder, moister climates. Darker (black and dark brown) forms of *A. lusitanicus* and *A. ater rufus* were found to occur at altitudes of 500 m and more by Chevallier (1969), while Chevallier (1976) provided further evidence for the selection by climatic factors of particular colour phenotypes in these species. Luther (1915) found that in *D. reticulatum* and *D. agreste* there was a simple Mendelian inheritance of body pigmentation, with pigmentation being dominant to albinism. A similar condition was described by Ikeda (1937) for the philomycid slug, *M. bilineatus*. Getz (1962) showed that in *A. subfuscus* dark body coloration was dominant to a creamy buff colour. Williamson (1959b) also concluded that body pigmentation in *A. ater* was determined by simple Mendelian inheritance with three loci being involved. Body pigmentation was controlled by two alleles at a single locus, with the allele for black melanin (M) being dominant to that for brown melanin (m). The allele for full pigmentation (F), in which the melanin was evenly distributed over the upper surface of the body in the adult slug, was dominant to that for streak (F^s), where the melanin was confined to the mid-dorsal area, while both alleles were dominant to white (f), in which melanin was found only in the grooves of the foot fringe and on the tentacles. The presence of longitudinal bands in juveniles was controlled by the third locus with two alleles, and longitudinal banding (U) was dominant to the absence of banding (u). Evans (1977) examined the degree of variation in body coloration in the *A. ater agg.* and found no clear separation between the forms. Evans suggested that the wide range of overlap in body coloration between *A. ater ater* and *A. ater rufus* indicated their close taxonomic relationship although this could have arisen from hybridization between the two forms. Evans (1977) also showed that body coloration and mottling in *L. flavus* was under polygenic control, with dark body coloration possibly dominant to light coloration.

5.6.3 Polymorphism

In addition to the colour polymorphism in *A. ater* described in Section 5.6.2, a number of other polymorphisms have been described for different species of slug. Lupu (1977) described polymorphism in body form and coloration of *L. cinereoniger* and in the reproductive tract of four species of *Deroceras*, including aphallic forms of *D. laeve*. A number of biochemical and immunological techniques have been employed in the study of polymorphism and taxonomy of slugs and snails. Details of the techniques used have been described by Wright (1974). Electrophoretic techniques have been used in a number of investigations in slug taxonomy. Grossu and Tesio (1971) were able to confirm the distinct taxonomic status of the morphologically similar species, *Limax zilchi* Grossu and Lupu and *L. cinereoniger* by comparing the blood proteins and tissue esterases of these slugs. The relationships between species within the genus *Deroceras* were examined by Grossu and Tesio (1975) and the systematics of the *A. hortensis* complex were clarified by Dolan and Fleming (1988), both using isozyme patterns. Grossu (1977) reviewed the use of electrophoretic methods in the study of polymorphism in gastropods. Burnet (1972) used electrophoretic methods to study isozyme polymorphism in two enzyme systems in the *A. ater agg.* Samples of *A. ater ater*, endemic to the British Isles, were found to be monomorphic for the enzymes. However, populations of *A. ater rufus* from localities in England and the Netherlands were polymorphic for the different enzyme systems. Burnet (1972) concluded that localized populations of *A. ater rufus* in Britain might have originated from individuals accidentally imported from continental Europe, providing support for the conclusions of Ellis (1969) discussed in Section 1.4.4. Evans (1977, 1985) compared isozymes of *L. flavus* with those of the closely related species, *L. maculatus* and also of *A. ater ater* and *A. ater rufus*. There was a low level of similarity between the two species of *Limax*, supporting the view that these were separate species. However, Evans found that the two forms of *A. ater* were very similar in terms of isozymes.

Backeljau (1985b) estimated the genetic similarity between and within the *A. hortensis agg.* and *A. intermedius* by means of isoelectric focused esterase patterns in digestive gland fragments. The three species of the *A. hortensis agg.* were shown to constitute three clearly different gene pools and could therefore be regarded as reproductively isolated and distinct biological species. Since *A. owenii* and *A. intermedius* showed monomorphic esterase patterns, Backeljau (1985b) suggested that they reproduced by self-fertilization or apomictic parthenogenesis. *A. hortensis s.s.* and *A. distinctus* showed polymorphic patterns and so were clearly outcrossing species. Backeljau concluded that *A. owenii* had a mixed breeding system as spermatophores and mating behaviour had been described for this species. However, Foltz *et al.* (1982b), using electrophoresis and biochemical staining to analyse genetic variety in enzymes, had previously found no evidence that the breeding system of *A owenii* involved anything other than outcrossing with polymorphic and heterozygous local populations which were in Hardy-Weinberg equilibrium. Backeljau *et al.* (1988) described an albumen gland protein polymorphism in *A. owenii* using vertical polyacrylamide gel electrophoresis (PAGE). Other species of the *A. hortensis agg.*, together with *A. intermedius* and *A. fagophilus*, were found to be monomorphic with respect to albumen gland proteins determined by PAGE. Backeljau *et al.* suggested that, as this polymorphism might distinguish populations, geographically isolated populations of *A. owenii* could represent distinct forms or genetic strains. An investigation into genetic variation in *D. reticulatum* in Northern Ireland by Fleming (1989) showed that seven out of 19 enzyme loci examined were polymorphic.

5.6.4 Breeding systems

McCracken and Selander (1980) used electrophoretic methods to investigate genetic variation in 14 species of slug in the eastern USA and showed that the majority of species were polymorphic at their structural gene loci, with high levels of heterozygosity and Hardy-Weinberg proportions of genotypes present in local populations. However, six species showed no heterozygosity and included monogenic strains, evidence that self-fertilization was the normal breeding system for those species. In populations of the slug *D. laeve*, genotypic frequencies were shown to deviate markedly from Hardy-Weinberg proportions as a consequence of self-fertilization (Foltz *et al.*, 1982a). A number of studies have been made of genetic population structure and breeding systems in slugs and the results have already been discussed in Section 5.2.3. The genetically determined albinism of the integument of *D. reticulatum*, described by Luther (1915) (see Section 5.6.2), was used as a marker in an estimation of the degree of homozygosity in generations of three experimental populations differing by the degree of inbreeding (Matekin and Pakhorukova, 1980). It was possible to relate differences in various quantitative biological parameters, such as fecundity and growth rate, to differences in the genotypic structure of the populations. The environmental conditions in the cultures selected in favour of albinistic individuals, but this selection was accompanied by an increase in homozygosity of the populations, associated with a rise in abnormalities during embryogenesis and the death of young slugs.

6

Nervous system, sensory structures and neurosecretion

6.1 NERVOUS SYSTEM

6.1.1 General account

The detailed structure of the pulmonate nervous system has been described by Bullock and Horridge (1965). Most of the available information is based on the snail *Helix* and Kerkut and Walker (1975) dealt mainly with the nervous system of the snail *H. aspersa* in their review of the nervous system of pulmonates. The central nervous system of slugs is similar to that of *Helix* and has been described and illustrated for *Vaginula solea* d'Orbigny (Coifmann, 1934), for *A. ater* (Smith, 1967), for *D. reticulatum* (Runham and Hunter, 1970) and for the Athoracophoridae (Tillier, 1984). It consists of nine ganglia and their connectives which have become fused into a circumoesophageal ring. This ring lies immediately behind the buccal mass and can be divided into three parts: the supraoesophageal ganglionic mass, the suboesophageal ganglionic mass and the pedal ganglia. Slight differences occur between different groups of slugs in the length and thickness of the connectives and in the extent to which the suboesophageal ganglia are fused. For example, in some athoracophorid slugs the fusion is less complete and the outline of four suboesophageal ganglionic masses can be distinguished (Tillier, 1984). The arrangement of the circumoesophageal ganglia is shown in diagrammatic form in Fig. 6.1.

The supraoesophageal ganglionic mass consists of a pair of cerebral ganglia connected by a cerebral commissure. Each cerebral ganglion tends to be lobed externally in pulmonates (Bullock and Horridge, 1965) and Kerkut and Walker (1975) described three lobes in *Helix*, the procerebrum (lateral lobe), the mesocerebrum and the metacerebrum (postcerebrum). These lobes are also present in slugs (Smith, 1967) and, according to Bullock and Horridge (1965), the procerebrum represents an invagination in the embryo (the cerebral tube), which in *Testacella* persists as a conspicuous cavity. A pair of cerebrobuccal connectives run forwards from the cerebral ganglia to the paired buccal ganglia which lie on the dorsal surface of the buccal mass.

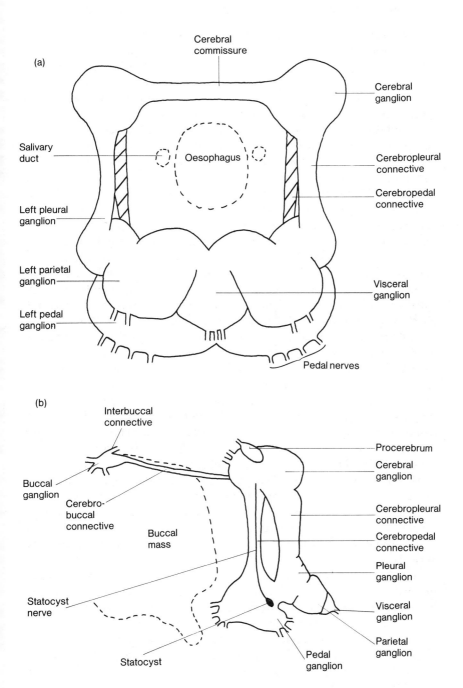

Fig. 6.1 Diagram of circumoesophageal ganglia from (a) posterior and (b) left-hand side.

The tentacular or digitate ganglion is linked to the procerebrum by the tentacular nerve. The cerebral ganglia are linked to the pleural ganglia of the suboesophageal ganglionic mass and to the paired pedal ganglia by two pairs of connectives. The suboesophageal ganglionic mass contains a central visceral ganglion, flanked by a pair of parietal ganglia and then paired pleural ganglia. These ganglia generally form a fused mass in slugs.

In addition to the cerebrobuccal connectives and tentacular nerves, a number of peripheral nerves, mostly paired, run from the cerebral ganglia and these include the optic nerve to the eye on the optic tentacle, the statocyst nerve located on the pedal ganglion and several nerves running to the buccal region, including the lips (Smith, 1967; Runham and Hunter, 1970). A single penial nerve runs from the cerebral ganglion to the penis and, in *Helix*, this nerve originates from cell bodies in the pedal ganglia, the axons run through the cerebropedal connective and emerge from the cerebral ganglion (Kerkut and Walker, 1975). The buccal ganglia are connected by an interbuccal connective (Smith, 1967) and nerves run from the buccal ganglia to the buccal mass musculature, the radula, the salivary glands and ducts and the oesophagus while the posterior gastric nerve runs to the stomach and digestive gland (Smith, 1967; Runham and Hunter, 1970). Welsford and Prior (1989) showed that increased oesophageal activity during feeding was mediated via the gastric nerves. The fine structure of the salivary nerve was described by Beltz and Gelperin (1979). It contained over 3000 axon profiles and these axons innervated the musculature of the salivary duct and also branched within the gland. The intestinal nerve from the visceral ganglion supplies branches to the kidney, digestive gland, heart (cardiac nerve), reproductive tract and intestine with a branch, the anal nerve, running to the rectum and the secondary ureter (Smith, 1967; Runham and Hunter, 1970; Garner, 1974). The posterior salivary gland nerve is also a branch of the intestinal nerve (Runham, 1975). At least 12 different types of identified neurons were found to be related to the nerve from the visceral ganglion in the slug *I. fruhstorferi* (Kim and Chang, 1987). The parietal ganglion supplies nerves to the skin and muscles of the body wall and mantle while nerves from the pedal ganglia supply the skin of the head area and the pedal musculature.

Veratti (in Bullock and Horridge, 1965) described a nerve plexus covering the wall of the gut in *Limax* throughout its length. This stomatogastric plexus is found generally in gastropods (Bullock and Horridge, 1965). The cells are scattered singly and in clusters, both at intersections and between them and may be unipolar, bipolar and multipolar. The plexus is denser in some places where accumulations of cells may be large enough to be regarded as ganglia although Bullock and Horridge (1965) considered that these accumulations were not sufficiently constant to receive a name.

Roach (1968) described how three distinct layers of nervous tissue were present in the wall of the gut of *A. ater*. These were probably interconnected and were joined to the central nervous system by the gastric and intestinal nerves. Characteristic patterns of spontaneous muscular activity were shown to occur in all regions of the alimentary tract (Section 3.1.6). Reaction of this rhythmic muscular activity in the gut to drugs, such as acetylcholine and 5-hydroxytryptamine, indicated that this gut activity was regulated by nervous coordination. Roach suggested that rhythmic activity in the gut was controlled or coordinated by the stomatogastric plexus and that, while activity might be influenced by the central nervous system, this appeared to be of relatively little importance in *A. ater*. The pedal nerves in gastropods branch as they reach the sole of the foot and then pass abruptly into a coarse meshed pedal plexus (Bullock and Horridge, 1965). Biedermann (1906) described this plexus in *Limax*. On either side of the midline of the sole of the foot there was a nerve cord with ganglia spaced at regular intervals along its length. These ganglia were linked to those of the opposite side by cross bridges, giving a ladder-like appearance to the plexus. The cross bridges were also linked by fine, cell-free longitudinal strands parallel to the two main cords. Laterally, this ladder, like the central plexus, was continuous with an irregular wide meshed plexus at the sides of the foot and in the body wall. Chatwin (1973) found that the pedal plexus in *D. reticulatum* was almost identical to that of *Limax* described by Biedermann (1906).

6.1.2 Structure of ganglia and connectives

The histology of the pulmonate ganglion was reviewed by Bullock and Horridge (1965) and the histology of the ganglia in the snail *Helix* was described in detail by Kerkut and Walker (1975). Runham and Hunter (1970) described the structure of the ganglion in *D. reticulatum*. The ganglion was covered by a thick sheath (the epineural sheath) consisting of muscle cells, collagen and other connective tissue. Some cells appeared to contain material which was secretory and Kerkut and Walker (1975) showed that, in *Helix*, the epineural sheath had a number of globular cells containing glycogen, with the glycogen content increasing during the winter. In *Helix* a thin inner layer of connective tissue was also present in the epineural sheath (Kerkut and Walker, 1975) and the vascular system penetrated the epineural sheath, with blood draining back through the sheath into the sinuses of the body cavity. In the ganglion of *D. reticulatum* the nerve cell bodies were arranged peripherally while the centre of the ganglion was filled with a dense network of interconnecting axon fibres (the neuropil). The nervous tissue was interspersed with supporting or glial cells, some of which contained glycogen while others formed sheath-like structures surrounding the nerve cells (Runham and Hunter, 1970). The cell bodies of the ganglion cells varied in size and included a few giant cells with diameters of up to 800 μm and exceptionally large nuclei (Bullock and Horridge, 1965).

The giant cells were relatively constant in number and in *Arion* and *Limax* there were one or two giant cells in each pleural ganglion, nine or ten in the parietal ganglion and similar numbers in the visceral and pedal ganglia while the buccal ganglion contained a single giant cell. One constant giant cell was found on the dorsal periphery of the metacerebrum. In addition to the giant cells, the ganglion contained ordinary ganglion cells and very small chromatin-rich cells with a diameter of 10 μm or less (Bullock and Horridge, 1965). The cells of the procerebrum were the smallest and most densely packed. About 25–30 giant neurons were present in the circumoesophageal ganglia of *L. alte* together with advanced type multipolar heteropolar neurons similar to those in vertebrates (Babu *et al.*, 1977). Turner (1966) showed that the arrangement of most of the larger cells and groupings of cells in the ganglia of *A. columbianus* was constant from slug to slug. Thus the basic stereotyped responses of the slug reflected stereotyped anatomical patterns in the central nervous system and Turner concluded that the central nervous system had a physiologically and anatomically patterned neuropil as opposed to a randomly functioning one. Some of the neuron pathways in the ganglia and connectives of *A. columbianus* had been previously described by Turner and Nevius (1951). A single conspicuous giant cell was present in each metacerebral ganglion of *L. maximus, L. flavus, D. reticulatum, T. budapestensis, A. ater, A. hortensis* and *A. californicus* (Senseman and Gelperin, 1974). Senseman and Gelperin described how this cerebral giant cell had processes which emerged from the ipsialateral cerebral ganglion only in the cerebrobuccal connective or in the external lip nerve. In *A. californicus* the processes in the cerebrobuccal connective made excitatory monosynaptic connections with the two largest neurons in the buccal ganglion and in *Limax* and *Ariolimax* a more medially located, large buccal neuron was also excited by the cerebral giant cell. Senseman and Gelperin (1974) found some differences in the morphology and physiology of the cerebral giant cell between different species of slug but concluded that they represented homologous neurons. Sejnowski *et al.* (1980) and Reingold *et al.* (1981) described a technique for mapping metabolically active neurons during spontaneous and driven electrical activity in the slug brain, involving the use of tritium-labelled 2-deoxyglucose autoradiography.

The detailed structure of the gastropod nerve fibre has been described by Bullock and Horridge (1965) and Turner and Nevius (1951) included a description of a nerve in a study of the nervous system of *A. columbianus*. The nerve was bounded by a thick outer sheath or perineurium and a pattern of radially arranged septa, projecting inwards from the perineurium, was characteristic of the gastropod nerve fibre. Myelotropic behaviour in the visceral connective of *Arion* was described by Bullock and Horridge (1965), but the fibres were not truly myelinated like those of the vertebrate nerve. Estimates for the conduction velocity of slug nerves include values of 0.64 m s^{-1} at 19.8°C for a fibre diameter of 35 μm and 0.35–0.44 m s^{-1} for the pedal nerve, both in *Ariolimax*, and 1.24 m s^{-1} for *Limax* pedal nerve (Bullock and Horridge, 1965). Gerschenfeld (in Kerkut and Walker, 1975) described the fine structure of the synapses in the ganglia of two snails and the slug *V. solea*. The synaptic junctions were always between axons (axoaxonal synapses) and never between axons and nerve cell bodies (axosomatic synapses). However, Zs-Nagy and Sakharov (1969) described axosomatic synapses in the *Limax* procerebrum. Postsynaptic membranes were mainly located on small finger-like axonal branches while the presynaptic endings were either of the terminal type or *en passant*. Gerschenfeld noted the presence of vesicles at the synaptic endings. The fine structure of neuromuscular junctions in the optic tentacle of *V. soleiformis* was described by Barrantes (1970). Barrantes found two distinct types of granulated vesicles in addition to clear synaptic vesicles. Since these occurred in separate nerve endings which showed other morphological differences, it was suggested that there were two different types of neuromuscular junctions. The fine structure of neuromuscular junctions in the optic tentacle of *L. flavus* was described by Rogers (1968).

Paired fast salivary burster neurons with axons running along the paired salivary nerves were described by Prior and Gelperin (1977). These were autoactive motoneurons and each innervated the ipsilateral salivary duct musculature. The basic pattern of endogenous salivary burster activity was composed of cyclical bursts of impulses, which during expression of the feeding motor programme became synchronized with the protraction phase of this programme (Prior and Gelperin, 1977; Gelperin *et al.*, 1978). This synchronization was due to electronic coupling between the salivary burster neurons and protractor neurons in the buccal ganglia, resulting in salivary burster activity being modified by the synaptic drive of the feeding motor programme (Prior and Grega, 1982). The effects of temperature acclimation on the endogenous activity of the salivary burster neurons and the effects of short-term temperature changes on their interaction with other neuronal elements in the salivary-feeding system were investigated by Prior and Grega (1982). They found that the activity of the salivary burster neurons showed temperature acclimation and had characteristic cold and warm blockade temperatures. Acclimatization of burster activity was a slow response to the mean ambient temperature. Prior and Grega also showed that there was increased synchrony of activity between the right and left salivary burster neurons at low temperatures. Since a reduction in the efficiency of synaptic interactions could interrupt synchronized motor patterns at low temperatures in chemically mediated synapses, a temperature-dependent increase in electronic coupling could facilitate the maintenance of synchrony in motor output. They concluded that this kind of temperature compensation could be an adaptation to short-term exposure to low temperatures at night, which were often close to the cold block temperature for chemically mediated synapses. The feedback from salivary duct stretch receptors was shown to modulate the activity of the salivary burster neurons thus giving the duct musculature reflex control over duct contraction (Beltz and Gelperin, 1980a, b). The salivary burster had both sensory and motor endings and Beltz and Gelperin suggested that some of the inputs to the buccal neurons were probably also involved in the activation of the feeding motor programme. A second autoactive salivary neuron was described from each buccal ganglion of *L. maximus* by Copeland and Gelperin (1983) and this neuron had axons in the ipsilateral and contralateral salivary nerves. This bilateral salivary neuron was described as a slow bursting neuron and was presynaptic to some of the secretory acinar cells of the salivary gland, increasing salivary flow during the feeding motor programme. The bilateral salivary neuron was affected synaptically by the metacerebral giant cell. Goldring *et al.* (1983) showed that the salivary glands of *A. columbianus* were composed of three morphologically distinct cell types and that these could be characterized by differences in their electrical activity. They suggested that different secretory products might be released from these distinct cell types but only under a specific set of conditions associated with changes in membrane potential.

6.1.3 Neurotransmitters

The transmission of nerve impulses at the synapse generally requires a chemical intermediate. The action potential in the presynaptic cell triggers the release of a neurotransmitter from vesicles at the axon terminal and this substance mediates transmission of information between the presynaptic and postsynaptic parts of the synapse. Several neurotransmitters have been identified from pulmonate nervous systems including acetylcholine, serotonin or 5-hydroxytryptamine (5-HT) and dopamine. Kerkut and Walker (1975) concluded that acetylcholine was both an excitatory and inhibitory transmitter substance between neurons in the pulmonate central nervous system, 5-HT was an excitatory transmitter and dopamine was an inhibitory central transmitter agent. The salivary burster neuron is a cholinergic motoneuron located in the buccal ganglion of *L. maximus*. Its axon runs down the salivary nerve and initiates contractions of the salivary duct. Barry and Gelperin (1982a) showed that exogenous choline augmented transmission at the isolated cholinergic synapse between the salivary burster neuron and the salivary duct muscle. Exogenous choline was taken up by the presynaptic terminals of the salivary burster neuron and converted to acetylcholine. The concentration of choline in the plasma of slugs fed on choline-enriched and choline-deficient diets varied with the availability of choline in the diet and was significantly higher in animals fed on the choline-enriched diet (Barry and Gelperin, 1982b). Since measurements of transmission made at the isolated cholinergic synapse showed that it was sensitive to fluctuations in the external choline concentration within the physiological range, Barry and Gelperin (1982b) concluded that the synapse in intact slugs would be sensitive to the fluctuations in exogenous choline which could occur with a change in diet. A study of acetylcholine turnover at the synapse between the salivary burster neuron and the salivary duct muscle showed that the salivary burster neuron was partially dependent upon choline reuptake for maintained acetylcholine synthesis and release. However, during high-frequency tonic firing, the salivary burster neuron was primarily dependent on choline reuptake for acetylcholine synthesis and release (Barry and Gelperin, 1984). Changes in neuronal acetylcholine concentration due to an increased dietary intake of choline have been shown to augment learning and memory functions (Sahley et al., 1986) and this is discussed in Section 7.1.6. The presence of an acetylcholinesterase in the circumoesophageal ganglia and the muscular foot of L. maximus has been demonstrated by Pessah and Sokolove (1983a).

Rosser (1982) recorded the presence of peptidergic neurons in the pedal ganglia of D. reticulatum and suggested that they might have a transmitter function in the pedal nerves innervating the musculature of the foot. Dopamine and serotonin (5-HT) are important neurotransmitters in a range of molluscan nervous systems (Rózsa, 1984). Cottrell and Osborne (1970) isolated serotonin from the giant nerve cells in the metacerebrum of L. maximus. They estimated the serotonin content of the cytoplasm of the cell soma to be 4×10^{-3} M. Osborne (1971) and Wieland and Gelperin (1983) also demonstrated the presence of dopamine and serotonin containing cells in the nervous system of L. maximus. Wieland et al. (1987) showed that dopamine and serotonin were localized to particular cell groups and fibre tracts within and between the cerebral and buccal ganglia of L. maximus. They concluded that, since these ganglia controlled feeding behaviour, which was readily modified by associative learning, then these substances might be involved in this process. Evidence to support this conclusion was provided by Wieland et al. (1989) who showed that the release of dopamine and serotonin from isolated Limax cerebral and buccal ganglia was stimulated by exposure to high concentrations of K^+ ions and that feeding motor programme responsiveness was suppressed reversibly by whole ganglion high K^+ treatment. Serotonin was shown to be present in the procerebrum of L. maximus and elicited action potentials from procerebral neurons in vitro (Yamane et al., 1989). Yamane et al. described several biochemical changes which were triggered by adding serotonin to the isolated procerebral lobe, including a rapid synthesis of cAMP. They suggested that serotonin-stimulated events might have direct consequences for intercellular communication in the odour-processing network of the procerebral lobe.

Delaney and Gelperin (1986) showed that L. maximus acquired an aversion to the taste and probably the odour of diets lacking an essential amino acid (Section 7.1.6). Serotonin depends on the amino acid tryptophan as a precursor and Gietzen et al. (1987) found that, when a tryptophan devoid diet was fed to L. maximus, the cumulative food intake was reduced by 50% over three weeks. They also discovered that slugs fed on the tryptophan devoid diet were more active and were able to right themselves more rapidly, when placed on their side, than control slugs and that this increased mobility was not simply the result of food deprivation. Thus two behavioural variables which might be influenced by serotonin were altered in slugs fed a tryptophan devoid diet. However, neither brain serotonin nor plasma tryptophan concentrations differed between tryptophan devoid and control animals. Furthermore, the observed depression in food intake did not appear to be due to measurable changes in brain serotonin or plasma tryptophan concentrations. Cooke and Gelperin (1988a) described how both the tetrapeptide FMRFamide and the amino acid neurotransmitter GABA exerted dose-dependent inhibitory effects on the expression of the feeding motor programme of the *in vivo* lip–brain preparation when added to the bath saline in physiological amounts. Since these substances could be involved in the generation or regulation of feeding behaviour and in food-aversion learning, the distribution of FMRFamide-like substances and of GABA-like substances in the nervous system of *L. maximus* was studied by Cooke and Gelperin (1988a, b) using immunohistochemical methods. Cooke and Gelperin (1988a) suggested that the large number, extensive distribution and great range of size of FMRFamide-like immunoreactive cell bodies and the wide distribution of immunoreactive nerve fibres indicated that FMRFamide-like peptides might serve several different functions in the nervous system of this slug. Experiments by Krajniak *et al.* indicated that FMRFamide-like peptides were involved in the regulation of reproduction and digestive activities in *L. maximus*. Since GABA-like immunoreactive fibres were not found in any peripheral nerves, Cooke and Gelperin (1988b) suggested that GABA acted as a central neurotransmitter in the slug.

6.2 SENSORY STRUCTURES

6.2.1 Tentacles

The location of the two pairs of tentacle has been described in Section 2.1. The structure of the slug tentacle has been described by Grassé (1968) and by Bullock and Horridge (1965). The tentacle is a hollow muscular tube and can be withdrawn by inversion except in the Soleolifera where it can only be contracted. Three movement components were shown to be involved in the withdrawal of *A. subfuscus* tentacles by Wondrak (1977). The tentacle retractor muscle is inserted towards the tip of the tentacle and runs back to the floor of the pallial complex. The fine structure of the tentacle retractor muscle in *Limax* was described by Reger and Fitzgerald (1981). The retractor muscle cells were distributed as a network beneath the epithelium of the tentacle and showed specializations which reflected the large surface area changes which occurred in these muscle cells. A study of the tentacle retractor muscle of *A. columbianus* by Chan and Moffett (1982a, b) showed that six identifiable cerebral motoneurons mediated the retraction of the optic tentacles of the slug. No evidence of a peripheral nerve plexus was found in the retractor muscle.

The cavity of the tentacle contains the large tentacular nerve from the procerebrum to the tentacular ganglion near the enlarged tip of the tentacle and a smaller optic nerve from the postcerebrum which serves the eye, also located near to the tip of the tentacle. Moquin-Tandon (in Taylor, 1902–1907) demonstrated that the tentacles of the slug were the site of olfaction. Kittel (1956) and Chase and Croll (1981) showed that distance chemoreceptors on the posterior (optic) tentacles were mainly involved in anemotaxis and movement along a concentration gradient while the anterior tentacles and the oral lappets were mainly concerned with contact chemoreception and were involved in trail following. Olfactory orientation and trail following are described in Sections 7.2.1 and 7.2.2. The primary receptor cell bodies are arranged in groups at the base of the sensory epithelium at the distal end of the tentacle. These cells have distal hair-like processes which penetrate the epithelial cells so that their sensory endings lie at the surface. Olfactory chemoreceptors have been described and illustrated by Grassé (1968) from *Limax* tentacles. The olfactory function of the tentacular ganglion was demonstrated by Gelperin (1974). The fine structure of the epidermis of the optic tentacle in *L. flavus* was examined by Kataoka (1976), who described the distal processes of the receptor cells. The nerve elements associated with these receptors are complex and include the large, subdivided tentacular ganglion, reflecting the importance of these olfactory receptors (Bullock and Horridge, 1965). Egan and Gelperin (1981) found that olfactory fibre tracts in *L. maximus* terminated in the cerebral ganglion, arborizing near the metacerebral giant cell. The metacerebral giant cell was shown to fire in bursts. Electrical stimulation of the olfactory nerve elicited complex postsynaptic potentials in the ipsilateral metacerebral giant cell that disrupted its firing rate and also altered the firing pattern of the contralateral metacerebral giant cell. Egan and Gelperin (1981) also found that olfactory stimulation failed to activate the feeding neural network of which the metacerebral giant cell was part although Chang and Gelperin (1976) had previously demonstrated metacerebral giant cell modulation of the chemoreceptively elicited feeding motor programme *in vitro*.

The structure of the eye of *D. reticulatum* was described by Newell and Newell (1968) and resembles that of *Helix*. Kerkut and Walker (1975) wrote an extensive review of literature on the structure of the pulmonate eye, and compared the structure of the eyes of *Helix* and *Deroceras*. The eye was located near the tip of the extended optic tentacle beneath a thin layer of collagen and a transparent area of epidermis. It was a closed vesicle, ovoid in shape and with a wall made up of a single layer of cells. The cells at the front of the eye were larger and transparent, had a high refractive index of 1.4 and functioned as a cornea, refracting light entering the eye. The eye had a spherical lens and between this and the retina was a layer of vitreous humour. The retina at the back of the eye was formed by two types of cells, photoreceptor cells and pigment cells. The photoreceptors were large columnar cells and they were relatively few in number. They reached a height of about 45 μm towards the centre of the retina but were shorter towards the sides of the eye. Each photoreceptor cell had a projection, the rod, at the apex and this projected into the vitreous humour. The rod had a fringe of radially arranged microtubules or microvilli, mostly arranged at right angles to the light. The pigment cells, which surrounded the photoreceptors, were more numerous and were packed with melanin granules in the inner part of the cell. A nerve fibre ran from the base of each photoreceptor cell to form a part of the optic nerve. Newell and Newell (1968) concluded that, since the number of photoreceptors was few, the eye would produce a poor retinal image but was structurally adapted for detecting changes in light intensity and for operating at night.

Newell and Newell (1968) also described an accessory retina in the eye of *D. reticulatum*. This was a diverticulum from the main retina on its ventral side and so below the optic axis of the eye. Two types of cell were present in the lining of this diverticulum, a few receptor cells (probably only 12) and a number of supporting cells which were unpigmented, unlike those of the main retina. Newell and Newell suggested that this accessory retina might be an infrared receptor. Kerkut and Walker (1975) described how slugs would turn and walk away from a black body heat emitter and how they lost this ability to avoid a radiant heat source if the ends of the tentacles were removed. When the tentacle was withdrawn, the eye was rotated so that the accessory retina was displaced to lie under the outer surface of the tentacle, placing it nearer to an outside source of radiation. Eakin and Brandenburger (1975) found that the eyes of *A. californicus* and *L. maximus* were similar to those of *D. reticulatum* and *L. flavus*. They compared the fine structure of the photoreceptors in the eye of the light-tolerant *A. californicus* with that of the nocturnal slug, *L. maximus*. They described two types of photoreceptor cell. Type I had an apical rod from which long microvilli extended into the vitreous humour while Type II was characterized by having only short terminal microvilli on the apical surface. They found several differences between the Type I photoreceptor cells of *A. californicus* and *L. maximus* including the distribution of the vesicles thought to carry the photopigment or its precursor to the microvilli. In the light-tolerant slug, these vesicles were aggregated distally beneath the microvilli, while in the light-avoiding slug the vesicles were aggregated basally near the nuclei. The fine structure of Type I and Type II photoreceptor cells has also been described for *L. flavus* by Kataoka (1975).

Suzuki *et al.* (1979) found that, during dark adaptation in *L. flavus*, the retinal sensitivity increased in two stages with a break in the dark adaptation curve. They suggested that the presence of the two phases in the *L. flavus* retina might represent the activities of the two types (Types I and II) of photoreceptor cell. This duplex system also suggested that the range of light intensity perceived by the slug was possibly wider than previously thought. Kataoka (1977) showed that the accessory retina of *L. flavus* also consisted of Type I and Type II photoreceptor cells and supporting cells. The Type II cells contained no photic vesicles in the cytoplasm and lacked apical rods although many microvilli were present on the apical surface of the cell. Axons from both types of photoreceptor cell joined with those from the main retina to form the optic nerve. The supporting cells resembled the corneal cells in structure. The corneal cells, beneath the tentacular epithelium, consisted of simple columnar epithelial cells, which lacked many cell organelles although they were packed with glycogen. In a detailed study of the accessory retina of *L. flavus*, Tamamaki (1989) used intracellular recordings to show that the λ_{max} for the spectral sensitivity curve of Type I photoreceptor cells from both accessory and main retinae was 460 nm. A similar value had been reported by Suzuki *et al.* (1979) for the light-adapted main retina of *L. flavus*, using recordings from the optic nerve. Eakin *et al.* (1980) found that Type I cells of the retina of the eye in the athoracophorid slug, *A. bitentaculatus*, had unusually long rods, bearing thousands of microvilli, extending from the distal ends of the cell towards the lens. They considered that this hypertrophy of the photoreceptors was an adaptation to nocturnal life as this slug forages in trees at night.

6.2.2 Other sensory structures

(a) Lips

Walker (1969) showed that there were several small lips around the mouth of *D. reticulatum* together with a pair of lateral mouth lobes. Runham (1975) described how the surface of the mouth lobes was covered by a highly specialized sensory epithelium connected to Semper's organ (see Section 6.3.2). Kieckebusch (in Stephenson, 1979) showed that chemoreceptors were present on the lip lobes of *H. pomatia* and Stephenson (1979) suggested that this was probably also true for slugs. He also found that mucus secreted on to the surface of the skin tended to flow along channels between the skin tubercles towards the tentacles and lips. Stephenson suggested that when the lip lobes approached a moist leaf surface, surface tension would cause the mucus and moisture film on the leaf to unite to form an 'aqueous bridge', enabling solutes from the leaf surface to flow across and come into contact with chemoreceptors on the lips and mouth lobes. Chemoreceptors on the lips of *L. maximus* have been described by Salanki and Bay (1975), Chang and Gelperin (1980) and Culligan and Gelperin (1983).

(b) Statocysts

The statocysts lie on each side of the circumoesophageal ganglia close to the pedal ganglion or between the pedal and pleural ganglia. They are mainly concerned with gravity perception and orientation. The statocyst is spherical, filled with liquid (statolymph) which contains many small calcareous statoliths. The statoliths ranged from between 150 and 200 in number in *A. ater* (Grassé, 1968). The structure of the statocyst of *L. alte* was described by Kulkarni (1977) while Wolff (1969) described the fine structure of the statocysts of *A. ater*, *L. maximus* and *L. flavus*. The statocyst was bounded by an outer sheath of connective tissue and within this was a sensory epithelium, containing two types of cell. The sense cells were flattened giant cells bearing sensory hairs against which the statoliths rubbed, and between these were smaller cells with microvilli. There were 11–15 hair cells, each having a single axon in the statocyst nerve which ran to the cerebral ganglia. Experiments described by Wolf (1970a) confirmed that hair cells were the receptor cells of the statocyst and showed that maximal impulse frequency occurred when the slug was in the upside-down position. Wolff (1970b) used a preparation of the circumoesophageal ganglion ring to record the electrical responses from a single statocyst nerve of *Arion* and *Limax*. There was an increase in activity in the nerve when the preparation was rotated and this activity was maintained when the connection between the ganglia and statocyst was cut and recordings made from the statocyst side. If the recording was made on the ganglion side of the cut, activity was still recorded when the brain was rotated. However, this stopped when the other statocyst was removed. Wolff (1970b) concluded that activity from the statocysts entered the brain and that some of this was relayed as efferent activity to the statocyst on the opposite side.

(c) Other sensory structures

Primary receptor cells (Section 6.2.1) occur at or near the epithelial surface in most parts of the body of the slug. Bullock and Horridge (1965) described how functionally identified taste receptor cells, olfactory receptor cells and photoreceptor cells were all found on the skin of the optic and anterior tentacles, the mouth lobes, and at four loci on the foot of the snail, *H. pomatia*, although their densities varied widely between the different locations. Wallis and Wright (1971) identified primary receptor cells at the tentacle tip of *A. ater* which were extra sensitive to tactile stimuli and they found that both phasic and tonic receptors were present. They described two types of receptor cell: one in which the dendrite terminated in a cup, from the centre of which arose a group of cilia; and the other in which the nerve ending terminated in a hillock, from which cilia arose. Wright (1972) estimated that the optic tentacle of *A. ater* had an average of 150 000 dendrites per square millimetre and that at least one of the four types of free nerve endings identified was probably mechanoreceptive. Newell (1977) described the structure of primary receptors with small tufts of cilia which were present in the dorsal epithelium and the foot epithelium of *A. hortensis* and *D. reticulatum*. McCrone and Sokolove (1979) showed that the optic tentacles were not essential for the photoperiodic induction of maturation in *L. maximus*. Some kind of extraocular light receptor was, therefore, involved in the photoperiodic response and McCrone and Sokolove suggested that this was located somewhere within the central nervous system of the slug.

Internal receptors are also present in slugs. Senseman (1978) described how feeding bouts in *A. californicus* were terminated by post-ingestional feedback which indicted the presence of proprioceptors in the gut. The evidence for the existence of proprioceptors in the gut of slugs was reviewed by Gelperin and Reingold (1981). Dobson and Bailey (1982) suggested that in *D. reticulatum* crop fullness could be a suitable homoiostat for the termination of feeding and that this might be mediated by stretch receptors in the crop wall. Proprioceptors in the buccal muscles of *L. maximus* may reflect the hardness of food (Gelperin *et al.*, 1978).

6.3 ENDOCRINOLOGY

6.3.1 Neurosecretion

In their review of endocrinology in pulmonates, Boer and Joosse (1975) discussed the definition of neurosecretory cells. Although this term could be applied to cells having the characteristics of neurons and also exhibiting signs of glandular activity, Boer and Joosse considered that it was necessary to define the criteria in more detail, including the use of ultrastructural evidence which would make possible a correct identification of neurosecretion including the distinction between neurotransmitters and neurosecretions. In *M. gagates* the secretory processes were already initiated at the time of hatching (Quattrini, 1962) while in *A. ater* secretion in the buccal ganglia began when the slugs were two weeks old and in the cerebral ganglia at puberty (van Mol, 1962). Quattrini (1966) described the ultrastructure of a group of neurosecretory cells from the left parietal ganglia of *M. gagates* and also described the way secretory material was transported along nerve fibres. Smith (1967) examined the relationship between neurosecretory changes in the circumoesophageal ganglia and maturation of the reproductive tract of *A. ater*. He used a histological technique (paraldehyde–fuchsin) to identify groups of neurosecretory cells and mapped their nerve cell bodies and nerve fibres for each of the three main ganglion masses of the circumoesophageal ring (supraoesophageal,

pleuro-parieto-visceral and pedal ganglia) at each of the stages in the development of the gonad (Section 5.1.3). Complex patterns of neurosecretory
activity were recorded in the supraoesophageal and pleuro-parieto-visceral
ganglion masses but little activity was found in the pedal ganglion mass. The
buccal ganglion also showed a great deal of neurosecretory activity.

Smith (1967) found that the only major change in the pattern of
neurosecretion that could be associated with reproduction took place at the
mid and late spermatozoa stages when large amounts of secretion suddenly
appeared in groups of small cells in the pleuro-parieto-visceral ganglion
mass. The secretion was found only in slugs in which the female glands had
begun to mature, it rose to a maximum by the late spermatozoa stage when
copulation occurred and then decreased rapidly until little was present by the
late oocyte stage. Wijdenes *et al.* (1980) also used a histological technique
(alcian blue/alcian yellow) to identify neurosecretory cells in the ganglia of
D. reticulatum, A. hortensis and the snail, *H. aspersa*. They described a
number of different cell types from all three species and concluded that the
neurosecretory systems of the three species were essentially similar although
some differences were apparent. Van Minnen and Sokolove (1981) used the
same technique to identify neurosecretory cells in the ganglia of *L. maximus*.
Stainable cells were identified in the cerebral, pleural, parietal and buccal
ganglia and in the visceral ganglion but not in the pedal ganglia. They
occurred as single cells or in groups of up to 100 cells. The neurosecretory
system was similar to that described by Wijdenes *et al.* (1980) for *D.
reticulatum* and *A. hortensis*. Wijdenes and Runham (1977) showed that
when the medial neurosecretory cells of the cerebral ganglia were destroyed,
body growth in *D. reticulatum* stopped although the differentiation of the
reproductive tract was only delayed and not blocked. They suggested that
the medial neurosecretory cells produced a growth hormone with a specific
somatotrophic effect. McCrone *et al.* (1981) found that, in the absence of
cerebral ganglia, suboesophageal ganglion mass explants in *L. maximus*
appeared to secrete a factor which inhibited somatic growth. Van Minnen *et
al.* (1983) showed that homogenates of part of the right parietal ganglion
containing large secretory cells ['M' cells of Van Minnen and Sokolove
(1981)] could induce egg laying in mature *L. maximus*. Marchand *et al.*
(1984) used immunocytological tests to demonstrate the presence of a
somatostatin-like substance in the cerebral ganglia of *L. maximus*. In the
cerebral ganglia the right Z-area cells, responsible for the synthesis of the
maturation hormone, were strongly somatostatin-positive, suggesting a
similarity between the maturation hormone and the somatostatin-like
substance contained in the Z-area cells.

Kulkarni *et al.* (1983) examined the effect of temperature on neurosecretion in the cerebral ganglia of the slug *Semperula maculata* (Semper). Slugs
acclimated at high temperatures (32°C, 34°C, 36°C) showed a reduction in
neurosecretory activity while slugs acclimated at low temperatures (10°C,

15°C) showed an increase in neurosecretory activity. The structure of neurohaemal areas in pulmonates was described by Grassé (1968) and Martoja (1972). These areas consisted of an accumulation of axon terminals in the neighbourhood of small blood sinuses. Two neurohaemal areas were associated with the cerebral ganglia, including one which was associated with the dorsal body.

6.3.2 Endocrine structures

(a) Tentacle

The optic tentacles have been shown to have a role in the endocrine control of reproduction in slugs (Pelluet and Lane, 1961; Pelluet, 1964) and this is discussed in Section 5.5. The source of the tentacular hormone was described by Lane (1964b) as the collar cells which were arranged around the base of the tentacular ganglion. Nagabhushanam and Kulkarni (1971c) suggested that three types of cell in the optic tentacle of L. alte were endocrine cells, including the collar cells and lateral cells in the dermatomuscular layer. One of the two cell types present showed an annual cycle of activity which was correlated with the annual reproductive cycle of L. alte. Takeda (1982) described the collar cells in six species of slug as large ovoid cells, about 60 μm in diameter. These cells were intimately connected with the tentacular ganglion, sending fine cytoplasmic processes into the ganglion. The collar cells underwent histological changes during the year in relation to reproduction. Since hormonal activity in the tentacle was restricted to the collar cells, Takeda (1982) concluded that these were the source of the tentacular hormone and suggested the use of the term optic gland to describe the collar cell group. Takeda et al. (1987) described the fine structure of the optic gland cells of L. marginata, showing that many large granules were present in the cytoplasm. These were considered to contain hormonal material relating to reproduction. Takeda et al. (1987) found that during the breeding season the medial neurosecretory cells of the cerebral ganglia were actively producing and releasing secretory materials. Since neurosecretory axons penetrated into the optic gland cells and the axons of the medial neurosecretory cells extended to the tentacular ganglion near the optic gland, Takeda suggested that the optic gland was controlled by a neurohormone originating from the medial neurosecretory cells of the brain.

(b) Dorsal bodies

The dorsal bodies are composed of groups of cells which extend over the surface of the cerebral ganglia and down to the pleural ganglia (Runham and Hunter, 1970). Boer and Joosse (1975) concluded from their review of the pulmonate dorsal bodies that they produced secretory granules and that their secretion was probably released by exocytosis into the loose connective tissue surrounding the groups of dorsal body cells. The dorsal body cells showed hyperplasia during the spring and were probably endocrine organs. Wijdenes and Runham (1976) suggested that the dorsal bodies in *D. reticulatum* produced a hormone which promoted maturation of the oocytes and the female accessory sex organs. The dorsal bodies of *L. maximus* were shown to be the source of the galactogen synthesis stimulating factor by Van Minnen and Sokolove (1984). Further details of the role of the dorsal bodies in the control of reproduction are given in Section 5.5.

(c) Arterial gland

The arterial gland of *D. reticulatum* consisted of irregularly shaped masses of opaque whitish tissue situated irregularly along the distal portion of the cephalic artery and along its branches (Laryea, 1969). Two types of granules were present in the cells and their contents were released into intracellular ducts which connected with intercellular channels, leading directly to the edge of the gland. There was no clear relationship between the size or histology of the gland and reproductive development although Runham and Hunter (1970) stated that, when reproduction had been completed, the breakdown of this gland preceded the death of the slug. Laryea (1969) also recorded the presence of an arterial gland in *D. caruanae* and *L. flavus*. The function of this gland remains uncertain.

(d) Semper's organ

This lobed mass of grey or whitish tissue is located between the buccal mass and the anterior tentacle but separated from the digestive tract. It has a rich blood supply and is innervated by a branch of the anterior tentacle nerve (Grassé, 1968). Lane (1964a) suggested that the function of Semper's organ in *Limax* and *Helix* might be endocrine but Golding (1974) described how it resembled an exocrine gland in ultrastructure and no endocrine function has been demonstrated.

(e) Cerebral gland

The paired cerebral glands represent the stalks of the ectodermal invaginations in the embryo which give rise to the procerebra of the cerebral ganglia and each gland is closely associated with the procerebrum. In *A. ater* the glandular activity of the cerebral gland varied with the different periods in the life cycle. The gland consisted of extra- and intracerebral parts, and during the first four months after hatching the intracerebral part was well developed and showed active secretion while in adult slugs this part had almost disappeared. The extracerebral part of the gland became active in the fourth month and remained so throughout the reproductive period (van Mol, 1961, 1967). Smith (1967) found a possible relationship between the cephalic gland secretion and reproductive activity. However, the role of the cerebral gland as an endocrine gland remains unknown.

(f) Gonad

There is evidence from experimental work that the gonad of slugs produces one or more hormones (see Section 5.5). However, in a review of pulmonate endocrinology, Boer and Joosse (1975) concluded that there was no histological evidence that the pulmonate gonad produced hormones.

6.4 CONTROL OF WATER BALANCE

Hughes and Kerkut (1956) measured the electrical activity in the isolated pedal ganglion of *D. reticulatum*. This activity consisted of potentials of characteristic shape and size, with a frequency of about 30 min^{-1}. No correlation was observed between frequency and any rhythmic activity of the slug. This spontaneous activity was affected by changes in the concentration of the surrounding medium. Increased concentration of the medium decreased activity while dilution increased activity. Hughes and Kerkut (1956) concluded that this effect was due to a change in osmotic potential rather than a change in the concentration of individual ions. Koshtoyants and Rózsa (1961) demonstrated that this effect could be caused by the presence of osmoreceptors on the sole of the foot of the snail, *H. pomatia*. They also found that an increase in activity of the pedal ganglion occurred within 0.5–1.0 min of being treated with water. Adaptation occurred after 10–15 min. Hughes and Kerkut (1956) and Kerkut and Taylor (1956) suggested that the increased activity induced by the effect of humidity was due to changes in the concentration of the blood. Changes in electrical activity in the pedal and cerebral ganglia were recorded in *A. ater* and *L. maximus* when the osmoreceptors of the foot sole were stimulated by solutions of various concentrations (Rózsa, 1962a, b). Rózsa found that this pattern of activity was closely related to the activity of the whole animal and concluded that slugs might react by the use of peripheral osmoreceptors to changes in the humidity of the environment as well as by central osmoreception. These findings, based on peripheral osmoreceptors, contradicted those of Hughes and Kerkut (1956) and Kerkut and Taylor (1956) described above.

Dainton (1954) concluded that, while the relative humidity of the atmosphere had no direct effect on slug activity, it could limit the duration of activity by influencing the water content of the body. Although slugs with high water content were not inherently active, they responded more readily to other stimuli than slugs with a low water content. Prior (1981) and Prior *et al.* (1983) described a sequence of behaviours during the desiccation of *Limax*, aimed at minimizing further water loss and corresponding to levels of body hydration (see Section 4.2.5). These included changes in the frequency of pneumostome closures, a decline in locomotor activity and finally a general contraction of the body. Prior (1981) suggested that changes in body hydration would result in the exposure of the nervous system to both osmotic and ionic stress. Prior described the responses of neurones to osmotic stress and concluded that they were more sensitive to variation in osmotic potential than in ionic concentration, thus agreeing with the conclusions of Hughes and Kerkut (1956).

The salivary burster neuron is an autoactive motoneuron whose cyclical bursting activity initiates contractions of the salivary duct. Prior (1981) described how this neuron was particularly sensitive to hyposmotic stress and how this sensitivity could be attributed to osmotic rather than ionic stress. While the salivary motor system was transiently modified by hyposmotic stress, it was also capable of rapid adaptation to the stress. Prior showed also that the cyclical synaptic input to a buccal neuron was particularly sensitive to hyposmotic stress. He concluded that, since osmotic stress could initiate the coordinated expression of a central motor programme, it was possible that variation in osmotic potential resulting from a change in body hydration could directly initiate the protective behaviour patterns that were related to hydration stress. The level of body hydration is known to affect feeding responsiveness. Prior (1983a) described how, in *L. maculatus* and *L. valentiana*, air dehydration to 70–60% of initial body weight abolished feeding. Feeding responsiveness was rapidly restored after slugs were allowed to rehydrate provided the blood was diluted to at least 180–190 mOsmol kg^{-1} water (equivalent to 70–75% of initial body weight). Prior showed that this hydration modulation of the feeding response was mediated by blood osmotic potential. Phifer and Prior (1985) suggested that the change in feeding response associated with changes in the osmotic potential of the blood, might be mediated by direct osmotic effects on the central nervous system because slugs lack a blood–brain barrier. Phifer and Prior found that *L. maximus* showed a similar dehydration-induced reduction in feeding responsiveness to that shown by Prior (1983a) for *L. maculatus* and *L. valentiana*. In an isolated central nervous system–lip preparation, the duration of the Feeding Motor Programme (FMP) could be reduced by increasing the osmolality of the saline bath although the pattern of the motor programme was not affected. The osmolality of saline which could modify the FMP corresponded to that of the blood of a slug dehydrated to 65–70% of initial body weight. This level of osmolality was similar to that which initiated contact-rehydration behaviour in slugs (Prior, 1989) (see Section 4.2.5). Phifer and Prior (1985) found also that the neural network underlying the FMP adapted to hyperosmotic saline and that the duration of the FMP bouts gradually returned to normal levels after long-term exposure of 6–8 hours.

In *L. maximus* contact-rehydration behaviour was initiated when the osmolality of the blood was increased to about 200 mOsmol kg^{-1} water by either air-dehydration or injection of a hyperosmotic mannitol solution. This behaviour could be terminated by either contact-rehydration or by injection of dilute saline (Prior, 1984). The possibility that water balance in slugs might be controlled by hormones was investigated by Makra and Prior (1983). They found that injection of both arginine vasotocin and angiotensin II into hydrated slugs could induce water uptake and concluded that these hormones or related peptides might be involved in the control of water balance in slugs. Makra and Prior (1985) confirmed that the vertebrate vasoconstrictor hormone angiotensin II could initiate contact-rehydration behaviour in *L. maximus*. They concluded that release of an 'angiotensin II-like' hormone might be involved in the control of both initiation of this behaviour sequence and the increase in integumental water absorption that are characteristic of contact-rehydration. The response of the pneumostome rhythm to dehydration has been described in Section 4.2.5. An investigation into the control of pneumostome rhythm in *L. maximus* by Dickinson *et al.* (1988) showed that this behaviour was initiated and controlled partly by the osmolality of the blood and partly by a peptide closely related to arginine vasotocin and arginine vasopressin. Since an increase in blood osmolality by mannitol injections did not reduce the area of the pneumostome opening, although a reduction could be induced either by natural dehydration or by the peptides arginine vasotocin and arginine vasopressin, Dickinson *et al.* (1988) suggested that at least two systems might be involved in the overall control of the pneumostome. They also found that the hormone angiotensin II was not involved in the control of the pneumostome although it could initiate contact-rehydration behaviour. Thus although peptide hormones were involved in water regulatory behaviour, no single hormone controlled all such responses. Extracts of cerebral and pleuropedal ganglia from *A. columbianus* and *L. maximus* were found to contain vasotocin-like or vasopressin-like materials. There was no evidence, however, that these materials participated as neurotransmitters or hormones in the regulation of water balance and the immunoreactivity and antidiuretic activity of ganglion extracts did not appear to change during dehydration and rehydration (Sawyer *et al.*, 1984).

Banta *et al.* (1990) described water-orientation behaviour in *L. maximus*, consisting of an increase in locomotor activity combined with a preference for the moist arm of a Y-maze. Water-orientation behaviour was exhibited when slugs were dehydrated to 70% of their initial body weight. It could also be elicited by injecting 10^{-1} M arginine vasotocin into fully hydrated slugs. Banta *et al.* showed that the effects of arginine vasotocin and dehydration were additive since injection of 10^{-7} M arginine vasotocin into slugs dehydrated to only 80% initial body weight also elicited water-orientation behaviour. Injection of angiotensin II or arginine vasopressin failed to elicit water-orientation behaviour. Banta *et al.* (1990) suggested that a dehydration-induced increase in blood ionic concentration was also involved in the initiation of water orientation.

6.5 CONTROL OF LOCOMOTION

Kunkel (1903) found that if whole *L. tenellus* were cut into small pieces, the isolated sections of foot sole exhibited regular wave patterns. These pedal waves have been described and their role in locomotion discussed in Section 2.4. Electrical stimulation of the pedal nerves of *Limax* was shown by Biedermann (1906) to vary the intensity of the pedal waves when the muscles of the slug were completely relaxed. Biedermann also found that the quiescent isolated foot of *Limax*, without pedal ganglia, would produce waves after continued mechanical stimulation, indicating some dependence on continuous peripheral stimulation and access of receptors to the sole plexus. The autonomic state of the foot sole in *Limax* was, therefore, subject to both peripheral and central influence and Crozier and Federighi (1925b) concluded from further experiments with *Limax* that the pedal plexus, although under the control of the central nervous system, was an independent peripheral reflex mechanism. *Arion* is more dependent upon the pedal ganglia and Bullock and Horridge (1965) described how, although pedal waves were stopped when all the pedal nerves were served, a few were sufficient to maintain these waves. Ten Cate (1923) showed that, when a crawling *A. ater* was cut into between two and four pieces, the head section showed normal movement of the sole while the other sections, which were separated from the brain, showed irregular locomotory movements on the sole.

Chatwin (1973) demonstrated that the foot sole of *D. reticulatum* could produce pedal waves after the central nervous system had been removed. Stimulation of the pedal nerves indicated the presence of inhibitory and excitatory fibres which influenced the basic autonomic wave pattern controlled by the peripheral nerve plexus. These observations supported previous suggestions that the locomotory wave was under central nervous control and that the basic wave pattern required central control to preserve its integrity. Prior and Gelperin (1974) also concluded that neural control underlying pedal wave activity in *L. maximus* was a function of a peripheral nerve plexus capable of generating the basic pedal wave pattern but which required initiation and modulation by the central nervous system for normal operation. Broyles and Sokolove (1978) studied the recovery of the pedal wave following transection of the nerves running backwards from the pedal ganglion. Initially, the portion of sole posterior to the cut showed no pedal waves, while pedal waves anterior to the cut had a reduced wave velocity and frequency. After a variable period of time (3–21 days), depending on the position of the transection, fully formed pedal waves reappeared in the denervated region of the sole. Broyles and Sokolove (1978) found no evidence for reinnervation in the foot and suggested that coordinated pedal waves could be maintained on parts of the foot which lacked direct connections with the central nervous system by the spread of excitation through the peripheral nerve plexus of the foot.

6.6 CONTROL OF REPRODUCTION

The endocrine regulation of reproduction in slugs has been described in Section 5.5.

6.7 CONTROL OF FEEDING

Senseman (1978) suggested that two processes, sensory adaptation and post-ingestional feedback, were sufficient to account for the short-term control of food intake in *A. californicus*. In a study of the Feeding Motor Programme (FMP) in *L. maximus in vivo* and *in vitro*, Gelperin et al. (1978) found that the central neural network controlling coordinated rhythmic activity in feeding neurons was located in the buccal and cerebral ganglia and that both sets of ganglia produced rhythmic components of the FMP. They demonstrated that aspects of *in vivo* feeding behaviour were retained and identifiable in an *in vitro* preparation of the lips, cerebral ganglia and buccal ganglia and that repetitive chemostimulation of the lips yielded reproducible bouts of FMP. In these preparations, the ring of tissue formed by the lips and mouth lobes was isolated from surrounding tissue while preserving the innervation of the lips from the cerebral ganglia. The ring was then cut mid-dorsally and mid-ventrally and each half-lip isolated in a separate perfusion chamber, leaving the nerve intact. It was shown that the expression of the FMP *in vitro* could be strictly controlled by defined chemical stimuli. Using this *in vitro* preparation, Delaney and Gelperin (1987) identified a group of cerebral to buccal interneurons which had powerful initiating and modulating effects in fictive feeding rhythms. These interneurons received synaptic input from lip mechanoreceptors and chemoreceptors, feedback from the buccal ganglion during fictive feeding and strong inhibiting input from those stimuli which elicit withdrawal in the whole animal, such as electric shock of the foot. Further details of this work were given by Delaney and Gelperin (1990a), who described between 14 and 18 neurons with somata in the cerebral ganglia and axons projecting to the buccal ganglia via the cerebrobuccal connectives. They also examined the role of these neurons in the initiation and modulation of fictive feeding using the *in vitro* preparation (Delaney and Gelperin, 1990b). Further studies on the integration of feeding-related sensory inputs by these cerebral interneurons were described by Delaney and Gelperin (1990c). However, in spite of the central neural control of rhythmic feeding movements, meal duration and bite frequency may vary within a meal (Section 3.1.7).

Reingold and Gelperin (1980) examined several aspects of the variation of the FMP in *L. maximus*, using intact slugs and also using the isolated 'lip-brain' preparations described by Gelperin *et al.* (1978). Reingold and Gelperin showed that intact slugs showed a higher frequency of bite cycle (one protraction–retraction cycle of the buccal mass) on soft food compared with hard food when the load on the buccal mass was altered by varying the hardness of their food. In *in vitro* preparations, where the load on the buccal muscles was increased by attached weights, the cycle frequency of FMP triggered by food extracts was increased when the buccal muscles were unloaded compared with FMP when muscles were loaded. When the chemostimulant concentration of the food was increased, the duration of feeding in both intact animals and in *in vitro* preparations was also increased. Inflation of the crop in *in vitro* preparations caused an early termination of feeding activity together with a decreased FMP cycle frequency supporting the suggestions of previous authors that proprioceptive (i.e. post-ingestional) feedback from the gut was responsible for meal termination (Sections 7.1.6 and 3.1.7).

Rapid food-aversion learning has been demonstrated in intact slugs (Section 7.1.6). Chang and Gelperin (1980) showed that the isolated lip–brain preparations described by Gelperin *et al.* (1978) were capable of rapid taste-aversion learning. When lip chemostimulation by attractive food extracts was paired with lip chemostimulation by bitter plant secondary substances, the isolated brain was found to suppress selectively its neural response to one food extract while remaining responsive to another. Isolated brains were able to learn after only one or two trials and retained the learning for 6–8 hours compared with 7–21 days in the intact animal. However, Chang and Gelperin suggested that improved techniques might allow longer lasting learning by isolated brains. One-trial associative learning by the lip–brain preparation from *L. maximus* was demonstrated by Culligan and Gelperin (1983). In previous training experiments both right and left lips had received stimuli simultaneously. Culligan and Gelperin found that learning could occur if one lip was used to apply stimuli during training while the opposite lip was used to apply stimuli during testing. Learning also occurred if the food extract and the bitter taste substance, quinidine, were applied during training to opposite lips. These results indicated that the synaptic alteration due to learning occurred in the central nervous system rather than at the sensory periphery. Gelperin and Culligan (1984) investigated whether it was possible to read out a memory state neurophysiologically that had been entered into the neural network behaviourally. This would test whether whole animal learning and isolated brain learning were using the same memory storage network. Intact *L. maximus* were trained and then tested for a learned aversion to an attractive food extract. Isolated lip–brain preparations from these slugs were then assessed for feeding motor programme responses to lip-applied food extracts. About 50% of the isolated lip–brain preparations derived from previously conditioned slugs expressed *in vitro* the learning acquired *in vivo*. Gelperin and Culligan (1984) found that lip–brain preparations which did not express the learning *in vitro* tended to show a general depression of function within the feeding control network and suggested that an improved life-support system for the lip–brain preparation might have improved the performance of the preparations.

Further information on the FMP was provided by Phifer and Prior (1985). Electrical stimulation of the anterior lip nerve in the isolated lip–brain preparation of *L. maximus* elicited FMP bouts which consisted of coordinated bursts of action potentials in buccal roots 1 and 2, the salivary nerve and the gastric nerve of the buccal ganglion. Phifer and Prior described how these bursts of activity drove radula protraction, radula retraction, salivary duct contraction and oesophageal peristalsis respectively and were, therefore, correlates of biting and swallowing in the intact slug. A bite cycle in the lip–central nervous system preparation consisted of action potential bursts in the buccal nerves that corresponded to one protraction–retraction cycle of the buccal mass, and a single FMP bout normally comprised between 20 and 60 individual bite cycles and could last several minutes after electrical stimulation of the anterior lip nerve was terminated. The small cardioactive peptide B (SCP_B) was first described by Lloyd (1978) for its actions on the heart of the snail, *H. aspersa*. Prior and Watson (1988) described how SCP_B had an excitatory effect on both buccal neurons and musculature in several molluscs. Prior and Watson investigated the effect of SCP_B on the activity of specified buccal neurons and the expression of the FMP in *L. maximus*. They found that, in an isolated central nervous system preparation, superfusion with SCP_B resulted in an increase in the burst frequency of the fast salivary burster neuron, while having no effect on the activity of the bilateral salivary neuron, another burster neuron. The response of the fast salivary burster neuron showed no indication of desensitization, even after long-term exposure. The membrane responses which lay behind the ability of SCP_B to increase the burst frequency of the fast salivary burster neuron were examined by Hess and Prior (1989), who confirmed that SCP_B had profound effects on the bursting activity of this neuron. Prior and Watson (1988) showed that, in the presence of SCP_B, a previously subthreshold stimulus to the lip nerve could initiate the full FMP. They concluded that the responsiveness of the initiation process of the FMP was enhanced by SCP_B.

SCP_B-like immunoreactive material was localized by Prior and Watson (1988) to several neurons within the buccal ganglia of *L. maximus*, including the large laterally positioned B1. Activation of neuron B1 has been shown to mimic the excitatory effects of exogenous SCP_B on other central neurons (Prior and Delaney, 1986) and on peripheral effectors (Welsford and Prior, 1987). These observations on the involvement of SCP_B in the control of feeding in *L. maximus* were all derived from *in vitro* experiments. The behavioural effects of injection of SCP_B on intact *L. maximus* were investigated by Schagene *et al.* (1989). They found that SCP_B alone was capable of initiating the sequence of behavioural reponses leading up to feeding (Section 3.1.1) which was normally initiated by chemosensory cues, although cyclical radula movements were not observed. They concluded that SCP_B did not directly elicit the FMP in *L. maximus* but rather that it increased the responsiveness of the FMP to chemosensory stimulation (cp. Prior and Watson, 1988), suggesting that an SCP_B-like peptidergic system might be involved in the control of feeding behaviour in *L. maximus*. Krajniak *et al.* (1989) suggested that SCP_B might also be involved in the regulation of digestive activity in *L. maximus*, particularly by modifying the muscular tone of the crop. Welsford and Prior (1989) suggested that neuron B1 might mediate FMP-induced alterations in crop activity, possibly by the release of SCPs in this slug.

6.8 CARDIOREGULATION

Molluscs have myogenic hearts, with pacemaker action potentials being initiated within the cardiac muscle fibres, and the heart is capable of beating rhythmically and spontaneously in isolation (Jones, 1983). Jones discussed the location of the pacemaker in gastropods and suggested that there was insufficient evidence for concluding that the pacemaker was diffuse and that the atrioventricular junction might be important in this respect. MacKay and Gelperin (1972) described the anatomy and innervation of the heart in *L. maximus*. They showed that the heart could be reflexively modulated by tactile stimulation of the body surface. The cardiac nerves ran to the heart through the venous and aortic connections (MacKay and Gelperin, 1972), although it was difficult to determine where they terminated in the heart. In several snails, including *Helix*, there was a high density of nerve endings at the auricular end of the ventricle and in the atrioventricular region (Jones, 1983). MacKay and Gelperin (1972) identified cardioexcitatory and cardioinhibitory areas of the visceral ganglion and demonstrated modulation of heart rate by single cells. Cardioexcitation was serotoninergic and cardioinhibition was cholinergic, although both acetylcholine and serotonin had biphasic effects, depending on their concentration. The threshold for both acetylcholine and serotonin was about 10^{-7} M and temporary diastolic arrest occurred with 5×10^{-5}M serotonin (MacKay and Gelperin, 1972).

Both dopamine and noradrenaline were also shown to be cardioactive in *L. maximus* by MacKay and Gelperin (1972).

Duval (1983) investigated the effects of temperature, body weight, activity and cardiac denervation on heart rate of slugs. The effect of body weight on heart rate was discussed in Section 4.1. The rate of heart beat was directly related to temperature in *L. maculatus* while no significant difference was recorded between the heart rate of *D. reticulatum* at rest and when crawling. An increased heart rate of 16% was recorded for *L. maximus*, when crawling, by MacKay and Gelperin (1972). An increased rate of about 30% was recorded at the onset of feeding activity in *D. reticulatum* and Duval (1983) suggested that protrusion of the buccal mass required increased heart rate to raise the blood presure. Duval considered it unlikely that this was related to digestion because the rate increased before food entered the buccal mass. Cardiac denervation by cutting the branches of the visceral nerve innervating the heart in *D. reticulatum* resulted in an increased heart rate of about 15%. The role of the central nervous system in the modulation of heart activity induced by feeding was investigated by Grega and Prior (1985) for the slug, *L. maximus*. They found that an increase in heart rate following a meal was mediated in part by the central nervous system, and in part was a direct response of the heart musculature. The central nervous system mediated an immediate response to proprioceptive input from stretching of the crop, while the heart musculature responded directly to increased blood glucose concentration following ingestion of food. The latter response was demonstrated in both intact slugs and in isolated perfused hearts. Activation of the FMP in *L. maximus* was shown by Welsford and Prior (1991) to increase the force of ventricular contractions in heart–central nervous system preparations. They concluded that SCPs were involved in feeding related changes in heart activity and that neuron B1 was involved in the control of this effect.

Grega and Prior (1986) examined the response to dehydration of cardiac activity in *L. maximus*. The responses of both intact slugs and isolated heart–central nervous system preparations to increased blood osmolality were investigated. In intact slugs, heart rate increased in response to progressive dehydration in air and to increases in blood osmolality due to injections of hyperosmotic solutions. In isolated preparations, exposure of the heart alone to hyperosmotic saline caused a decrease in heart rate, while exposure of only the central nervous system resulted in an increase in heart rate. Grega and Prior (1986) concluded that the increase in heart rate observed in intact air-dehydrated slugs was primarily mediated by the central nervous system.

7

Behaviour

7.1 FEEDING BEHAVIOUR

7.1.1 Range of foods

Most slugs are omnivorous, Taylor (1902–1907) for example, described how
M. gagates rejected only 22 out of 195 different foods offered to it while *A.
ater* refused 39 out of 197 foods. Frömming (1954) provided a great deal of
information about the range of plants accepted by slugs and their preferences
and described *A. ater* as a polyphagous species. When *A. hortensis*, *A.
circumscriptus* and *A. intermedius* were offered leaves and flowers from 40
wild plants, each species attacked at least 90% of the plants with *A. hortensis*
feeding on a part of every plant offered. The food preferences of each of the
Arion species were generally similar. A rather different result was obtained
when four species of *Limax* were offered material from 29 ornamental plants.
While *L. maximus* and *L. marginata* attacked over two-thirds of the plants,
L. flavus and *L. tenellus* were more restricted in their feeding behaviour.
Limax flavus fed only on about a third of the plants offered and restricted
itself to the bulb or stem in most instances while *L. tenellus* caused slight
damage to only four of the plant species offered. Taylor concluded that
L. flavus did not generally eat green plant material and that *L. tenellus*, a
rather local species living in old woodland, fed almost entirely on fungi.

  Getz (1959) offered leaves of 45 plant species to *D. laeve*, *D. reticulatum*
and *A. circumscriptus*. *Deroceras laeve* showed a more varied choice of food
(18 selected out of 45 species) than either *D. reticulatum* or *A. circumscriptus*
(10 selected out of 45 species). The range of foods eaten under natural
conditions vary widely within a single species, depending on the food
available to the slug. Neijzing and Zeven (1976) described how *D. laeve* fed
on the anthers of *Streptocarpus* flowers. Pallant (1969, 1972) found, from an
examination of crop contents and faeces, that the diet of *D. reticulatum*
consisted mainly of fresh plant material, the dicotyledons *Ranunculus
repens* Linnaeus and *Urtica dioica* Linnaeus from deciduous woodland and
grasses, particularly *Holcus lanatus* Linnaeus, from rough grassland. An
examination of gut contents of slugs from arable ground by Hunter (1968c)
showed that the surface living *D. reticulatum* had a greater tendency to
feed on green plant material than the soil-dwelling species, *A. hortensis*
and *T. budapestensis* and Airey (1986) demonstrated these preferences

experimentally. Lutman (1978) found that *D. reticulatum* tended to eat fresh plant material while other surface-living slugs on rough grassland took senescent or dead material. However, Jennings and Barkham (1975b) concluded from faecal analyses that the diet of all eight species of slug examined from deciduous woodland, including *D. reticulatum*, consisted largely of dead higher plant material or the underground parts of plants. Seasonal changes in diet probably reflect changes in the availability of food. Jennings and Barkham suggested that an increase in the amount of fungal material, consumed by five species during the autumn, was the result of a seasonal increase in availability of fungal food combined with mild damp nights that encouraged feeding above ground.

Slugs have long been associated with damage to the fruiting bodies of the larger fungi. Taylor (1902–1907) suggested that fungi were the main food of the larger *Limax* spp. while Oldham (1922) described *L. tenellus* as feeding solely on fungi. Slugs feed on a wide range of basidiomycetes, including those poisonous to man, and are particularly fond of the death-cap, *Amanita phalloides* Secretan (Ramsbottom, 1953). The cap of the related *Amanita muscaria* Persoon is frequently damaged by slugs yet Richter (1980b) found that slugs feeding on immature fruiting bodies of this species died and suggested that toxic alkaloids, intended to deter grazing, might be responsible. Frömming (1954) compared the acceptability of 23 basidio-mycete fungi to *L. maximus* and *D. reticulatum* and found that, while *L. maximus* fed on all the fungus species offered, *D. reticulatum* rejected over a third of them and consumed relatively less of the fungal food than *L. maximums*. Although *D. reticulatum* refused *Polyporus squamosus* Fries, Taylor (1902–1907) reported that *D. laeve* fed on this species. Duthoit (1964) found that six slug species consumed both the puff-ball *Lycoperdon* sp. and wheat seeds when offered a choice between the two. The larger ascomycetes are also eaten and Taylor described how *L. cinereoniger* fed on *Peziza* spp. and *Morchella esculenta* Linnaeus. Webb (1988) found that although *T. sowerbyi*, *T. budapestensis*, *L. flavus* and *L. maximus* readily ate *Auricularia auricula-judae* (Fries), *A. hortensis* would accept this fungus only in the absence of alternative food and that *D. reticulatum* declined to feed on it. Preference tests, using *A. ater*, *A. hortensis* and *A. fasciatus*, showed that more plant tissue infected by rust fungi (*Puccinia* spp. and *Coelosporium* sp.) was consumed when compared with healthy tissue (Ramsell and Paul, 1990), although *D. reticulatum* showed no preference for infected plant tissue.

Slugs and snails feed less frequently on mosses and liverworts although they feed readily on lichens (Chatfield, 1976b). Pallant (1969, 1972) found moss leaves in crop contents and faeces of *D. reticulatum* from both grassland and woodland, while Jennings and Barkham (1975b) showed that three species of moss were acceptable to six species of woodland slug, although they were less palatable than many angiosperms. Although parts of mosses are eaten, particularly the immature capsule and the protenema, these plants are less attractive to slugs than higher plants. This is partly because of the physical barrier presented by the cell wall, and partly because phenolic compounds are present, particularly in older leaves, and act as feeding deterrents (Davidson and Longton, 1987; Davidson *et al.*, 1989). The palatability of 16 species of lichen to three slugs, *D. laeve*, *L. maximus* and *L. flavus*, and two snails was compared by Frömming (1954). *Limax flavus* refused all the lichens, *L. maximus* fed on seven species while *D. laeve* accepted 11 species, including those species eaten by *L. maximus*. The growth form of the lichen did not appear to influence its acceptability to slugs. Frömming showed that *A. subfuscus*, *A. circumscriptus* and *A. intermedius* would feed on the apothecium of *Cladonia* spp. but only in the absence of alternative foods. *Arion fasciatus*, *A. hortensis* and *A. intermedius* also fed readily on the horsetail *Equisetum arvense* Linnaeus. There are few references to slugs feeding on ferns although Frömming (1954) included several pictures of ornamental ferns damaged by *D. reticulatum*, *D. laeve* and *A. intermedius*. Jennings and Barkham (1975b) found that bracken, *Pteridium aquilinum* (Linnaeus), was eaten by several species of slug, particularly *A. ater* and Richter (1979) stated that this fern formed an important part of the diet of *A. columbianus* in coniferous forests in the USA. Richter also found that *A. columbianus* included a gymnosperm, *Pseudotsuga menziesii* (Mirbel) in its diet while Tan (1987) reported slug damage to seedlings of the fir *Abies alba* Mill. However, Harper (1988) considered that *A. columbianus* would starve rather than feed on *Sequoia sempervirens* (Lamb) seedlings.

According to Taylor, in addition to plant material, slugs will also feed on carrion and even living invertebrates, including snails and insects. The prey of slugs has included cecidomyiid galls (Leatherdale, 1955) and mealy bugs (Coccoidea) eaten by *D. laeve* (Quick, 1951). Mienis (1989) also recorded this slug feeding on wax scale insects (Coccoidea) in Israel. When slugs are confined with other slugs, in the absence of food, they may resort to cannibalism (Taylor, 1902–1907) and Chatfield (1976b) described how *A. ater* fed readily on freshly killed slugs and even on narcotized slugs, including *L. flavus*. Dirzo (1980) observed cannibalism in *A. ater* and *A. subfuscus* in laboratory cultures. However, Landauer and Cardullo (1983) showed that *L. flavus* avoided lettuce treated with an aqueous extract of conspecifics, and Niemelä *et al.* (1988) demonstrated a similar feeding aversion in laboratory feeding trials. In these trials, *D. agreste* preferred untreated lettuce to lettuce treated with a series of aqueous extracts prepared from the slugs *D. agreste*, *L. cinereoniger*, *Arion* sp. and the snail *Bradybaena fruticum* (Müller). This feeding aversion was most marked for extracts of *Deroceras* and the closely related *Limax* and was weaker for the taxonomically distant genera *Arion* and *Bradybaena*.

Although some earthworm chaetae and arthropod remains found in the crop contents of *D. reticulatum* by Pallant (1972) may have been fortuitously ingested with plant material, some of this material was probably obtained by feeding on dead invertebrates. Lovatt and Black (1920) found that *D. reticulatum* readily fed on injured and dead earthworms. An increase in the proportion of earthworm and arthropod remains eaten by five slug species (*A. ater, A. fasciatus, A. hortensis, A. intermedius* and *A. subfuscus*) from deciduous woodland in late spring was noted by Jennings and Barkham (1975b). Pitchford (1956) recorded *D. reticulatum* feeding on the flesh of a dead rabbit while *A. ater* will feed on dead hedgehogs and birds killed by cars (Newell, 1971). Tischler (1976) found that in rainy weather *A. ater* was attracted to carrion by smell. According to Taylor (1902–1907), slugs are also attracted to bones where they extract the gelatine and some slug species, especially *A. ater*, are coprophagus. Lutman (1978) confirmed that *A. ater* fed on sheep faeces. Cabaret and Vendroux (1986) tested the response of four terrestrial gastropods, including *A. ater*, to herbivore faeces and found that fresh faeces were not attractive while the faeces of sheep and goats were less repulsive than those of cattle or horses. *Arion ater* will also feed on dog faeces (Weiss, 1981). Mienis (1988) listed several instances where *L. flavus* and *L. maximus* were regularly attracted to milk.

The diet of *G. maculosus* was initially considered to be restricted to lichens and liverworts under natural conditions (Taylor, 1902–1907) but according to Platts and Speight (1988) this slug feeds on a wide range of lichens, mosses, liverworts and fungi and in captivity will eat most vegetable material. Quick (1960) stated that in captivity *G. maculosus* had been observed to attack small snails. Slugs of the genus *Testacella* are entirely carnivorous and Taylor listed earthworms, slugs, snails and centipedes among their food although he stated that they would not feed on dead animal material. *Testacella maugei* readily fed on other *Testacella* spp. in the absence of alternative food and Stubbs (1934) recorded how a *T. maugei* ate the tail end of a *T. sowerbyi* in captivity. Stokes (1958) described feeding behaviour in *T. scutulum*. Slugs fed both on the soil surface or below it, worms being seized by either end or in the middle. Feeding was spasmodic and the time taken to eat a worm varied from less than an hour to several days, the worm being either partly or entirely consumed. Several days usually elapsed after a meal before another worm was captured. If slugs were disturbed during a meal they often regurgitated their prey. Taylor described how when a *Testacella* seized an earthworm emerging from its burrow at the soil surface, the worm would retract into its burrow dragging the slug with it. The latter was able to attenuate its body sufficiently to enable it to retain its hold on the prey.

7.1.2 Food choice

It is clear that, with the exception of the carnivorous species such as

Testacella spp., slugs will consume a wide range of foods and Cameron (1939) described them as indiscriminate feeders with a low level of sensory response. However, subsequent investigations have shown that slugs are capable of a high degree of selection in their feeding behaviour. The technique of presenting slugs with whole plants or parts of plants, either separately or in a choice situation, has formed the basis of several investigations into food preferences. Getz (1959) offered leaves of 45 different plants separately to *A. circumscriptus, D. reticulatum* and *D. laeve* and found that about 28% were very acceptable, 25% were relatively acceptable and 47% were rejected. Duval (1971) demonstrated that many common agricultural weeds were acceptable to *D. reticulatum* by offering leaves of 26 species separately to slugs, and was able to rank their acceptability to the slug into four categories: eaten readily 39%, eaten less readily 19%, eaten occasionally 15% and virtually rejected 27%. Two grasses included in the test were eaten readily although they were chosen for their coarse leaves. When this experiment was repeated for *A. hortensis* the corresponding values for the above categories were 15, 29, 19 and 37% (Duval, 1973). The extent of feeding by slugs on single leaf discs was estimated by Jennings and Barkham (1975b) using a scale based on the proportion of leaf consumed by the slug. Three categories of woodland plant material were used, i.e. fresh tree leaves, tree leaf litter and leaves from the ground flora. *Arion intermedius, A. hortensis* and *A. fasciatus*, species which normally remain near the ground to feed, had different food preferences and accepted less green plant material than *A. ater, A. subfuscus* and *D. reticulatum*, which climb food plants readily when conditions are favourable. There was wide variation in the acceptability of leaves of different trees and herbaceous plants to slugs although bryophytes all gave low acceptability scores.

Pallant (1969) offered *D. reticulatum* a choice of leaves of two or more woodland plant species simultaneously and estimated the amount of feeding on a four-point scale. The most readily accepted foods, *R. repens* and *U. dioica*, were also the most frequently recorded plants in the gut contents of this slug from woodland. Duthoit (1964) offered wheat, barley and oat seeds to five species of slugs, both as single foods and in a choice situation. The damage to oat grains by each species was negligible while *D. reticulatum* showed a marked preference for wheat grains compared with barley, whether or not presented with a choice. The barley grain is enclosed within a husk while the wheat grain is naked, and Port and Port (1986) reported that removal of the barley husk made the seed more palatable to *D. reticulatum*. Port and Port also stated that, while there was no evidence for feeding preferences between the cultivars of bread wheat (*Triticum aestivum* Linnaeus), differences have been demonstrated between four species of *Triticum* in favour of species with a high total-nitrogen content. Dirzo (1980) offered a choice between mature leaf discs of test plants and leaf discs of lettuce, used as a reference material, to rank the acceptability of 30 plant species to *D. caruanae*. He used an acceptability index based on the quotient of leaf area of test material consumed/leaf area of lettuce consumed to compare the acceptability of the different plants to the slug. About 43% of plant species tested, including the grass *Poa annua* Linnaeus, were rejected by the slugs. The acceptability index was generally high for annual plants, for species with soft thin leaves and for those not known to contain secondary compounds while hairiness did not appear to influence acceptability. The use of a relative acceptability index, similar to that of Dirzo (1980), and an absolute acceptability index, based on the mean area of test plant leaf consumed, was compared by Rathcke (1985) in an investigation into food choice by three species of slugs. Both indices showed similar statistical trends and similar rank orders. In experiments comparing the quantity of test plant material consumed by *D. reticulatum* with the quantity of reference plant consumed in choice situations, Richardson and Whittaker (1982) demonstrated that the resulting rank order of acceptability is altered when different reference species are used. A similar result was obtained when consumption was measured as dry weight rather than leaf area consumed.

Mølgaard (1986a) discussed the use of indices to measure the palatability of plants to slugs and described both an eating index and, derived from this, a preference value which allowed for the time taken by the observations. The feeding behaviour of *D. caruanae* varies according to the plant species offered to the slug (Dirzo and Harper, 1980). Slugs ate whole, or parts of, leaves of a preferred species, *Capsella bursa-pastoris* (Linnaeus), and rarely killed the plants. When starving, they fed on the rejected species *P. annua* but chewed through the young shoots at ground level, cutting them down but not consuming the shoots and many of the damaged plants of *Poa* died. Dirzo (1980) suggested that silica deposits in the leaf margins of *P. annua* hinder access to the tissues of this grass.

According to Wadham and Wynn Parry (1981), rice, *Oryza sativa* Linnaeus, is probably the heaviest accumulator of silicon among the grasses (Gramineae) and they showed experimentally that the grazing of rice leaves by *D. reticulatum* could be reduced by an increase in the level of silicon accumulation in the plant. Another example of a mechanical deterrent to feeding by slugs and snails was the presence of raphides in the leaves of *Arum maculatum* Linnaeus and *Endymion nonscriptus* (Linnaeus) (Knight, 1964). Latex, a viscous, milky secretion produced by laticiferous plants, acts as an effective feeding deterrent to many herbivores, including slugs, both chemically, by toxins, and mechanically, by coagulating on exposure to air and thus hindering feeding (Dussourd and Eisner, 1987). A number of specialist insects feed on laticiferous plants, by draining the latex through cuts in leaf veins before feeding. Dussourd and Eisner showed that *D. reticulatum* readily fed on regions of the leaves of a laticiferous plant, *Asclepias syriaca*, that had been drained of latex by artificial vein cutting.

Byers and Bierlein (1982) offered three species of slugs, *D. laeve, D. reti-culatum* and *A. fasciatus*, choices between growing seedlings of different ages of cultivars of three legumes, red clover, alfalfa and birdsfoot trefoil. They found that the youngest seedlings at the cotyledon stage were preferred to the later stages and that cultivars of birdsfoot trefoil were least preferred by slugs. Horrill and Richards (1986) provided an explanation for this preference for younger seedlings and this is discussed below. Yamashita *et al.* (1979) reported that a species of *Deroceras* was a serious pest of Kenya white clover (*Trifolium semipilosum* Fres.) cv. Safari in Queensland. In laboratory acceptability tests, they showed that this legume was readily accepted by both *Deroceras* sp. and *Lehmannia nyctelia* Bourguignat. However, these slugs refused 14 species of tropical grasses that were also offered to them although temperate pasture grasses are eaten by slugs (e.g. Pallant, 1972). Field observations by Yamashita *et al.* showed that *T. semipilosum* was preferentially grazed compared with the naturalized white clover, *Trifolium repens* Linnaeus, and it was suggested that this difference was due to the lower hydrocyanic acid content of the former. Both *T. repens* and *Lotus corniculatus* Linnaeus (birdsfoot trefoil) are polymorphic for the presence or absence of cyanogenic β-glucosides. Those plants, which contain both cyanogenic glucosides and the appropriate β-glucosidase enzyme capable of hydrolysing the glucosides and releasing hydrocyanic acid gas after mechanical damage to the stems or leaves, are termed cyanogenic, while those not doing so are acyanogenic. Jones (1962) found that in wild populations of *L. corniculatus*, acyanogenic plants were selected by *D. reticulatum* and demonstrated experimentally that *A. ater* fed selectively on acyanogenic *L. corniculatus*. However, when the acyanogenic plant had been consumed then the slug would begin to eat the cyanogenic plant. Jones (1962) suggested that an advantage was conferred upon cyanogenic *L. corniculatus* because the evolution of hydrocyanic acid gas deterred herbivores such as slugs. The interaction between the polymorphism of cyanogenesis in *L. corniculatus* and *T. repens* and various animals, including slugs, was reviewed by Jones (1973a) and he listed *A ater, A. hortensis, A. subfuscus* and *D. reticulatum* as slugs which differentially feed on acyanogenic forms of *L. corniculatus*.

The evidence that cyanogenesis in *T. repens* acts as a defence against herbivores is conflicting. Bishop and Korn (1969) failed to find evidence for selective feeding on acyanogenic *T. repens* by *D. reticulatum* or the snail *H. aspersa* while Crawford-Sidebotham (1972a) showed that differential feeding by six slug and seven snail species on acyanogenic rather than cyanogenic forms was more marked in *L. corniculatus* than in *T. repens* and that the different species of gastropod did not behave in the same way. Differences in the behaviour of several slug species were demonstrated by choice chamber experiments with cyanogenic and acyanogenic *T. repens*

(Angseesing, 1974; Angseesing and Angseesing, 1973). *Arion ater* became more selective against cyanogenic morphs with increased time and also showed decreased selectivity when its food sources were changed, while *D. reticulatum* showed some degree of selectivity, but *A. subfuscus* was not selective. When differences in the acceptability of individual plants of *T. repens* to the slugs *D. caruanae, D. reticulatum, A. ater* and the snail *H. aspersa* were investigated by Dirzo and Harper (1982a, b), they found that cyanogenesis markedly reduced but did not totally prevent damage to the plants. This effect was emphasized in the field where, in areas of high density of slugs and snails, cyanogenic morphs were over-represented, while in areas of low mollusc densities, acyanogenic morphs were over-represented. However, Dirzo and Harper (1982b) emphasized that mollusc grazing was only one of the several forces affecting the cyanogenesis polymorphism in *T. repens*. Knight *et al.* (1978) found that damage by *Limax* spp. was higher in acyanogenic than in cyanogenic morphs in field plantings of *T. repens* in Virginia, USA. Crawford-Sidebotham (1972a) suggested that the selection of cyanogenic plants under natural conditions was likely to operate on seedlings rather than mature plants, because only a small degree of damage would greatly reduce the fitness of a seedling, compared with that of a plant. This effect was demonstrated by Horrill and Richards (1986) who found that the amount of releasable cyanide in cyanogenic seedlings of *T. repens* increased from 5 days to 35 days of age. Over this period, levels of cyanide decreased in cotyledons but increased in stems and young leaves. *Arion hortensis*, introduced into mixed populations consisting of either mainly cyanogenic and or mainly acyanogenic seedlings of *T. repens*, did not discriminate between morphs with respect to grazing and no difference in the level of damage was detected at 5 days' age. However, for older seedlings, very few grazed cyanogenic seedlings suffered lethal damage while most grazed acyanogenic seedlings were lethally damaged. Horrill and Richards concluded that older seedlings of cyanogenic *T. repens* were protected from lethal grazing by slugs and that this could influence the frequency of cyanogenic morphs in populations more than discriminatory grazing on mature plants. They also showed that *A. hortensis* was unable to differentiate between cyanogenic and acyanogenic seedlings until it grazed the plant tissue, when mechanical damage by grazing released hydrocyanic acid.

Experiments by Burgess and Emos (1987) demonstrated that, while all six populations of *D. reticulatum* from permanent pasture sites containing white clover showed preferential grazing of acyanogenic morphs of *T. repens* the degree of preference varied significantly among populations. Slugs from sites with a low frequency of cyanogenic morphs showed a greater degree of selective eating of acyanogenic morphs than slugs taken from a site having a high frequency of cyanogenic morphs. Burgess and Emos suggested that these differences in selectivity were caused by differences in the rate of initiation of feeding on cyanogenic morphs and by differences in the extent of damage once feeding had been initiated. The degree of preferential grazing shown by *D. reticulatum* might be influenced by the frequency of encounters with cyanogenic clover morphs. As this frequency increased, slugs capable of incorporating a higher proportion of cyanogenic clover in their diet would be favoured since a greater proportion of food available could be utilized, thus reducing search time. Slugs taken from populations at sites with a high frequency of cyanogenic clover might, therefore, show a reduced preference for acyanogenic morphs of *T. repens*. The hypothesis that relatively unacceptable plants are more acceptable to slugs from the same habitat than to slugs from other populations not normally exposed to these plants, was rejected by Whelan (1982a) on the basis of choice tests using *A. subfuscus* and *D. caruanae*.

7.1.3 Design of feeding assays

Investigations into the basis of food selection by slugs have involved either an assessment of their food preferences or the estimation of the degree of acceptability of a particular plant measured by some kind of acceptability index. Techniques involving the presentation of slugs with whole plants or parts of plants have already been discussed (e.g. Pallant, 1969; Dirzo, 1980). However, it is often difficult to determine which leaf or stem characteristics influence the rejection or acceptance of potential food items; for example physical structures such as hairs or hardness, may mask the effects of secondary plant chemicals and other substances. These characteristics may be unevenly distributed even within a single plant, showing for example differences between young and old leaves. Thus feeding experiments may be confounded by the different responses of slugs to conflicting stimuli. Several methods have been devised for testing the effect of both plant extracts and individual chemicals on feeding behaviour in slugs. These have usually involved presenting the substance on or in an inert material. Stephenson (in Johnson, 1968) presented slugs with chemicals and potato extracts incorporated into filter and chromatography papers. However, Frain (1981) found that feeding activity on treated paper was too low, that measurement of the area eaten was difficult and that the paper often absorbed mucus. Whelan (1982a) used a calcium alginate gel, which was insoluble in water, to present plant extracts to slugs. The gel was formulated to be attractive to slugs by including wheat germ and bran. This gel had the advantage that it could be prepared without excessive heating, which might denature extracts incorporated into it. However, Frain (1981) found that this type of gel was unattractive to slugs.

A number of investigations have involved the use of agar discs in feeding assays with slugs as this acts as an inert carrier with no attractiveness of its own while extracts or chemicals under investigation can be incorporated into it (e.g. Senseman, 1977; Frain, 1981; Storey, 1985). Atkin (1979) used a base comprising equal parts of agar and cellulose. However, agar has the disadvantage that it is a warm setting gel and so volatile and heat degradable substances are more difficult to incorporate into it although these can often be added as the gel cools. The hardness of agar-based foods could be changed by adjusting the agar concentration although there was little variation in hardness within a batch of agar (Senseman, 1978). Storey (1985) found that the form in which the agar discs were presented to slugs influenced feeding behaviour and factors such as disc thickness, the presence of free water and moisture content of discs needed to be taken into account. Frain (1981) found that the water content of agar discs varied during experiments, making direct weight measurements of food eaten unreliable. She scored each disc on a scale of 0–5 and subsequently converted scores to weight losses and obtained three measures of feeding activity: feeding score (average amount of food consumed by slugs), percent response (percentage of slugs which fed on a disc of a particular food) and the ratio comparing consumption of one food item with another. Whelan (1982a), using a choice test, compared gels containing test plant extract, with a control gel containing lettuce extract. Results were expressed as an acceptability index which compared the ratio of area of test gel eaten to the total area of gels eaten in each experiment, the index ranging from 0 (no test gel eaten) to 1 (only test gel eaten). Storey (1985) compared three methods of assessing feeding: counting the number of discs damaged, allocating a score of between 1 and 10 to the percentage of each disc eaten and weighing each disc before and after the experiment. She found the scoring method to be most suitable whilst the weight change of individual discs was subject to considerable error. Frain (1981) showed that, under experimental conditions, slugs habituate to a monotonous diet so that it becomes gradually less attractive and this needs to be taken into account in the design of feeding assays.

7.1.4 Food selection

It is clear that slugs are not indiscriminate feeders but actively select certain foods in preference to others. The basis of food selection can be broadly defined by two concepts, the presence of inhibiting substances and the presence of feeding stimulants, although the two are not necessarily mutually exclusive. Lawrey (1983) tested two hypotheses which might explain the grazing preferences of the slug *Pallifera varia* Hubricht for particular lichen species. The preference hypothesis, that slugs select lichens of highest food quality, was tested by measuring the concentration of essential elements in lichens eliciting high and low preference by slugs. The avoidance hypothesis, that preference by slugs actually results from the rejection of unpalatable lichens, was tested by offering slugs choices of paper discs impregnated with secondary products of unpalatable lichens. Lawrey found that preferred lichens had significantly lower concentrations of essential elements and that avoided lichens produced secondary compounds that inhibited grazing by the slug. He concluded that the avoidance hypothesis provided the better explanation for the grazing behaviour of *P. varia*, with choice by the slug being simply the result of avoidance of those lichen species that produce inhibitory substances and that lichens with the highest concentrations of essential elements were most likely to produce defensive compounds.

Plants contain a wide variety of secondary chemicals including nitrogen compounds such as alkaloids and glycosides, terpenoids, phenolics including tanins, and others. The role of these substances has been extensively investigated but their function is difficult to define. Rhoades and Cates (1976) considered it unlikely that secondary substances have a role in the normal metabolic processes of plants or that they are of nutritional significance to herbivores in the same way as primary substances. A defensive role against herbivores has been generally ascribed to these secondary substances, and the interaction between slugs and cyanogenesis in *L. corniculatus* (Jones, 1962) was one of the first examples of the defensive role of secondary substances in plants against feeding by molluscs. Rhoades and Cates (1976) classified secondary substances formed by plants into two groups: those which are toxic and act on the metabolic processes of the herbivore and those which reduce the availability of plant nutrients by interfering with digestion and whose effect is quantitative, i.e. dependent upon concentration. Some secondary substances may act in both ways. Harbourne (1977) confirmed that secondary compounds produced by plants may have a defensive role and stated that synergism sometimes occurred when mixtures of closely related compounds were produced by plants.

Although in excess of ten thousand substances are involved in plant–animal interactions (Harborne, 1977), relatively few of these, apart from the cyanogenic glycosides described previously, have been examined in relation to slugs. Terpenes, which occur mainly in the leaves of aromatic plants, are released when the leaf is damaged. Rice *et al.* (1978) found that the palatability of leaves of the mint, *Satureja douglassii*, to *Ariolimax dolichophallus* was affected by the monoterpenoid composition of the plant. Gouyon *et al.* (1983) compared the acceptability of different monoterpene compositions of *Thymus vulgaris* Linnaeus plants to *D. reticulatum* in a choice situation. They showed that the different terpenoid compositions provided very different defences against the slugs and that the behaviour of individual animals was determined by several factors including the past experience of the slug. The monoterpenoid alcohol, geraniol, from leaves of the scented geranium (*Pelargonium graveolens* Linnaeus), showed antifeedant activity when tested with slugs (Airey *et al.*, 1989). Glucosinolates in seeds of the oilseed rape (*Brassica* sp.) serve to protect seedlings from slug damage (Glen *et al.*, 1990), while a metabolite of the glucosinolate gluconasturtin, present in root extracts from the horseradish (*Armoracia rusticana* Gaertn.), protects this plant against slug attack (Airey et al., 1989).

Mølgaard (1986a) suggested that there was a general pattern in food plant selection by land molluscs. Plants generally avoided were those containing condensed tannins and polyphenolics, such as caffeic acid in *Plantago* spp., while those containing potent constituents against mammals were often readily selected by molluscs. Two subspecies of *Plantago major* Linnaeus, ssp. *major* and ssp. *pleiosperma*, occur in Denmark (Mølgaard, 1986b). The caffeic acid ester with glucose occurs in leaves of both subspecies, while the ester with rhamnose occurs only in ssp. *pleiosperma*. Both *A. ater* and *D. reticulatum* showed a higher preference for ssp. *major* leaves than for ssp. *pleiosperma* in choice tests, indicating that the rhamnose ester was the more efficient feeding deterrent. An investigation of frequency-dependent grazing by *D. reticulatum* (Cottam, 1985), demonstrated that this slug showed a strong preference for the white clover, *T. repens*, over the grass, *Dactylis glomerata* Linnaeus in different combinations of the two species. *Dactylis* is unpalatable to moluscs (Grime *et al.*, 1968) probably due to the silica content of the leaves (Wadham and Wynn Parry, 1981). However, despite these preferences, the slug also grazed anti-apostatically on both species, i.e. selecting those which were less common. This may simply be the result of slugs sampling the available foods, even those which are less palatable, or, it may be that slugs require a varied food intake (Frain and Newell, 1982) so that uncommon and often unpalatable foods may be eaten.

The term coevolution has been used to describe the interdependent interactions which evolve between plants and animals. If a plant develops a secondary chemical that deters the majority of herbivores, any herbivore which develops a physiological means of coping with the deterrent can obtain an uncontested food supply. Although slugs are generalized feeders, Dirzo (1980) described them as 'acceptability-moderated generalists' with a hierarchy of accepted foods and he considered that their coevolutionary specialization was at an early stage of development although Dirzo and Harper (1982b) suggested that the mollusc–white clover interaction represented a possible coevolutionary pathway. It is clear, from the examples given here, that plant species which have evolved polymorphisms for selected secondary compounds (eg. Jones, 1962; Mølgaard, 1986b) are afforded some protection from grazing by slugs. However, this protection is achieved at the expense of the energy resources of the plant, putting it at a competitive disadvantage in some habitats. Populations of wild ginger, *Asarum caudatum* Lindl., in western Washington are polymorphic for growth rate, seed production and palatability to the slug *A. columbianus* (Cates, 1975). In habitats where slugs were not as abundant, Cates found that individuals allocating more energy to growth rate and seed production and less to the production of antiherbivore compounds were more abundant.

He suggested that, in the absence of grazing, palatable and faster growing plants would be at a competitive advantage over slower growing protected individuals, although the fitness of the latter would be increased under grazing pressure. Rai and Tripathi (1985) demonstrated how preferential feeding by the slug, *Mariaella dussumieri* Gray, and some insects was able to reduce the competitive ability of one annual species of *Galinsoga* against another species of the same genus. The differential acceptability of plants to slugs was considered by Crawford-Sidebotham (1971) to be an important determinant in the dynamics of some plant communities.

Slugs have been used as examples of generalist feeders to test predictions based on theoretical models of plant–herbivore coevolution (e.g. Cates and Orians, 1975). Early successional plant species tend to be less available to herbivores both in time and space, and Cates and Orians suggested that these plants would make a smaller commitment to antiherbivore defence compounds and would be more attractive to herbivores. They demonstrated this by means of choice tests involving 100 plant species and using the slugs *A. columbianus* and *A. ater*. Early successional annuals were more palatable than early successional perennials which, in turn, were more palatable than later successional species. A similar conclusion was reached by Reader and Southwood (1981) using five generalized herbivores including *D. reticulatum*. Cates and Orians also found no correlation between palatability and evolutionary association of the slugs with the plant species, thus confirming the conclusion of Dirzo (1980) that their coevolutionary specialization was at an early stage of development. This conclusion was also supported in choice tests made with the slugs *A. subfuscus*, *A. fasciatus* and *D. reticulatum* by Rathcke (1985). Rathcke rejected the coevolution hypothesis, that plant choices by slugs would tend to be unique for each species, as it coevolves with local plants, because the feeding responses of the three slug species examined were similar. However, Rathcke's results supported the herbivore-adaptation hypothesis, that feeding choices are determined by adaptations of generalist herbivores to plants available in their local habitats, and there was some support for the plant-defence hypothesis, that plants readily available to herbivores have a greater defence commitment and are not as acceptable as less available plants. Observations of several plant species by Maiorana (1981) suggested that plants incur greater herbivore damage in shaded than in nearby sunny areas. However, choice tests presenting leaves of sun and shade plants to several herbivores including a slug, *Arion sp.*, were inconclusive, with the slug showing a preference for shade leaves of some species and sun leaves of other species.

Although the avoidance hypothesis, that food selection is based largely on the presence of inhibiting substances, is strongly supported by the evidence reviewed here, feeding stimulants undoubtedly play some part in food selection by slugs. Hsiao (1972) considered that primary plant chemicals common to all plants, such as amino acids and sugars, are phagostimulants which serve to initiate feeding in insects, and Stephenson (1979) found that this was true for *D. reticulatum*. Senseman (1977) showed that starch was a potent feeding stimulant for the slug *A. californicus* and that slugs could distinguish between foods on the basis of their starch concentrations, preferring higher levels of starch. He suggested that the stimulatory effect of the starch might be due to the action of salivary enzymes hydrolysing starch into low-molecular weight sugars such as maltose, glucose and maltotriose and demonstrated that these sugars had the ability to stimulate feeding although at a more reduced level than starch. Senseman also considered that casein (milk protein) was as effective as starch as a feeding stimulant. Although the amino acid proline acts as a feeding stimulant for snails, Bright *et al.* (1982) found no evidence that a barley mutant, having very high proline levels in the leaves, was more susceptible than other cultivars to several pests including *D. reticulatum*.

Taylor (1902–1907) described how *A. hortensis*, *D. reticulatum* and *T. sowerbyi* were strongly attracted to beer placed in shallow saucers on the ground. Beer has long been known to attract slugs and Smith and Boswell (1970) reviewed its use by gardeners to control slugs. They reported that both *D. reticulatum* and *S. plebeia* Fisher were strongly attracted to fresh beer. Further tests, using only *D. reticulatum*, showed that, while stale beer, unfermented grape juice and *Drosophila* fermented bait were attractants, wine, ethanol, methanol, dimalt, vinegar and water had little or no attraction for slugs. Both *A. hortensis* and D. reticulatum preferred dark beers to pale beers and lagers (Thomas, 1988) while unfiltered beers were preferred to filtered beers of the same type, suggesting that yeast was an attractant to slugs. Anon (1990) described how slugs preferred alcohol-free lager to a range of alcoholic beverages. Selim (1979) attempted to isolate the attractive components of beer and suggested three possible attractants: acetoin, dihydroxyacetone and diacetyl, although complete beer remained more attractive than these chemicals alone. Selim also found that the attractiveness of beer was largely removed by treating beer with reagents known to react selectively with aldehydes and ketones. Frain (1981) attempted to identify the attractant component of beer by isolating samples of beer and its by-products at each stage in the brewing process and testing their attractiveness to slugs. She concluded that the attractiveness of beer to slugs lay in the sediment, for the unrefined fermentation product (pressed yeast) proved to be more attractive than the beer supernatant. Pressed yeast is a mixture of yeast cells and wort which is pressed out of the vats after fermentation. Frain also suggested that beer may lose some of its attractiveness during the refining process as by-products formed at a later stage in the fermenting process were less palatable to slugs than the pressed yeast. Thus spoilage chemicals formed might mask the attractive substances which stimulate feeding. The chemical diacetyl, suggested by Selim as a possible attractant component of beer, is one of many chemicals which are regarded as spoilage substances in beers and Frain found that diacetyl was not an attractant to slugs.

7.1.5 Varietal susceptibility of potatoes to slug damage

Thomas (1947) first reported that tubers of different potato cultivars varied in their susceptibility to attack by slugs. This was confirmed in the Terrington NIAB trials (Gould, 1965; Winfield et al., 1967) and by other field trials in different regions (Pinder, 1969; Warley, 1970; Atkin, 1979; South, 1973a; Airey, 1986a). Table 11.1 compares damage to different cultivars of potato from 1947 onwards. Stephenson and Bardner (1976) summarized the NIAB trial data for varietal susceptibility in potatoes using a 'susceptibility index' for the leading cultivars. This index for any particular cultivar was calculated as the ratio of the percentage of tubers of that cultivar damaged compared with the percentage of tubers of the cv. Majestic damaged. The cv. Majestic was chosen as a standard because it is less susceptible to slug attack. The most resistant cv. Stormont Enterprise had a susceptibility index of 0.232, cv. Majestic had an index of 1 and the most susceptible cultivar, Maris Piper, had an index of 4.02, showing that there were about 17 times as many slug-damaged tubers of Maris Piper compared to Stormont Enterprise.

Field trials of 15 potato cultivars (South, 1973a) showed that first early cultivars were generally more resistant to slug attack and that this persisted even when they were grown as maincrop potatoes. The first early cv. Epicure was an exception to this, damage to this potato being more than six times as great compared to the least susceptible cultivar, Sharpes Express, another first early cultivar. Thus the normal earlier lifting date for first early cultivars does not completely explain their resistance to slug damage and a feeding preference by slugs for certain cultivars is indicated. Damage to maincrop potatoes is generally light during the early summer but increases rapidly from August onwards. The reasons for this are considered in Chapter 11 but Thomas (1947) reported that little damage occurred until the potato tuber was mature and this view was supported by Pinder (1969), Warley (1970) and Wareing (1982). Pinder showed that tubers of earlier planted cv. Majestic received significantly more damage than tubers from seed planted a month earlier and thus maturing later. However, Airey (1986a) found that tuber damage began early in the summer although extracts from tuber skins became less unacceptable as the season progressed and slugs attacked larger (older) tubers in preference to smaller (newer) tubers. Atkin (1979) showed that differences were present between cultivars regardless of their state of development or the lifting date.

Stephenson (in Johnson, 1965) suggested that slugs may locate susceptible cultivars in the soil by detecting water-soluble diffusates from the tubers.

However, this seems unlikely as South (1973a) found no significant difference in numbers of slugs in soil surrounding four different potato cultivars, although these varied in their susceptibility to slug attack, and Warley (1970) obtained a similar result. Airey (1988a) showed by laboratory tests that *D. reticulatum* did not respond to, or discriminate between, root diffusates of different cultivars. Despite previous suggestions that *D. reticulatum* fed only on cut or damaged tubers (e.g. Thomas, 1947), Storey (1985) showed that *D. reticulatum* is a primary feeder capable of causing as much damage to tubers as other species such as *A. hortensis* and *T. budapestensis*. Airey (1987b) demonstrated that six slug species, *D. reticulatum*, *D. caruanae*, *T. budapestensis*, *A. hortensis*, *A. fasciatus* and *A. ater*, were capable of damaging intact potato tubers while *L. maximus* and *L. pseudoflavus* were deterred by the tuber skin. However, *L. maximus* will feed on potato slices (e.g. Gelperin, 1974). Interactions between species appeared to reduce, rather than enhance levels of damage (Airey, 1987a) and Airey concluded that this might be due to agonistic behaviour between the species.

The differential susceptibility of potato cultivars to slug attack persists when isolated potato tubers are compared in the laboratory (Warley, 1970; Atkin, 1979; Wareing, 1982; Storey, 1985; Airey, 1986a). Storey found that slug damage to tubers was characterized by a small entry hole in the skin and a large cavity eaten out of the cortical tissues although this rarely extended into the pith of the tuber, suggesting that the cortical tissue of potato tubers is preferred to the skin or the pith tissues. She also found that while differences between cultivars are shown by the skin and cortex of the potato tuber, these are not present in the pith tissues, and Airey (1986a) came to a similar conclusion. Warley (1970) found no evidence that physical characteristics of tubers, such as skin thickness, skin strength and crude fibre content, were related to differences in susceptibility between cultivars and this was confirmed by Atkin (1979), who demonstrated that penetrometer measurements on the skin of different cultivars varied from year to year and that there was no correlation between resistance to penetration and susceptibility to slug attack. South (1973a) found, from an examination of scanning electron micrographs of the skin of potato tubers, that the surface of susceptible cultivars tended to become more irregular and broken up while that of less susceptible cultivars remained smooth. He suggested that breaking up of the cork periderm, perhaps at maturity, might render the cultivar more susceptible to attack by the slug radula. However, Storey (1985) found that the skin characteristics were highly variable, being influenced by environmental factors such as soil type, and did not provide an adequate explanation of differences in susceptibility between potato cultivars although these characteristics might contribute towards the differences.

Hunter *et al.* (1968) found that leaves and stems of cv. Maris Piper were more extensively damaged than those of cv. Majestic, the order of susceptibility being smilar to that found among potato tubers, and both Storey (1985) and Airey (1986a) described similar differences in the susceptibility of leaves from different cultivars, both from field trials and in the laboratory. These observations suggest that a chemical substance is involved in the differential susceptibility of potato cultivars rather than physical characteristics of the skin or other tissues. It seems more likely that slugs discriminate between potato cultivars by contact chemoreception (taste) rather than distance chemorecepton (smell) as both Atkin (1979) and Wareing (1982) demonstrated that slugs showed no preference for the odours of potatoes.

Pinder (1974) suggested that biochemical changes in tubers at maturity rendered them more susceptible to slug damage while Warley (1970) considered that a feeding inhibiting substance was more likely to be present in more resistant cultivars than an attractant substance in more susceptible cultivars. Both Storey (1985) and Airey (1986a) suggested that such a chemical was probably located in the skin and cortex rather than the pith of the tuber. Atkin (1979) concluded that the starch fraction of potatoes was not a phagostimulant for slugs although Storey considered that the starch fraction used by him was too low to elicit a feeding response. Storey identified starch as the major phagostimulant within potato tubers, differences between the starch characteristics of potato cultivars playing an important role in determining their palatability to slugs. She suggested that differences in the palatability of potato starches might be due to differences in the starch composition, i.e. the amylose-amylopectin ratio, less susceptible cultivars having higher amylose concentrations. Alternatively, differences in palatability might be due to the hydrolysis products of different starches when degraded by slug enzymes.

In feeding trials, Senseman (1977) found that maltose, glucose and maltotriose stimulated feeding by slugs but at a more reduced level than starch, while Frain (1981) found that a sucrose solution was not attractive enough to maintain feeding in *D. reticulatum*. Warley (1970) found that sugar levels in potato tubers, both as reducing sugars and as sucrose, were highest in less susceptible cultivars although Storey (1985) demonstrated the variable nature of sugar concentrations within potato cultivars. It seems unlikely that tubers with higher sugar levels would be more attractive to slugs as lawrey (1983) showed that plants with the highest concentrations of nutrients tended to have high levels of deterrent secondary plant compounds. Furthermore, immature tubers have been shown to contain higher levels of sugars (Warley, 1970; Storey, 1985) and these are less susceptible to slugs. While Wareing (1982) and Storey (1985) found that different sugars can stimulae slugs to feed, Storey considered that the concentration of these sugars was crucial and it was likely that sugar concentrations in tubers were too low to explain the susceptibility of potatoes.

The plant secondary compounds most frequently associated with plant-herbivore interactions in the Solanaceae are alkaloids. Specialist herbivores, for example the Colorado beetle (*Leptinotarsa decemlineata* Say), have become adapted to only single species, for example the potato, and can only cope with those alkaloids produced by that plant (Edwards and Wratten, 1980). Alkaloids have been shown to be major contributors to the resistance of *Solanum* spp, in several insect species. The glycoalkaloids of cultivated potatoes are nitrogen-containing steroids with a glycosidic linkage. They have a bitter taste and may be toxic to both animals and man. Storey (1985) showed that total glycoalkaloid levels varied between cultivars, with the lowest concentrations being present in susceptible potatoes. Storey found that the major glycoalkaloids present in potatoes – solanine and chaconine – acted as feeding deterrents at concentrations similar to those found in less susceptible cultivars. She showed that feeding on susceptible cultivars could be reduced if levels of glycoalkaloids were artificially increased. Phenolic compounds are present in all higher plants and may act as defence compounds by interfering with the nutritional value of the plant to a herbivore, for example by acting as enzyme inhibitors. Chlorogenic acid, the main phenolic compound in potatoes, stimulated feeding by *A. hortensis* and *T. budapestensis* but reduced feeding by *D. reticulatum*, when presented together with sucrose (Stephenson in Johnson, 1969). However, the concentration of chlorogenic acid was greatest in least susceptible cultivars and this was confirmed by Storey (1985). Storey reported that, while low concentrations of chlorogenic acid had a stimulatory effect on feeding, slight increases in chlorogenic acid levels caused an inhibitory effect on feeding. Thus, in susceptible cultivars, low levels of chlorogenic acid acted as feeding stimulants while, in resistant cultivars, the higher levels recorded would inhibit feeding. Atkin (1979) suggested that, as proteolytic enzymes were inhibited by potato proteins, proteolytic activity in the slug crop might be inhibited by feeding on resistent cultivars of potato as these had a high total-protein content. Results of bioassays in which slugs were fed chemical fractions from potato tubers in agar-based diets showed that low molecular weight compounds were essential for the expression of resistence of cultivars to slug attack (Johnston *et al.*, 1989). They also suggested that to have a significant effect on slug growth, glycoalkaloid concentrations would have to be higher than those normally present in potatoes. However, they did not examine the possible effect of lower levels of glycoalkaloids on slug-feeding behaviour. Johnston *et al.* concluded that resistance was probably due to the interaction of low molecular weight and high molecular weight compounds, possibly involving the enzymic oxidation of phenolics to quinones which in turn might inhibit slug digestion.

7.1.6   Learning and feeding behaviour

Slugs as generalized herbivores are faced with the problem of assessing both the nutritional value and potential toxicity of foods. If slugs were capable of learning to avoid certain foods due to unpleasant tasting or harmful compounds within them, this mechanism would determine food selection, and the presence of secondary compounds would be an effective plant defence mechanism against slugs. Whelan (1982b) showed that slugs usually fed on novel foods, even when they were unacceptable, but once eaten they were avoided during a second trial, suggesting that a learning process has occurred. Bailey (1989) found that slugs usually fed on the first item of food found, but often ignored items of food encountered later. However, if food was scarce, slugs fed almost every time. He showed that feeding was most intense during the first two and a half hours after feeding began. Later meals were both shorter and less regular. Starved slugs differed from fed slugs mainly by taking a second meal shortly after the first and consuming twice as much food as fed slugs. Feeding behaviour in *L. maximus* was described by Gelperin (1975). Foraging slugs were attracted to food odours via olfactory receptors located on the superior (optic) tentacles. When the food source was located, the slug explored the surface of the food using both pairs of tentacles and, if the odour signalled a potential food, the lips were everted so that their gustatory receptors were in contact with the food. If the food was judged acceptable then radular rasping and ingestion followed lip contact while unacceptable food was usually rejected after lip contact. Slugs normally depend therefore on both olfactory and gustatory cues to detect their food. Gouyon *et al.* (1983) described this feeding behaviour by a model based on two parameters: the probability of attacking a given plant, which is related to odour, and the probability of continuing to eat an attached plant, which is related to taste. Gouyon *et al.* showed that *D. reticulatum* learnt rapidly to recognize which terpene chemotypes of *Thymus vulgaris* they preferred. It is now known that slugs are capable of associating the independently sensed olfactory and gustotory cues from a food source and can learn to avoid olfactory cues that singal an unacceptable food and thus partition their foraging time more effectively.

Rapid food-aversion learning in *L. maximus* was demonstrated by Gelperin (1975). Slugs learnt to avoid a novel palatable food when CO_2 poisoning was paired with ingestion of the new food. Avoidance was most commonly associated with complete rejection of the unsafe food, based on olfactory cues alone. Some animals learnt food avoidance in a single trial and remembered without error for three weeks. One-trial associative learning was demonstrated in *L. maximus* by Sahley *et al.* (1981a). They showed that, by using a Pavlovian-like conditioning procedure, the slug's normal preference for odours produced by potato and carrot could be markedly reduced by pairing a brief feeding experience on potato or carrot (conditioned stimulus) with exposure to a saturated solution of quinidine sulphate (unconditioned stimulus), a bitter tasting plant extract. These experiments, in which *L. maximus* was capable of associating the odour and taste properties of food, provide a demonstration of associative learning first-order conditioning. In a second group of experiments, Sahley *et al.* (1981b) showed that the food odour preference of *L. maximus* could be modified by a second-order conditioning procedure. Slugs that had received carrot odour-quinidine pairings followed by potato odour-carrot odour pairings, subsequently displayed a reduced preference for potato odour in comparison to slugs which had received only a single pairing. Thus the preference of slugs for potato odour was reduced even though potato odour was never directly paired with the quinidine sulphate. *Limax maximus*, therefore, appeared to demonstrate a behavioural strategy, second-order conditioning, similar to that observed in vertebrates. Sahley *et al.* (1981b) also showed that a blocking effect influenced associative learning in the food odour preferences of *L. maximus*, again emphasizing the similarity between associative learning in *Limax* and vertebrate learning. Sahley *et al.* (1981c) discussed how knowledge of food aversion learning in slugs had demonstrated that invertebrate systems could serve as useful models in which to study the neural basis of associative learning. Gelperin (1986) reviewed work that had been done on the highly developed learning ability of *L. maximus* for discriminating food odours and tastes. He described a computer model (LIMAX) which allowed performance of taste-taste associative conditioning experiments on-line. Using this model, Hopfield and Gelperin (1989) predicted that *Limax* would be unable to learn to avoid a compound odour stimulus without also learning to avoid its components. However, they found that slugs (*L. maximus*) could learn a strong aversion to a combined odour while the two components odours remained strongly attractive when presented individually. Most studies of associative learning in slugs have involved the use of aversive reinforcers. However, Sahley *et al.* (1990) demonstrated the appetitive learning ability of *L. maximus*, where an initially aversive odour was made attractive by repeatedly pairing it with a very attractive chemostimulant as the unconditioned stimulus.

In an investigation into the short-term control of food intake by *A. californicus*, Senseman (1978) showed that the rate of ingestion of an agar-based food was inversely related to the hardness of that food and, under conditions in which food intake was restricted by the hardness of the food, meal duration and size were directly related to the chemostimulant intensity of the food, with meal size being inversely proportional to meal duration. Senseman derived a 'feeding equation' to predict meal duration over a range of chemostimulant concentrations and food hardness. He concluded that two processes, sensory adaptation and post-ingestional feedback, were sufficient to account for the short-term control of food intake in *Ariolimax* and emphasized the need to assess the rate of decay of the feedback process as well as the recovery of the chemosensory receptors when investigating conditioned food aversion in slugs. Dobson and Bailey (1982) demonstrated that a strong positive stimulus to these chemoreceptors could overcome post-ingestional feedback from the crop in *D. reticulatum*, Delaney and Gelperin (1986) considered the means by which generalist herbivores, such as slugs, select their food when they are faced with varied and seasonally changing food resources. They found, by feeding amino acid deficient diets to *L. maximus*, that slugs acquire, post-ingestively, an aversion to the taste and probably the odour of diets lacking an essential amino acid as the result of associative learning. However, removing non-essential amino acids does not cause an aversion. Similar results were obtained by Lee and Chang (1986) for the slug *Incilaria fruhstorferi deiseniana*. Delaney and Gelperin reported that injection of the missing essential amino acid into the haemocoele one hour after the completion of a meal blocked the development of a learned aversion. Thus slugs presented with a large number of potential foods, some nutritionally better than others, could sample the foods available and post-ingestively evaluate their nutritional quality. This information could then be used to pattern future foraging. Increased dietary intake of choline elevates blood choline and brain acetylcholine levels and this change in neuronal acetylcholine concentration may augment learning and memory functions. Sahley (1985) showed that, while initial learning of a food-avoidance task was not augmented by a high choline diet fed to the slug *L. maximus*, the duration of memory retention was prolonged. This supports the involvement of cholinergic synapses in the memory retention mechanism.

Administration of opioid agonists, such as morphine sulphate, to the slug *L. maximus* resulted in significant, dose-dependent increases in the ingestive responses of food-deprived slugs and in the initiation of feeding in satiated animals although low doses attenuated feeding (Kavaliers *et al.*, 1984, 1986). These effects could be blocked by opiate antagonists, such as naloxone hydrochloride, and these opiate antagonists alone cause a significant decrease in the feeding of free-feeding slugs. Thus opiates are involved in the control of the feeding behaviour of *L. maximus* and, since their effect is similar to that in mammals, this suggests the early evolutionary development of opioid involvement in the control of feeding (Kavaliers *et al.*, 1986). Kavaliers *et al.* (1985) showed that magnetic stimuli reduced the ingestive effects of the opiate agonist, morphine sulphate, in free-feeding and food-deprived *L. maximus*. Kavaliers and Hirst (1985) demonstrated that exposure of *L. maximus* to stress by pinching the posterior portion of the body (tail-pinch stress) caused feeding responses similar to those observed after treatment with exogenous opiates and that these were blocked by the opiate antagonist, naloxone hydrochloride. Analgesic effects which increased the thermal nociceptive threshold of the slug were also caused by tail-pinch stress. Kavaliers and Hirst suggested that exposure to stress increases endogenous opiate activity in slugs in a manner similar to that reported in mammals. Dalton and Widdowson (1989) also found evidence that a physical stressor could produce analgesia which might be due to the release of endogenous opiates. When tail-pinch stress was applied to *A. ater*, there was a significant increase in the response time when the slug was tested for the foot-lifting response on a hotplate. This analgesia could be completely reversed by injection of opiate antagonists while analgesia could also be elicited by injection into the food of β-endorphin and enkephalin analogues.

7.2 CHEMORECEPTION

7.2.1 Olfactory orientation

Moquin-Tandon (in Taylor, 1902–1907) described how the movements of *L. maximus* were continuously orientated towards food sources such as a damaged apple. The capacity for orientation was lost when the tentacles were amputated, suggesting that the tentacles were the site of olfaction. Adams (in Taylor, 1902–1907) illustrated how acute the olfactory sense was by describing how a *L. maximus* crawled directly towards a plate of food six feet away on a lawn at night and how the slug reorientated itself as the plate was moved away three times, and Taylor recorded that *L. cinereoniger* had a similar capacity for orientation towards food. However, Bailey (1989) has suggested that many food items are found by random encounter. Kittel (1956) compared the olfactory behaviour of *A. ater* and *L. cinereoniger* by testing their ability to detect the presence of various fungi in a field. Both species reacted to olfactory stimuli from fungi by first crawling in the direction that they were set down and then, after stopping and moving the tentacles from side to side, slugs moved towards the food but in a spiral path rather than in a straight line. The distance of attraction varied with the species of fungus but *A. ater* was able to detect the stinkhorn (*Phallus impudicus* Persoon) at 120 cm or more. The posterior (optic) tentacles determined the direction of movement until the slug got to within a few centimetres of the fungus when the anterior tentacles and oral lappets were applied to the surface of the food. If the posterior tentacles were removed, the distance at which fungi could be detected was reduced to a few centimetres. Removal of both pairs of tentacles meant that food could only be detected if placed directly in front of the mouth. Kittel concluded that the posterior tentacles were distance chemoreceptors while the anterior tentacles and oral lappets were mainly contact chemoreceptors. Ricou (1961) described how some fungi with a strong odour, such as *P. impudicus*, could attract slugs at distances of up to 7 metres. She also described how, since the olfactory receptors of *L. maximus* were sensitive to a dilution of $1:10^6$ of mustard gas, a recommendation was made during the first world war that the slug be used as a gas detector in the trenches.

Kittel's observations were confirmed by the investigations of Chase and Croll (1981) into differences in function of the anterior and posterior pairs of tentacles in olfactory orientation of the snail *Achatina fulica*. They showed that the pairs of tentacles had different functions. The anterior tentacles were mainly involved in trail following while the posterior tentacles were involved in anemotaxis and movement up a concentration gradient. Anemotaxis required only a single tentacle making successive discriminations (i.e. klinotaxis) while orientation to a concentration gradient required simultaneous bilateral comparisons by both tentacles (i.e. tropotaxis). Stephenson (1979) concluded from a study of the sense organs associated with feeding behaviour in *D. reticulatum* that the acceptance or rejection of a potential food plant relied on gustatory cues. Stephenson (1978) described a bioassay, based on a Y-tube olfactometer, in which the responses of slugs to plant extracts or solutions of chemicals could be evaluated while Pickett and Stephenson (1980) compared Y-tube olfactometry and a trail-following bioassay. The Y-tube was found to be unsuitable although the trail-following technique proved successful and was used to investigate which plant volatiles and components influenced feeding behaviour in *D. reticulatum*.

7.2.2 Trail following and homing behaviour

Miles (1969) described a technique for recording slug trails, using lightly sanded, water saturated paper. Hamilton (1977) discussed the use of mucous trails in gastropod orientation studies and reviewed the various techniques used for the detection and recording of these trails. Hamilton also derived an index of directness which could be used in conjunction with trails to assess the degree of directed orientation shown by an animal. Cook (1977) treated the mucous trails of *L. maculatus* with a suspension of powdered calcium carbonate which stuck to the trail and then sprayed them with lacquer to produce a permanent record of the movements of slugs in laboratory experiments. Trail-following behaviour is a widespread phenomenon among gastropods and Wells and Buckley (1972) concluded that it was a general feature of gastropod behaviour although terrestrial gastropods had been neglected until relatively recently. The anterior tentacles in *A. fulica* are mainly responsible for trail following according to Chase and Croll (1981). However, Ushadevi and Krishnamoorthy (1980) concluded that the posterior tentacles of *Mariella dussumieri* (Gray) regulated trail following. Cook (1985) found, for *L. maculatus*, that while there was no change in rates of locomotion before and during trail following there were decreases in tentacle activity during trail following. Bilateral amputation of tentacles brought about a general reduction in the speed of locomotion and prevented the decrease in tentacle activity during trail following. Cook concluded from tentacle amputation experiments that the posterior tentacles were concerned with the detection and identification of trails which were subsequently followed and that the anterior tentacles controlled the behaviour of the slug on the trail.

Cook (1977) investigated trail following in *L. maculatus* and found that each individual followed its own trails and those of conspecific slugs further and more frequently than would be predicted on a random basis. The direction in which the trail was followed was related to the angle at which the trail was approached. *Limax maculatus* would not follow trails of *D. reticulatum* or *T. budapestensis* but showed a high frequency of turning on to the trails of *L. flavus*, with which it was sympatric. *Limax flavus* behaved in a similar way to *L. maculatus*. Although both *L. flavus* and *L. maculatus* confused each other's trails at the point at which the decision to follow was taken, once on the trail, slugs only continued to follow if the marker trail was laid by a member of the same species, an initial period of exposure to the trail being required to make a discrimination. Trail-following by *Limax* only followed on about 30% of the occasions when a trail was encountered and they rarely followed trails for more than 20 cm. In studies using time-lapse cinephotography and video recording, Wareing (1986) found that *D. reticulatum* generally followed the direction in which the first trail was laid and, in over half of the cases, following led to the second slug catching up with the first, resulting in courtship. Wareing suggested that this type of orientated behaviour enabled a slug to find a mate. Older trails (up to three days) were not followed by *D. reticulatum* although *L. maculatus* was capable of following trails which had dried out and were up to three days old (Cook, 1979a). Dundee *et al.* (1975) suggested that *Veronicella* spp. aggregated in response to a mucus-borne pheromone while Cameron and Carter (1979) suggested that inhibitory pheromones in mucus might be responsible for the differing reactions of slugs and snails, such as *L. maculatus* and *L. flavus* (Cook, 1977), to closely related species and less closely related species. Landauer *et al.* (in Rollo and Wellington, 1979) found evidence for an alarm pheromone in mucous trails of *Lehmannia valentiana* while Ushadevi and Krishnamoorthy (1980) also suggested that a pheromone in the mucus of the slug *M. dussumieri* might be responsible for trail following.

Homing behaviour was described by Taylor (1902–1907) for *L. maximus*, *L. cinereoniger* and *L. flavus*. He included a diagram of a mucous trail, 20 feet or more in length, produced by *L. maximus* on a wall. This originated from and returned to the same crevice but the return track was distinct from the outward one, only crossing it at one point. Ingram and Adolph (1943) described how *A. columbianus* established a home site by excavating a depression in the soil and then foraged over an area extending at least 4.5 metres from this home. The home site was readily identified by an accumulation of faeces and converging mucous trails. Edelstram and Palmer (1950) showed that adult *H. pomatia* returned in late summer to their earlier winter quarters with a considerable degree of precision and suggested that smell combined with memory were the guiding senses. When large numbers of *D. reticulatum* were released at a single location on undisturbed arable land (South, 1965), their dispersive behaviour was similar to the 'homing type' described by Dobzhansky and Wright (1943) where, after some initial dispersive movement, they established a home range. Moens *et al.* (1966) found that when *D. reticulatum* were released into a strange environment, they moved over large distances initially and then, when they found suitable sheltered sites, they rested there for some days. Newell (1966a, b) showed, by the use of time-lapse cine photography, that *D. reticulatum* returned to a specific 'home' after their nightly activity. However, Duval (1972) reported that *D. reticulatum* repeatedly changed its daytime resting place rather than establishing a permanent home base. Gelperin (1974) found that *L. maximus* could return to a home site from a distance of over 90 cm by a direct route. He concluded that mucous trail following and vision were not involved in this homing behaviour and found that in the presence of a low-velocity wind, homing occurred upwind. Disconnection of the digitate ganglion in the posterior tentacle which is concerned with olfaction, from the central nervous system eliminated homing thus demonstrating the olfactory basis of homing behaviour.

Laboratory experiments on *L. maculatus* showed that this slug consistently homed without following trails, although failure to find a home would often result in a period of trail following (Cook, 1979a, b). He interpreted this behaviour by suggesting that two pheromones were involved: a volatile pheromone, serving as an olfactory beacon, involved in the distance chemoreception of home; and a second, non-volatile, pheromone, detected only by contact with the mucus, and responsible for trail following. Thus *L. maculatus* homed mainly by using an olfactory beacon but with a trail-following capacity held in reserve. Unlike Edelstram and Palmer (1950), Cook concluded that memory was not involved in homing, a dual pheromone mechanism being sufficient to account for the homing phenomenon. The consistent return to the same site in laboratory experiments was considered by Cook (1980) to be an artefact, resulting from the provision of only one home. He investigated homing in *L. maculatus* under less confined conditions in the field using time-lapse cinematography and confirmed that distance chemoreception, rheotaxis and trail following adequately accounted for homing in this species. However, he found that individual slugs did not have a specific home but rather returned to a site that was previously occupied by members of the same species. Thus homes were species specific rather than individual specific. Rollo and Wellington (1977) suggested that the faeces of *L. maximus* might provide the olfactory cue for homing in that species.

Eight slug species (*A. columbianus*, *A. ater*, *A. circumscriptus*, *A. hortensis*, *A. subfuscus*, *D. caruanae*, *D. reticulatum* and *L. maximus*) showed an ability to relocate artificial shelters from distances of more than a metre away in the field (Rollo and Wellington, 1981). When it was raining or windy, slugs approached shelters in a spiralling manner but in calm, dry weather their path was direct. This difference in behaviour was similar to that recorded by Kittel (1956) and Duval (1970a) for olfactory orientation to food. *Ariolimax columbianus* chose shelters occupied by conspecifics more readily than unoccupied shelters and travelled some distance between occupied shelters, ignoring unoccupied shelters on the route. Some slugs consistently returned to the same shelters over long periods. During hot or dry weather, individuals of several species showed a strong attachment to particular shelters while there was more movement among shelters during cool, moist conditions. Rollo and Wellington also confirmed previous conclusions (e.g. Cook, 1980) about the importance of distance chemoreception and trail following in homing, using *L. maximus*, *D. reticulatum* and *A. ater*. However, they emphasized that trail following also serves other purposes, such as territorial behaviour and locating mates. Rollo and Wellington found that *A. ater* preferentially homed to shelters containing carrot-derived faeces as opposed to those containing only carrots. This suggested that a homing pheromone was present either in faeces or in mucus deposited on the faeces.

Both *L. maculatus* and *L. flavus* form large packed aggregations in sheltered locations, for example under logs and stones (Chatfield, 1976a; Evans, 1978). This type of aggregation, involving frequent intimate contact between individuals, is distinct from the aggregated distribution, where animals rarely come into contact, which results from the combined effect of physical environmental factors and a lack of dispersal and was described by South (1965). Cook (1981a) described how *L. maculatus* and *L. flavus* were packed closely together in their daytime resting sites, with large areas of their flanks in contact. He described these dense aggregations as 'huddles'. The benefit appeared to be a physiological one brought about by a net reduction in surface area and may be related to water conservation, thermoregulation or some other metabolic function. Huddles formed when two populations of conspecifics were mixed and also formed from mixed populations of *L. flavus* and *L. maculatus*. They formed more frequently and involved more contact in dry compared with moist conditions and, since two slugs making close contact reduced their evaporation rate by 34% compared with unhuddled slugs, Cook (1981a) concluded that huddles were non-social aggregations whose prime function was to conserve water. In a comparative study of aggregation in four *Limax* spp., Cook (1981b) found that the compactness of the huddles could be described by a series:

$$L.\ maculatus = L.\ flavus > L.\ maximus > L.\ marginata$$

where *L. maculatus* and *L. flavus* moved until they made maximal contact with other slugs, forming close huddles. *Limax maximus* also moved to make contact but these were minimal and the slugs did not pack together, while

L. marginata tended to move away from one other. Cook concluded that huddling enabled slugs to make greater use of daytime resting sites and to maintain their occupation even when sharing the site with more aggressive species such as *L. maximus*. Similar huddling and postural adjustment, in relation to dry conditions, has been described by Waite (1988) for *D. reticulatum*. Chelazzi *et al.* (1988) confirmed the importance of distance chemoreception in the homing of *L. flavus* and showed in choice tests that this species headed preferentially for group shelters rather than individual shelters.

7.3 AGONISTIC AND DISPERSIVE BEHAVIOUR

7.3.1 Agonistic behaviour

Although aggression in slugs has been reported by a number of authors (e.g. Taylor, 1902–1907; Quick, 1960), the first detailed accounts were given by Rollo and Wellington (1977, 1979). Slugs were first seen to be aggressive during the late juvenile stage as they became sexually mature. The most aggressive species was *L. maximus* and this species attacked both conspecifics and other species, in some cases fatal attacks occurring. Barker and McGhie (1984) have described how this species often inflicted severe body wounds through repeated biting. Rollo and Wellington (1977, 1979) also found that *A. subfuscus* was nearly as aggressive as *L. maximus* although it was less of a threat to most other species because it moved more slowly. *Deroceras caruanae* and *D. laeve* showed much higher levels of aggression than *D. reticulatum*, which was seldom aggressive. However, Airey (1987a) reported aggression and cannibalism in *D. reticulatum*. Rollo and Wellington described how a high-density population of *D. laeve* contained many individuals with severe wounds typical of those resulting from aggressive interactions while *D. caruanae* was capable of driving the much larger *A. columbianus* from a feeding area. Rollo and Wellington also found little aggressive behaviour in *A. columbianus* whilst aggressiveness in *A. ater* was mainly associated with courtship. Limited observations on *A. hortensis*, *A. circumscriptus* and *A. intermedius* suggested that these species were not particularly aggressive.

Rollo and Wellington (1979) found that aggressiveness varied seasonally, occurring mainly during summer when hot, dry weather reduced the numbers of shelters and availability of food. There was little aggression in winter when slugs were frequently aggregated. Since slugs avoided shelters occupied by large aggressors and shelters closest to food were usually selected, Rollo and Wellington concluded that agonistic behaviour improved the chances of acquiring shelter and food simultaneously. Airey (1987a) described how interspecific interactions between slugs tended to reduce damage in a potato crop while feeding by one member of a species could stimulate feeding by conspecifics, thus increasing damage by a single species.

The mode of aggression, which tended to be similar for all species, was described in detail by Rollo and Wellington (1979). Attack usually involved repeatedly biting the victim, sometimes adopting a 'rear-and-lunge' sequence. Slugs with aggressive intent also followed fresh mucous trails. Various responses to attack were recorded, including 'tail wagging', withdrawal of head and tentacles under the mantle, mantle flapping, the secretion of large quantities of mucus and moving away from the area. When *A. ater* was attacked it contracted until almost hemispherical and then rocked from side to side, a form of behaviour described by Taylor (1902–1097) as a response to being startled or being irritated. This form of defensive behaviour, which may be effective against predatory beetles, was less effective against other slugs, which were not deterred by mucus or by the difficulty of biting a hemisphere, and *A. ater* soon moved away to escape attack. Richter (1980a) described how the conspicuous caudal mucus plug in *A. columbianus* had an important role in defence against predators and antagonistic slugs (*Limax* sp.) by fouling the mouthparts or by providing a temporary distraction during which time *Ariolimax* escapes. Deyrup-Olsen *et al.* (1986) described the autotomy escape response of the slug *Prophysaon foliolatum* (Gould), aimed at predators rather than other slugs, where muscle contraction severed the posterior section of the foot from the rest of the body.

7.3.2 Dispersive behaviour

Hamilton and Wellington (1981a) investigated the effects of changing the food supply and population density on the dispersive movements of *A. ater* and *A. columbianus*. High slug density and sparse, poor-quality food induced rapid dispersal by *A. ater* but neither slug density nor amount or quality of food had any significant effect on the dispersal of *A. columbianus*. There were significant seasonal changes in the rate of dispersal, mortality, and weight of dispersing *Arion* while only the mortality of *Ariolimax* changed significantly. When the effects of density and food on the nocturnal behaviour of these slugs were examined, Hamilton and Wellington (1981b) showed that density did not affect the level of activity, short-term movement, resting or feeding in *A. ater*, although *A. columbianus* was more active and moved, rested and fed more frequently when slug density was high. Hamilton and Wellington (1981b) concluded that while the nocturnal activity of *A. ater* appeared to be mainly food orientated, that of *A. columbianus* was more responsive to slug density. Seasonal changes in levels of activity and behaviour were shown by *Arion* but not by *Ariolimax*. Rollo (1983b) showed that *A. ater* and *L. maximus* showed a definite dispersive phase associated with female-phase maturation. This was absent from *A. columbianus*. Rollo suggested that this dispersal would be advantageous in transient field environments and this was the reason it was absent from the forest-dwelling slug, *A. columbianus*. Similar results to those of Hamilton and Wellington were obtained from an investigation into the effect of food deprivation on the locomotor activity of *D. reticulatum* and *T. budapestensis*. The surface-living *D. reticulatum* showed a marked increase in locomotor activity when deprived of food, while the locomotor activity of the soil-dwelling *T. budapestensis* was reduced (Airey, 1987b). Airey concluded that different slug species adopted different strategies in their attempt to survive periods of food deprivation. Hogan (1985) recorded that *D. reticulatum* would travel up to 12 m in a single night in the absence of food. However, Bailey (1989) showed that in an enclosed arena, neither food abundance nor area of arena influenced the distance moved each night by *D. reticulatum* (mean 4.6 ± 3.1 m).

7.4 LOCOMOTOR ACTIVITY

7.4.1 Timing of activity

Slugs are most active at night and different species of slug reach their peaks of nocturnal activity at different seasons of the year (Barnes and Weil, 1944). The initiation time of this activity varies with the season (Barnes and Weil, 1942) while individual species differ in the timing of their nocturnal activity (Barnes and Weil, 1945; White, 1959b). White suggested that *D. reticulatum* became less active later in the night, that *T. budapestensis* was most active around midnight and that *A. hortensis* became most active late at night. Barry (1969) found that the activity of slugs, mainly *D. reticulatum*, in Ohio cornfields began at dusk and gradually increased until it reached a peak between four and six hours after dark. Daxl (1969) showed how the timing of activity differed for each of four slug species. *Deroceras reticulatum* was periodically active with three peaks during the night, the last peak occurring just before sunrise, while *D. laeva* showed little activity during the night and had a single peak of activity about two hours after sunrise. The activity of *L. flavus* reached a peak soon after it began about two hours after sunset, and this level of activity was maintained for six or seven hours until shortly after sunrise. The activity of *A. hortensis* showed a steady rise through the night, reaching a maximum about two hours after sunrise. Daxl concluded that the diurnal activity of slugs was regulated by a complex of endogenous and exogenous factors, particularly light, temperature and changing seasons.

7.4.2 Endogenous rhythms

It is clear from the previous section that the locomotor activity of slugs shows daily, i.e. circadian, rhythmicity but it is necessary to know whether these rhythms are subject to endogenous or exogenous control. Slugs kept under conditions of constant temperature in darkness show a persistent but steadily deteriorating diurnal rhythm of activity (Dainton, 1954a). Lewis (1969b) found that, under conditions of constant humidity, constant temperature and darkness, A. ater exhibited rhythmic locomotor activity in which the period of the rhythm was less than 24 hours. However, the onset of activity advanced each day so that the period of the rhythm eventually decreased to about 20 hours before it broke down. This rhythm would not entrain to an 18.5-hour light cycle nor follow a 6-hour phase advance in a 24-hour light cycle. Lewis concluded that the rhythm was a truly endogenous one and suggested that its adaptive significance was due to A. ater remaining under cover during the daytime where it was unable to estimate the time of day. Premature emergence might result in exposure to unfavourable environmental conditions. Sokolove et al. (1977) demonstrated an endogenous circadian locomotor rhythm in L. maximus, using 'running wheels'. In constant darkness at a constant temperature, the locomotor activity free-ran with a period of about 24 hours. This period tended to be shorter in constant light. The circadian rhythm was entrained to light–dark cycles (L:D 12 : 12 or L:D 16 : 8) and could be phase shifted by changes in the L:D cycle. The period of the rhythm was temperature compensated with a Q_{10} approximately equal to 1.00. Two distinct peaks in feeding activity in a 24-hour cycle were described by Raut and Panigarhi (1990) for L. alte. They concluded that the feeding rhythm was under endogenous control.

Morton (1979) showed that D. caruanae possessed an endogenous cycle of activity entrained by the normal 24-hour day and night cycle. The slug remained inactive under shelter during daylight hours and, as dusk approached, a phase of locomotor activity brought the slug into contact with food. As dawn approached slugs returned to their shelter. Morton also recorded an endogenous rhythm in gut activity which was correlated with this cycle (see Section 3.1.5). The rhythm of locomotor activity persisted under constant illumination although it was out of phase with the normal light–dark sequence. In total darkness, however, slugs lost their normal rhythm of activity. Beiswanger et al. (1981) compared the locomotor activity of blinded slugs and control slugs and showed that the eyes on the posterior tentacles of L. maximus and L. flavus were not essential for maintenance of rhythmicity or for L:D entrainment. In total darkness the active period in the free-running cycle projected back to the dark phase of the previous light–dark cycle. Blinded slugs remained entrainable to white-light L:D cycles and could follow a 6- or 12-hour phase shift. Unlike intact slugs, however, they could not entrain to L:D cycles with wavelengths greater than 600 nm. These results indicated the existence of extraocular photoreceptors capable of mediating entrainment and relatively insensitive to red light. Beiswanger et al. (1981) concluded that the eyes provide photic input to the slug circadian system although they are not essential components of the circadian clock that controls locomotor activity.

In the endogenous circadian rhythms described above, there was evidence that light cycles normally acted as zeitgebers for the entrainment of these rhythms. Pavan and Babu (1979) showed that light acted as a zeitgeber for locomotor rhythms in the slug *L. alte*, while in constant light and dark conditions the slugs exhibited a continuous pattern of activity. Endogenous rhythms in the acetylcholine–acetylcholinesterase system of the central nervous system and foot muscle of *L. alte* were demonstrated by Pavan *et al.* (1982). Under an L:D 12 : 12 regime, these rhythms were in phase with one another. Pavan *et al.* suggested that the ACh and AChE rhythmic phase synchronized the motor activity of *L. alte*. Ford and Cook (1987) showed that the endogenous circadian activity of the slug *L. maculatus* was entrained by L:D cycles of 24 hours, temperature cycles of 23 hours and a combined temperature and light cycle of 23 hours. They concluded that the role of temperature fluctuations in the control of activity of *L. maculatus* appears to be in the enhancement of the rhythms which are primarily regulated by changing photoperiod. In constant darkness activity was expressed as a free-running rhythm with a period of 24.7 hours while, in continuous light, rhythmicity was weak with a periodicity of less than 24 hours. A light cycle of 19 hours weakly entrained the locomotor rhythm but failed to entrain the feeding rhythm. Since feeding and locomotor rhythms could free-run with different periodicities, it was suggested that the endogenous circadian clock was a multi-oscillator system. In an investigation into entrainment to light pulses using *L. maculatus*, Ford and Cook (1988) showed that a 30-minute pulse of light (10 000 lux) under otherwise constant dark conditions was usually accompanied by a phase-dependent phase shift in the free-running locomotor rhythm of the slug and a change in the free-running period. In the field, the activity pattern appeared to be limited by low temperatures, low humidity and high winds (Cook and Ford, 1989).

Little is known about endogenous circannual rhythms in slugs. Segal (1960) considered that the presence of an endogenous circannual rhythm in *L. flavus* was indicated by the temperature independence and almost precise initiation of egg laying under conditions of constant temperature (20°C) and photoperiod (L:D 11:13). A low temperature (10°C) with constant photoperiod appeared to shorten the period of egg laying without shifting the phase. Vianey-Liaud (1975) found that rearing *M. gagates* in constant darkness had no effect on growth in the life cycle and suggested that growth was regulated by an endogenous process. Daxl (1969) briefly described seasonal activity in six slug species, including *Boettgerilla vermiformis*, in terms of numbers of slugs active but without allowing for the effect of changes in the life cycles of these slugs on numbers. However, Barnes and Weil (1944) found that the numbers of slugs active at night reached an annual peak which was different for each species, suggesting an annual rhythm in the life cycle. *Helix aspersa* exhibited a circannual cycle of activity when subjected to a constant 12-hour photoperiod, with a free-running period considerably shorter than a year (Bailey, 1981). Under a normal seasonal lighting regime, entry into hibernative by the snail *H. aspersa* was controlled by a circannual rhythm entrained by photoperiod and was not directly related to temperature. South (1989a) concluded that the annual life cycle of *A. intermedius* was entrained by light, with a rapid stage of growth beginning each year in mid or late March and coinciding with lengthening days as the photoperiod (L:D) reached 12 : 12 hours. This rapid growth was independent of the prevailing temperature. South also considered that photoperiod had little influence on the timing of the life cycle of *D. reticulatum*, an opportunist species capable of growing and reproducing over a wide range of environmental conditions.

7.4.3 Physical environmental factors

Despite the evidence for the existence of endogenous circadian rhythms in slug behaviour, their role has not been universally accepted. Dainton and Wright (1985) and Dainton (1989) have maintained that it is unnecessary to postulate endogenous rhythms to explain activity and that this is instead a direct response to changes in weather conditions, particularly temperature.

Many of the early reports of the effect of selected physical factors such as temperature and humidity on slug activity (e.g. Dainton, 1943) were based on brief laboratory experiments carried out usually during the daytime when the slugs would normally be inactive. Their relevance to the behaviour of animals under natural conditions and over long periods of time is doubtful. The observations of Barnes and Weil (1945) that the optimum conditions for slug activity are a warm still night with plenty of surface moisture, based on the numbers of slugs active at night, relates to the amplitude of the activity rhythm and do not apply to the endogenous rhythm.

Barnes and Weil (1945) found that slug activity on the surface of the soil ceased at freezing temperatures and White (1959b) described how numbers of slugs caught during night searching began to decrease as soon as the temperature fell below 4.4°C. Poulicek and Voss-Foucart (1980) reported that, in Belgium, adult *D. reticulatum* hibernated from October until March while Getz (1959) also considered that these slugs hibernated during the winter in Michigan (USA). However, South (1989a) found that this species remained active during the winter months in England under normal conditions. Mellanby (1961) showed that some slugs maintained their activity at low temperatures, with *D. reticulatum* moving and feeding normally at 0.8°C although *A. hortensis* was seldom spontaneously active below 5°C. *Tandonia budapestensis* was active at 5°C but inactive at 0.8°C and Mellanby concluded that slugs were well adapted to life at low temperatures. Barnes and Weil (1945) found that when moderate evening temperatures were followed by an early morning frost, activity occurred during the first half of the night but ceased during the latter half. Activity was also reduced during heavy rain and by strong winds. Lack of surface moisture appeared to limit activity and rainless periods in the spring had less effect on limiting activity than summer droughts when the soil became very dry. Young and Port (1991) described how reduced water availability on the surface of the soil limited both the distance travelled by *D. reticulatum* and the time this species spent foraging. Dainton (1954a) investigated the effect of temperature and humidity on the induction of activity in *D. reticulatum*. She found that atmospheric moisture had no direct effect on activity and that, between temperatures of 4 and 20°C, activity was induced by falling temperature and suppressed by rising temperature, with changes as small as 0.1°C being effective. Thus activity would be induced by a fall in temperature after a rain shower and during the late evening. This effect was reversed at temperatures between 20 and 30°C when activity was induced by rising temperatures with the result that slugs would tend to move away from a region of high temperature. Dainton (1989) concluded that the daily rhythm of activity simply followed the normal diurnal temperature rhythm except that continuous rain reduced activity. Dainton (1954a) also showed that, in a temperature gradient, *D. reticulatum* aggregated about a preferred temperature of 17–18°C.

Dainton (1954b) examined the effect of light and air currents on activity and found that, although an increase in light intensity initially caused a burst of activity, natural light cycles did not control the timing of the active phase. Air currents induced activity in a resting animal and increased the speed of locomotion. Air currents directed at the head or tentacles resulted in a klinotactic response with the animal turning and moving away from the stimulus. The adaptive significance of this kind of behaviour has been discussed in Section 4.1.3(a). Dainton and Wright (1985) showed that a falling temperature stimulated activity for *A. ater* and that this occurred at temperatures below 22°C while Dainton (1989), in confirming her previous conclusions, also showed how falling temperatures after heavy rain showers in summer could explain the exceptional daytime activity of slugs under these conditions.

Fluctuating temperatures were shown to be important in initiating locomotor activity in *P. carolinianus*, *L. valentiana* and *D. reticulatum* by Karlin (1961). Diminishing light intensity did not initiate activity although light intensity could restrict activity induced by temperature. Both Webley (1964) and Hunter (1968a) showed that feeding activity in common slug species was dependent on temperature. Webley (1964) concluded that relative humidity had little effect on activity while Hunter (1968c) found that activity was greater at 100% than 95% relative humidity. It would appear from these results that the timing of activity in some slugs, particularly *D. reticulatum*, is controlled largely by environmental temperature fluctuations and that light and possibly humidity are of little importance. However, Newell (1966b) showed a closer relationship between surface crawling activity and diurnal light changes rather than with temperature for *D.*

reticulatum. Lewis (1969a) demonstrated that temperature changes in the laboratory had no effect on the timing of locomotor activity for *A. ater* and concluded that field activity was not controlled by natural temperature rhythms. Lewis was able to synchronize activity in *A. ater* with 24-hour light cycles, after having first destroyed or weakened their original endogenous rhythm by keeping animals at 10°C in constant darkness for several weeks. Since activity was readily synchronized with 24-hour artificial light rhythms, Lewis concluded that natural light was the most important environmental factor involved in the control of activity in this species.

Using regression analysis, Crawford-Sidebotham (1972b) examined the relationship between numbers of slugs active at night and four weather parameters: temperature, relative humidity, vapour pressure and vapour pressure deficit. He demonstrated that the activity of *A. hortensis*, *A. subfuscus*, *A. lusitanicus* and *T. budapestensis* was best related to temperature and to vapour pressure deficit. There was evidence that the same was true for *D. reticulatum*. Activity tended to increase with increasing temperature, but to decrease with increasing vapour pressure deficit and so might be related to the general drying capacity of the air. Vapour pressure deficit exerts its influence on slug activity because the rate of evaporation of water from the surface of a slug is directly proportional to it. Crawford-Sidebotham described how attempts at constructing a model to predict the active numbers of a particular species had met with limited success, although at higher levels of activity the deviation from the prediction became less. Rollo (1982) developed an empirical model to describe the key factors regulating activity in *L. maximus*, using curvilinear regression models. He also developed a 'limit-type' model based on the endogenous clock and climatic factors. Models using lag-weather data did not explain as much variability as did concurrent weather. The most important factors included were time of day (i.e. circadian rhythm), light intensity, changes in light intensity and surface temperature. Other key factors included shelter temperature, temperature gradients, length of the night and time of sunset. Age and degree of hydration of slugs were also key factors in some instances. Rollo showed that a model incorporating weather thresholds explained about 83% of the variability in the activity of *L. maximus* over the season.

Hogan (1985) investigated the relationship between foraging activity and microclimate in the field using multivariate techniques. The activity of *D. reticulatum* was found to be related to temperature (air, soil and subsoil temperatures) and, to a lesser extent, relative humidity. Laboratory studies with actographs showed that, at constant tempertures between 0 and 24°C, the locomotor and feeding activity of *D. reticulatum* increased with increasing temperature to an optimum and thereafter declined (Wareing and Bailey, 1985). The optimal temperature for locomotor activity in short daylengths in winter was 13°C but this shifted to 17°C by early summer while the optimal temperature for feeding activity remained at 14°C.

Wareing and Bailey also confirmed the endogenous nature of the rhythm of locomotor activity. Cycling temperatures produced more locomotor activity and less feeding in 24 hours than constant temperatures, even when the temperature was higher at night. Cooling stimulated and warming inhibited activity during the period of change between cycling temperatures and, when cooling occurred at dusk, over half of the normal 24-hour activity occurred in the next three hours. Greater responses were produced by faster rates of cooling but these could result in lower overall activity because the slugs remained longer at lower temperatures. The stimulating effect of cooling on activity recorded by Wareing and Bailey recalls the increased activity induced by falling temperatures that was observed by Dainton (1954a). The effect of microclimate on slug activity in the field was investigated by Young and Port (1989), using infrared time-lapse video photography. Microclimate was expressed in terms of seven variables (humidity, windspeed, soil moisture content and four measures of temperature) recorded on a microclimatic level in relation to activity. Air temperature, soil surface temperature, windspeed, humidity and soil moisture content were each found to be correlated with activity. These factors were then used to formulate a 'limit-type' model to predict the activity of slugs in the field. This model accurately predicted activity down to a low threshold level. Cook and Ford (1989) described a threshold model for describing activity. This incorporated a mathematical description of daily and annual rhythms and three-dimensional representations of the variation of activity with time of day and with each environmental factor. Rollo (1989) discussed experimental considerations and analytical methodologies for investigating molluscan activity, including the construction of threshold models.

This review shows that there is an extensive literature attempting to relate weather to slug activity and the earlier work, particularly, often ignored the existence of endogenous rhythms. Rollo (1982) summarized the relationship between the two by stating that, in the field, weather acts mainly on the amplitude of the rhythm while the period and phase of the endogenous rhythm respond mainly to more predictable zeitgebers such as photoperiod. Cook and Ford (1989) expressed the similar view that the role of environmental variables in the control of laboratory activity was as a series of factors which restricted the time available for movement rather than as initiators of movement. Thus the endogenous rhythm determined when a slug was ready to begin activity, i.e. the period and phase of the rhythm, while the weather determined how that readiness was expressed, i.e. the amplitude of the rhythm. In some instances, the endogenous rhythm may be modified as a result of the physiological condition of the slug. For example, Hess and Prior (1985) found that, under dry conditions, L. maximus remained active well beyond the time that activity would have ceased for this species.

8

Life cycles

8.1 DEVELOPMENT OF THE EGG

Early records for the developmental period of the slug egg tend to be unreliable as the environmental conditions are not generally specified. According to Taylor (1902-1907) the eggs of most *Limax* spp. take about a month to hatch, but rather longer in the case of *Limax flavus* (40-60 days). Similar periods are required by eggs of *A. ater* (30-50 days), *A. hortensis* (20-40 days) and *A. intermedius* (3 weeks). Eggs of *D. reticulatum* hatch in three to four weeks while those of the related *D. laeve* may take longer (20-40 days). *Milax gagates* eggs hatch in less than a month. Taylor gives no information about environmental conditions except that these hatching times vary with weather conditions. The developmental period for *T. haliotidea* of 10-12 days is shorter than for other slugs while 20-36 days for *T. scutulum* and *T. maugei* is much shorter than the figure of 144-201 days recorded by Barnes and Stokes (1951) for *T. scutulum* kept in a cellar as near natural conditions as possible. The developmental period of six to eight weeks given by Taylor for *Geomalacus maculosus* was derived from Simroth in Taylor (1902-1907) and Rogers (1900) and appears to be the only record for this species. The eggs of *T. sowerbyi* take about four weeks to hatch (Miles *et al.*, 1931).

Künkel (1916) demonstrated that egg development was temperature dependent, being most rapid at temperature between 18 and 25°C. For all 12 species examined, the shortest developmental period was generally between three to four weeks, although slightly less for the smaller species of *Arion*. At lower temperatures, however, development could take as long as three or four months in *A. ater*, *A. subfuscus*, *L. marginata*, *L. tenellus* and *D. reticulatum* and the developmental period varied widely within a single batch of eggs. Frömming (1954) showed for *D. laeve* that egg development was considerably delayed at even lower temperatures (−3 to 4°C) and most embryos died. In this species, the majority of eggs hatched in the early morning. A developmental zero for this species of 4°C was confirmed by Stern (1979).

The effect of temperature on egg development under controlled conditions was investigated for *D. reticulatum* by Bachrach and Cardot (1924), Carrick (1942), Judge (1972) and Pakhorukova and Matekin (1977). Judge demonstrated a curvilinear relationship between temperature and mean time taken for 50% of eggs to hatch while Pakhorukova and Matekin found that a third-degree polynomial fitted these types of data, which approximates to a hyperbola with the developmental period increasing above an optimum of 20 or 21°C. The lowest temperature at which development was completed was 4.4°C (Judge, 1972) although the developmental period was prolonged, varying from 175 to 227 days. At 0°C all embryos died within six days (Carrick, 1942). The mean time taken for eggs to hatch exceeded 100 days at 5 and 6°C but decreased rapidly to between 27 and 33 days at 15°C and to between 18 and 22 days at 20 and 21°C. Even at these temperatures, however, there was an appreciable difference between maximum and minimum hatching times, e.g. 15–22 days at 20°C (Carrick, 1942) and 19–35 days at 21°C (Judge, 1972). Both Carrick and Bachrach and Cardot found that eggs failed to complete their development at temperatures above 21°C and, while both Judge and Pakhorukova and Matekin found that eggs hatched at temperatures of 26.6 and 28°C respectively, at these temperatures the developmental period increased and fewer eggs hatched. Carrick applied the concept of day-degrees to the developmental period of *D. reticulatum* but with little success.

A comparison of the effect of temperature on developmental period for the related species *D. reticulatum* and *D. agreste* by Pakhorukova and Matekin (1977) showed that the rate of embryonic development was considerably slower for *D. agreste*, the difference being most marked in the optimum temperature range for these species. The developmental zero is probably about 4°C for most temperate slugs. The marked differences recorded by Dmitrieva (1969) in length of developmental period for individuals of *D. agreste* from different geographical areas have not been observed for *D. reticulatum*. Interspecific differences were shown between four sympatric species of *Limax* by Cook and Radford (1988) from a comparison of the effects of temperature on egg development over the range 5 to 20°C. *Limax maximus* showed the shortest developmental period while *L. marginata* eggs took a significantly longer time to hatch at 15°C, with a higher embryo mortality, and failed to hatch altogether at 20°C. However, eggs of this species and of *L. maculatus* showed a higher survival rate at 5°C and, for the latter species, at 10°C. The ecological significance of these interspecific differences is discussed in Chapter 10. The developmental period for *L. maximus* eggs agrees with that of other authors, for example, Sokolove and McCrone (1978) give a period of about one month at 15°C. However, the extended hatching period of 8.7 days at 20°C contrasts with the observation by Prior (1983b) that all eggs in a batch hatch within about 24 hours at a temperature of 18–20°C.

Hunter (1968a) found that, under controlled conditions, the developmental period for eggs of *A. hortensis* ranged from two weeks at 20°C, four and a half weeks at 10°C and 14 weeks at 5°C. Similar values were found for *T. budapestensis*, ranging from three weeks at 20°C, ten weeks at 10°C and 18 weeks at 7.5°C, with little development taking place at 5°C. Eggs of *M. gagates* laid in autumn and incubated at laboratory temperatures hatched in 26–37 days (Focardi and Quattrini, 1972) while spring-laid eggs under similar conditions hatched a little earlier, in 20–25 days (Quattrini, 1970), although this difference might have been due to temperature differences in the laboratory. The development of the egg of tropical species of slug is similarly temperature dependent. Raut and Panigarhi (1988a) found that eggs of *L. alte* failed to hatch at temperatures of 15°C and below and at 35°C. The mean developmental period ranged from 20.92 days at 20°C to 13.35 days at 30°C with a corresponding percentage hatch from 72.6 to 81.6%. The maximum number of eggs hatched at room temperature, which ranged from 19 to 35°C.

The developmental period of the egg under natural conditions varies with the seasons. Hunter (1968a) found that eggs of *A. hortensis* took from three months in winter to only one month in summer to hatch. Eggs of *T. budapestensis* laid in December hatched in about four months compared with little more than a month for eggs laid in April. Development of eggs laid during winter months was delayed so that they tended to develop at the same time as eggs laid in spring, resulting in a peak hatch in late spring. South (1989b) showed that eggs of *A. intermedius* laid in August hatched within a month compared with three months for eggs laid in October. *Deroceras reticulatum* eggs laid in autumn took between five and six months to hatch while eggs laid in late spring took between two and a half months and four months to hatch. Carrick (1938) recorded an even shorter incubation period of only 22–36 days in August and September for this species while the longest period was 96 days during winter and early spring. He found that egg mortality increased as the period of development increased in winter. However, Dmitrieva (1969) considered that *D. reticulatum* eggs were rarely killed during the very cold winters of the Leningrad region provided they were protected by an adequate layer of snow. Under these conditions eggs were laid and began development in August and September. Development was arrested when the temperature fell below a developmental zero of 5°C during late autumn, winter and early spring and development was completed between April and June, giving an extended developmental period of nine or ten months for the egg of this species.

8.2 POSTEMBRYONIC DEVELOPMENT

Newly hatched slugs vary in weight from 2 to 3 mg in *Deroceras* spp. and the smaller *Arion* spp. up to 20 mg or more in *A. ater* and the larger *Limax* spp. Fully grown individuals, when extended, measure from 20 to 25 mm in length, weighing little more than 1 g (*A. intermedius*), up to 100–200 mm or more in length with weights of 20–30 g or more in *A. ater* and the larger

Limax species (Frömming, 1954; Quick, 1960). There is considerable intraspecific variation in the size of slugs at hatching and during their subsequent development, even amongst individuals hatched from the same batch of eggs. Peake (1978) showed that this variation was common in pulmonates generally and suggested that variations in the nutritional content of individual eggs might be an important factor in determining future growth patterns. For this reason, Prior (1983b) concluded that body weight was an unreliable measure of absolute or relative age. South (1982) described how the coefficient of variation for body weight in groups of 20 *D. reticulatum* ranged from 25–30% at the beginning of the juvenile phase to 40–50% at the end of that phase. Each group originated from a single egg batch and was reared under constant conditions. Similar values for *A. intermedius* were 20–25% and 30–35%. A similar high degree of variation in growth rate was described by Shibata and Rollo (1988) for offspring from unmated *D. laeve*, a species with low genetic variability (see Section 5.6.4). Shibata and Rollo concluded that variation in growth rate did not require high genetic polymorphism. They also found that the diet of the parent slug and, therefore, egg quality did not influence the growth rate of the young slug. Similarly, the early nutritional experience of the young slug did not influence the growth rate. Since individuals reared in isolation showed similar variation, intraspecific competition was unlikely to be responsible for this variation. Variation in egg size, however, seemed to account for the variation in growth rates, with larger eggs producing slower-growing individuals which matured later. Shibata and Rollo (1988) suggested that this variation might be related to resource supply by ensuring that mature slugs were available to breed when conditions were favourable and by reducing competition.

Abeloos (1944) made a detailed study of postembryonic development in *A. ater, A. subfuscus, A. hortensis, A. intermedius* and *A. circumscriptus* at 20°C and with an abundant mixed diet. He described three phases of growth in the development of the first four species: an initial phase of rapid growth (infantile phase) followed by one of slow growth (juvenile phase) and, finally a phase with little or no growth during which the slugs matured and laid eggs (mature phase). The development of the reproductive system was closely related to these growth phases. The gonad and reproductive tract remained very small during the infantile stage, developed rapidly during the juvenile stage and functioned after the beginning of the mature stage. The term prepuberty was applied to the transition between infantile and juvenile phases while the change from juvenile to mature phases was termed puberty. The mature phase was sometimes accompanied by a loss of weight and the reproductive tract later showed senescent changes similar to those described by Szabó and Szabó (1934c). *Limax maximus, L. cinereoniger* and *L. flavus* showed growth patterns similar to those described for *Arion* spp. (Abeloos, 1943). The infantile phase was similar in all three species with an absolute growth rate of 0.01–0.02 (log-weight (mg) increase per day) which was less than the corresponding figure for the four species of *Arion* (0.03). The juvenile growth rate was 0.005 for *L. maximus* and *L. flavus* compared with 0.005–0.01 in the *Arion* spp. The growth rate during the juvenile phase was slower for *L. cinereoniger*, prolonging this stage, with the result that egg laying was delayed until the second year of life. *Limax flavus* eggs contain larger quantities of albumen and, as a result, embryonic development is extended, so that the young slug at hatching is relatively larger than in other *Limax*. Since the infantile growth rate is similar and the slug attains the same weight at the end of the phase as the other *Limax* species, then the infantile phase is shortened. The juvenile phase was also relatively shorter so that the first eggs were laid 7–10 months after hatching. In comparison with *L. maximus*, sexual maturity was delayed in *L. cinereoniger* and accelerated in *L. flavus* with growth continuing after maturity in the latter species. Thus, unlike most *Arion* species which show considerable uniformity in their growth patterns, species of *Limax* show a series tending towards the reduction of the infantile phase of development in *L. flavus*.

There was one exception to the growth pattern described by Abeloos (1943) within the genus *Arion*. *Arion circumscriptus* showed only two growth phases, an initial rapid one associated with rapid development of the hermaphrodite gland and a second one of slower growth during which egg laying took place. The growth rate for the first stage was considerably less than that of the first (infantile) phase of other *Arion* spp. and was more similar to that of other juvenile phases. Abeloos concluded that the two growth phases in *A. circumscriptus* were equivalent to the juvenile and mature phases of other *Arion* spp., the infantile phase having been suppressed by the earlier development of the reproductive system. He had observed a similar type of development in the genus *Deroceras* (*D. agreste* and *D. laeve*) and in milacid slugs but gave no further details. However, Vianey-Liaud (1975) described three phases in the development of *M. gagates* and classified slugs into three groups on the basis of their growth pattern:

(1) *A. ater, A. subfuscus, A. hortensis, A. intermedius, L. maximus, L. cinereoniger* and *M. gagates*. Characterized by the existence of three growth phases where the mature phase showed little or no growth or even a loss in body weight.
(2) *L. flavus*. Characterized by the existence of three growth phases but showing an increase in body weight during the mature phase.
(3) *A. circumscriptus, D. agreste* and *D. laeve*. Characterized by an absence of an infantile phase and by an increase of weight during the mature phase, this increase being more marked for *Deroceras* spp.

However, Stern (1979) described two juvenile phases in the development of *D. laeve* corresponding to the infantile and juvenile phases of *Arion* spp.,

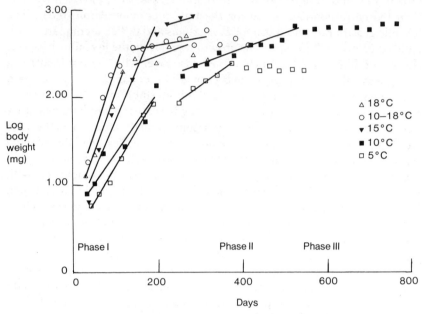

Fig. 8.1 Growth curves for *Deroceras reticulatum* at five temperatures showing growth phases I–III (after South, 1982).

while Runham and Laryea suggested that the stages of growth in *D. reticulatum* were similar to those found in most arionid slugs.

The growth rates and life cycles of *D. reticulatum* and *A. intermedius* were compared by South (1982) at several constant temperatures between 5 and 26°C and at a temperature which alternated between 10°C (16 hours) and 18°C (8 hours) daily. The results are summarized in Figs 8.1 and 8.2. The growth pattern for *A. intermedius* was similar to that described by Abeloos (1944), although the growth rates were rather less than those recorded by him for this species. The growth pattern for *D. reticulatum* appeared to be similar but phase III was poorly developed except at lower temperatures, and since egg laying occurred consistently in the second phase of growth, the latter one corresponded with the mature phase of growth in *A. intermedius*. *Deroceras reticulatum* thus shows a similar growth pattern to that of *A. circumscriptus* (Abeloos, 1944) with the infantile phase being suppressed. The relationship between phase I growth rate and temperature for *D. reticulatum* approximated to a hyperbola, with the maximum rate of development occurring at 18°C and minimum values at 5 and 26°C. Dmitrieva (1969) found that 17–19°C were the most favourable temperatures for growth in a study of growth rates in *D. reticulatum*. The preferred

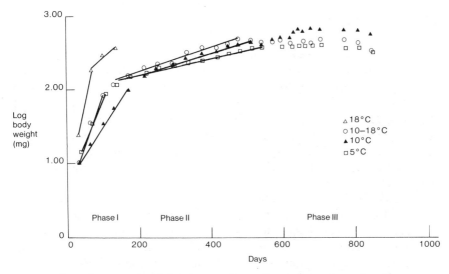

Fig. 8.2 Growth curves for *Arion intermedius* at four temperatures showing growth phases I–III (after South, 1982).

temperature for this species in a temperature gradient was also found to be 17–18°C by Dainton (1954a). The association between temperature and growth rate for *A. intermedius* was less clear although maximum growth also occurred at 18°C. Chevallier (1982) considered these to be the optimum temperatures for terrestrial gastropods of Atlantic Europe generally. Temperatures of 24°C and above were unfavourable for both species, growth rates were reduced, no eggs were laid and slugs did not reach maturity.

The effect of the alternating temperature on the development of *A. intermedius* was similar to that of the corresponding constant temperature of 12.7°C. However, for *D. reticulatum*, development was more rapid than would be expected for a constant temperature of 12.7°C with a higher phase I growth rate than at any constant temperature. This might be due to a higher level of activity caused by temperature changes since Dainton (1954a) showed that changes of temperature below 20°C stimulated activity for this species. South concluded that, for *D. reticulatum*, most life-cycle parameters were clearly temperature-sensitive processes, this close dependence being emphasized by the enhanced effect of the alternating temperatures while *A. intermedius* appeared to be less sensitive to the effect of temperature change. *Deroceras reticulatum* was also capable of growth and reproduction over a wider range of temperatures than *A. intermedius* under controlled conditions.

8.3 LIFE SPANS AND LIFE TABLES

The recorded life spans of slugs vary widely, depending on whether they relate to slugs reared under protected conditions or in the wild and whether they represent the maximum recorded age or mean life span. Comfort (1957) summarized data for slugs kept in captivity. These show life spans of between 11 and 13 months for the smaller *Arion* spp., extending to 18 months for *A. ater.* Life spans of several years are recorded for *Limax* spp. with a maximum of five years for *L. cinereoniger* and a life span in excess of 570 days for marked *T. scutulum* and *T. haliotidea* recovered from the field. Oldham (1942) recorded a life span of six and a half years for a single captive *G. maculosus.* Szabó and Szabó (1929) gave a maximum recorded age of 14 months for laboratory reared *D. agreste.* Künkel (1916) found that some slugs and snails reared separately had a longer life span than those that were allowed to mate. Thus *A. ater,* kept isolated, lived for 16–18 months compared with 12–14 months with those allowed to mate. Szabó and Szabó (1936) showed that delaying mating prolonged the life span of *D. agreste* and suggested that self-fertilization, which results in few eggs being laid, is less exhausting to the animal than sexual reproduction. South (1982) showed that life span under controlled conditions was directly related to temperature for *D. reticulatum,* with a mean life span ranging from 527.3 days at 5°C to 71.6 days at 26°C. However, although the mean life span for *A. intermedius* was reduced at higher temperatures – for example, from 938 days at 5°C to 95 days at 24°C – the relationship was less clear as the mean life span was similar at 5, 10 and 12.5°C (mean of alternating 10 and 18°C) and was then considerably reduced at 18°C. The maximum recorded ages for these species of about two and a half years contrast with the annual life cycle of these species under natural conditions. Since the maximum body weights for these species and their growth rates were also dependent on the temperature at which slugs were reared, caution is needed when extrapolating laboratory data for use in the field.

A life table and survivorship curve for *D. agreste* was constructed by Pearl and Miner (1935) and based on the data of Szabó and Szabó (1929). They obtained a slightly sigmoid curve approaching the Type II curve of Pearl (1928) and indicating a mortality rate which remains almost constant until near the end of the life span. The ratio of total life span to mean duration of life is about 3 : 1 for this type of curve, some individuals living three times as long as the population average. Survivorship curves for *D. reticulatum* (South, 1982), based on the survival of slugs at different temperatures after hatching, are more similar to the convex Type I curve of Pearl (1928). Mortality in this type of curve is low throughout life until old age, when death of the slugs takes place over a relatively short period. The ratio of total life span to mean duration of life might be expected to approach unity and

South obtained values in most instances ranging from 1.07 : 1 to 1.46 : 1. This pattern of survival might be expected in the protected conditions of a laboratory culture. The survival of slugs belonging to separate generations of *D. reticulatum* and *A. intermedius* on permanent pasture (South, 1989b) showed a more or less constant mortality rate throughout life, similar to the results of Pearl and Miner (1935). This does, however, indicate that there is a considerable mortality rate during the egg and very young stages under natural conditions.

The changes that take place during senescence were reviewed and discussed by Szabó (1935). Outward changes include a hardening of the body wall accompanied by a darkening in colour. The outline of the internal shell, located beneath the mantle, becomes more obvious. The buccal mass may protrude from the mouth and a similar evagination of the reproductive organs may occur if the animal lays eggs at this stage. Ulceration of the body wall occurs more frequently than in younger animals. These features have been observed in *D. agreste, L. flavus, L. cinereoniger, A. circumscriptus* and *A. subfuscus* (Szabó and Szabó, 1930a, 1931a; Szabó, 1935a). Internally, the changes associated with senescence vary with species. In *D. agreste*, senescence may be accompanied by loss of weight (Szabó and Szabó, 1929). The most obvious changes take place in the digestive gland. Loss of weight is accompanied by a reduction in the cytoplasm of cells lining the acini of the digestive gland. The resulting atrophy of the digestive gland is accompanied by an increase in the amount of connective tissue in that gland. Where there is no atrophy of the digestive gland, the amount of connective tissue within the lobes still increases and eventually interferes with the functioning of the gland. This proliferation of connective tissue was considered the major cause of death at senescence (Szabó and Szabó, 1931b). Szabó and Szabó (1930b) concluded that since the increase in connective tissue begins early before any atrophy is observed, it is unlikely that atrophy causes an increase in connective tissue. A similar proliferation of connective tissue, usually accompanied by atrophy of the acini in the gonad, has been described for *A. agreste* by Szabó and Szabó (1936). Similar changes have been recorded for other species and a full account of postembryonic changes in the gonad of slugs has been given in Chapter 5.

Structural changes due to senescence in *L. flavus* were investigated by Szabó and Szabó (1931b), while Szabó and Szabó (1934a) compared these with those of *D. agreste*. Proliferation of connective tissue in the digestive gland, gonad and other parts of the body occurred with ageing in *L. flavus* but to a lesser extent than in *D. agreste*. A more important sign of senescence in *L. flavus* was the deposition of a brownish-black pigment, particularly in the ganglion cells of the nervous system. This pigment accumulated in the ganglion cells, causing their death and resulting in the atrophy of organs innervated by the ganglia. Szabó and Szabó found that the accumulation of pigment in *L. flavus* was more rapid than the rate of proliferation of connective tissue and concluded that, although both factors were operating, the accumulation of pigment was the primary cause of death by senescence in *L. flavus*. These two factors worked independently of one another. There was little or no pigment in the nerve cells of *D. agreste, D. reticulatum* and *D. laeve* although increased pigmentation with age was also recorded for *A. circumscriptus, A. subfuscus* and other molluscs (Szabó and Szabó, 1931b; Szabó, 1935). It has already been noted that the life span of slugs can be lengthened if they are prevented from mating. Examination of the hermaphrodite gland of older unmated individuals of *D. agreste*, which had not laid eggs, showed that undifferentiated cells of the germinal epithelium (Polimanti cells of Szabó and Szabó), which were few and scattered in young animals, had proliferated and formed a distinct tissue. Where this was present, signs of senescence such as atrophy in other organs of the body were absent (Szabó and Szabó, 1936). They concluded that the Polimanti cells served to lengthen the life span of slugs, possibly through an endocrine effect. Fournié and Chétail (1982b) suggested that an overload of calcium stored in the body and the hardness of the integument, consequences of a sclerosis of the calcium cells, may be the cause of death in *D. reticulatum* and also of the frequent ruptures which preceded death itself (see Section 3.5).

South (1982) described a well-developed postreproductive period in the life cycle of *A. intermedius* during which slugs showed clear signs of senescence. There was no clear postreproductive period for *D. reticulatum*, although the extended mature phase at lower temperatures indicated some postreproductive survival at these temperatures. Egg-laying continued until all slugs had died although slugs dissected near the end of their life span showed signs of senescence. Carrick (1938) described how for this species, death was not preceded by any outward sign of senility and that egg laying continued until a few days before death at laboratory temperatures. Künkel (1916) reached similar conclusions and Dmitrieva (1969) showed that *D. reticulatum* stopped feeding and died at the end of the reproductive period. Focardi and Quattrini (1972) found that death in *M. gagates* occurred within two or three weeks after the last eggs were laid, rapidly following a marked loss of weight due to dehydration during the last two or three days of life. A similar weight loss was recorded for *D. reticulatum* by South (1982).

8.4 FIELD STUDIES

Taylor (1902–1907), Künkel (1916) and other early authors suggested that members of the genus *Arion* had an annual life cycle. This was confirmed for *A. ater* (Chevallier, 1971; Jennings, 1975; Rollo *et al.*, 1983a), *A. lusitanicus* (Chevallier, 1971), *A. subfuscus* (Chevallier, 1971; Beyer and Saari, 1978; Chichester and Getz, 1973; Jennings, 1975), *A. hortensis* (Hunter, 1968b),

A. intermedius (Jennings, 1975; South, 1989a) and *A. circumscriptus* (Burenkov, 1977). A summary of the life cycle of *A. intermedius* on permanent pasture is shown in Fig. 8.3. The timing of the cycle varies considerably within some species such as *A. hortensis* and hatching has been recorded at various times between mid-summer and mid-winter. This difference may be for geographical reasons (Beyer and Saari, 1978) although Jennings (1975) found that newly hatched individuals of both *A. hortensis* and *A. fasciatus* were present for most of the year in the same wood and suggested this might have been due to particularly favourable climatic conditions. Davies (1979) recorded specific differences in the timing of the appearance of mature slugs within the *A. hortensis* complex with *A. distinctus* maturing later and *A. owenii* maturing earlier than *A. hortensis* s.s. South (1989a) found that the onset of egg laying in *A. intermedius* was delayed until the spring by an exceptionally dry summer. Laviolette (1950c) recorded a similar delay in the maturation of *A. ater* while Chevallier (1971) found that a few individuals of *A. ater*, *A. lusitanicus* and *A. subfuscus* took two years to complete their life cycle when adverse weather conditions retarded their growth in the first year. The body weight of these slugs at

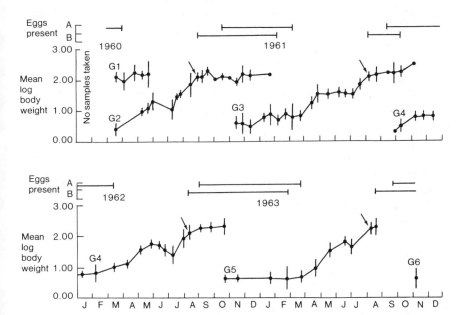

Fig. 8.3 Summary of life cycle of *Arion intermedius*. Presence of eggs indicated by continuous lines for late developmental stages (A) and early developmental stages (B). Body weights shown as mean log weights with confidence limits. Arrows show first appearance of mature-unmated slugs. G1–G5: generations of *A. intermedius*. (After South, 1989a.)

maturity was then greater than the usual weight of one-year-old animals. Similar delays have been recorded for *A. hortensis* in northern Britain (Hunter, 1968b) and *A. circumscriptus* in the Moscow region (Burenkov, 1977).

Published information on the life cycle of *Deroceras reticulatum* shows some discrepancies, as Carrick (1938) demonstrated from his review of early records for the breeding season of this species. *Deroceras reticulatum* has two main breeding seasons each year with young slugs hatching in late spring and early autumn, although eggs may be found throughout the year (Miles *et al.*, 1931; Karlin and Naegele, 1958). It has been suggested, either, that this species bred continuously throughout the year at a rate which depended on weather conditions and reached a peak in spring (Bett, 1960), or, that there were two successive generations each year (Hunter, 1968b) with a spring generation hatching in May and giving rise to an autumn generation hatching in late September, the adults having a life span of between five and seven months. Bett suggested that the shortest life span might be only three or four months in the summer with newly hatched slugs in May and June maturing in late July or August. Judge (1972) suggested a similar subannual life cycle for *D. reticulatum* but this kind of cycle seems unlikely because Bett's conclusions were based on an estimated developmental period for the egg of only 2.5 weeks derived from the laboratory studies of Carrick (1938). Hunter and Symonds (1970) considered that development was slower under natural conditions and that the life span was probably nine months or more in length. This conclusion has been supported in laboratory studies (South, 1982) and in the field (South, 1989a) where the maximum life span of *D. reticulatum* was estimated at between 9 and 12 months and the shortest developmental period for the egg was 2.5 months even during the summer. Thus two complete generations in a year are unlikely as overwintering eggs may take up to five months or more to hatch. Burenkov (1977) observed two breeding seasons in the Moscow region in autumn and spring although he considered that *D. reticulatum* was a univoltine species, the spring eggs being laid by slugs which had not laid eggs in the previous season.

Hunter and Symonds (1970) showed that a population of *D. reticulatum* on uncultivated ground near Cambridge could be divided into two overlapping sets of generations with a generation interval of about nine months. South (1989a) found a similar pattern (Fig. 8.4) on permanent pasture in Northumberland but with a generation interval of about 12 months. Successive generations could be divided into two overlapping sets: set A (Generations I, III, V, VII and IX in Fig. 8.4) and set B (Generations II, IV, VI and VIII in Fig. 8.4). Young set A slugs hatched in late summer, developed during the winter months, matured and laid eggs late in the following spring and survived into the summer months, while young set B slugs first appeared in late spring, matured and laid eggs in late autumn and survived until early in the following spring. Such an arrangement would

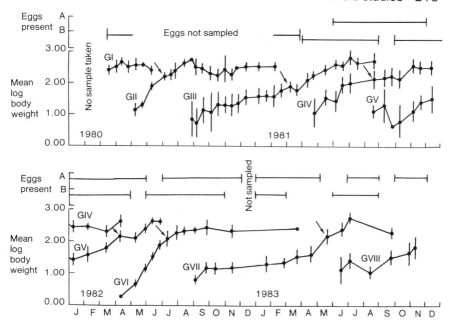

Fig. 8.4 Summary of life cycle of *Deroceras reticulatum*. Presence of eggs indicated by continuous lines for late developmental stages (A) and early developmental stages (B). Body weights shown as mean log weights with confidence limits. Arrows show first appearance of mature-unmated slugs. GI–GVIII: generations of *D. reticulatum*. (After South, 1989a.)

explain the spring and autumn peaks in numbers of young slugs noted by previous authors. Duval and Banville (1989) described similar overlapping sets of generations of *D. reticulatum* from two Quebec strawberry fields, although at other similar but more intensively cultivated sites populations of this species were univoltine. They suggested that intensive cultivation and molluscicide treatment breaks down overlapping and leads to a single population.

South (1989a) considered it unlikely that set A or set B slugs represented distinct races of *D. reticulatum* since there was some variation in the timing of each generation. The extended hatching period for the eggs meant that a few slugs probably mature early or rather late and so may contribute to the preceding or succeeding generations of the alternate set. Dmitrieva (1969) showed that a population of *D. reticulatum* living under the more extreme climatic conditions of the Leningrad region consisted of a single generation in each year with an overwintering egg stage, which corresponds to the life cycle of set B slugs (South, 1989a). There was some variation in this pattern, however, and in two years out of the three examined, as the result of particularly favourable climatic conditions during the egg-laying period, some early eggs hatched and a second generation of slugs, corresponding to set A, appeared at the beginning of the autumn. It seems likely that, as the result of the prolonged cold winter, these young slugs were either killed or their development delayed so that they became indistinguishable from the slugs hatching in the spring. Warley (1970), working in the south-east of Scotland, came to a similar conclusion for *D. reticulatum*. An otherwise continuous breeding cycle was restricted by the cold conditions of winter and early spring, during which time hatching was negligible. The development of separate generations thus appears to depend on climatic conditions and under suitable conditions, *D. reticulatum* might breed continuously throughout the year. Both Pilsbury (1948) and Karlin and Naegele (1958) reported that this occurred in the USA when weather conditions permitted. In Northumberland continuous breeding was prevented by the slower rate of development of eggs (set B) and young slugs (set A) during the winter months and by some reduction in the activity of set B slugs during dry periods in the summer months. South (1989a) suggested that the pattern of generations was probably consistent within a single geographical region. *Deroceras laeve* appears to have a similar life cycle and Burenkov (1977) described it as a univoltine species in the Moscow region while Boag and Wishart (1982) found that in Alberta it was an annual species which overwintered as immature slugs. Chevallier (1973a) described the life cycle of *D. caruanae* as being similar to that of *D. reticulatum*, with eggs being found throughout the year. Reygrobellet (in Chevallier, 1973a) described growth in *D. caruanae* as being more rapid than in *D. reticulatum*, with the optimum rate of growth at 20°C. *Tandonia budapestensis* has a biennial life cycle, hatching during May to August and maturing during its second autumn and winter (Hunter, 1968b). *Milax gagates* also follows a similar cycle with slugs reproducing only once during their lifetime, either in autumn or spring (Galangau, 1964; Focardi and Quattrini, 1972). Slugs of the genera *Arion*, *Deroceras* and *Milax* belong to the annual and biennial group of molluscs which are semelparous, i.e. breeding once only during their life cycle. In contrast, the longer-lived *Limax* spp. are a pluriennial, iteroparous species, having a specific lifespan of two or three years depending on the species and with two or three egg-laying periods in their life cycle (Künkel, 1916). The one exception to this pattern appears to be *L. tenellus* which has an annual cycle. Szabó and Szabó (1934b) described an individual *L. flavus* which lived for four years and laid eggs during each autumn. The timing of the life cycle in some pluriennial *Limax* spp. is regulated by photoperiod. Both Barker and McGhie (1984) in New Zealand, and Sokolove and McCrone (1978) in the USA, demonstrated an annual cycle of changes in the reproductive system of *L. maximus*. Spermatogenesis was induced by the transfer from short days to long days in the spring and early summer, and this was followed by oogenesis during the summer and fertilization followed by egg laying in the autumn. In *L. flavus*, however,

Segal (1960) found that the annual cycle persisted in the absence of external seasonal changes, including photoperiod. *Ariolimax columbianus* is another pluriennial slug with a life span of four to six years and an annual breeding period in the autumn (Rollo *et al.*, 1983a). In a study of four species of *Limax*, Cook and Radford (1988) showed that *L. flavus*, *L. maximus* and *L. maculatus* were polyvoltine, semelparous species but found that *L. marginata* was a univoltine, iteroparous species with an annual cycle, with few adults surviving into the following spring in contrast to the pluriennial cycle described by Künkel (1916) for this species.

The growth pattern for *A. intermedius* in the field differed from that described under controlled conditions (South, 1982, 1989a). Growth in the field was slow initially (Fig. 8.3) but increased rapidly in late March, and this increase was maintained until mature slugs first appeared in August. This transition to the mature phase of the life cycle (Fig. 8.3) corresponds to the *puberté* of Abeloos (1944) and there was little growth after this point. A similar pattern of growth was shown by *A. ater*, *A. lusitanicus* and *A. subfuscus* under natural conditions (Chevallier, 1971). Chevallier divided the infantile growth phase into (i) an initial stage of slow winter growth (*croissance infantile hivernale*) which was dependent on temperature, followed by (ii) a stage of rapid spring growth (*croissance juvenile pre-estivale*). He then described a *phase juvenile estivale*, corresponding to the juvenile phase of Abeloos (1944), which ended at puberty when slugs matured and were ready to copulate.

Two stages of growth were distinguished by South (1989a) for *A. intermedius*, and these are shown in Fig. 8.3. They represent (a) the first stage of slow winter growth corresponding to the first part of the infantile phase and (b) the second part of rapid spring growth similar to that described by Chevallier and corresponding to the second part of the infantile phase together with a part of the juvenile phase. Thus the potentially rapid growth rate for the infantile phase of this species under laboratory conditions (South, 1982) was initially suppressed in the field by low temperatures, particularly in the long cold winter of 1962/63. Hunter (1968b), working on arable land within a few miles of the study area, recorded similar delays at that time for *D. reticulatum*, *A. hortensis* and *T. budapestensis*. South (1989a) found no statistical correlation, however, between mean air temperature and growth rates for either first or second stages and concluded that growth in *A. intermedius* was less sensitive to the effect of temperature change. He suggested that the life cycle might be synchronized by photoperiod since the rapid stage of growth each year coincided with lengthening days as the photoperiod (L:D) reached 12 : 12 hours. The second growth stage in *A. intermedius* was usually delayed in late spring by dry conditions when growth ceased, body weight was reduced and the slug entered a quiescent state. This pause in development resulted in egg laying being delayed until late winter and early spring in one exceptionally dry year.

Young *A. hortensis* appear earlier that *A. intermedius* and Hunter (1968b) found that young *A. hortensis* confined in cultures in the field in August grew rapidly during their first autumn, slowly through the winter and then rapidly during the following spring. Similar rapid growth in the autumn was prevented for *A. intermedius* by the later egg-hatching date although the growth of *A. hortensis* was subsequently delayed by low winter temperatures in a similar way to *A. intermedius*.

A single immature growth stage was described by South (1989a) for *D. reticulatum* under natural conditions (Fig. 8.4). This was equivalent to the juvenile phase of most other slugs since the infantile phase is suppressed in this species (South, 1982). Growth in *D. reticulatum* continued throughout the winter months except under exceptionally cold conditions and was not usually delayed by dry conditions in spring. There was no evidence that the life cycle of *D. reticulatum* was synchronized by seasonal changes in photoperiod, which is not surprising since this species will breed whenever environmental conditions are favourable. Growth rates for overwintering juvenile slugs (set A above) were less than the corresponding value for slugs kept at a constant temperature of 5°C (South, 1982) although the mean temperatures were similar to this value. Slugs in the field, however, were exposed to fluctuating temperatures which included long periods when the temperature remained below 5°C. Poulicek *et al.* (1980) found that, in Belgium, adult *D. reticulatum* hibernate from October until March, and Getz (1959) recorded that adults hibernate during the winter in Michigan (USA). However, Mellanby (1961) described how, at 0.8°C, *D. reticulatum* remained active and fed normally and this was confirmed by Wareing and Bailey (1985), although the rate of feeding activity was considerably reduced at 0°C. South (1989a) recorded this species feeding on grass under a deep layer of snow. Crawford-Sidebotham (1972b) concluded that activity in *D. reticulatum* is not restricted to the same extent by low temperatures as it is in other species of slug. The summer generations of juvenile slugs (set B above) developed at significantly higher rates than set A slugs and growth rates were higher than the corresponding value for slugs kept at a constant temperature of 10°C (South, 1982) although mean temperatures experienced by these slugs were similar to this temperature. The fluctuating temperatures experienced by slugs in the field probably resemble cycling temperatures more closely than constant temperatures and these higher growth rates may be explained by the increased activity which results from cycling temperatures under laboratory conditions (South, 1982, 1989a).

Runham and Laryea (1968) described the development of the reproductive system of *D. reticulatum* and concluded that the stages in maturation of the system were related to the growth phases in the life cycle of the slug. Details of these stages were given in Chapter 5. South (1989a) related these stages to the growth phases of *D. reticulatum* taken from permanent pasture. The gland of very small slugs (less than 30 mg body weight) was at the 'undifferentiated stage' of Runham and Laryea. The point at which the gland could first be weighed accurately but was still relatively small (0.4–1.1% of body weight) corresponded to the beginning of the 'spermatocyte stage'. The rapid increase in size of the gland to between 3.3 and 5.5% of body weight by the end of the immature stage represented the 'spermatid stage'. Spermatozoa appeared in the hermaphrodite duct at the end of the 'spermatid stage' and so the gland of the mature unmated slug represented the 'early spermatozoan stage' of development when the gland reached its maximum size although its size relative to body weight remained the same. There was a subsequent reduction in size of the gland to between 1.6 and 2.6% of body weight due to the loss of spermatozoa and to egg laying by mature mated slugs and this corresponded to the 'oocyte' stage of Runham and Laryea. A small number of mature slugs (about 10%) found with considerably reduced hermaphrodite glands (less than 1% of body weight), showing signs of atrophy, represented the 'postreproductive stage'. The small number of these slugs supports the conclusion of South (1982) that egg laying continues almost up to death and that the postreproductive stage is relatively short except at low temperatures.

9

Predators, parasites and disease

9.1 INTRODUCTION

The study of predators and parasites of molluscs has long been of academic interest although early records for slug predation and parasitism are largely restricted to the anecdotal information contained in comprehensive works on slugs and snails, such as Taylor (1902–1907). More recently, the possible use of a range of other organisms in the biological control of pest species of slug has attracted considerable attention and the potential for exploitation of these organisms suggests that they may have an important role in the future control of slugs (Port and Port, 1986). It is difficult to obtain evidence from direct observations in the field but this has been done successfully for birds (e.g. Feare *et al.*, 1974). Slugs have also been offered to a range of potential predators in the laboratory (e.g. Feare *et al.*, 1974; Poivre, 1972; Tod, 1973) although acceptance does not show whether slugs form part of the normal prey of a specific predator. One difficulty in the study of the predation of slugs is the identification of mollusc remains in the gut of a potential predator, particularly in fluid feeders such as many carabid beetles (Davies, 1953). Digestion of the slug is often rapid, particularly in birds, and the only structures which resist digestion are the jaw, radula and, very occasionally, fragments of the internal shell. The jaws of slugs, particularly *D. reticulatum*, can, however, be identified and the weight of individuals of this species consumed by the predator can be estimated (South, 1980). Immunological techniques have been used successfully to determine the presence or absence of slug remains in the gut of predatory carabid beetles and of ragionid fly larvae (Tod, 1973; Sergeeva, 1982, 1984).

9.2 PREDATORS

9.2.1 Vertebrates

Mead (1961, 1979) carried out an extensive review of the predators and parasites of terrestrial gastropods, with particular reference to the giant African snail *Achatina fulica*. His list of vertebrate predators included amphibians, reptiles, birds and mammals although he suggested that some instances of slug and snail consumption by mammals were probably incidental or accidental. He described how the giant Central American toad,

Bufo marinus, was introduced into Guam from Hawaii in 1937 in an attempt to control the large black pest slug, *Veronicella leydigi*. This early attempt at biological control was very effective but there were many indirect effects, not all of which were advantageous.

Slug predation has been recorded for all five groups of vertebrates, including fish. Elliott (1967) included the slug *A. circumscriptus* among the terrestrial invertebrate component of the diet of the trout (*Salmo trutta*) in a Dartmoor stream. While frogs and toads have long been cited as important predators of slugs and snails (e.g. Noel, 1891; Taylor, 1902–1907), there is a general lack of quantitative information on the subject with most references coming from books on gardening and herpetology or based on the feeding behaviour of captive amphibians. Taylor considered that frogs and toads were great enemies of slugs, particularly the genus *Limax*. The use of the toad *Bufo marinus* in the control of slugs has been referred to previously while Grassé (1968) included *Bufo* and *Ceratophrys* as predators of *Veronicella tuberculata*. Pilsbury in Taylor (1902–1907) described how *Limax* sp. was eaten by salamanders.

Slugs are included in the diet of several species of colubrid snake in North America. These include the redbelly snake (*Storeria occipitomaculata*), whose seasonal activity follows that of their main food resource of slugs (Semlitsch and Moran, 1984), the brown snake (*S. dekayi*) which feeds mainly on slugs but also ate earthworms (Catling and Freedman, 1980a, b) and two species of garter snake *Thamnophis ordinoides* and *T. elegans* (Gregory, 1978) whose diet is a little more varied. An analysis of a slug-eating response by newborn *T. elegans* demonstrated geographic variation in behaviour with a genetic basis (Arnold, 1981). Two conspecific populations of *T. elegans*, representing a coastal and an inland geographic race, were tested for their response to different prey including the slug *Ariolimax californicus*. The coastal snakes were terrestrial foragers, preying mainly on slugs, while the inland snakes were aquatic foragers with fish and amphibia as prey. Although both populations were polymorphic for slug-eating tendency, a slug-eating morph predominated in the coastal population while a slug-refusing morph predominated in the inland population. These differences in feeding behaviour were stable. The slug-refusing morph starving to death if only offered slugs. Arnold (1981) showed that in mollusc-rich environments in coastal California, where the slug-eating habit evolved, slugs constituted 90% of the diet.

Burghardt (1968) showed that when newborn young of three sympatric species of water snakes (*Natrix* spp.) were tested with water extracts of a range of small animals, none of the species responded to extracts from slugs or other terrestrial animals. Slug eating may be more widespread among colubrid snakes, Broadley and Cock (1975) described a snake known as the slug-eater (*Duberria lutrix*) from Zimbabwe. Taylor (1902–1907) included the slow-worm *Anguis fragilis* among the predators of slugs and stated that this lizard would feed readily on *D. reticulatum*, in captivity taking four or five or more at one meal, while refusing to eat *A. hortensis* because of its sticky mucus and tougher skin. Poivre (1972) found that, in addition to earthworms, slow-worms fed on slugs and insect larvae in north-eastern France. He offered both young *A. ater* and *D. reticulatum* to slow-worms and confirmed Taylor's observation that *D. reticulatum* was the preferred prey while the sticky mucus of *A. ater* hindered the feeding of the slow-worm, although it still succeeded in eating the *Arion*.

The available information on bird predation tends to be contradictory (Runham and Hunter, 1970). Wild and Lawson (1937), in their review of the enemies of British slugs and snails, listed 24 bird species recorded feeding on slugs. However, Kleiner (in Mead, 1961) demonstrated from an extensive examination of bird gut contents that, while snail remains were frequent and occurred in a wide range of species, their relative bulk was small in comparison with the total food taken. He found that while a wide variety of birds will consume snails, in some instances these may have been taken accidentally into the gut. This is probably equally true for slugs and, in addition, since digestion in birds is rapid and their remains may have been overlooked, it is difficult to judge the relative importance of slugs as a component of the diet of birds. Boycott (1934) considered that most bird species reject larger slugs such as *L. maximus* and *A. ater* as uneatable. Pitchford (1954) described how a blackbird (*Turdus merula*) cleaned the mucus from an *A. subfuscus* by dragging it through the ashes of an ash path before eating it. Slugs have been recorded in the diet of starlings (*Sturnus vulgaris*) by Collinge (1921) (6.5% of the total bulk of food) and Schermer (1958), although Dunnett (1956) did not consider them as a significant source of food in his study of the autumn and winter food of starlings. Schrey (1981) observed that gastropods comprised about 20% of the stomach contents of starlings on Heligoland, but three-quarters of these gastropod remains were identifiable snails, and although the unidentified portion may have included slug remains, these probably represented only a small proportion of the starling's diet. South (1980) found that starlings feeding on permanent pasture readily took *D. reticulatum* and that larger individuals were selected. The starling's method of feeding, by probing the turf, meant that surface-living slugs, such as *D. reticulatum*, were readily available to these birds.

Taylor (1902–1907) recorded the rook (*Corvus frugilegus*) as a predator of *D. reticulatum* although Lockie (1956) did not record slugs during his investigation into the food of the rook, the carrion crow (*Corvus corone*) and the jackdaw (*Corvus monedula*). However, Holyoak (1968) recovered slugs from the gizzards of jackdaws and rooks while Feare *et al.* (1974) showed that they formed a very small proportion of the diet of rooks for a limited period of the year. South (1989b) found no slugs in the gut contents of a small sample of rooks, jackdaws, crows, lapwing (*Vanellus vanellus*), redwing (*Turdus musicus*), fieldfare (*Turdus pilaris*), common gull (*Larus canus*) and black-headed gull (*Larus ridibundus*) which had been feeding on permanent pasture. Cuendet (1983) also found that slugs were not included in the gut contents of black-headed gulls feeding on farmland. Grassé (1968) included vultures and owls as predators of arionid slugs. Taylor (1902–1907) stated that poultry, especially ducks and geese, were partial to slugs although, while readily eating *A. subfuscus*, they rejected the larger slug *A. ater*. Curtis (1860) and Anon. (1905) recommended the introduction of a flock of ducks for slug control. A wide range of other bird species were recorded in their review by Wild and Lawson (1937) as predators of slugs including, fieldfare (feeding on *D. reticulatum*), redwing (on *Limax* and *Arion*), blackbird (*Turdus merula*) (on *Limax*), thrush (*Turdus ericetorum*) (on *Limax* and *A. ater*), red grouse (*Lagopus scoticus*) (on *D. reticulatum* and *A. hortensis*), other game birds, three species of owl and several species of charadriiform birds. Koval (1976) included *L. flavus* as a component of the diet of the wryneck (*Jynx torquilla*) in gardens in the Dnieper area (USSR). Slugs have also been recorded from the gut contents of mature rock doves (*Columba livia*) in the USSR (Kotov, 1978).

Corbet and Southern (1977) included slugs as a component of the diets of hedgehogs (*Erinaceus europaeus*), common shrews (*Sorex araneus*), moles (*Talpa europaea*) and badgers (*Meles meles*) although shrews showed a dislike for molluscs in laboratory tests. However, MacDougall (1942), Raw (1966) and South (1980) all concluded that slugs were not frequently taken by moles. Morris (1983) confirmed that slugs are a frequent component of hedgehog gut contents. Boycott (1934) reported that *A. ater* was a favourite food of the hedgehog and was also eaten by badgers. However, Herter (1938) observed that while hedgehogs ate *Deroceras* they rejected larger slugs of the genera *Arion* and *Limax*. Dimelow (1963) found from a series of preference tests that while captive hedgehogs would eat *Milax* spp., *Limax* spp., *Deroceras* spp. and some *Arion* spp., they rejected *A. ater*, *A. fasciatus* and *A. subfuscus* on account of their tougher skin. Size was not important since hedgehogs readily tackled *L. maximus* 15 cm in length and mucus did not appear to be a deterrent, as both *T. budapestensis* and *A. hortensis* were consumed despite the mucus they secreted. Taylor (1902–1907) considered *D. reticulatum* to be a favourite food of hedgehogs in the wild, while Brockie (1959) found that this species was eaten frequently by hedgehogs in New Zealand. South (1980) recovered slug jaws, mainly belonging to *D. reticulatum*, from hedgehog faeces at different times of the year.

Pernetta (1976) and South (1980) concluded that slugs were not important components of the diets of the common shrew and the pygmy shrew (*Sorex minutus*), although slug remains occur fairly regularly in the gut contents of the common shrew in Britain. Skaren (1978) showed that while the least shrew (*Sorex minutissimus*) fed readily in captivity, it refused slugs, snails and earthworms even when hungry. Mead (1961) concluded that other species of shrew are among the worst enemies of land molluscs. In a study of the foods of six species of sympatric shrews in New Brunswick, slugs comprised a major component of the food of two species, *Blarina brevicauda* and *Sorex palustris* (Whitaker and French, 1984). Slugs were among the major foods of the vagrant shrew (*Sorex vagrans*) when it was feeding in relatively ungrazed mountain meadows in Oregon but they were replaced by other invertebrates in areas subjected to increased grazing (Whitaker *et al.*, 1983). The suggested reason for this change was that the trampling and compression resulting from grazing resulted in a decrease in populations of some invertebrates, including slugs.

The Malayan insectivores *Echinosorex gymnurus* and *Suncus murinus* frequently feed on snails and slugs (Liat, 1966). Liat also showed that molluscs, including the slug *Microparmarion malayanus*, were the preferred food of *Rattus jalorensis* and *R. argentiventer* although other *Rattus* spp. were less inclined to feed on snails and slugs. There seems little doubt that many of the instances where gastropod remains have been found in the gut contents of a wide range of mammals represent incidental or accidental consumption (Mead, 1961). This is confirmed by the occurrence of parasites in many mammals, including ungulates, where the vector is either a slug or a snail.

9.2.2 Invertebrates

Mead (1961, 1979) included Coleoptera (Carabidae, Lampyridae and Drilidae), Diptera (Calliphoridae, Sciomyzidae and Phoridae), mites and other gastropods among the invertebrate predators of slugs and snails. The terricolous triclad flatworm *Geoplana septemlineata*, a predator of terrestrial snails and endemic to Hawaii, was reported by Mead (1979) to feed on the introduced slug, *D. reticulatum*. Mead (1979) described how another large terricolous triclad from New Guinea, which fed on *Achatina* and other snails, was also a predator of slugs. The latter, after attack by the flatworm, developed blisters on their backs before dying. Records of flatworm predators of terrestrial molluscs in New Zealand were reviewed by Barker (1989b), who described how another triclad, *Geoplana ventrilineata*, was a predator of introduced and endemic slugs in New Zealand.

Stephenson and Knutson (1966) listed more than 46 species of invertebrate parasites and predators associated with slugs. Predators listed as the more important natural enemies of slugs included two species of lampyrid beetles, six species of Diptera, including four sciomyzid fly larvae and four species of carabid beetles. According to Baronio (1974), the families Lampyridae and Drilidae (Coleoptera) and Sciomyzidae (Diptera) feed almost exclusively on slugs and snails. He listed several other families of insects which contain malacophagous species including the Phoridae and Calliphoridae (Diptera) and the Carabidae and Staphilinidae (Coleoptera). Grassé (1968) recorded cockroaches feeding on eggs of the slug *Veronicella*.

In a review of beetles as predators, Crowson (1981) stated that beetles with adaptations for feeding on gastropods were known from both Adephaga (Carabidae) and Polyphaga (Drilidae, Lampyridae and Silphidae), although the adaptations described were mainly aimed at dealing with the snail shell. Ingram (1946) recorded the cychrinid *Scaphinotus interruptus* feeding on the slugs *D. reticulatum* and *M. gagates* while Larochelle (1972) included slugs among the principal prey of many species of Cychrini. Altieri *et al.* (1982) demonstrated that a significant reduction in numbers of *L. maximus* and *H. aspersa* could be achieved by releasing a number of adult *Scaphinotus striatopunctatus* on horticultural land.

In an early reference to beetles as predators of slugs, Taylor (1902–1907) recorded that *L. marginata* was liable to be attacked by a species of *Carabus* which tore open the skin and fed on the viscera. *Carabus violaceus* has been observed attacking in the middle of the body and killing *A. hortensis* (Moore, 1934) and killing and carrying off *D. reticulatum* and *M. gagates* using its mandibles (Tomlin, 1935). The closely related carabid *Calosoma frigidum* is a predator of *A. ater* and other slugs, although when given a choice it prefers *A. ater* (Poulin and O'Neil, 1969). Davies (1953) examined the crop contents of 84 carabid beetles. He noted that since some carabids partially predigested their food, the crop contained only liquids. Thus slug and snail remains could only be detected when fragments of the radula has been injested. The only radulae recovered consisted of one from 18 *P. madidus* examined and two from 12 *A. parallelepipedus*. Cornic (1973) developed a method for identifying slug remains in carabid gut contents and showed that *Pterostichus vulgaris* and *Ophonus rufipes* included slugs in their diet in an apple orchard whenever slugs were active on the surface of the soil during moist weather. At these times slugs made up between 25 and 50% of *Pterostichus* gut contents. A third carabid, *Harpalus affinis*, fed almost exclusively on plant material.

In a series of feeding tests, Stephenson (in Johnson, 1965) offered live slugs to seven common species of carabid beetle and demonstrated that some but not all of the larger polyphagous carabids will feed on slugs. Three species (*C. violaceus*, *Pterosticus melanarius* and *Abax parallelepipedus*) readily ate slugs while the other four were reluctant to feed even if starved. Warley (1970) found that, out of four species of carabid, only *Feronia madida* was able to feed successfully on adult *D. reticulatum*, although the other species readily consumed newly hatched slugs. Tod (1970, 1973) developed a serological method involving the use of rabbit antisera sensitized to *D. reticulatum* tissues. Although not specific to slug tissue, the antiserum was specific to molluscan tissues. Fourteen species of carabid beetle were shown to have eaten material of mollusc origin under field conditions. Tod also concluded that the larger the species of carabid, the more likely it was to eat slugs. Larger beetles were more capable of killing slugs because of their greater strength and strongly developed mouthparts. This enabled them to hold the slugs firmly in their mouthparts, away from the head so that they were not hindered by the slug's mucus. The largest beetles, such as *Carabus* spp., were strong enough to cut the slug in two while smaller beetles had to struggle with the slug and only occasionally succeeded in killing it. Some smaller species might be scavengers feeding on dead slug material. The results also suggested that laboratory feeding trials could be unreliable. Beetles feeding most frequently on slugs included *Carabus violaceus, C. nemoralis, C. problematicus, Pterostichus niger, Calathus fuscipes* and *Cychrus caraboides,* the latter species probably feeding exclusively on slugs at the site investigated. An autotomy escape response in the slug *P. foliolatum* was described by Deyrup-Olsen *et al.* (1986). They suggested that the autotomy section, which was packed with glycogen cells, might offer a significant reward to a predator such as a carabid beetle. The amount of slug material consumed per day was measured by Symondson (1989) for the carabid *A. parallelepipedus* feeding on *D. reticulatum* under laboratory conditions. Female beetles consumed significantly more material per day than male beetles, i.e. 73 mg per day compared with 44 mg, although there was no difference in the kill rates. He showed that, on small enclosed plots, *A. parallelepipedus* was able to make significant reductions in numbers of *D. reticulatum* and the possible significance of this form of biological control is discussed in Chapter 12.

The larvae of glow-worms and fireflies (Lampyridae) feed mainly on slugs and snails. Mead (1961) reviewed the use of lampyrid beetles in the control of *Achatina fulica* and described their feeding behaviour. Fincher (1947) reported that the glow-worm *Lampyris noctiluca* fed on *A. ater.* The larvae of *L. noctiluca* and *Phausis spendidula* also attack *A. ater, A. subfuscus, A. hortensis* and *A. fasciatus* (Schwalb, 1960). Crowson (1981) noted that glow-worm larvae detect and follow the slime trails of their prey. The way glow-worm larvae attack slugs and snails was described by Anon. (1920a), including the occasional use of a toxin to immobilize their prey. Extracts of the midgut and haemolymph of larvae of the fireflies (*Photuris* spp.) produced an inhibitory effect on the heart of *L. maximus* (Copeland, 1981) and may contribute to the paralysing toxin which is injected by the larvae into their snail prey. Larvae of the related Drilidae also feed on slugs and snails (Crawshay, 1903; Bayford, 1906; Baronio, 1974). Tod (1973) found that adults of three species of staphylinid beetles (*Philonthus decorus* (Gravenhorst), *Staphylinus aeneocephalus* Degeer and *Quedius lateralis* (Gravenhorst)), together with the silphid *Phosphuga atrata* (adults and larvae) and a cantharid beetle, had eaten mollusc material while Warley (1970) described how staphilinid larvae fed on *D. reticulatum* eggs. *Ocypus olens,* a staphylinid which preys on snails, has also been observed feeding on the carnivorous slug *T. maugei* (Humphreys, 1982) and the slug *L. valentiana* (Orth *et al.,* 1975). Crowson (1981) described how both larvae and adults of *Phosphuga* accumulate a solution of ammonia in a rectal pouch and that this is then poured out to kill or immobilize the prey. Larvae of the beetle *Prosternon tessellatum* (L.) (Elateridae) were observed to attack and consume slugs and earthworms in the field although they preferred vegetable food in the laboratory (Evans, 1971). Wheater (1989) showed that *Carabus caraboides, C. problematicus, C. violaceus* and *Staphylinus abbreviatus* responded to slug mucus when searching for prey, using receptors located on the terminal ends of their palps.

Mead (1979) concluded that, in the case of the Diptera, since many of the recorded instances were for flies emerging from the dead bodies of slugs and snails, the fly larvae were often saprophagus rather than parasitic or predatory. Beaver (1972) recorded 14 species of saprophagous Diptera belonging to eight families, including Muscidae, Calliphoridae, Phoridae and Psychodidae, breeding in dead snails (*Cepaea nemoralis*). Kühlhorn (1986) recorded 178 species of Diptera of 29 families trapped on carcasses of *A. ater*, the majority belonging to the Calliphoridae or the Muscidae which are generally associated with carrion and other decaying material. An experimental study of the colonization of carcasses of *L. maximus* and *Philomycus carolinianus* (Kneidel, 1983, 1984) showed a lack of competition and a high species diversity among the Diptera which included larvae of the phorid *Megaselia scalaris*. These observations support Mead's conclusions that many of the flies associated with slugs and snails are frequently saprophagus species. However, larvae of other species of *Megaselia* have been recorded as predators of slug eggs. These include *M. aequalis* feeding on eggs of *D. laeve* (Robinson and Foote, 1968) and male and female *M. ciliata* emerging from damaged eggs of *Deroceras* sp. (Disney, 1976). Robinson (1965) also recorded an unidentified species of phorid fly feeding on the embryo in eggs of *D. laeve*. The larvae of the calliphorid fly *Melinda cognata* are parasites of the snail *Helicella virgata* (Keilin, 1919) while larvae of the house fly *Musca domestica* have been observed feeding on live snails (Anon., 1920b). Keilin also recorded an instance where the flesh-fly *Sarcophaga melanura* deposited larvae on the back of *A. hortensis*, the larvae then penetrating the tissues, killing and liquefying the slug. Sergeeva (1984) demonstrated, using serological techniques, that slugs form a minor component of the diet of rhagionid fly larvae.

The Sciomyzidae are small to medium-sized cyclorrhaphan flies which are found in damp and marshy places. The larvae are predators or parasitoids of slugs and snails. The family includes both aquatic and terrestrial species and, since some aquatic snails are intermediate hosts of schistosomiasis and other parasites, extensive research has been carried out on the biology of sciomyzid flies. The possible value of sciomyzid larvae as snail predators in the control of trematode disease was reviewed by Berg (1964). Larvae of the aquatic species are active predators, rapidly killing their prey and destroying many snails. A few terrestrial species are parasitoids while many others show both parasitoid and predatory tendencies. Foote (1963) and Trelka and

Foote (1970) described three North American sciomyzids whose larvae were host specific on slugs. First and second instar larvae of *Tetanocera plebeia* Loew live under the mantle of the host while older larvae are predatory on slugs. First instar larvae of *T. valida* Loew feed in the invaginated eye tentacle of the host while older larvae force their way into the mouth of active slugs killing them within 8–24 hours. The newly hatched larvae of *T. plebeia* and *T. valida* are host specific for *D. laeve* and *D. reticulatum*. First instar larvae of *T. clara* Loew move into the mouth of the slugs *Pallifera dorsalis* (Binney) and small *Philomycus cavolinianus* (Bosc.), remaining there until the third instar. The third instar larvae are predatory and invade the mouth of their prey but do not kill the slug until several hours later. Larvae of all three species consume between two and eight slugs during their development.

Knutson *et al.* (1965) described the biology of a fourth species of slug-killing sciomyzid, *Tetanocera elata* (Meig.), which is distributed widely in northern and central Europe where it occupies a broad range of habitats. Oviposition occurred even in the absence of slug and the first instar larva remained near the empty egg cases until contact with a host as it crawled past. The first instar larva was host specific to *D. reticulatum* and *D. laeve* and when offered alternative slug hosts, *A. ater*, *A. circumscriptus* and *L. tenellus*, these were refused. In nature it was found only on *D. reticulatum* and an infestation of 14% was recorded from a sample of this slug. The larva penetrated beneath the edge of the mantle until only the posterior spiracles remained exposed, and it remained there feeding on mucus until early in the second instar when the host generally died. Beaver (1972) confirmed the host specificity of *T. elata* and that the first instar fed on mucus but found that it did not at first penetrate the skin. Knutson *et al.* found that the larva behaved as a predator from late second instar onwards, pursuing its prey, and they suggested that it then immobilized the prey by lacerating nervous tissue in the slug's head by the use of its mouthparts. Between four and nine slugs were killed by each larva before it pupated. During the predatory stage, larvae were not host specific and killed and ate *A. fasciatus*, *A. intermedius*, *L. flavus*, *L. tenellus*, *T. budapestensis* and *T. sowerbyi*. However, they refused to eat *A. ater*, *A. hortensis* and *A. subfuscus*. Most slugs harboured from one to five or even ten larvae and the timing of the end of the parasitoid stage varied slightly according to the number of larvae infesting the host.

Reidenbach *et al.* (1989) reported that the larvae of four sciomyzid flies, belonging to the genus *Euthycera*, attacked slugs. These were *E. cribrata* (Rondani), *E. stichospila* (Czerni), *E. chaerophylli* (Fabricius) and *E. arcuata*. In addition, *Limnia unguicornis* (Scopoli) larvae fed on slugs in the laboratory. Laboratory observations of predation by *E. cribrata* larvae on *D. reticulatum* showed that, although they destroyed similar numbers of slugs to *T. elata*, their feeding behaviour differed in that the larvae continued to feed on the tissues of the slug for some time after its death. The degree of host specificity varies widely within the Sciomyzidae and Knutson *et al.* (1970) demonstrated that larvae of the snail-killing species *Salticella fasciata*, which feeds normally on helicid snails, will feed in the laboratory on a wide range of dead snails and on dead *D. reticulatum* as well as non-mollusc remains including freshly killed woodlice and earthworms. This species can also complete its larval development from hatching to pupation feeding solely on dead snails. Behavioural studies of third instar larvae of *T. plebeia* and *T. elata* by Trelka and Berg (1977) indicated that both species immobilize slugs by inserting their mouthhooks and injecting a toxic secretion into the posterolateral part of the slug. The toxin causes immobilization and cessation of locomotor waves. The chemical nature of the toxic component and its site of action has been described by Trelka (1972). Trelka and Berg also found that the predatory larvae of *T. plebeia* follow fresh mucous trails but only after recent contact with a slug, and that they often follow them in the wrong direction.

The trombidiform mite *Riccardoella limacum*, which is commonly found on the surface of the skin of slugs, was first described by Réaumur in 1710 as an 'insecte des limaçons' according to Turk and Phillips (1946) who described the life cycle of the mite. Copulation takes place on the host with eggs being most frequently laid in the mantle cavity of the host although they were also found embedded in mucus on the skin. Females were occasionally viviparous. A brief larval stage was followed by two nymphal stages, the protonymph and deutonymph. The deutonymph was neotenous, contained the eggs and was oviparous or viviparous. This was followed by a final moult to the adult stage which was not common and never contained eggs. However, Baker (1970a) found no evidence for neoteny in the deutonymph and showed that adult females contained mature eggs and were oviparous with no evidence of ovoviviparity or viviparity. Adult males and females were also present in much larger numbers than other active stages. Baker also described a third nymphal stage, the tritonymph, in the life cycle. The mite has a wide host range and Turk and Phillips listed 31 species of slug and snail on which it had been found. Slug hosts included three *Testacella* spp., five *Arion* spp. (*A. circumscriptus, A. hortensis, A. intermedius, A. subfuscus* and *A. ater*), three species of *Milax, D. reticulatum* and four *Limax* species (*L. cinereoniger, L. maximus, L. flavus* and *L. marginata*). Snail hosts included several helicid and zonitid snails and the operculate *Pomatias elegans*. Baker (1970a) found that the average number of mites found on slugs taken in field collections from Middlesex and Surrey at different times of the year ranged from four to 11. In Northumberland, the number of slugs infested varied according to the host species (White, 1959a). The degree of infestation was greater on *T. budapestensis* (39.1%) and *A. ater* (29.4%) and least on *D. reticulatum* and *A. circumscriptus* (about 2% of slugs examined) with 20 mites being found on a single *L. maximus*.

Turk and Phillips (1946) found that mites frequently leave their host and run about in the vicinity of the slug. They then locate their host by following the mucous trail which is generally present. When given a choice between *L. flavus* and zonitid snails, *Riccardoella limacum* chose the slug even when initially placed on the snails. They suggested that the mite fed either on mucus or, alternatively, away from their host. Baker (1970b) demonstrated the absence of slug mucus and the presence of slug amoebocytes in the gut of *Riccardoella*, showing that the slug mite is a blood feeder, obtaining its food by piercing the tissues of its host with a feeding tube.

Other invertebrate predators of slugs include the centipede *Scolopendra subspinipes* Leach, feeding on *Veronicella leydigii* (Lawrence, 1939), one centipede and one harvestman eating mollusc material (Tod, 1973) and the slug *Testacella maugei* feeding on *T. sowerbyi* in captivity (Stubbs, 1934). *Testacella* sp. have been found dead after being observed feeding on slugs which were dying after eating metaldehyde baits.

9.3 PARASITES

9.3.1 Platyhelminthes

Slugs are known to be intermediate hosts of helminth parasites of domestic and wild animals. Intermediate hosts of the lesser liver fluke *Dicrocoelium dendriticum*, a parasite of ruminants and pigs, include the slugs *A. subfuscus*, *L. tenellus* and *D. reticulatum* (Chevallier, 1973b). Krissinger (1984) included the slugs *D. reticulatum* and *Lehmannia poirieri* among land molluscs which could act as experimental intermediate hosts of the dicrocoeliid *Lutztrema monenteron*, a parasite of passerine birds. In his review of the literature relating to the trematode genus *Brachylaemus*, Foster (1954) included the slugs *Philomycus bilineatus*, *D. reticulatum*, *A. ater* and *D. laeve* as intermediate hosts of this geographically widespread genus which occur as adults in birds and some mammals including hedgehog, badger and some small mammals. *Tandonia sowerbyi* and *A. lusitanicus* were also intermediate hosts for brachylaemids (Cragg *et al.*, 1957), while Villella (1953) included *D. laeve* among several intermediate hosts for *Entosiphonus thompsoni* Sinitsin, a brachylaemid parasite of shrews and mice in Michigan. Sporocysts and metacercariae of *Brachylaima fuscatum* (Rudolphi), a parasite of doves (*Streptopelia* sp.), were reported from kidneys of *Deroceras* sp. by Shamsuddin and Al-Barrak (1988). Foster (1958a) described seasonal changes in infestation of *D. reticulatum* and *T. sowerbyi* by brachylaemid metacercariae with a peak incidence in June and July and lowest incidence in January to March and recorded both primary (spring) and secondary (autumn) infestations of the slugs. He also suggested that *T. sowerbyi* was the natural intermediate host of the parasite, supporting this by demonstrating that parasitism in *D. reticulatum* led to a host mortality related to the degree of infestation, while *T. sowerbyi* withstood the damaging effects of the parasite (Foster, 1958b). Metacercariae were located only in the kidney and infestation led to widespread necrosis of that organ in both species. Lupu (1970) found that infestation by intermediate stages of a trematode parasite in *L. tenellus* resulted in degeneration of the reproductive system causing sterility. Stadnichenko (1969) described this effect in more detail for several gastropod species.

Taylor (1902–1907) recorded the presence of a cestode cyst in the pulmonary cavity of an *A. ater* and reported that a species of *Limax* was an intermediate host of the small poultry tapeworm. Wetzels (in Frömming, 1954) cited *L. flavus* as an intermediate host of *Davainea proglottina* (Davaine). Grassé and Rovelli (in Brown, 1933) fed onchospheres of *D. proglottina* to *L. maximus*, *L. flavus* and *D. reticulatum* and then fed the resulting cysticercoids to hens where they produced mature forms. Brown described a technique for infecting *D. reticulatum* with onchospheres which passed through the gut wall and developed in the body cavity or digestive gland, producing cysticercoids. Chevallier (1973b) included milacid slugs among the hosts of *D. proglottina* and listed *D. reticulatum*, *L. marginatus* and arionid slugs as intermediate hosts of *Raillietina bonini*, a tapeworm found in pigeons. Grassé (1968) cited *A. ater* as the intermediate host of the cestodes *Hymenolepis multiformis* and *Anomotaenia arionis* as well as *Anomotaenia subterranea* which is a parasite of shrews. The life cycles and ecology of two cestodes *Choanotaenia crassiscolex* and *C. estavarensis*, from the common shrew *Sorex araneus* L. and water shrew *Neomys fodiens* (Pennant) respectively, were described by Jourdane (1972, 1977). Their natural intermediate host was *A. lusitanicus*, the cysticoids being located in the renal and pulmonary epithelium, but *D. reticulatum* was able to act as an intermediate host in the laboratory. The morphology and histochemistry of the cysticercoid of *Rodentotaenia crassiscolex*, located within a capsule of muscular and connective tissue in *D. reticulatum*, was described by Valkounova and Prokopic (1978, 1979).

9.3.2 Nematoda

Nematodes have been recorded as parasites of slugs and their eggs on numerous occasions. Some free-living soil nematodes are also capable of living as facultative parasites on slugs depending on environmental conditions. Arias and Crowell (1963) recorded the free-living rhabditid genera *Rhabditis*, *Parogrolaimus* and *Diplogaster* as parasites of *D. reticulatum* under culture conditions. Larvae of the nematode *Alloionema appendiculata* Schneider are facultative parasites of *Arion* spp., living in the foot muscles and in the lumen of the intestine. In a review of the rhabditid genus *Angiostoma* Dujardin, Spiridonov (1985) listed three species known to live as parasites in the gut of slugs and described a fourth species, *Angiostoma asamati*, from the crop of the slug *Turcomilax ferganus* (Simroth). The other species were *Angiostoma limacis* Dujardin, from *A. ater*, *A. subfuscus* and other *Arion* spp., *Angiostoma stammeri* Mengert, from *L. cinereoniger* and *L. maximus*, and *Angiostoma dentifera* Mengert, from *L. cinereoniger* and *A. subfuscus*. The species of *Angiostoma* are widely distributed within their hosts, and Taylor (1902–1907) records these nematodes (as *Leptodera* spp.) from the foot, intestine and salivary glands of *A. ater* and the rectum, salivary glands and foot of *L. maximus*. He also recorded the nematode *Cosmocercoides* from *Limax* eggs. *Angiostoma limacis* is an obligate parasite of slugs although the 1st and 2nd larval stages are free-living (Johnson, 1978). This species shows seasonal changes in the degree of infestation in *A. hortensis*.

Ogren (1959a) recorded the nematode *Cosmocercoides dukae* (Holl.) from *Deroceras* sp. and *D. gracile* (Rafinesque) and showed that both adults and larvae were located in the mantle cavity (Ogren, 1959b). The female *C. dukae* deposited 3rd stage larvae in the mantle cavity and these then appeared in slime trails left by slugs or snails (Anderson, 1960). Larvae were found in the reproductive tract and in eggs of slugs, in the latter case moving into the mantle cavity of the embryo which subsequently hatched. Anderson also recorded *C. dukae* from the rectum of amphibians but considered these were derived from eating infested slugs or snails. However, Vanderburgh and Anderson (1987a) demonstrated that two species of *Cosmocercoides* were involved and retained *C. dukae* as the name of the species occurring in slugs. Hunter (1967b) recorded a *Cosmocercoides* sp. in the gut of *A. hortensis*, *D. reticulatum* and *T. budapestensis* with a significantly greater proportion of *A. hortensis* being parasitized than in other species. The seasonal changes in infestation by *C. dukae* in a natural population of *D. laeve* were described by Vanderburgh and Anderson (1987b). Transmission of the nematode occurred in late spring and early summer before the death of the adult slugs in July. Transmission to young slugs continued into the autumn. Sections of slugs experimentally infested by nematode larvae showed that the larvae penetrated slug tissues as they move to the mantle cavity. *Arion fasciatus* and *A. hortensis* were also recorded as hosts of *C. dukae*.

Both adult and larval stages in the life cycles of the nematodes described occur in slugs. Slugs are also intermediate hosts to the larvae of many nematode parasites of vertebrates. Chevallier (1973b) listed *Deroceras* sp., *L. flavus* and *A. circumscriptus* among the intermediate hosts of the cat lungworm *Aelurostrongylus abstrusus*. Likharev and Rammel'meier (1952) included *Arion* spp. and *Deroceras* spp. as intermediate hosts of *Crenosoma vulpis*, a lungworm of cats, dogs and other carnivores together with

A. intermedius as an intermediate host of *Filaroides branchialis* (Werner), also a lung parasite of smaller carnivores. Guilhon (1963) demonstrated that *A. ater* acted as an intermediate host of *Angiostrongylus vasorum*, a lungworm parasite of dogs. Third stage infective larvae were also recorded from unidentified slugs associated with infested dogs (Simpson and Neal, 1982). *Angiostrongylus vasorum* was found to be capable of developing to the 3rd stage infective larva in *A. lusitanicus, A. hortensis, D. reticulatum, L. flavus* and nine species of snail under laboratory conditions (Guilhon and Cens, 1973). Several species of slug and snail can act as intermediate hosts of the rat lungworm, *Angiostrongylus cantonensis*, which may be transmitted to man where it can cause eosinophilic meningoencephalitis (Linaweaver, 1966; Eckert and Laemler, 1972). In some instances of eosinophilic meningitis, the cause of infection could be traced to the consumption of raw terrestrial slugs (Kuberski and Wallace, 1979; Sharma *et al.*, 1981).

Naturally occurring slugs found infected with *A. cantonensis* include the veronicellid slugs *Laevicaulis alte* and *Vaginulus plebeius* and the limacid *D. laeve* on Pacific islands (Wallace and Rosen, 1969; Ash, 1976), three unidentified species of slug on Okinawa (Intermill *et al.*, 1972), *L. alte* in India (Limaye *et al.*, 1988) and *L. alte* and *L. marginatus* in Japan (Noda *et al.*, 1982). The level of infestation was particularly high in the veronicellid slugs and Ash (1976) considered these slugs to be the chief reservoir of infection for feral rats. However, Uchikawa *et al.* (1987) showed experimentally that the rat *Rattus rattus* preferred smaller molluscs such as the slug *L. marginatus* and the young snail *A. fulica* to the larger veronicellid *L. alte* and adult *A. fulica*. Thus smaller molluscan intermediate hosts, whose level of infestation with *Angiostrongylus* larvae is low, probably play a more important role in the transmission of the parasite in nature than the more heavily infested larger hosts such as *L. alte*, which may contain more than 1000 *A. cantonensis* larvae per slug (Wallace and Rosen, 1969). At lower temperatures, a greater proportion of larvae developed to the 3rd stage in *L. marginatus* than in snails although larvae did not develop into further stages at temperatures of 10°C or below (Hori *et al.*, 1985). Experimental infection of the slug *L. marginatus* and several species of snail by exposure to rat faeces containing 1st stage *A. cantonensis* larvae, showed that, while several snail species became infected by larvae, the best results for development to 3rd stage infective larvae were found in *L. marginatus* and the snail *Bradybaena oceania* (Hori *et al.*, 1976). Ko (1978) recorded the occurrence of *A. cantonensis* in the heart of a spider monkey, *Ateles geoffroyi*, which had died of an unknown cause. He suggested that the monkey had acquired the infection by accidentally or deliberately consuming slugs or snails from the cage area. Eckert and Laemler (1972) reported an instance of angiostrongylosis in Switzerland where the intermediate host was *A. ater*. They also recorded several new experimental intermediate hosts, including *D. reticulatum* and *D. laeve*. Andrews (1989) described how veronicellid slugs, especially *S. plebeia*, were the principal intermediate hosts of *Angiostrongylus costaricensis* (Morera and Cespedes), the cause of human abdominal angiostrongyliasis in Central America. People become infected when they consumed food contaminated by slugs or infected mucus.

Slugs are intermediate hosts of several other lungworm parasites of ruminant hosts. The slugs *D. reticulatum*, *D. agreste*, *A. subfuscus* and *A. circumscriptus* were found to be among the most suitable intermediate hosts of protostrongylid parasites of sheep, in Poland by Urban (1980). The natural levels of infestation of slugs and snails by protostrongylid larvae in the Rabat region of Morocco were investigated by Cabaret *et al.* (1980). The slugs, represented by *Limax* sp. (mainly *L. maximus*) and *M. gagates*, formed about 10% of the gastropods present. *Milax gagates* showed one of the two highest levels of infestation by nematode larvae. *Limax* sp. acted as a host to *Muellerius capillaris* Müller and *Neostrongylus linearis* while *M. gagates* was host to a *Protostrongylus* sp. In southern Armenia, Davtyan (in Likharev and Rammel'meier, 1952) listed six limacid slugs, including *L. flavus*, as obligatory hosts of the lungworm *M. capillaris*, although the slugs did not become infected by *Synthetocaulus* spp., also a lungworm parasite of sheep and goats. Slugs were occasionally infected by larvae of *Cystocaulus nigrescens* (Jerke), but these rarely reached the infective stage. First stage larvae of *M. capillaris* from sheep tissue were allowed to develop to the infective stage in slugs (*Deroceras* sp.) and snails, which were then fed to mice, guinea pigs and rabbits (Beresford-Jones, 1966). Living 3rd and 4th stage larva were subsequently recovered from these mammals.

A comparison of the suitability of the different parameters used to estimate the degree of infection of molluscs by protostrongylid larvae was made by Cabaret (1982a). Cabaret (1982b) demonstrated that goats could be infected with *M. capillaris* by grazing them on pasture previously contaminated with pieces of dead infected *A. ater*. In an investigation of the infection of lambs on pasture by the lungworm *Muellerius capillaris* in Morocco, Cabaret and Pandey (1986) found that infection of lambs was mainly due to the injestion of grass contaminated with larvae liberated by the death of the slugs. The slug intermediate hosts were *M. gagates* and *L. maximus*, *M. gagates* again having a higher level of infestation than *L. maximus*. The parasitism of slugs reached a maximum in February and March. The early reaction of the slug to infection by larvae of *M. capillaris* was investigated by Cabaret and Weber (1987). An early cellular reaction resulted in the formation of a granuloma but this did not seem to interfere with the larva. Chevallier (1973b) also recorded *D. reticulatum*, *D. laeve*, *A. subfuscus* and *T. sowerbyi* as intermediate hosts of *M. capillaris*. Rose (1960) found that *D. reticulatum* acted as an intermediate host for the sheep lungworm *Cystocaulus ocreatus* while *Deroceras* spp. were shown capable of transmitting infective larvae of the trichostrongylid *Cooperia oncophora* on pasture by Gronvold and Nansen (1984).

The slug *D. laeve* is an important intermediate host of the muscle worm *Parelaphostrongylus odocoilei*, a parasite of the mule deer (*Odocoileus hemionus*) (Platt in Boag and Wishart, 1982) and of the meningeal worm *Parelaphostrongylus tenuis*, a parasite of the white-tailed deer (*Odocoileus virginianus*) (Maze and Johnstone, 1986). Lankester and Anderson (1968) showed that larvae of *P. tenuis* developed most rapidly in *D. reticulatum* while in the field they found that this slug, together with *D. laeve* and *A. circumscriptus*, were infected with larvae. Rowley *et al.* (1987) found that, in addition to *D. laeve*, the slug *Philomycus carolinianus* was also an intermediate host of *P. tenuis*. Experiments by Platt and Samuel (1984) showed that the predominant mode of entry of 1st stage larvae of *P. odocoilei* was direct penetration of the ventral foot epithelium of *D. laeve* while low levels of larval entry into *D. reticulatum* indicated that this slug was an unsuitable intermediate host for *P. odocoilei*. Examination of a large number of gastropods from Jasper National Park, Alberta, by Samuel *et al.* (1985) confirmed that *D. laeve* was the most important host for *P. odocoilei*. They also showed that mule deer became infected as fawns in autumn by accidentally injesting infected slugs. Skorping and Halvorsen (1980) tested the susceptibility of slugs and snails to experimental infection with *Elaphostrongylus rangiferi* Mitskevich, a meningeal parasite of the reindeer, *Rangifer tarandus* (Linnaeus). While all species were susceptible to infection, they found considerable variation in the incidence and intensity of infection and in the development of the larval nematode. Among the slugs, the most rapid development was in *A. silvaticus* and *D. laeve*, development was slower in *D. reticulatum* while few larvae developed and at a very slow rate in *A. subfuscus* and *A. hortensis*. *D. reticulatum* and *A. hortensis* were also found by experiment to be intermediate hosts of the strongyloid *Morerastrongylus andersoni* (Petter) by Petter (1974).

Hobmaier (1941) showed that larvae of the nematode *Skrjabingylus chitwoodorum* Hill, a parasite of skunks, could develop to the infective stage in several slugs including *D. reticulatum*, *L. maximus*, *L. cinereoniger*, *L. flavus* and *Milax* sp. Dubnitskii (1956) successfully transmitted a parasite of the frontal sinuses of mustelids, *Skrjabingylus nasicola* (Leuckart), to mink (*Mustela vison* Schreber) by feeding them with infected *D. reticulatum*. However, the diet of most mustelids in nature is mainly composed of small vertebrates and experiments by Hansson (1967) showed that the shrew *Sorex araneus* (Linnaeus) could act as a paratenic host for *S. nasicola*, since molluscs are regularly eaten by *S. araneus*. In the laboratory the mouse *Apodemus sylvaticus* (Linnaeus) and the vole *Clethrionomys glareolus* (Schreber) were successfully used by Weber and Mermod (1985) as paratenic hosts to infect the ferret (*Mustela putorius furo*) with *S. nasicola*. They estimated that when *D. reticulatum* were exposed to 100 1st stage larvae, only a mean of 77.3 penetrated the slug and that only 23.8 of these larvae reached the infectious 3rd stage. Slugs have been cited as the intermediate hosts of the gapeworm *Syngamus trachea*, a nematode parasite of domestic poultry (Taylor, 1935). Asitinskaya and Shulepova (1977) demonstrated by experiments that when ascarid eggs were injested by the slugs *D. reticulatum* and *D. sturanyi* about 10% of them were destroyed through digestion. Airey *et al*. (1984) recorded the nematode *Diplogaster* sp. from slugs.

9.3.3 Protozoa

Several sporozoan parasites have been recorded from slugs. Pellerdy (1965) included a reference to the coccidians *Isospora incerta* Schneider from *L. cinereoniger* and *Pfeifferinella impudica* Léger and Hollande from *L. marginatus*. Grassé (1968) listed the coccidians, *Klossia loosi*, from the kidney of *Limax* sp. and *Arion* sp., and *Isospora nana*, from the intestine of slugs. Parasites recorded in an extensive survey of slug pest species (Airey *et al*., 1984) included a microsporidian which occurred in 30% of all *D. reticulatum* seen. Where populations were dense, the level of infection was as high as 100%. This microsporidian infects only the intestinal epithelium of the host and its effect on the slug, which includes a 40% reduction in fecundity, a 30% reduction in growth and a 23% reduction in longevity, was described by Jones (1985). The possible use of this parasite, which appears to be specific to *D. reticulatum*, as a biological control agent of this slug was suggested by Jones and Selman (1984) and the species, *Microsporidium novacastriensis*, was fully described by Jones and Selman (1985). Boettger (in Mead, 1979) described abnormalities in *A. ater* which might have had a gregarine (sporozoan) etiology.

Although several ciliate species have been found in slugs, their parasitic status is generally uncertain, with the exception of the species of *Tetrahymena*. Pelseener (1928) recorded the ciliate *Concophthirus steenstrupi* (Stein) from *D. reticulatum* and *A. ater*. Reynolds (1936) considered that *Colpoda steni* Maupas, found in the digestive gland and pulmonary cavity of *D. reticulatum*, was a facultative parasite although Brooks (1968) described this species as a harmless species occurring in the gut of slugs and snails. Windsor (1959) found *C. steni* in the digestive gland of 20% of the *D. reticulatum* and 50% of the *L. marginatus* collected from the field.

Warren (1932) described *Paraglaucoma limacis* Warren (now *Tetrahymena limacis*) as a commensal from the digestive gland of *D. reticulatum*. Borden (1948) considered this ciliate, which occurred in about 7% of *D. reticulatum* examined, to be a facultative parasite. Kozloff (1956a) found *Tetrahymena pyriformis* (Ehrenberg) in the intestine and digestive gland of *D. reticulatum* and recorded *T. limacis* (Warren) from the slug *Prophysaon andersoni* (Kozloff, 1956b). *Tetrahymena rostrata* (Kahl) was also recorded from the kidney of *D. reticulatum* (Kozloff, 1957). Cysts of *T. rostrata* were found repeatedly in faeces of *D. reticulatum* even after weeks of feeding the slug on well-washed lettuce by Thompson (1958). This ciliate could be cultured experimentally in the body cavity of insects and Thompson suggested that insects may serve as natural hosts for *T. rostrata* although it was normally a free-living protozoan. Most records for *Tetrahymena* spp. as parasites of slugs were initially from North America, but Corliss *et al.* (1962) found *T. limacis* in the blood, digestive gland and gut of British examples of *T. budapestensis*. Arias and Crowell (1963) found that in addition to the various internal organs, *T. limacis* was present in the subcutaneous tissue of *D. reticulatum*. Unlike previous authors (e.g. Kozloff, 1956b), Arias and Crowell considered that heavy infestations of this species would bring about the death of the host. The role of *T. limacis* and *T. rostrata* as parasites of *D. reticulatum* was investigated in a comprehensive account of the biology of these ciliates by Brooks (1968). He found that *T. limacis* was confined to the lumen of the gut, including the digestive gland, and that it had little effect on the host. Infection with *T. rostrata*, however, was marked by extensive damage and inflammation in the kidney and pulmonary region accentuated by a rapid multiplication of amoebocytes around the site of infection. These amoebocytes formed masses around the parasites, often obstructing the blood vessels. The ciliate eventually became distributed throughout the body including the connective tissue of the body wall. Slug eggs became infected with *T. rostrata* during their formation in the reproductive tract. Embryos became infected via the mouth and sometimes succumbed to the infection. New hosts for both species of *Tetrahymena* included the slug *M. gracilis*. Michelson (1971) showed that contrary to previous observations *T. limacis* was distributed throughout the tissues and organs of the host, even in visible papules on the external epithelium. The parasite elicited a marked pathological response including fibrotic encapsulation of the ciliates, necrosis, tissue destruction and metaplasia. Michelson suggested that *T. limacis* was an obligate, rather than facultative parasite of molluscs and that slugs were infected via the respiratory aperture rather than orally. Runham *et al.* (1973) reported a high incidence of infection with *Tetrahymena* in cultures of *D. reticulatum* collected from the wild. Godan (1983) listed several new hosts for the species of *Tetrahymena* and described the life cycle of these ciliates. *Deroceras reticulatum* infected with *T. rostrata* showed characteristic symptoms. These included a swollen mantle and the elongation and constriction of the posterior end of the body. The tissue response to infection by *Tetrahymena* described previously appears to be a general response to foreign bodies (see Section 9.5). Brooks (1968) described the role of amoebocytes in the response to infection by *T. rostrata*.

9.4 DISEASE

Mead (1961) considered that the most neglected aspect in the entire field of malacological biology was the study of the role of microorganisms in molluscan symbiosis and pathology. This was confirmed in the review of bacteria and viruses in molluscs by Godan (1983). Few, if any, viruses have been described from slugs. A virus-like particle was described by David *et al.* (1977) from the digestive gland of living *D. reticulatum* and from several other tissues in dead and moribund slugs. However, similar particles from slugs were shown to be carbohydrate reserves of galactogen and glycogen (Kassanis *et al.*, 1984). Nicholas (1984) described virus-like particles from the epithelial cells of the sarcobellum of *D. reticulatum* and suggested that they might be related to the Herpetoviridae. These particles were not found in other parts of the reproductive tract. Mead (1961, 1979) described the phenomenon of population decline in *A. fulica*, whereby populations of this snail, after increasing out of all proportion during the first few years after introduction, showed a decline in numbers to more normal levels. The condition was characterized by leukodermic lesions in the tentacles and head region of the snails caused by destruction of melanophores in a small area of the dermis, and there was evidence that the bacterium *Aeromonas hydrophila* (Chester) was associated with this condition. Runham (1989) described how bacterial infections, particularly *Aeromonas*, had been recorded in snails used in snail farming while Chevallier (1979) reported that *Pseudomonas* caused disease on snail farms. Slug populations rarely reach densities comparable with the densities of newly introduced *Achatina* populations or snails kept under culture conditions, and this kind of disease has not been generally recorded for slugs. However, Raut and Mandal (1986) have reported fatal leukodermic lesions from the body surface of the slug *Laevicaulis alte* (Férussac). The etiological agent of this condition was unknown but the symptoms were similar to those found in *A. fulica*. Mead (1979) also reported that the slug *L. flavus* had been used as a living vehicle to transport the etiological agent associated with population decline in a viable state and that they showed symptoms typical of the condition. A fatal blistering disease of *Veronicella ameghini* Gambetta, triggered by laboratory conditions, has been described by de Gravelle (in Mead, 1979) but the etiology of this condition is unknown. A pathogenic disease of introduced veronicellid slugs was described by Dundee (1977) from the USA and he suggested that this disease might be used as a biological control agent.

Few fungal infections have been recorded from slugs although Arias and Crowell (1963) identified a species of *Fusarium* associated with dead slugs in laboratory cultures. Chevallier (1979) has recorded this fungus as an egg parasite of *H. aspersa*. Fungal infections frequently develop in slug eggs, however – especially under laboratory culture conditions. Tervet and Esslemont (1938) found that a fungus, *Verticillium chlamydosporium*

Goddard, was parasitic on the eggs of *D. reticulatum*. They found a relatively high degree of infection of eggs even in those collected from the field. However, although Lovatt and Black (1920) found widespread fungal infections in their slug cultures, they considered that fungi were of minor importance under natural conditions. South (1989) found that an average of only 8–9% of *D. reticulatum* eggs and 6–7% of *A. intermedius* eggs taken over two years from permanent pasture contained fungal mycelium or spores. He concluded that the fungus was most likely an obligate saprobe rather than a parasite, developing in eggs where the embryo had already died. Arias and Crowell (1963) also recorded a fungus *Arthrobotrys* preventing the development of eggs.

Concern has been expressed about the possible role of slugs as carriers of microorganisms which are human pathogens, especially as some slugs, such as *L. flavus*, tend to be synanthropic species. Elliott (1969) demonstrated that coliform bacteria, including possible human pathogens, predominated in isolates from the skin surface and intestine of *L. maximus*. Weiss (1981) showed that, as a result of the coprophagus habits of *A. ater*, this slug acted as a carrier of different *Salmonella* serotypes between dog faeces and hedgehog predators of the slug. A number of species of Diptera were shown by Kühlhorn (1986) to transmit different serotypes of *Salmonella*, together with eggs of the nematodes *Ascaris lumbricoides* Linnaeus and *Trichuris trichiura* Linnaeus, from the body surface, faeces and dead bodies of *A. ater* to human foodstuffs and fodder for domestic animals. The slugs probably acquired the parasites as the result of their coprophagus habits.

Little is known about immune responses in slugs or snails. Curtis and Cowden (1978) investigated the responsiveness of the slug *L. maximus* to injections of fluorescein- and rhodamine-conjugated immunogens. They found that these were rapidly incorporated and eliminated by cells of the slug. However, these foreign proteins were not eliminated by circulating amoebocytes, but rather, by adventitial cells of the aorta and other blood vessels. Lectins are proteins or glycoproteins of non-immune origin that agglutinate cells or precipitate complex carbohydrates. They may be concerned with disease resistance. Various agglutinins or lectins have been extracted from snails and slugs, particularly from the eggs of aquatic snails and the albumen gland of *Helix* spp. and other snails. Several of these show considerable specificity for particular human blood groups (Uhlenbruck, 1969). However, their occurrence is rather erratic and Wright (1974) suggested that their presence or absence in different populations of the same species might serve as genetic markers. Kothbauer (1970) demonstrated that an agglutinin from the albumen gland of *H. pomatia* conferred immunobiological protection on the living egg and also retarded the decay of dead eggs. He concluded that, while the main purpose of agglutinins extracted from the bodies or eggs of ten snail species was that of egg protectins, they might have other biological functions in addition to this

(Kothbauer, 1970). Krüpe and Pieper (1966) examined seven species of arionid and limacid slugs but found haemagglutinins in only *A. circumscriptus* (anti-A). A lectin present in the mucus of *A. ater* was described by Habets *et al.* (1979). In addition to its agglutinating and precipitating properties, the purified lectin possessed proteinase-inhibiting properties.

Pemberton (1970) reported the presence of non-specific agglutinins for human erythrocytes in saline extracts of whole *L. flavus*. Miller *et al.* (1982) found that an agglutinin from this extract, *Limax flavus* agglutinin (LFA), was highly specific for sialic acid residues of glycoproteins and described the purification and macromolecular properties of this lectin. A number of applications have been found for LFA, which is now commercially available. Theolis *et al.* (1984) used the lectins concanavalin A (Con A) and LFA in the course of an investigation into the characterization of antigenic determinants of human apolipoprotein B. The development of a cytochemical affinity technique for the demonstration of sialic acid residues in various rat tissues by light and electron microscopy was described by Roth *et al.* (1984). LFA was applied to tissue sections and fetuin–gold complexes were subsequently used to visualize the tissue-bound lectin. Similar lectin–gold complexes have been used to detect specific monosaccharide residues in mouse brain microblood vessels using a number of site specific lectins, including LFA (Vorbrodt *et al.*, 1986). Tsuyama *et al.* (1986) used LFA and other lectins in a study of the ultrachemistry of glycosylation sites of rat colonic mucous cells while Kasinath and Singh (1987) and Wagner and Roth (1988) described how LFA could be used to detect sialic acid on the surface of cultured glomerular epithelial cells and during glomerulus development in the newborn rat kidney. Marban-Mendoza *et al.* (1987) obtained significant control of root gall formation by the tomato root knot nematode *Meloidogyne incognita* (Kofoid and White) using applications of the lectins Con A and LFA, although LFA gave poorer control than Con A.

9.5 ABNORMALITIES AND INJURIES

Humpbacked slugs have been reported from time to time in the literature, the earliest record being that of Simroth (1905) who described a hump-like curvature of the posterior part of the mantle of an *Arion* sp. In some instances, for example *L. flavus* (Szabó and Szabó, 1931c, 1934b), these have been shown to be true neoplasms. In other instances, for example *A. ater* (Boettger, 1956), the cause was confused by pathological changes associated with a gregarine infection. The status of these abnormalities has been discussed by Sparks (1972). Other hump-like conditions have been described for *L. flavus* (Frömming, 1954) and *D. reticulatum* (Frömming *et al.*, 1961). The 'hump' or dermal sac contained viscera displaced upwards from the body cavity and was referred to as a hernial sac. The authors suggested that this was brought about during hatching as the slug attempted to force its way through too small an opening in the egg shell. However, Segal (1963) showed that a similar hernia in *L. flavus* was the result of a developmental abnormality in the diaphragm, separating the pulmonary cavity from the body cavity, rather than a difficulty in hatching. Segal concluded that this abnormality was temperature dependent and that its appearance was predictable at lower temperatures. South (1989b) described a condition in *D. reticulatum* where slugs were found with a slit either in the mantle, with the shell often missing, or behind the mantle or with a tight constriction across the tail region. They appeared to survive this damage which was recorded only at times of environmental stress due either to very cold weather or very dry weather. In a detailed study of the structural and metabolic changes following wounding in *A. hortensis*, Dyson (1964) demonstrated that wounds heal readily and the death rate is low if slugs are kept at a cool temperature after wounding. The tissue response to an injection of a charcoal suspension into the subintegumentary tissue of *L. poirieri* was described by Arcadi (1968). The charcoal rapidly induced the tissue to form a capsule, composed of cells which appeared to be fibroblasts. Amoebocytes or epithelioid cells then invaded, engulfed and surrounded the charcoal particles.

10

Ecology

10.1 ESTIMATION OF POPULATIONS

10.1.1 Choice of methods

The methods for estimating the size of slug populations may be divided into three types. Absolute estimates are expressed as numbers per unit area of habitat (i.e. density) and give the most reliable estimate of the size of the population. However, they tend to be time consuming and laborious. Relative estimates are usually related to some measure of slug activity, such as the catch per unit effort (e.g. time searching) or numbers obtained by some form of trapping. Since slug activity is largely governed by weather, relative estimates measure a variable and unknown fraction of the population. However, they still allow comparisons to be made of the spatial and temporal distributions of slug populations. Relative methods usually require only simple equipment and can be carried out more rapidly and enable more extensive investigations to be made. Finally, population indices can be calculated based on some artefact of the population. In the case of slugs, for example, the extent of damage to groups of wheat grains or potato tubers has been used. It is sometimes possible to relate relative estimates and population indices to absolute numbers, using regression analysis, provided the estimates are made simultaneously. Relative estimates or population indices may then be used for subsequent estimates, converting them to absolute numbers using correction factors.

10.1.2 Relative methods

Most relative methods used for sampling slug populations have employed either trapping or baiting or a combination of both. Timed searches of the surface of the soil, usually at night, have also been used. These methods are essentially a measure of slug activity and Webley (1964) estimated that weather accounted for about 70% of daily variation in slug numbers taken by baiting. The relationship between weather and slug activity has been discussed in Chapter 7. Webley also found that 12% of the variation in catch at baits was due to specific differences in slugs and Crawford-Sidebotham (1970) confirmed this differential susceptibility of slug species to baits.

(a) Surface searching

A surface searching method, based on the catch per unit effort, was described by Barnes and Weil (1944). The distribution of slugs in gardens was determined by night searching using a powerful flashlamp and a sampling route was defined, which included a range of habitats. The sampling route was then followed for a fixed period of time, collecting, or simply recording, numbers of all slugs found without disturbing the foliage, etc. Barnes and Weil found that this method yielded reproducible results and it was subsequently used by White (1959b), Bett (1960) and Hunter (1968a). There are several disadvantages to this method. Lighter coloured slugs tend to be recorded in higher proportions than darker species and smaller species or immature slugs also tend to be overlooked. Furthermore, van den Bruel and Moens (1958a) showed that surface searching varied in efficiency with slug activity and tended to exaggerate the relative proportions of species which did not burrow extensively, for example *D. reticulatum*. Crawford-Sidebotham (1972) found that this method was suitable for studying the effect of weather on slug activity at the soil surface but emphasized that it was necessary to develop a 'hunting image' before commencing a sample. Time searches were also used by Paul (1975) to examine the distribution of slugs and snails in woodland. Relative densities of *A. columbianus* have been estimated by counting individuals sighted along transects in coniferous forests throughout the sampling season (Richter, 1979).

(b) Trapping

Trapping usually involves the collection of slugs from under units of cover, for example: boards, sacking, stones, leaves (Miles *et al.*, 1931; Getz, 1959; Jensen and Corbin, 1966; Hunter, 1968a; Judge, 1972; Beyer and Saari, 1978); inverted squares of turf (Thomas, 1944); and inverted wooden boxes (Howitt, 1961), all of these being laid on the ground. The refuge trap described by Schrim and Byers (1980) consisted of a cylindrical hole dug in the soil, providing a dark, damp refuge for slugs to congregate, covered by a foil-wrapped piece of fibreglass to reflect solar radiation and prevent the temperature in the hole from rising. South (1964), using roof tiles laid on the ground, showed that trapping was not a reliable method for estimating slug numbers on grassland and that this method exaggerated the proportion of mature slugs present. He found no significant difference between numbers from traps made of different materials. Hunter (1968a) obtained similar results on arable ground, showing that trapping exaggerated the proportion of surface-dwelling slugs, e.g. *D. reticulatum*, present when compared with soil washing. Lyth (1972) compared trapping and baiting and found that the former returned too few slugs to be of any value. However, Hommay and Lavanceau (1986) successfully compared the relative abundance of different slug species in 12 *départements* of France using trap samples. The traps were of two different types. One type consisted of non-degradable mats covered on the lower surface with plastic while the upper surface was coated with aluminium to act as a heat shield. These traps were baited with slug pellets and, during dry conditions, the soil under the trap was first watered. The other type of trap consisted of cardboard squares, ridged on the lower surface and with black plastic sheet covering the upper surface. No bait was used and the soil was not watered. Van den Bruel and Moens (1958a) used a different kind of trap, counting the slugs immobilized while crossing a band 0.50 m wide of DNOC treated soil. Water traps, plastic dishes containing water with a wetting agent sunk into the ground with their rims flush with the surface, were used by Nelson (1971) in a survey of moorland invertebrates and these successfully trapped *A. ater* and *D. reticulatum*. The method of removal trapping or collecting (South-wood, 1978), where estimates of the total population are made using the rate at which the catches on successive nights decline, was used by Cook and Radford (1988) to estimate populations of four species of *Limax*.

(c) Baits

The numbers of slugs killed by toxic baits have been used to estimate slug populations. Bran baits incorporating metaldehyde have been used in many investigations including Barnes and Weil (1942), Thomas (1944), van den Bruel and Moens (1958a), Gould (1962a) and Webley (1964). Baits based on methiocarb have also been used (eg. Gould and Webley, 1972). These baits are either left exposed or covered by small tiles and Webley (1963) found that more slugs were trapped with exposed baits while, if baits were covered, more slugs were caught under black glass than clear glass. Van den Bruel and Moens (1958a) concluded that both baits and trapping with DNOC were suitable relative methods provided that birds could be prevented from removing the dead slugs, although they found that baits, like surface searching, have a differential attraction for different species. South (1973b) came to a similar conclusion in a comparison of relative and absolute sampling methods. Glen and Wiltshire (1986) found that methiocarb baits provided an estimate of the number of larger slugs in the top 10 cm of soil but gave no indication of numbers of the smaller slugs which often cause considerable damage to crops. Beer, in small containers sunk level with the surface of the soil, is also an effective bait for slugs, which fall in and drown (Smith and Boswell, 1970). Combinations of traps and baits have also been used, for example metaldehyde baits covered by (i) black-painted glass (Thomas, 1944) and (ii) linoleum tiles (Beyer and Saari, 1977).

(d) Population indices – feeding activity

Estimates of slug-feeding activity have been made by exposing known quantities of foods – for example, wheat grains on terylene net (Duthoit, 1961) or potatoes in wire baskets (Stephenson, 1967) – to slug attack in the field. Although there are difficulties in the use of these techniques – for example, other animals consuming the foodstuffs – the results have been used successfully to measure feeding activity at different levels in the soil. Hunter (1969a) stuck groups of ten wheat grains to the base of disposable Petri dishes, which were then perforated to allow slugs entry but to exclude larger animals such as mice and birds. These results of these techniques can be quantified and expressed as indices of feeding activity.

(e) Conclusions

Although relative methods do not provide an accurate estimate of slug density, they are suitable for particular tasks. Warley (1970) compared several relative methods with the results of flooding, an absolute method. These methods included uncovered baits (commercial slug pellets), baits covered by tins, tins alone and wooden boards with and without baits. He concluded that, for practical purposes, three categories of adult slug density might be distinguished, i.e. low, medium and high, and that relative methods are probably suitable indicators of the category to which a particular population belongs. These categories could then be related to the probable scale of slug damage that might be expected in a crop. Gould and Webley (1972) successfully used three relative methods to compare the effects of methiocarb and metaldehyde pellets applied to plots before drilling winter wheat. The methods used were (a) counts of dead slugs on the soil surface after treatment with pellets, (b) perforated plastic Petri dishes, each with ten wheat grains stuck to the base, activity being measured by the number of grains damaged after several days, and (c) methiocarb baits covered by tiles. Lyth (1972) found that baiting gave satisfactory results for comparing numbers of slugs in neighbouring areas of woodland, coppice and grassland.

10.1.3 Absolute methods

(a) Surface searching of a defined area

South (1964) found that counting slugs by searching the surface of grass-land *in situ*, contained by 30 cm by 30 cm quadrats, significantly under-estimated numbers. Surface searching of similar quadrats in a potato field underestimated numbers of all species except the surface-living *D. reticulatum* and *D. laeve*, numbers of which were exaggerated (South, 1973b). The extent to which slugs burrow depends upon the soil conditions and varies with species. Few slugs were found more than 3 or 4 cm below the surface of rough grassland except during very dry weather (South, 1964), while a much larger proportion burrowed into the soil of an arable plot (Hunter, 1966). Mordan (1973) found that surface searching 30 cm × 30 cm quadrats, including the removal of the vegetation and disturbance of the soil surface, gave satisfactory results in a study of woodland slugs and snails. Thus this method is generally only applicable where the soil is firm, for example on grassland. Mordan found that 75% and between 72 and 85% of slugs could be recovered by searching quadrats of deciduous woodland and rough grassland respectively at all seasons, and this recovery efficiency was checked by soil washing. South (1964) found that chemical repellants, such as St Ives fluid used for leatherjackets (Dawson, 1932) or formalin used for earthworms (Raw, 1959), poured on to the surface of quadrats to bring slugs to the surface, were unsuitable because they killed rather than repelled slugs. van den Drift (1951) tested searching combined with dry sieving litter *in situ* in a beech forest but considered it to be unsuitable for sampling invertebrates although several *A. subfuscus* were included among the animals recovered.

(b) Hand sorting

In the comparison of relative methods, referred to previously, with an absolute method for estimating populations, van den Bruel and Moens (1958b) concluded that the absolute method used, hand sorting 0.3 m cubes of soil, gave the most reliable results for arable and grassland samples. However, Mordan (1973) concluded that hand sorting was no more efficient than the surface searching of quadrats for woodland samples. South (1964) found that small slugs (< 100 mg), known to be present, were not recovered when turves were dry-sieved and hand sorted. Phillipson (1983) combined hand sorting soil cores with infrared heat extraction of litter samples in a study of slug populations in beech woodland.

(c) Soil washing and flotation

The flotation method described by Vágvölgyi (1952) for snails is unsuitable for slugs as samples must be dried prior to flotation. Williamson (1959a) used a wet-sieving technique to recover molluscs from woodland litter. However, the method was ineffective if soil was included in the sample and was therefore unsuitable for slugs. An adaptation of Williamson's wet-sieving method, combined with flotation, was used by Bishop (1981) in an investigation of slug and snail faunas from wetland habitats. A soil-washing method for extracting animals from samples was first described by Salt and Hollick (1944) and modified by Raw (1951). A modification of this method for the extraction of slugs and eggs was described by South (1964). Soil

Table 10.1 Percentage recovery rates for slugs and eggs using soil washing.

Author	Species				
	1	2	3	4	5
South (1964)	between 98 and 100%			–	–
Hunter (1968)					
(i) >12.5 mg	100	–	–	100	100
(ii) <12.5 mg	63	–	–	67	86
(iii) eggs	91	–	–	15	100
Warley (1970)					
(i) >25 mg	100	–	100	100	–
(ii) <25 mg	95	–	100	100	–
(iii) eggs	96	–	98	65	–
Rollo and Ellis (1974)					
eggs only	97*	–	98	94	–

Species: 1,*D. reticulatum*;   2,*A. intermedius*;   3,*A fasciatus*;
4,*A. hortensis*;   5,*T. budapestensis*.
* *Deroceras* spp.

samples were broken down, using a water jet, on a bank of three 38 × 38 cm graduated sieves, the finest being 12-mesh to the centimetre. Each sieve with its residue was then immersed in magnesium sulphate solution (relative density 1.17–1.20). The organic material floated at the surface and slugs and eggs were removed. The extraction efficiency of this method was checked by the addition of known numbers of slugs and eggs before washing. This method has been used by several authors, including Hunter (1968a), Pinder (1969) and Warley (1970) and examples of the recovery rates obtained are compared in Table 10.1. Results generally show a high recovery rate for larger (> 25 mg) slugs and in two instances, South (1964) and Warley (1970), for eggs and small slugs. Both Hunter (1968a) and Warley (1970) suggested that the low recovery rate for *A. hortensis* eggs was due to their being broken up by the water jet. Hunter also found that slugs lost weight, mainly through the loss of mucus, during soil washing. However, soil washing is probably the most accurate technique available for estimating slug numbers and is the most suitable method for recovering eggs. Rollo and Ellis (1974) described a faster mechanical process, preceded by the flooding process described below, for washing soil samples and recovering slugs and eggs. Although laborious and requiring special facilities, soil washing is a useful technique when used in conjunction with other methods (South, 1965) or to calibrate or check the efficiency of other methods for estimating populations, particularly relative methods.

(d) Flooding

Baunacke (1913) and Piéron (1928) demonstrated that individuals of *D. reticulatum* moved upwards when immersed in water. This upwards movement is particularly noticeable on flooded grassland (South, 1964), even newly hatched slugs escaping from water 30 cm deep within a few hours. As a result of these observations, South (1964) described a method for the extraction of slugs from turf samples, called the 'Cold Water Process' (CWP). Sampling units of turf were placed on edge in plastic bins with tight-fitting lids and progressively flooded during a period of four or five days. Slugs present were driven up to the top of the turf where they were collected. This method was initially used for turf samples but Hunter (1968a) and Warley (1970) modified the technique for arable soil samples. Modifications of the flooding technique have also been used for samples from arable soils by Glen *et al.* (1984) and Glen and Wiltshire (1986). The flooding technique has also been used for woodland soil samples, after first removing and searching the litter (Jennings and Barkham, 1975a), for abandoned farmland (Beyer and Saari, 1978) and for montane grassland (Lutman, 1978). Table 10.2 compares the percentage recovery rates obtained by several authors using the CWP. In each instance, slugs and eggs not extracted were recovered by soil washing. The recovery rates were consistently high with one exception, and this was corrected by modifying the CWP for arable samples (South, 1964). Flooding techniques are less laborious, give satisfactory

Table 10.2 Percentage recovery rates for slugs using the cold water flooding technique.

Author	Habitat	Species				
		1	2	3	4	5
South (1964)	Turf	99	–	81	94	–
	Arable (coarse tilth)	92	–	54	80	–
	Arable (fine tilth)	100	89	–	–	100
Hunter (1968)	Arable (fine tilth)	92	88	–	–	89
Warley (1970)	Turf	99	97	–	98	–
	Arable	96	89	–	97	–
Jennings and Barkham (1975)	Woodland soil	–	95	91	–	–

Species: 1, *D. reticulatum*; 2, *A. hortensis*; 3, *A. intermedius*; 4, *A. fasciatus*; 5, *T. budapestensis*.

estimates of slug number and slugs are recovered in good condition. They do not, however, recover eggs and take rather a long time to complete.

(e) Hot water process

Milne *et al.* (1958) found that leatherjackets and other animals, including slugs, could be expelled by raising hot water through turves. They called this method the 'Hot Water Process' (HWP). The slug repelling efficiency of this method was assessed by South (1964), who found that it was less efficient than the CWP. Warley (1970) used a modified form of this technique and, for turf samples, obtained similar percentage recovery rates to those for cold water flooding. However, his results for arable samples were rather lower than corresponding rates for cold water flooding. The HWP is not really suitable for arable samples since slugs recovered are usually damaged by the heat and eggs are not recovered.

(f) Defined-area trapping

A quantitative method for estimating slug density on pasture *in situ*, the defined-area trap, has been described by Ferguson *et al.* (1989) and Ferguson and Hanks (1990). This involved delimiting a sampling unit of 0.1 m² of turf with the defined-area trap and trapping out the slugs contained within the defined area over several days. Ferguson *et al.* (1989) demonstrated that this method gave results comparable with other soil sampling methods but with savings in time and labour and it provided a non-destructive means of sampling for slugs. However, it was only applicable to sampling grassland. Byers *et al.* (1989) compared the use of defined-area traps and refuge traps (see Section 10.1.2(b)) with the aim of predicting slug density from refuge trap data. They concluded that although the refuge trap, a relative method, was unsuitable for estimating slug density, it could be used to make relative comparisons between molluscicide treatments.

(g) Conclusions

Soil washing is the most efficient absolute method for estimating slug numbers and has the added advantage of recovering eggs. The 'Cold Water Process' was more efficient than the 'Hot Water Process' and recoveries of slugs by the CWP were often similar to those obtained by soil washing. The defined-area trap also gave comparable results to the CWP for grassland samples. However, the CWP and the defined-area trap, unlike soil washing, do not recover eggs. Hand sorting and surface searching are considerably less efficient and may miss whole sections of the population although, under certain conditions, they can give satisfactory results (e.g. Mordan, 1973).

10.1.4 Capture–recapture

The capture–recapture technique provides an alternative absolute method to those based on counts of animals within fixed units of habitat. Snails can be marked by painting their shells and this technique has been applied to *T. haliotidea* and *T. scutulum*, by painting the small external shell of these slugs with cellulose acetate paint (Barnes and Stokes, 1951). They found that this method of marking was partly successful but could not determine how many individuals retained the marks, although marked individuals were recovered for several months after marking.

Deroceras (Agriolimax), *Arion* and *Helix* have been successfully labelled by feeding them on food containing the isotopes ^{32}P and ^{131}I (Fretter, 1952) and similar techniques were employed by François *et al.* (1965) and by Moens *et al.* (1965) also using ^{32}P. Marked slugs could be located in the field using a Geiger-Müller detector. Moens *et al.* (1966) recorded the daily movements of *D. reticulatum* in the field by labelling the slugs with ^{182}Ta wire inserted under the vestigial shell. Newell (1965) made population estimates for *D. reticulatum* by the capture–recapture method using ^{32}P- labelled slugs. Newell found that marked slugs could be identified for over a month after marking, although the proportion of marked animals recaptured fell from 13.4% after 7 days to 3.3% after 21 days. This method gave rather wide 95% confidence limits, ranging from ±32.3 to ±57.0%. François *et al.* (1968) made a number of estimates of *D. reticulatum* populations in small $3\,m \times 3\,m$ experimental plots in a clover field over periods ranging from 12 to 17 days, also using ^{32}P-labelled slugs, and obtained 95% confidence limits ranging from ±23 to ±57%. Experimental plots were delimited by the use of a copper wire fence as a barrier to reduce the movement of slugs between plots. The use of this type of barrier was described by Moens *et al.* (1967). Stephenson and Dibley (1975) also described how an electric fence could be used to retain slugs in outdoor enclosures.

Slugs have also been marked by feeding them on food containing neutral red, a vital dye (Newell, 1965; South, 1965). This dye stained the digestive gland and the colour was easily seen through the foot, even in darkly pigmented individuals, and lasted up to three weeks in the field. Neutral red marking was used successfully by South (1965), for *D. reticulatum*, and Pinder (1969) to study dispersal in slugs. Hunter (1968a), using neutral red marked *D. reticulatum*, *A. hortensis* and *T. budapestensis*, found that the capture–recapture method tended to underestimate slug numbers. Methylene blue crystals, introduced under the integument of several larger species of *Arion*, formed coloured patches which lasted for about six weeks and slugs marked in this way were used for the estimation of population densities and in studying dispersal by Müller and Ohnesorge (1985). Freeze branding applied to the mantle was found by Richter (1976) to be a suitable method for individually marking the large slugs, *A. ater* and *A. colum-bianus*, and the small slugs, *D. laeve* and *P. andersoni*, although the pigment pattern of *L. maximus* made it difficult to distinguish the brand after a week on this species. The brand was applied using irons of copper and staple wire cooled by immersion in liquid nitrogen. The area of contact remained very dark compared with the surrounding tissues for the first month on other species, after which it slowly faded, although in the field *A. ater* and *A. columbianus* retained their individual marks for some months.

Rollo (1983b) used freeze branding to monitor the growth and movements of *A. columbianus*, *A. ater* and *L. maximus* in field cages although the rapid healing of brands prevented reliable recognition of individuals. Cook and Radford (1988) used freeze branding in a study of the mobility of *L. maculatus* between and within different stone wall sites, finding that the brands became indistinct after about three weeks. Hogan and Steele (1986) described a dye-marking technique, using a Panjet dental inoculator, for marking the integument of slugs and compared it with the freeze-branding technique. The blue dye, Alcian Blue, was preferred for its greater visibility and least damage occurred when slugs were marked at the posterior end of the body. Hogan and Steele found that survival was significantly higher after marking with the Panjet when compared to freeze branding and that marks persisted for at least two months. However, freeze branding enabled slugs to be marked on an individual basis, which was not possible using the Panjet which was limited by the number of different dyes available and sites suitable for injection. Both freeze branding and the Panjet can be used in large-scale field experiments, unlike previous techniques.

10.1.5 Size of sample

Van den Bruel and Moens (1958b), South (1964), Hunter (1968a) and Pinder (1969) all found that a sampling unit of 30 cm × 30 cm was most suitable for sampling arable and agricultural grassland. This size of unit has also been used for woodland (Mordan, 1973) and montane grassland (Lutman, 1978).

Warley (1970) obtained satisfactory results on arable with a smaller 23 cm × 20 cm unit. Bishop (1981), Phillipson (1983) and Glen et al. (1984) found that 25 cm × 25 cm sampling units were suitable for wetland habitats, woodland and arable soils respectively. Beyer and Saari (1978) used 23 cm × 30 cm units for abandoned farmland. Cores 27 cm diameter were used by Rollo and Ellis (1974) for sampling arable land and about 15 cm diameter by Phillipson (1983) for sampling woodland litter. The defined-area trap (Ferguson et al., 1989) enclosed a circular sampling unit 35.7 cm in diameter on pasture. Smaller sampling units, cores 10.1 and 6.3 cm diameter, were tested by South (1964) and Hunter (1968a) who found that units of this size generally failed to provide sufficient slugs per sampling unit. In addition, too many smaller sampling units are required to equal the precision arising from the use of 30 cm X 30 cm units (South, 1964).

The depth of a sampling unit will depend on the vertical distribution of slugs and this is fully discussed in Section 10.2.2. South (1964) found a sampling depth of 10 cm adequate for grassland and, for most of the time, in a field of winter wheat on a light soil, where only *D. reticulatum* was present. It was necessary, however, to sample down to the bottom of the furrow (27 cm) immediately after ploughing. Hunter (in Runham and Hunter, 1970) also showed that this species was mainly present in the top 7.6 cm of soil on arable ground for much of the year but that *T. budapestensis* and *A. hortensis* occurred deeper in the soil. Few, if any, slugs were found below a depth of 30 cm and van den Bruel and Moens (1958b), Hunter (1968a) and Pinder (1969) adopted a sampling depth of 30 cm for arable soils. Warley (1970) used a sampling depth of 10–12.5 cm on grassland and 25 cm on arable land and he concluded that the latter depth could be halved (i.e. 12.5 cm) for samples taken before ploughing. Both Glen *et al.* (1984) and Rollo and Ellis (1974) found a depth of 10 cm to be adequate for arable land and Lutman (1978) used this depth for samples from montane grassland. In more stable environments units have been taken to a lesser depth; Mordan (1973) used a sampling depth of 7.5 cm in woodland while Bishop (1981) used sampling units 3 cm deep for wetland habitats.

A sampling depth of 10 cm on grassland, and between 25 and 30 cm on arable, appears to be adequate for sampling slugs although under certain conditions this depth could be considerably reduced. The most suitable size of sampling unit appears to be between 0.10 and 0.05 m^2. Above this size, units become unmanageable, and below it, too few slugs are generally recovered for the statistical analysis of data. Finally, the sample size, i.e. number of sampling units, is generally limited by the available facilities but for regular sampling between 12 and 20 units appear to give satisfactory results. Hunter (1968a), Pinder (1969) and Warley (1970) divided sampling units horizontally into subunits to provide information on the vertical distribution of slugs.

10.2 SLUG POPULATIONS

10.2.1 Size of populations

Some examples of the size of slug populations in arable, grassland and woodland habitats are shown in Tables 10.3, 10.4, 10.5. With the exception of South (1989b), these estimates are probably for slugs only and so the actual size of population would be higher. While these estimates give some idea of the size of slug populations, there are considerable variations in density with time, reflecting the life cycle of the slug, and between generations (Lutman, 1978; South, 1989b). Warley (1970) suggested that the

Table 10.3 Population densities of slugs from arable habitats.

Habitat and species	Locality	No. m^{-2}	Extraction method	Reference
Arable (young wheat)	–		Metaldehyde baits	Thomas (1944)
D. reticulatum		148[b]		
Arable (sweet potatoes)	Trinidad		Metaldehyde baits	McCalan in Stephenson (1967)
Total slugs		3.3		
Arable	Northumberland		Flooding + soil washing	Hunter (1968a)
D. reticulatum		59[b]		
A. hortensis		39[b]		
T. budapestensis		37[b]		
Arable (wheat)	Leningrad, USSR		Unknown	Dmitrieva (1969)
D. reticulatum		20, 100, 40[a]		
Gardens				
D. reticulatum		60, 140[a]		
Garden	East Africa		Capture–recapture	Coe (1971)
Trichotoxon copleyi		2.8		
Arable (winter wheat)	Oxfordshire	up to 180	Flooding	Glen et al. (1984)
Total slugs				Newell (1967)
Arable (undersown stubble)				
D. reticulatum		18[b]		
Arable (ploughed, stubble)	East Lothian		Flooding or heat treatment	Warley (1970)
D. reticulatum		32.3[b]		
A. hortensis		18.3[b]		
A. fasciatus		2.2[b]		
Arable (potatoes)	East, West and Midlothian			

Table 10.3 Cont'd

Habitat and species	Locality	No. m^{-2}	Extraction method	Reference
D. reticulatum		20.5[b] ⎫	Mean of 4 crops	South (1973a)
A. hortensis		35.7[b] ⎬		
A. fasciatus	Kent	7.8[b] ⎭		
Arable (potatoes)			Flooding	
D. reticulatum		60.4[b]		
A. hortensis		210.4[b]		
T. budapestensis		65.6[b]		
A. fasciatus		17.7[b]		
Allotments	Leeds		Flooding	Atkinson et al. (1979)
Total slugs		47.6		
Arable (winter barley)	Bristol		Flooding	Wiltshire and Glen (1989)
Total slugs		200		
Arable (winter wheat)				
D. reticulatum		53.6		
Total slugs		57.1		

[a] Maximum numbers for two successive years.
[b] Approximate number – original number not expressed per square metre.

Table 10.4 Population densities of slugs from grassland and wetland habitats.

Habitat and species	Locality	No. m^{-2}	Extraction method	Reference
Rye-grass	–		Copper sulphate	Carrick (1938)
D. reticulatum		12[c]		
Rough grassland	Midlothian		Flooding/heat treatment	Warley (1970)
D. reticulatum		66.0[a,c]		
A. hortensis		37.7[a,c]		
A. fasciatus		1.7[a,c]		
Grass paddock	East Lothian			
D. reticulatum		23.7[c]		
A. intermedius		77.5[c]		
A. fasciatus		1.1[c]		
Permanent pasture	Northumberland		Flooding + soil washing	South (1989b)
D. reticulatum		42.4[a,c]		
A. intermedius		19.2[a,c]		
A. fasciatus		4.0[a,c]		
Pasture	New Zealand		Defined-area trap	Ferguson et al. (1989)
D. reticulatum		8.8–43.3		
Solidago–Aster field	NY State, USA		Flooding	Beyer and Saari (1978)
A. subfuscus		5.0[a]		
Clover fields	Belgium		Capture-recapture	François et al. (1968)
D. reticulatum		26, 13–44[c]		

Table 10.4 Cont'd

Habitat and species	Locality	No. m^{-2}	Extraction method	Reference
Agrostis–Festuca grassland	Llyn Llydaw, N. Wales		Flooding	Lutman (1978)
D. reticulatum		11.6, 11.9[b]		
D. laeve		0.1, 2.0[b]		
A. intermedius		12.5, 14.3[b]		
A. fasciatus		1.0, 0.8[b]		
A. subfuscus		0.7, 0.6[b]		
A. ater		0.0, 0.4[b]		
Fen	Wicken Fen		Wet sieving + flotation	Bishop (1981)
A. ater		0.8		
A. intermedius		0.8–3.2		
D. laeve		0.8–1.6		
D. reticulatum		1.6		
Fen and freshwater marsh	East Anglia			
A. intermedius		0.2–2.8		
D. agreste		0.8		
D. laeve		0.2–4.0		
D. reticulatum		0.4		

Juncus effusus – Sphagnum mires	Central Highlands, Scotland	
A. ater		0.2–0.4
A. intermedius		1.2–3.8
A. subfuscus		0.2
D. agreste		0.2
D. laeve		0.6–1.6
D. reticulatum		0.2
Molinia – Myrica mires	Central Highlands, Scotland	
A. intermedius		0.2–2.2
A. subfuscus		0.2

[a] Mean annual numbers.
[b] Mean annual numbers for two or three successive years.
[c] Approximate number – original number not expressed per square metre.

Table 10.5 Population densities of slugs from woodland habitats.

Habitat and species	Locality	No. m^{-2}	Extraction method	Reference
Beech forest	Netherlands		Tullgren funnel	Van den Drift (1951)
A. subfuscus		14.0[a]		
Ash-oak woodland	Huntingdonshire		Hand searching	Mordan (1973)
A. ater		1.1[a]		
A. fasciatus		<1		
A. hortensis		0.7[a]		
A. intermedius		4.9[a]		
L. maximus		<1		
Grassland in ash-oak woodland	Huntingdonshire			
A. ater		0.7[a]		
A. fasciatus		1.3[a]		
A. hortensis		0.2[a]		
A. intermedius		2.9[a]		
D. reticulatum		1.7[a]		
D. laeve		3.5[a]		
Mixed woodland	Norfolk		Flooding	Jennings and Barkham (1975a)
D. reticulatum		1.3[a]		
D. laeve		0.4[a]		
A. ater		1.2[a]		
A. fasciatus		6.1[a]		
A. hortensis		5.9[a]		

A. intermedius	19.1[a]		
A. subfuscus	0.8[a]		
L. marginatus	0.3[a]		
Acid woodlands		West Cork and Kerry	Hand sorting	Bishop (1977a)
A. ater	0.4–1.2		
A. circumscriptus s.s.	0.4		
A. intermedius	0.2–1.2		
A. subfuscus	0.2–0.8		
D. reticulatum	0.2		
G. maculosus	0.6		
L. marginatus	0.2		
L. cinereoniger	0.2		
Beech woodland		Oxfordshire	Infrared heat extractor	Phillipson (1983)
(i) litter			
A. hortensis	30.5		
D. reticulatum	1.0		
L. marginatus	0.1		
(ii) soil			Hand sorting
A. hortensis	1.6		
A. fasciatus	0.6		
D. reticulatum	0.3		

[a] Mean annual numbers.

ranges of mean slug densities on grassland and arable habitats might be classified as low, medium and high. A low *D. reticulatum* population, for example, would range from 5.4 to 12.9 slugs m^{-2}, a medium population from 12.9 to 48.4 slugs m^{-2} and a high population from 48.4 to 147.5 slugs m^{-2}. Comparable ranges for *A. hortensis* would be: low 1.1–9.7 slugs m^{-2}, medium 9.7–37.7 slugs m^{-2} and high 37.7–123.8 slugs m^{-2}.

10.2.2 Spatial distribution

(a) Vertical distribution

South (1964) showed that all the *D. reticulatum*, both slugs and eggs, in samples taken from grassland on a sandy loam were located within 2–3 cm of the soil surface. Similarly, for *A. intermedius*, 94% of slugs and all eggs were found at this level. Slugs of these two species were usually lying on the mat or feeding on sheltered grass and only entered loose soil during drought or severe frost. Eggs were found at the base of grass stems, under stones or in loose soil. *Arion fasciatus* was scarce and two out of the seven slugs recovered were taken from depths of between 2.5 and 7.5 cm, suggesting a difference between the vertical distribution of *A. fasciatus* and the other two species. Warley (1970) divided turf samples into a vegetation layer and a soil layer, down to 10 cm depth. He found that between 60 and 80% of *D. reticulatum* were located in the vegetation layer, except when the vegetation was frozen, while the proportion of *A. hortensis* (mean: 45%) and *A. fasciatus* (mean: 55%) was lower and showed wider variation. In a sample from another grassland site, 78% of *D. reticulatum* and 95% of *A. intermedius* were found in the vegetation layer. Carpenter *et al.* (1985) used a rhizotron situated in grassland to demonstrate that slugs may use earthworm burrows to move down into the soil, particularly in hot and dry or cold and wet weather. They found that up to a third of the slugs visible in burrows were located below a depth of 10 cm although most *D. reticulatum* were above this level. However, the proportion of the total population moving into burrows was unknown. Eggs of the worm-eating slug *T. scutulum* have been found at depths of up to 36 cm in garden soils indicating that this species can penetrate deeply in the soil (Barnes and Stokes, 1951).

In a wheat field, South (1964) showed that *D. reticulatum*, including eggs, were generally found in the upper 10 cm of soil except immediately after ploughing when slugs were found down to a depth of 15 cm, i.e. at the bottom of the furrow. Hunter (1966) also found significant differences between the vertical distribution of different species of slug in samples taken from arable ground over two years. Most *D. reticulatum* (83%) occurred in the top 7.5 cm of soil while fewer *A. hortensis* (62%) and *T. budapestensis* (50%) were found at that level. Few *D. reticulatum* (6%) were found below a depth of 15 cm and then, mainly after ploughing. A larger proportion of

A. hortensis and *T. budapestensis* were found at depths of between 7.5 and 22.5 cm. The proportion of *D. reticulatum* eggs in the top 7.5 cm of soil was significantly greater than the proportion of *A. hortensis* eggs, which in turn was significantly greater than the proportion of *T. budapestensis* eggs. Rollo and Ellis (1974) confirmed that 90% of *D. reticulatum* eggs were located in the first 10 cm of soil in cornfields. Stephenson (1966) found that young *T. budapestensis* moved below the soil surface to depths of up to 37 cm even where there were no cracks in the soil although, where there were natural cracks, these were used for movement. Stephenson observed the openings of burrows containing *T. budapestensis* in a field of medium heavy loam and also concluded that, in the absence of cracks in the soil, this slug could burrow by ingesting soil, even on heavy land. Both *A. hortensis* and *T. budapestensis* were shown to feed actively at depths of up to 36 cm in a potato field on a medium–heavy silt between June and September by Stephenson (1967). Pinder (1974) showed that in a potato crop on a loam soil significantly more *D. reticulatum* and *A. hortensis* were found in the upper 7.5 cm soil than in the lower soil and relatively few slugs were found below a depth of 15 cm.

Few slugs appear to penetrate to any depth in woodland soils although Mordan (1973) considered that slugs might move deeper into cracks in a heavy clay soil during dry weather. Phillipson (1983) showed that 95% of *A. hortensis* and 76% of *D. reticulatum* were recovered from woodland litter samples while the few *A. fasciatus* present were all located in the soil. The only *L. marginata* present were recovered from the litter samples. It seems unlikely that the larger woodland *Limax* spp. move into the soil but rather shelter under fallen branches, under bark and in decaying tree stumps (Taylor, 1902–1907). *Lehmannia marginata* is frequently found on trees, up to heights of four metres in wet weather and sheltering in crevices in the bark and between roots in dry weather (Quick, 1960). The forest-dwelling slug, *A. columbianus*, shelters under logs and in crevices and has even been found some distance underground in a disused burrow (Harper, 1988). Jennings and Barkham (1979) recorded the vertical distribution above the surface of woodland soils at which six species of slug fed. The mean heights ranged from 0.18 cm (*A. fasciatus*) to 8.12 cm (*A. ater*) above the soil surface. It was possible to distinguish two groups of slug on this basis, i.e. *A. ater*, *D. reticulatum* and *A. subfuscus* (3.89–1.82 cm) and *A. intermedius*, *A. hortensis* and *A. fasciatus* (0.18–0.74 cm) and Jennings and Barkham suggested that the reason for this separation was that the former group consumed larger quantities of green food than the other group.

(b) Horizontal distribution

The geographical distribution of British slugs has been discussed in Chapter 1. The local distribution is often 'patchy' (Barnes and Weil, 1944) and in adjacent gardens (Barnes and Weil, 1945) or even in adjoining parts of the same garden (White, 1959b; Lyth, 1972), slug faunas can be very different. South (1965) investigated the horizontal distribution of slugs using the 'coefficient of dispersion' first described by Blackman (1942). There was significant evidence that populations of *D. reticulatum*, *A. intermedius* and *A. fasciatus* were underdispersed (i.e. aggregated) on grassland but evidence for *D. reticulatum* from arable was inconclusive. Slug eggs occurred in batches and the batches themselves were underdispersed. For example, a single turf (15.2 × 15.2 cm) contained 310 *A. intermedius* eggs, representing an aggregation of at least 20 batches of eggs since there were rarely more than 15 eggs in a batch. The negative binomial distribution has been used as a model to describe this type of contagious distribution in invertebrates (Bliss and Fisher, 1953) and South (1965) found that frequency distributions for *D. reticulatum* and *A. intermedius* showed close agreement with the negative binomial distribution, providing additional evidence of aggregation in populations of these species. Using the coefficient of dispersion, Hunter (1966) found evidence for clumping in populations of *D. reticulatum*, *A. hortensis* and *T. budapestensis* from arable land. The highest coefficients were associated with hatching of the eggs when the degree of aggregation was more likely to be high. Mordan (1973) showed that *D. reticulatum* and *A. intermedius*, but not *D. laeve* and *A. fasciatus*, were aggregated in a grass field surrounded by oak–ash woodland, particularly when eggs were beginning to hatch. *Arion intermedius* from the oak–ash woodland were not generally aggregated except when newly hatched individuals were present (Mordan, 1973). The lack of evidence for an underdispersed distribution might have been due to the small numbers of slugs in some samples. However, Jennings and Barkham (1975a) found no evidence that individual populations of eight species of slug, including *A. intermedius*, were aggregated in mixed woodland and suggested that these slugs were usually randomly distributed or overdispersed.

Warley (1970), using the exponent k of the negative binomial distribution as a dispersion parameter, showed that newly hatched slugs of three species, *D. reticulatum*, *A. hortensis* and *A. fasciatus*, were strongly aggregated as they hatched from batches of eggs and that the degree of aggregation became less as they matured. The degree of aggregation in *D. reticulatum* increased in very dry and cold frosty weather as slugs found shelter. Warley obtained similar results in a potato crop. In common with many other terrestrial invertebrates, slugs generally show an aggregated type of dispersion which is most marked in newly hatched slugs emerging from batches of eggs. This type of aggregation, where slugs rarely come into contact with one another, generally results from the combined effect of physical environmental factors and a lack of dispersal from the egg batch (South, 1965). It is distinct from the huddling behaviour, involving frequent contact between individuals, already described in Chapter 7. Rollo (1983a) found that the aggressive slug,

L. maximus, had an overdispersed distribution in field cages due to their agonistic behaviour (see Section 7.3.1).

(c) Dispersal

South (1965) investigated the dispersive movements of *D. reticulatum* on undisturbed arable land by releasing large numbers of dye-marked slugs at a single location and measuring the distance travelled by marked slugs recovered in traps along lines radiating out from the release point. Under these conditions, the rate of increase of variance in distance travelled provides a measure of the rate of dispersal of the population (Dobzhansky and Wright, 1943) and South found that, although variance increased rapidly over the first two or three days, there was little further increase during the remainder of the time and no evidence that the changes in dispersal rate depended on weather conditions. South concluded that this dispersive behaviour was similar to the 'homing type' described by Dobzhansky and Wright (1943) where, after some initial dispersive movement, *D. reticulatum* established a home range. Homing behaviour in slugs has already been discussed in Section 7.2.2. The mean distances travelled by *D. reticulatum* in two experiments were 0.46 m in 5 days and 1.13 m in 7 days respectively. While slugs dispersed in several directions, there was an overall trend in dispersal downwind of the prevailing wind. Moens *et al.* (1966) also recorded two similar stages in dispersal behaviour for this species when it was released into a strange environment. Dye-marked *A. ater*, released on a lawn, vacated a circular area 20 m in diameter within 9–12 days and new slugs moved in to replace them (Müller and Ohnesorge, 1985). In this experiment, *A. ater* were recorded travelling distances of 3.5 and 12.8 m in a night on a lawn and of 1.1 and 5.8 m on asphalt. Coe (1971) demonstrated that the slug *Trichotoxon copleyi* Verdcourt travelled a maximum distance of 6–7 m over ten nights from its initial capture point. In laboratory experiments at temperatures between 15 and 20°C, Duval (1970a) found that *D. reticulatum* could travel a mean distance of 100 cm in 5–6 hours, twice the distance moved by *A. hortensis* (51 cm) and *T. budapestensis* (41 cm).

The dispersal of freeze-branded *L. maculatus*, living on stone walls, was recorded by Cook and Radford (1988) over a period of 22 days. Immigration or emigration was minimal in the population and no marked slug was recorded from a site 30 m distant. Near the end of the experiment, it was estimated that slugs had moved a mean distance of 3.2 m from the release point and only two slugs remained at the original release points, showing that slugs had not remained stationary. Both Miles *et al.* (1931) and Carrick (1938) suggested that slugs moved into crops from uncultivated headlands and Pinder (1974), using dye-marked *D. reticulatum*, showed that these slugs moved into an experimental potato crop from adjacent grassland in summer, possibly as the result of the increased cover afforded by the developing haulm. However, he considered that this movement was unlikely to be important on a field scale. Airey (1986a) found no evidence for a net movement of slugs into a potato crop while investigating the dispersal of dye-marked slugs. The available information on dispersal of slugs suggests that they disperse relatively little throughout life. Lloyd (1963) found that *Deroceras* sp. and *Arion* sp. living under bark of fallen trees, moved into the litter during severe weather. However, there was no evidence for large-scale diurnal movements independent of the weather. Barnes (1952) showed that *D. reticulatum* was unable to recolonize a garden until slugs were artificially introduced. Boycott (1934) considered that terrestrial molluscs were ill-adapted for dispersal, particularly with regard to colonization of new habitats. However, Eversham (1989) described how mammals, including foxes, cats and dogs, could disperse slugs that had become tangled in their fur.

(d) Factors determining spatial distribution

A detailed examination of the dispersal of *D. reticulatum* individuals within small areas showed that this slug was aggregated with respect to areas of (i) between 9 and 26 cm^2 and (ii) about 929 cm^2, but not with respect to intermediate areas (South, 1965). South concluded that the smaller aggregations represented individuals resting near to one another in sheltered positions, e.g. in hoof prints or under clods of soil. Since numbers of *D. reticulatum* were positively correlated with the distribution of tussocks of the grass, *Dactylis glomerata*, he suggested that, as a result of oviposition behaviour, more eggs of this species were laid in the shelter of *Dactylis* tussocks than elsewhere and that a greater proportion of these eggs survived dry weather. This produced an aggregated distribution which largely persisted because of the homing type of behaviour of this slug and could explain the larger sized aggregations. *Dactylis* tussocks provide a considerable degree of protection to animals sheltering in them (Luff, 1965) and the presence of adequate shelter appeared to be an important factor determining the distribution of slugs on grassland, with *A. fasciatus* being more dependent on soil moisture than either *D. reticulatum* or *A. intermedius*. Boycott (1934) and Getz (1959) reached similar conclusions about the distribution of *A. fasciatus*. Pallant (1974) found that *D. reticulatum* slugs and eggs and *A. intermedius* slugs were associated with tussocks of the grass, *Deschampsia caespitosa* (Linnaeus) in rough pasture rather than a cropped sward of other grasses. He concluded that tussocks provided slugs with protection from grazing animals and a more favourable microclimate, including lower temperatures, than the sward. There was no correlation between the distribution of slugs and that of *Ranunculus repens*, a favoured food plant of *D. reticulatum*. A similar result was obtained by Beyer and

Saari (1978) who showed that the distribution of *A. subfuscus* was associated with tufts of the grass, *Phleum pratense* Linnaeus. This grass, like *D. glomerata* and *D. caespitosa* previously, offered slugs a more favourable microclimate than the surrounding vegetation. François *et al.* (1968) demonstrated a gradient in numbers of *D. reticulatum* over a clover field in the direction of the confluence of a ditch and a stream.

Although there was no association between the resting sites of *D. reticulatum* and the distribution of the favoured food plants for this species in mixed woodland (Pallant, 1967), the distribution of both eggs and slug resting sites under logs was closely related to the growth of the ground flora, suggesting that the association was due to the shelter provided by the plants rather than the availability of food. Phillipson (1983) found similarly that most of the slugs, primarily *A. hortensis* and *D. reticulatum*, in a beech wood were found in areas where most shelter was provided by deep litter and ground vegetation. However, Beyer and Saari (1977), in a study of the effect of tree species on the distribution of slugs, showed that *A. fasciatus* and *D. laeve* were absent from stands of pine and spruce, except where a deciduous ground cover was present, while *A. subfuscus* was abundant in stands of both deciduous and coniferous trees. They concluded that *A. fasciatus* and *D. laeve* required a deciduous ground flora, possibly for food, while *A. subfuscus*, which according to Taylor (1902–1907) shows a preference for fungi, was able to obtain an adequate diet of fungi in coniferous areas where other ground flora was absent. Beyer and Saari (1977) found no evidence for specific associations between six slug species and 13 tree species.

In a field of potatoes, Warley (1970) demonstrated that *D. reticulatum* and *A. hortensis* were more numerous on the ridges where they were aggregated around potato plants while *A. fasciatus* was more numerous in the furrows rather than the ridges (Warley, 1970). Similar results for the distribution of *D. reticulatum* and *A. hortensis* in a potato crop were obtained by Pinder (1974) although Stephenson (1967) found that *A. hortensis* and *T. budapestensis* tended to be more numerous in the furrows rather than the ridges prior to tuber ripening. Stephenson concluded that furrows offered a moister habitat to slugs, since rain tended to be diverted to the furrow by the potato haulm, particularly when the amount of haulm was small during the drier part of the season. The choice of resting places for slugs in bare soil is influenced by the size of soil aggregates (Stephenson, 1975a), with *D. reticulatum* preferring a moderate or coarse soil to fine soil and Stephenson suggested that a response to contact stimuli might be partly responsible for this behaviour. Duval (1970a) had previously shown, in experiments with *D. reticulatum* and *A. hortensis*, that the spaces between aggregates, which were preferred as shelters, were such that a slug could insinuate itself completely into the cavity. She suggested that, in this position, a slug would make maximum contact with the soil, and so contact stimuli might be partly responsible for a slug remaining motionless once its requirements were satisfied.

Adult slugs preferred fine aggregates for egg-laying sites and in the field these sites were often found beneath larger lumps of soil left by cultivation (Stephenson, 1975a). On the basis of his choice tests, Stephenson concluded that, in the field, eggs of *D. reticulatum* would be laid in fine to medium (3–10 mm) aggregates, while newly hatched slugs would be able to find their preferred resting places in soil of medium coarse (10–12.5 mm) to coarse (>12.5 mm) aggregates in the vicinity of the egg-hatching site thus maintaining small discrete aggregations associated with larger lumps of soil. *Tandonia budapestensis* also prefers coarse-textured soils and in cultures, using a medium heavy slit, this species excavated cavities in the harder lumps of soil to deposit its eggs (Stephenson, 1966). Miles *et al.* (1931) and Carrick (1942) found that increased slug numbers were correlated with an increase in the soil organic matter content. Although decomposing plant material offers slugs a source of food, it seems likely that the increased water-holding capacity of soils with high organic matter is the main reason for this association. Barnes and Weil (1945) and South (1965) considered it unlikely that slugs were influenced by the organic matter content of garden soils or of grassland. However, in gardens, adequate alternative shelter is available to provide protection against water loss.

It may be concluded from this discussion that the horizontal distribution of slugs is, to a large extent, determined by the need for shelter, particularly from the effects of desiccation. Some species of slug have more exacting moisture requirements and it has been shown that this also influences their vertical distribution. The availability of food does not appear to exert much influence on the distribution of slugs and this is not surprising as most slugs are generalist herbivores. However, there may be a few exceptions, such as the distribution of some species in coniferous woodland. Boycott (1934), in an extensive review of the habitats of land molluscs in Britain, concluded that shelter was one of the main considerations determining the distribution of terrestrial molluscs, while food had little influence on distribution.

10.2.3 Distribution in time

(a) Introduction

Few attempts have made to study slug populations for any appreciable length of time. Where this has been attempted, the results have generally been expressed in terms of slug activity rather than absolute numbers (Barnes and Weil, 1944) or, where density has been used as a measure, results have been interpreted largely in terms of the life cycle of slugs and the factors responsible for the regulation of numbers have not been critically appraised (Hunter, 1966; Warley, 1970). Thus, while a great deal is known about the effect of climatic and other physical factors on the feeding and locomotor activity of slugs in both the laboratory and the field (Chapter 7), and some information is available on their effect on life cycles (Chapter 8), there is little information about the ways in which numbers are regulated under field conditions. In order to obtain this information it is necessary to distinguish between changes in numbers due to stages in the life cycle, such as increases due to egg laying, and changes resulting from adverse weather conditions, such as drought or severe frosts, and so numbers need to be compared for successive generations of slugs over several years.

(b) Climatic factors (density-independent factors)

Records of the total numbers of slugs trapped each year in a Manchester garden between 1925 and 1930 showed that adverse weather conditions had a marked effect on slug numbers but that the population rapidly recovered once conditions returned to normal (Miles *et al.*, 1931). Prolonged periods of intense cold during the winter and drought during the spring and early summer had an adverse effect on numbers while a mild wet season led to a marked increase in the slug population. Miles *et al.* found that *D. reticulatum* was better able to survive low temperatures than *T. sowerbyi*. They considered that predators had little effect on slug numbers.

Observations made on *D. reticulatum* populations on arable ground in the Leningrad region, between 1964 and 1966, enabled Dmitrieva (1969) to distinguish three critical periods in the year. The first was the autumn period of egg laying which began in mid-August and lasted until the beginning of September. Adequate moisture was available during this period but temperature was an important factor because egg laying, which began when midday air temperatures fell below 15°C, virtually ceased after air temperatures fell below 5°. The second critical period was the winter survival of the eggs and lasted from about October until April. Low temperatures were a critical factor at this time with mean minimum temperatures falling to below −20°C for much of the time. Dmitrieva found that *D. reticulatum* eggs could resist temperatures down to −11°C and that it was uncommon for eggs to be killed during the winter, provided that there was an adequate snow cover to insulate the soil against lower temperatures. The third critical period was the time when the egg developed and hatched in the spring after the mean air temperature increased beyond 5°C. This period began at the end of April and lasted until June. Temperatures were usually adequate at this time but, if the spring was dry, egg development was delayed and a smaller proportion of slugs hatched. Thus temperature and moisture were important factors in determining the numbers of slugs both directly and indirectly, by hindering the timing of the life cycle. Reference has already been made in Chapter 8 to the exceptional development of a

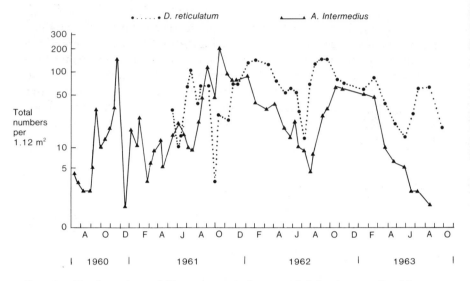

Fig. 10.1 Total numbers of *Deroceras reticulatum* and *Arion intermedius* (slugs and eggs) per 1.12 m² on each sampling date (after South, 1989b).

second generation in 1964 as the result of unusually favourable temperature and moisture conditions (Dmitrieva, 1969).

A population study of the slugs *D. reticulatum* and *A. intermedius* was carried out between 1960 and 1963 on permanent pasture on a farm in Northumberland (South, 1989b). Total numbers of both species (Fig. 10.1) showed fluctuations with peaks due to egg laying by successive generations. The sampling period included three complete generations of both species and examination of numbers for individual generations of *D. reticulatum* and *A. intermedius* showed a common pattern with an initial build-up of numbers, due to egg laying, followed by a rapid decline in the early stages and then a gradual fall in numbers through to maturity (Fig. 10.2). This pattern represents a considerable mortality in the egg and very young stages. Short-term changes in numbers could be generally explained by increases in numbers of eggs. In the long term, numbers of *D. reticulatum* were reduced by a prolonged summer drought although *A. intermedius* numbers were unaffected as these slugs aestivated during dry periods. There was evidence, however, that the consequent delay in development of the reproductive system of *A. intermedius* might have affected the size of the next generation by reducing the numbers of eggs laid. South (1989a) found that a drought in early summer hindered the development of the hermaphrodite gland of *D. reticulatum* so that the gland was smaller and fewer eggs were laid. Thus, in both species, the effect of a drought was carried forward to the next generation. Kennard (1923) recorded how semiarid conditions in 1921 reduced numbers of slugs but not snails and suggested that the reduction in

(a)

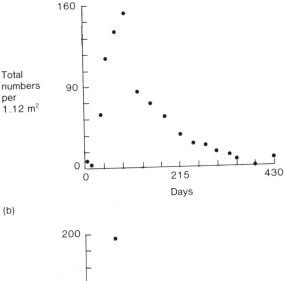

(b)

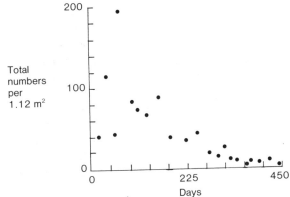

Fig. 10.2 Total numbers per 1.12 m² for (a) *Deroceras reticulatum* Generation VII and (b) *Arion intermedius* Generation IV (after South, 1989b).

numbers of *D. reticulatum* was probably due to the destruction of young slugs.

Mordan (1973) found that numbers of slugs in an ash–oak wood and in rough grassland surrounded by woodland were reduced by exceptionally dry periods during early autumn and in spring. Populations of *A. intermedius* and *D. laeve* were particularly affected in this way. A reduction in slug populations, especially for *A. intermedius*, *A. fasciatus* and *A. hortensis*, during dry summers was also recorded by Jennings and Barkham (1975a) from mixed woodland and they concluded that *A. ater* was better able to withstand dry conditions than the other slugs. South (1989b) found little evidence that cold weather affected numbers other than by delaying development, provided that adequate shelter, such as grass tussocks, was available. However, there was a significant reduction in numbers of *D. reticulatum* slugs on an adjacent barley stubble during the late autumn and early spring due to a period of very cold weather with severe night frosts before any snow had fallen. Slugs had congregated in the autumn under rows of straw left after combining and this cover had become consolidated so that its insulating properties were reduced. These observations were similar to those made by Dmitrieva (1969). A detailed assessment of egg mortality in the field showed that this was generally low for both species and less than 10% of eggs were dead when examined. Carrick (1938) found that egg mortality in this species increased at lower incubation temperatures and there was evidence that egg mortality increased during cold winters (see Section 10.2.3(c) below).

(c) Predators, parasites and disease (imperfectly density-dependent factors)

Although a wide range of slug predators has been recorded and slugs are hosts to many parasites (Chapter 9), there are few, if any indications that these exert any significant effect on slug populations, except under rather artificial conditions such as those reported by Symondson (1989). Although starlings (*Sturnus vulgaris* Linnaeus) fed on slugs on grassland (South, 1980), South (1989b) found no evidence that predators, parasites or disease caused any significant reduction in the populations of *D. reticulatum* and *A. intermedius*. He also concluded that a fungus present in some eggs was an obligate saprobe rather than a parasite. The incidence of this fungus in *D. reticulatum* eggs increased from 1.7% in January to 17.1% in March during the extended cold period in the winter of 1961–62 and South concluded that the fungus invaded eggs that had already been killed by a delay in development due to the exceptionally low temperatures.

(d) Competition (density-dependent factor)

The effect of competition on reproduction and mortality of three species of slug, *A. columbianus*, *A. ater* and *L. maximus*, was investigated using field cages containing monocultures and two-way combinations of these species (Rollo, 1983a). *Ariolimax columbianus* and *A. ater* are non-aggressive species while *L. maximus* is a highly aggressive species during the summer (see Section 7.3.1). *Ariolimax columbianus* is a native west North American species that evolved with no major molluscan competitor while the other species, introduced from Europe, occupy a much wider range of habitats and have penetrated the habitat of *A. columbianus*. Rollo concluded that, in comparison with other slugs, *A. columbianus* was strongly K-selected and probably evolved through intraspecific exploitative competition in predictable, stable forest habitats. Reproductive success and survivorship of

A. columbianus were adversely affected by the presence of either of the other slugs although the aggressive *L. maximus* had a far greater impact. *Arion ater* could be classified as a fugitive *r*-strategist species adapted to less predictable, unstable environments and a characteristic feature of a fugitive species is the rapid dispersal ability noted by Hamilton and Wellington (1981a) for this species (see Section 7.3.2). Rollo (1983a) also concluded that *L. maximus* was an α-selected species, i.e. an interference adapted species where the effect of aggression was more pronounced on competitors than intraspecifically.

The effect of competition on behavioural time budgets was investigated for the same three slug species by Rollo (1983b). He found that *A. ater* and *A. columbianus* had no detectable effect on one another's time budgets while *L. maximus* disrupted most categories of behaviour in these non-aggressive species, although its own time budget was not significantly changed. Rollo concluded that *L. maximus* was adapted to maximize behavioural disturbance of interspecific competitors while minimizing intra-specific effects. Rollo *et al.* (1983a) described and tested a comprehensive model which simulated the physiological and behavioural responses of *A. columbianus*, *A. ater* and *L. maximus* in variable environments. They treated the animals as physiological input–output devices linked with their environment by means of a behavioural control programme and found that the results could be accurately related to known life history parameters. These slugs have similar morphological designs but differ considerably in their life history tactics and behavioural time budgets and Rollo *et al.* (1983b) used the model described previously to study the effects of substituting alternative risk-management styles (i.e. making one species behave like another) in different environments. The simulations suggested that competition has shaped the time-budgeting tactics of *A. ater* and *L. maximus* and that each species performed better in harsh weather, using the behavioural programme of the other. *Limax maximus* is an aggressive species with a narrow nocturnal activity period and Rollo *et al.* (1983b) concluded that the concentrated activity of *L. maximus* may allow it to displace competitors more effectively, while the broader time span of activity in *A. ater* may be necessary to avoid *L. maximus*. Although the above experiments demonstrate that both intra- and interspecific competition may occur in some slug species under crowded conditions, the slug densities of $8 \, m^{-2}$ in the field cages described by Rollo (1983b) were considerably higher than those recorded from natural populations. For example, records for *A. ater* (Tables 10.4 and 10.5) show a range of between 0.2 and $1.2 \, m^{-2}$ while for *A. columbianus* a minimum density of $0.25 \, m^{-2}$ was estimated by Richter (1979) and for *L. maximus* a figure of less than one in $1.9 \, m^2$ was recorded by Mordan (1973).

Different species of slug, as generalist herbivores, will tend to share similar food resources and this suggests that interspecific competition might occur where densities are high, for example on arable ground, or, where species diversity is high, for example in woodland habitats. The Competitive Exclusion Principle states that if two competing species coexist in a stable environment then they do so as a result of niche differentiation. Jennings and Barkham (1979) found that the six most abundant slug species in mixed woodland could be separated into two groups on the single niche dimension of vertical feeding height, with one group, *A. ater. A. subfuscus* and *D. reticulatum* consuming larger quantities of green food than the other group, *A. intermedius*, *A. hortensis* and *A. fasciatus*. Although this separation was probably insufficient to avoid competition entirely, it seemed likely that the species also simultaneously differed along other unknown niche dimensions. Two additional species were clearly separated from the other six species because of their specialist requirements, *L. marginatus* feeding on epiphytes on trunks of trees and *D. laeve* being more tolerant of wetter soils. Hunter (1966) found that the vertical feeding heights of *D. reticulatum*, *A. hortensis* and *T. budapestensis* showed some separation from one another, with the latter species feeding deeper under ground than the other species, while *D. reticulatum* fed mainly on or above the soil surface. Cameron (1978) found substantial differences in the sites of activity of some coexisting species of land molluscs, including *L. marginatus* and *A. ater*, although the differences in resting sites were less marked. While these examples show some degree of niche separation, it seems likely, in view of the abundance of potential food, that interspecific competition would normally be avoided because adverse climatic conditions acted to keep numbers of coexisting species below the carrying capacity of the environment. An investigation of the ecology of four sympatric species of *Limax* was made by Cook and Radford (1988). They concluded that there was substantial niche separation between *L. maximus* and the other species, mainly based on its feeding preferences. The remaining species had a similar feeding ecology but differences in life cycles and temperature sensitivity distinguished *L. marginatus* from both *L. flavus* and *L. maculatus*. There was again no evidence for interspecific competition here, particularly as both *L. maculatus* and *L. flavus* form huddles together in their daytime resting sites (Cook, 1981).

(e) Conclusion

The review of the effect of density-independent climatic factors on slug populations suggests that numbers are not stable but are limited in most populations by the shortness of time during which they can increase before harsh weather, particularly the lack of moisture or very low temperatures, intervenes. Thus a period of unusually favourable weather will allow slug populations to increase to higher levels and this was demonstrated, for example, by the increase in numbers reported by Anon. (1988) after an unusually warm wet summer and by the exceptional development of a second generation of *D. reticulatum* reported by Dmitrieva (1969) and described previously (p. 267). Waldén (1981) considered the population dynamics of slugs in a particular locality to be, in most instances, a history of repeated local extinctions during dry years and rapid recolonization during wet years or seasons. This idea is supported by the views of Andrewartha and Birch (1954) and of Milne (1957, 1962), who concluded that, for most of the time, the control of increase in invertebrate numbers is due to the combined action of density-independent and imperfectly density-dependent environmental factors. The latter includes the effect of predators, parasites and disease and their effect on slug populations has already been discussed. South (1989b) suggested that while the mortality due to predators, parasites and disease was probably small in most slug populations, this factor could contribute to a general decline in numbers, particularly at critical times in the life cycle of the slugs.

There is no clear evidence from studies of natural slug populations that intraspecific competition has a significant effect on numbers. The highest densities of slugs are generally found on disturbed habitats such as arable land or agricultural grassland (Tables 10.3, 10.4) and are mainly of *D. reticulatum* or, less frequently, *A. hortensis* or *A. intermedius*. These are species which Rollo and Wellington (1977, 1979) considered were not particularly aggressive. *Deroceras reticulatum* can be regarded as an *r*-strategist species (South, 1982) and McCracken and Selander (1980) observed that while *D. reticulatum* was the most numerous and widespread introduced slug in North America, it was common in only a limited range of disturbed habitats such as horticultural areas. This suggests that, like other *r*-strategist species, it does not compete well with other species in more stable environments. This is confirmed by the much lower densities of this species in the more stable woodland habitats listed in Table 10.5. Thus it seems unlikely that competition acts as a regulating factor. In stable environments, such as woodland, the degree of niche separation appears to be sufficient to avoid interspecific competition while in unstable habitats, such as arable ground where populations tend to be higher, the species present are not particularly aggressive. Thus it seems unlikely that competition for food and shelter has any significant effect on natural slug populations. Boycott (1934) concluded that the quantity or quality of food had little influence on populations and this was confirmed by Runham and Hunter (1970).

10.3 COMMUNITIES

10.3.1 Introduction

A community is generally regarded as any assemblage of populations of living organisms which occur together in space and time. Such a grouping has generally developed as the result of the combined action of various environmental factors while interactions develop between the component species over a period of time. It is perhaps unrealistic to consider slug communities in isolation as they form part of a wider community and interact with other components, such as the plants which form a source of food. However, Boycott (1934) concluded that particular groupings of slugs and snails were found in different places and that these associations were such that experience enabled conchologists to predict, from the appearance of a place, which species they were likely to find in it. Boycott recognized three factors as important in determining the distribution of slugs and snails. These were the history of the fauna, the climate and the nature of the habitat, the latter usually referring to the presence or absence of shelter and calcareous material. He considered that the effects of food and competition were not important.

Thus it seems unlikely that slugs and snails form clearly defined associations or communities with interdependent bonds but rather that the habitat distribution of individual species is determined by their range of tolerance to particular environmental factors. Characteristic ecological groupings then arise from the coincidence of this range for several species. Boycott (1934) was able to classify some habitats on the basis of specific slug and snail faunas and this principle of ecological groupings was used successfully by Evans (1972) in the reconstruction of past environments from snail remains at archaeological sites. Evans classified the Limacidae as intermediate between woodland and open country species or catholic species while he considered the Arionidae as generally living in damp habitats. The slug faunas listed in Tables 10.3–10.5 suggest that species diversity is highest in relatively undisturbed woodland and wetland habitats while arable farming tends to reduce species diversity. However, total numbers of slugs tend to be highest on arable and managed grasslands.

10.3.2 Woodland slugs

Boycott (1934) considered that woodlands had no characteristic slug or snail fauna but provided the conditions which most slugs and snails required. Their fauna depended on their age, soil and on the degree of shelter and conservation of moisture afforded by them. Three slugs, *L. cinereoniger*, *L. tenellus* and *L. marginata*, could be regarded as woodland slugs although *L. marginatus* also occurred in some open habitats. Boycott also listed the slug species not found in woods. These were synanthropic species including *Testacella* spp., *L. flavus* and *M. gagates*. The species of *Milax* and *Tandonia* are occasionally found in woodland, e.g. *T. budapestensis* was recorded from woodlands in Surrey and Devon by Ellis (1964b). Although *L. flavus* is rarely found away from buildings (Cook and Radford, 1988), it has been recorded from similar woodland sites to those occupied by the closely related

L. maculatus which is a woodland species (Evans, 1978). Boycott (1934) concluded that the fauna of acid woodlands tended to be very poor both in species and numbers. While this is true for snails, the position with regard to slugs is less clear and Boycott concluded that most slugs were indifferent to lime, occurring freely in calcareous and non-calcareous places.

The distribution of woodland molluscs in the Netherlands was studied by Bruijns *et al.* (1959). The richest slug faunas were found in deciduous forests of oak, beech and poplar on eutrophic soils where the pH was greater than 7. Slug species found in these forests included *A. ater*, *A. subfuscus*, *A. circumscriptus*, *A. hortensis*, *A. intermedius*, *L. maximus*, *L. cinereoniger* and *L. marginata*. *Deroceras reticulatum* and *D. laeve* occurred exceptionally in this type of woodland while in southern Limburg, where soil and climatic conditions were especially favourable, *T. rustica* was also present. In contrast with these habitats, coniferous forests on poor sandy soils with a pH less than 7, contained few molluscs where the ground flora was sparse although *A. subfuscus*, *A. intermedius* and *L. tenellus* were present and fed on fungi. Where soil conditions were slightly better and some ground vegetation developed, *D. laeve* was also present. More slugs were present in oak–birch forests on these soils including *A. intermedius*, *A. subfuscus*, *L. tenellus*, *L. cinereoniger* and *D. laeve*, with *L. marginata* in addition on better soils. However, *L. maximus* did not occur in these acid woodlands. Finally, birch–fen woodland on sandy, peaty, poor acid soils, often with *Sphagnum* growth, contained only *A. intermedius* and *D. laeve*. Alder woodland on wet, peaty soils, with a high groundwater table, contained many molluscs including freshwater species. Slugs present included *A. ater*, *A. subfuscus*, *A. circumscriptus*, *D. reticulatum* and *D. laeve*. The same species, with the exception of *A. ater*, were present in birch–fen woodland growing under poorer conditions although only *A. intermedius* was numerous. This survey demonstrated that acid woodlands contain fewer slug species and also that slugs are less numerous there. The association of *L. marginatus* with better soils was supported by Valovirta (1968) who found that this species was confined to hyperite hills in central Finland where pH values exceeded 6.1.

Wäreborn (1969) examined the environmental factors influencing the distribution of land molluscs in five types of woodland habitat over 75 localities in a calcium carbonate free area of southern Sweden. Wäreborn suggested that the necessary supply of calcium for molluscs was found in the organic component of the soil, mainly the litter and foerna (fermentation) layers. Correlations were found between calcium content of the foerna and both numbers of mollusc species and their abundance. The availability of calcium salts in the tree leaves forming the litter depended on the tree species since calcium was less readily available from oak leaves than from leaves of ash, lime, maple and elm.

Moist, mixed woods, with a ground flora of *Vaccinium myrtillus*

Linnaeus, *Sphagnum* ssp. and *Polytrichum commune* Linnaeus and with a mean soil pH of 4.6, contained few species and low numbers of slugs. *Limax cinereoniger* was the most widely distributed species while *A. ater* and *A. subfuscus* were also present. The mean calcium content of the foerna was 7.1 parts 10^{-3}.

The drier, mixed coniferous woods with birch and *V. myrtillus* had a similar soil pH and calcium content. In addition to the previous species, *L. marginata*, *L. tenellus* and *A. silvaticus* were also present although these occurred at localities where calcium levels were slightly higher than average. *Limax cinereoniger* was again the most widely distributed species and numbers of slugs were generally low. Oakwoods, with a ground flora of *V. myrtillus*, the grass *Deschampsia flexuosa* (Linnaeus) and other heath plants, had a mean soil pH of 4.9 and a mean calcium content of 9.6 parts 10^{-3}. The fauna was similar to that of the mixed coniferous woods except that *L. tenellus* was absent, the distribution of *L. cinereoniger* was more restricted and *L. marginata* and *A. silvaticus* were more widely distributed than before.

These three groups of woodland were regarded by Wäreborn as oligotrophic habitats while the remaining two deciduous woodland types were regarded as meso-eutrophic habitats. These were a dry and a moist series of meadow wood habitats. They occurred on rocky hillsides and contained deciduous trees, including lime, ash, maple, elm, oak and birch. The ground flora was rich in herbs and grass while *V. myrtillus* and other heath plants were absent. The dry meadow wood habitats had a mean soil pH of 6.1 and a mean calcium content of 9.4 parts 10^{-3}. They had a more diverse slug fauna including *A. silvaticus*, *A. ater*, *A. subfuscus*, *A. circumscriptus s.s.*, *L. cinereoniger*, *L. marginata*, *A. fasciatus s.s.*, *L. tenellus* and *D. agreste*. *Arion silvaticus* was the most widely distributed slug, although *A. ater*, *A. subfuscus* and *A. circumscriptus s.s.* also occurred in a number of localities. The pH value of the moist meadow wood series of habitats was similar and the mean calcium content was 10.0 parts 10^{-3}. The slug fauna was similar to that of the dry meadow woodlands except that *D. agreste* was absent and the hygrophile species *D. laeve* was present. *Arion silvaticus* remained the most widely distributed slug and *A. ater*, *A. subfuscus*, *A. fasciatus s.s.* and *L. marginata* also widely distributed. The distribution of *L. cinereoniger* was more restricted in moist meadow woodlands. Wäreborn (1969) considered that *A. fasciatus s.s.* was a synanthropic species derived from nearby meadows. Wäreborn concluded that even in calcium carbonate free areas, the distribution of slugs is dependent on the calcium content of the organic matter derived from tree leaves, with some tree species providing more available calcium than others.

A study of the distribution of slugs and snails at 28 deciduous woodland sites in southern England and Wales was made by Mordan (1973). The woodlands included a wide range of soil types and tree species and soil pH values ranged from 3.6–8.2. He recorded a total of 12 species of slug (Table 10.6) although no more than seven species were found in any particular wood. Indices of similarity, based on total slug and snail faunas and calculated for each pair of sites, were classified by an agglomerative method of cluster analysis to produce a dendrogram showing a classification of the woodland sites. These sites could be divided into three main types: (a) a 'calcareous' group comprising mainly beechwoods on chalk with soil pH values ranging from 6.0 to 8.2, (b) a group on relatively base-rich soils with soil pH values ranging from 4.7 to 7.2, and (c) an acid or oligotrophic group with soil pH ranging from 3.6 to 5.0. The mean numbers of slug species associated with these groups were 4.7 (range 3–7), 4.3 (range 3–6) and 3.6 (range 0–5) respectively. Thus the numbers of slug species associated with each group did not differ markedly although there was a slight reduction in diversity with a lowering of pH values. The most widespread slug was *A. intermedius* (82% of sites), which was often present in relatively high numbers at both eutrophic and oligotrophic sites. Other common species included *A. ater* and *A. subfuscus* which were also found in acid woodlands. *Arion hortensis* was also widespread (61% of sites) although this species was not found in acid woodlands with a pH less than 4.2, while *A. fasciatus* showed a similar distribution. *Tandonia budapestensis* was found only at calcareous sites with a pH greater than 7.0, while *D. laeve* was only recorded from wet woodlands. *Deroceras caruanae* has a localized distribution in Britain and its restricted occurrence in woodland sites in Wales reflected its geographical distribution. Although *L. marginata*, *L. flavus*, *L. maximus* and *D. reticulatum* were only recorded in low numbers, they were distributed throughout the range of soil types.

The slug fauna of a mixed deciduous wood in Norfolk, located on neutral to slightly alkaline soils, consisted of eight species (Table 10.6) (Jennings and Barkham, 1975a). *Arion intermedius*, with a relative abundance of 55%, was the most abundant slug with *A. fasciatus* (18%) and *A. hortensis* (17%) the next most frequent species. *Arion intermedius* was also the most abundant slug in an ancient mixed oak–ash wood on a heavy boulder clay soil in Cambridgeshire (Paul, 1975). However, it avoided wet sites in the wood. *Arion circumscriptus s.s.* was also widespread and Paul regarded this species as typical of woodland in East Anglia. *Arion ater* was also widely distributed, but *A. subfuscus* was less frequent and this species was considered to be generally rare in East Anglia. The distribution of *A. hortensis* in the wood was patchy and Paul suggested that this was a synanthropic species which may have either dispersed into the wood from an adjoining railway or been introduced into the wood. Zeissler (1975, 1980) suggested that the presence of this species in two German forests was due to introductions, possibly from gardens. *Deroceras reticulatum* was common and, according to Paul (1978a), coppicing appeared to favour this species by

Table 10.6 Frequency of occurrence of species in European woodlands or woodland types.

Author	No. of woodlands or woodland types	A.i.	A.s.	A.f.	A.h.	A.a.	L.mx.	L.mar.	L.c.	L.t.	D.r.	D.l.
Kendall (1921)	3	3	2	1	2	2	1	2	–	–	2	2
Boycott (1934)	3	2	3	2	1	3	1	3	2	2	3	5
McMillan (1954)	2	1	1	2	2	2	1	–	–	–	2	2
Bruijns et al. (1959)	6	5	6	4	1	3	1	2	2	2	3	5
Wåreborn (1969)	5	–*	5	4	1	5	1	4	5	3	–	1
Mordan (1973)	28	23	10	9	17	13	6	2	–	–	6	3
Jennings and Barkham (1975a)	1	1	1	1	1	1	–	1	–	–	1	1
Paul (1975)	1	1	1	1	1	1	1	1	–	–	1	1
Bishop (1977a)	11	11	9	3	–	3	–	5	1	–	3	2
Bishop (1977b)	4	3	3	3	2	3	1	3	4	2	2	–
Paul (1978b)	14	11	1	11	13	11	8	1	–	–	10	7
Cameron (1978)	21	5	5	7	15	11	4	12	–	–	16	2
Waldén (1981)	15	*1	8	7	–	8	1	5	5	8	1	–
Phillipson (1983)	1	–	–	1	1	–	–	1	–	–	1	–
Tattersfield (1990)	38	30	14	34	29	32	23	35	20	2	25	13

Total occurrences (excluding Swedish sites)	133										
	96 (72%)	56 (42%)	79 (59%)	85 (64%)	85 (64%)	47 (35%)	68 (51%)	29 (22%)	8 (6%)	73 (55%)	38 (29%)
Total occurrences	153										
	97 (63%)	69 (45%)	90 (59%)	86 (56%)	98 (64%)	49 (32%)	77 (50%)	39 (26%)	19 (12%)	74 (48%)	39 (26%)

* Northern edge of range for *A. intermedius*.

A.i.: A. intermedius	*A.s.: A. subfuscus*	*A.h.: A. hortensis* agg.
A.a.: A. ater	*L.mx.: L. maximus*	*L.c.: L. cinereoniger*
L.t.: L. tenellus	*D.r.: D. reticulatum*	
Other species	*A.f.: A. fasciatus* agg.	
	L. mar.: L. marginata	
	D.l.: D. laeve	

McMillan (1954)   *T. budapestensis*, 1
Bruijns *et al.* (1959)   *T. rustica*, 1
Wäreborn (1969)   *D. agreste*, 1
Mordan (1973)   *T. budapestensis*, 3; *L. flavus*, 3; *D. caruanae*, 3
Bishop (1977a)   *T. budapestensis*, 1; *D. caruanae*, 1; *G. maculosus*, 1; *A. lusitanicus*, 1
Paul (1978b)   *T. budapestensis*, 2
Tattersfield (1990)   *T. budapestensis*, 7; *A. flagellus*: 1.

increasing the growth of herbaceous plants while Pallant (1969) found a correlation between the distribution of this species and the ground flora in woodlands. Paul (1975) considered that *A. subfuscus*, *A. intermedius* and *A. circumscriptus s.s.* were sensitive to dry conditions as they were absent from those areas of the wood which became dry in summer and where the ground flora was sparse. The hygrophilous slug *D. laeve* was common in the moister parts of the wood. Paul (1978b) investigated the suggestion that the mollusc fauna of ancient woodlands might resemble that described from fossil assemblages of the Atlantic Optimum period. However, slugs and snails rapidly colonized new additions to old woodlands and only those species classified by Paul as rare in East Anglia were absent from isolated modern woods in the area. Paul concluded from this study of slug and snail faunas of ancient woodlands in west Cambridgeshire, that four groups of molluscs could be distinguished. The first group, consisting of freshwater and hygrophilous species, included *D. laeve*. The second group comprised those species which were almost ubiquitous. Results showed that *A. hortensis* was present at 13 out of 14 sites and *A. ater* and *A. intermedius* were present at 11 out of 14 sites. *Arion hortensis* was described as a widespread species and not confined to woodland. *Arion circumscriptus s.s.* was included in the third group of molluscs, those common species which are characteristic of woodland in East Anglia. A final fourth group comprised somewhat rarer species more closely confined to woodland in East Anglia, although sometimes occurring in other habitats elsewhere in Britain. They included *L. maximus*, *L. marginata* and *A. subfuscus*.

Bishop (1977a) recorded 13 slug species (Table 10.6) from acidic woodlands in south-west Ireland, with a maximum number of seven species per site. The woodlands were located in an area of high rainfall and mild winters. Holly occurred at all sites while oak and ivy were widespread and *V. myrtillus* was often present. Bishop found a correlation between numbers of molluscs and litter volume, litter nitrogen and litter calcium. This result recalls the conclusions of Wäreborn (1969), who stressed the importance of litter calcium to acidic woodland molluscs, and of Phillipson (1983), who found that slug numbers were associated with litter depth. The acidic woodland sites had similar ranges of pH (3.8–6.1) and of litter calcium (3.0–11.1 parts 10^{-3}) to those described by Wäreborn (1969). The most widespread and abundant species was *A. intermedius* while *A. subfuscus* also occurred at most sites. Several species, including *A. silvaticus*, *D. caruanae* and *T. budapestensis*, were restricted to sites that had been disturbed and where exotic trees and shrubs had been planted within the native woodland. *Geomalacus maculosus*, found at one site only by Bishop, has a very restricted geographical distribution. In south-west Ireland it occurs in both woodland and open country sites on old red sandstone (Platts and Speight, 1988). Bishop (1980) found that the diversity of terrestrial molluscs at acidic woodland sites in the southern Alps in Italy was high and included 12 species of slug. However, the overall diversity of the mollusc fauna could not be explained by the distribution of vegetation and soil pH alone.

A survey of the habitats of mollusca in the central highlands of Scotland by Bishop (1977b) showed that pure stands of pine were virtually free of molluscs although *L. cinereoniger* could sometimes be found under logs there. However, Boycott (1934) recorded *L. cinereoniger*, *L. tenellus*, *L. marginata*, *A. ater*. *A. subfuscus* and *A. intermedius* from natural pine-woods in the Rothiemurchus Forest to the north of this area. Bishop (1977b) found that the slug fauna of the *Vaccinium*-rich birch wood association (pH 4.5–5.5) included *A. ater*, *A. intermedius*, *A. silvaticus*, *A. subfuscus*, *L. cinereoniger* and *L. marginata*. The herb-rich birch and oak wood association (pH 5.0–6.5) included these species together with *A. circumscriptus s.s.*, *A. fasciatus s.s.*, *A. hortensis*, *D. reticulatum* and *L. tenellus*. The slug fauna of mixed deciduous woodlands on fertile loams (pH 6.5–7.5) was even richer and included all the previous species with the addition of *L. maximus*.

From an investigation of the mollusc faunas of wooded talus and boulder slope habitats, Waldén (1981) described the following threshold pH values for slugs in south-west Sweden:

pH > 4.0   *A. subfuscus*
pH 4.5   *A. ater*, *A. silvaticus*, *A. intermedius*, *L. cinereoniger*,
            L. tenellus
pH < 5.0   *A. fasciatus s.s.*, *L. marginata*, *D. laeve*
pH 5.0   *A. circumscriptus s.s.*, *D. agreste*, *D. reticulatum*
pH > 5.0   *A. hortensis*

The distribution of these slugs in wooded talus and boulder slope habitats is summarized in Table 10.6. Up to seven species were found together at these sites, although those with a pH of 5 or less had only one species present. Waldén concluded that the rich microhabitat differentiation and good access to shelter of these habitats contributed to their high species diversity in both slugs and snails.

Cameron and Redfern (1972) and Cameron (1978) investigated the terrestrial slug and snail faunas of the Malham area in North Yorkshire and the distribution of slug species for the woodland habitats examined is summarized in Table 10.6. These woodlands ranged from areas of seminatural woodland on limestone rock and rich mixed deciduous woods planted on limestone soils to conifer plantations on acid, partly podzolized soils. Cameron (1978) described the woodland fauna as being characterized by high frequencies of *A. hortensis* and *L. marginata*, while these species and *L. maximus*, *A. fasciatus s.s.* and *A. silvaticus* were more or less absent from open sites. *Arion intermedius* was noticeably scarce in these woodlands, occurring at only five out of 21 sites. This is one of the most widespread species in British woodlands and the distribution in woodlands in the Malham area is unusual. However, the scarcity of this species in the woodlands of the Malham area has been confirmed by observations made during the spring over six successive years (South, unpublished).

The distribution of slugs and snails at 38 seminatural, deciduous woodlands in the south Pennines was examined by Tattersfield (1990). These woodlands were located either on limestone (mean soil pH range 7.0–8.0) or on millstone grit and shales (mean soil pH range 4.3–6.5). The total number of slug species was similar for both types of woodland. However, *A. hortensis, L. maximus, A. circumscriptus s.s.* and *A. fasciatus s.s.* were more frequent in limestone woods, while *A. intermedius, A. subfuscus* and *L. cinereoniger* were more frequent in acidic gritstone and shale woodlands. *Limax tenellus* was confined to these acidic woodlands. The hygrophilous slug *D. laeve* was also found only in the moister acidic woodlands. Several species, including *D. reticulatum, A. ater, L. marginata* and the third member of the *A. fasciatus agg., A. silvaticus,* showed no particular habitat preference. The synanthropic species, *T. budapestensis,* was recorded only from limestone woodlands and was most frequent in areas of woods associated with human disturbance. *Limax maximus* was also associated with areas of disturbance and its distribution was negatively associated with that of *L. cinereoniger. Limax cinereoniger* was present at most of the acidic woodland sites examined and Tattersfield (1990) concluded that its distribution was closely related to the moister environment presented by these woodlands.

In North America, the only slug recorded from woodlands in the Great Smoky Mountains (Tennessee and North Carolina) was a species of *Pallifera,* although this slug was widely distributed there (Getz, 1974). In north-eastern North America, both *Pallifera dorsalis* and *Philomycus carolinianus* were widespread species in coniferous and deciduous woodlands, although *P. dorsalis* was the more abundant species, also occurring in ecotonal habitats (Chichester and Getz, 1973). Another native species, *D. laeve,* was common in woods and Chichester and Getz considered this slug to be ecologically ubiquitous in the region. Several introduced European species of slug have also been found in woodlands in this area (Chichester and Getz, 1969). *Arion subfuscus* was primarily a woodland slug and was widespread in both coniferous and deciduous types. Other introduced species found in mainly mixed and deciduous woods included *A. fasciatus s.s., A. circumscriptus s.s., A. silvaticus, A. intermedius, A. hortensis* and *D. reticulatum.* These species also occurred over a wide range of other habitats and *A. fasciatus* was considered to be the most ubiquitous species of the genus.

In the same region, *A. subfuscus* was abundant in stands of both coniferous and deciduous trees although *A. fasciatus s.s.* and *D. laeve* were absent from stands of pine and spruce, except where a ground cover was present. These two species were common in stands of deciduous trees where smaller numbers of *A. circumscriptus s.s.*, *D. reticulatum* and *P. carolinianus* were also present (Beyer and Saari, 1977). Gleich and Gilbert (1976) found that in Maine, where *D. reticulatum* and *P. dorsalis* were the most common slugs, slugs were significantly less abundant in coniferous forests than in deciduous or mixed forests although this was not true for snail numbers. Slugs of the genus *Ariolimax* are indigenous to the coastal forests of western North America (Harper, 1988) and *A. columbianus* is the most abundant native mollusc in the Pacific north-west although the forest species of *Prophysaon* are also locally abundant (Rollo, 1983a). Most introduced species of slug have extended their range westwards in North America (Rollo and Wellington, 1975) and, although populations of *A. columbianus*, *L. maximus* and *A. ater* occur together in some habitats, *L. maximus* is usually found in fields rather than forests while *A. columbianus* mainly inhabits coniferous and deciduous woodland and *A. ater* occurs in both habitats (Rollo, 1983a).

The frequencies with which different slug species occur in published accounts of the slug faunas of various woodland habitats in Europe is summarized in Table 10.6. Because *A. intermedius* is at the northern edge of its range in southern Sweden and is absent from much of that area (Kerney and Cameron, 1979; Waldén, 1981), the first set of total frequencies omits the 20 Swedish sites. The most frequent slug species at the other woodland sites was *A. intermedius*, followed by *A. hortensis agg.*, *A. ater*, *A. fasciatus agg.* and *D. reticulatum*. These species, with the exception of *A. hortensis agg.* and *A. fasciatus agg.*, were described by Boycott (1934) as living 'anywhere' and being the least particular with regard to their requirements and they are not specifically woodland slugs. This supports Boycott's opinion that woodlands have no characteristic slug or snail fauna but provide the conditions most slugs and snails require. If the synanthropic Testacellidae and Milacidae are excluded, then most west European slugs have been recorded from woodlands at some time. While fewer slugs are found in acidic woodlands, the distribution of many species of slug appears to be less affected by low pH than that of snails. Wäreborn (1969) showed that sufficient calcium for slugs may be present in the litter of acidic woodlands and the list of threshold pH values for slugs given by Waldén (1981) shows that several slug species can live under acid conditions provided that adequate shelter and food is available. However, *A. hortensis* has the highest pH threshold (Waldén, 1981) and less acid conditions appear to be necessary for this slug (Boycott, 1934; Mordan, 1973). The hygrophilous *D. laeve* is a characteristic species of damp woodlands. The two species described by Boycott (1934) as confined to woodland, *L. cinereoniger* and *L. tenellus*, were not widely distributed in the woodlands examined, although they were more frequent at the Swedish sites, and they are associated with one another (Table 10.6). Bruijns *et al.* (1959) concluded that whenever *L. maximus* was introduced into woodland, then *L. cinereoniger* disappeared and a negative association between these species has been described by Tattersfield (1990). *Limax cinereoniger* and *L. tenellus* were described by Evans (1972) as anthropophobic species and they are largely confined to ancient woodlands. However, Tattersfield (1990) suggested that the distribution of *L. cinereoniger* might be more closely related, in some instances, to the available moisture in the woodlands. *Limax marginata* is a more common species and also occurs on rocky ground and stone walls. Although *A. subfuscus* was primarily a woodland slug in North America, in Europe it occurs in a much wider range of habitats although it is common in woodlands.

10.3.3 Other habitats

In contrast to woodland habitats, the slug faunas of agricultural land show a paucity of species (Tables 10.3 and 10.4). Arable farming tends to reduce species diversity although the pest species *D. reticulatum*, *A. hortensis* and *T. budapestensis* have successfully adapted to these conditions. *Deroceras reticulatum* is especially abundant on arable land and Evans (1972) cited archaeological evidence that agriculture has had a beneficial effect on this opportunist species. The reasons for this are not clear since, although this slug is widely distributed, numbers in natural habitats appear to be relatively low. However, this species is less exacting in water requirements than most other slugs (South, 1965; Lyth, 1972) and this, together with the absence of competition (see Section 10.2.3(d)), may be responsible for its success on arable land. Valovirta (1979) described *D. reticulatum* and *D. laeve* as pioneer species of the primary plant succession on an uplift archipelago in the Baltic, while *A. subfuscus* was characteristic of later successional stages (thickets and forest). A *Deroceras* sp. also successfully colonized bulldozed earth from construction excavations on a suburban site in Germany, numbers increasing from $5.8\,\mathrm{m}^{-2}$ after two years to $219.4\,\mathrm{m}^{-2}$ after four years (Strueve-Kusenberg, 1982). Matzke (1987) found that three species of slug (*A. fasciatus*, *A. hortensis* and *D. reticulatum*) were associated with land reclaimed from brown coal open-cast mining sites in east Germany.

Van den Bruel and Moens (1958a) compared the slug faunas of four different agricultural habitats in Belgium. The first type consisted of arable ground with a well-developed plant cover, including established clover fields and plots invaded by weeds. The soil in these fields contained a moderate amount of organic matter and was firm and difficult to penetrate.

The most abundant slug was *D. reticulatum*, making up between 50 and 80–99% of the population. Van den Bruel and Moens specifically noted the absence of *D. agreste* from these habitats. Other species present included *A. ater* and *A. subfuscus*, together with a few *A. circumscriptus* in the vegetation bordering the fields. Market gardens and vegetable gardens formed the second type of habitat and were located on easily worked, deep, slightly stony soils, rich in humus. Serious damage had been recorded here to potato tubers and other plant tissues below the soil. The most abundant slug was *A. hortensis*, comprising between 73 and 97% of the population. *Tandonia budapestensis* was the second most frequent slug (2–16% of total), while *M. gagates* and *T. sowerbyi* were also present. These subterranean species always formed at least 80% of the slug population, the remainder consisting of *A. ater*, *A. subfuscus* and *D. reticulatum*. Vegetable gardens where potato tuber damage was unknown, represented a third type of habitat. No *T. budapestensis* were found here although a few *M. gagates* and possibly *T. sowerbyi* were present. The slug fauna included an appreciable proportion of *A. hortensis* (25–50%) and *D. reticulatum* (27–65%), together with a few *A. subfuscus* (6–13%) and *A. ater* (3–5%). The fourth habitat type consisted of moist vegetation along the side of a stream in a valley. The luxuriant vegetation included grasses and *Urtica dioica* while pollarded willows and elder were also present. The only milacid slugs present were a few *M. gagates*. *Arion ater* and *D. reticulatum* were the most frequent slugs although *A. hortensis* was also common.

Hommay and Lavanceau (1986) recorded the relative abundance of the different species of slug found on arable ground in 12 French *départements*. They found that *D. reticulatum* was the most common slug (48.1–99.5%), with *A. hortensis* the second most frequent slug (0.3–46.1%). Other species were either absent or present in only small numbers except for a few isolated instances. Milacid slugs were conspicuously absent from samples except in Calvados in the north and Lot-et-Garonne in the south-west. Hommay and Lavanceau found that species diversity was higher (3–6 species) in western districts than in the east (generally one or two species) although other species were present in the area. For example, *A. subfuscus* was present over the entire area of the survey although it was recorded from arable ground only in the south-west.

The results of these surveys and from Table 10.3 lead to the conclusion that *D. reticulatum* and *A. hortensis* are the most common slugs on arable land in western Europe, and this also appears to be true where they have been introduced into North America (Chichester and Getz, 1969, 1973; Duval and Banville, 1989). Tables 10.3 and 10.4 also suggest that *D. reticulatum* is more widespread than *A. hortensis*, particularly in cereal fields and on agricultural grassland, and is the species most frequently cited as damaging cereal crops (Martin and Kelly, 1986). The status of *T. budapestensis* and other milacid slugs on arable land is less clear. They occur as pests in Britain (Table 10.3) and Belgium (van den Bruel and Moens, 1958a) and may be locally abundant (Barnes and Weil, 1944, 1945) but the survey by Hommay and Lavanceau (1986) indicated that their distribution was very localized. Van den Bruel and Moens (1958a) found that, as cultivated ground became more stable and the vegetation cover developed, other ubiquitous species such as *A. ater*,

A. subfuscus and possibly *A. circumscriptus agg.* became more numerous while *T. budapestensis* became less frequent. Van den Bruel and Moens concluded that serious damage to potato tubers occurred mainly when *T. budapestensis* was the commonest species or when it was associated with *A. hortensis* (i.e. in their second habitat type). When the main slug association was between *D. reticulatum* and *A. hortensis* (i.e. in their third habitat type) and few or no *T. budapestensis* were present, much less damage occurred. The reason for the absence of the *T. budapestensis* was uncertain but it seems likely that it requires a deeper, richer soil than the other species (Section 10.2.2). The distribution of pest species on arable ground is discussed further in Chapter 11.

Managed grasslands also present slugs with a difficult environment as grazing or repeated cutting reduces the available shelter and produces a drier microclimate. Slugs are less able to penetrate the firm soil of grassland. Species diversity is thus generally low. The slug fauna of intensively managed grassland which was subjected to heavy trampling comprised only *D. reticulatum* and *D. laeve* despite the fact that the ground water was regulated (Bruijns *et al.*, 1959). However, fields cut for hay supported a more diverse fauna, including *A. ater*, *A. subfuscus*, *A. intermedius* and *A. circumscriptus s.s.* On less intensively grazed grassland on a dry sandy loam, South (1964) found that only two species, *D. reticulatum* and *A. intermedius*, were common, although *A. fasciatus* was present in small numbers. The cover provided by the vegetation, particularly grass tussocks, met the needs of the first two species although the more exacting moisture requirements of *A. fasciatus* obliged it to seek shelter in the soil.

Bruijns *et al.* (1959) recorded only *D. reticulatum* from warmer dry calcareous grasslands in the Netherlands while they found *A. subfuscus* and *A. intermedius* on dry grassland on sandy soils although these did not occur in large numbers. Shelter seems to be crucial to these environments and Chichester and Getz (1969) recorded *D. reticulatum*, *D. laeve* and *A. fasciatus s.s.* as common in abandoned farmland in North America, with *D. reticulatum* in particular achieving very high population densities. Cameron (1978) studied the slug and snail fauna of a range of grasslands, mostly on rocky areas over limestone in the Malham area of the Pennines. These included areas grazed by sheep and limestone crags and pavements. In some instances, on grassland with no exposed rocks, the pH was as low as 5.5. Slug frequencies in these open habitats were generally low, although this was possibly an artefact of the sampling technique used. *Arion ater*, *D. reticulatum* and *A. intermedius* were the most widely distributed species, while *A. circumscriptus s.s.*, *A. subfuscus*, *A. hortensis* and *D. agreste* were also present. Only two species were found by Nelson (1971) on grassland further north in the Pennines. *Arion ater* was numerous on ungrazed grasslands while *D. reticulatum* was also widespread. Slugs were found more frequently on limestone grassland and associated with *Juncus squarrosus* and less frequently on alluvial *Agrostis–Festuca* grassland and *Calluna* moor while they were absent from valley bogs. However, Cameron (1978) recorded *A. ater* from a raised bog.

Lutman (1978) listed six species of slug from a montane grassland site in north Wales (Table 10.5) although only two species, *D. reticulatum* and *A. intermedius*, were common and the densities of these species were probably lower than at other grassland sites. *Deroceras reticulatum* was associated with herb-rich, somewhat disturbed grass while *A. intermedius* numbers were higher on fairly moist, undisturbed herb-rich swards. *Deroceras laeve* was restricted to *Nardus stricta* Linnaeus vegetation types which tended to retain more moisture. The montane zone was considered by Bishop (1977b) to lie above 750 m where habitats suitable for molluscs were few and an impoverished lowland slug fauna was present comprising *A. ater*, *A. intermedius*, *A. subfuscus* and *L. marginata*. A similar fauna, with the addition of *D. laeve*, was described for mountain habitats (250–600 m) on the Faroe Islands (Sølhoy, 1981). The vertical distribution of slugs on Scottish mountains was discussed by Dance (1972). The commonest species were similar to those of Bishop (1977b) except that *A. intermedius* was considered not to be as tolerant of adverse conditions as the other species. Dance listed the highest altitudes reached by these species as 961 m for *A. intermedius* (north Wales) and 1190 m for *A. subfuscus*, *A. ater* and *L. marginata* (Scotland). *Deroceras reticulatum* was not found above 610 m in Scotland but has been recorded at 778 m in Wales.

While lack of shelter appears to limit species diversity on managed and other dry grassland, more species are found in wetland habitats (Table 10.4). The most widespread slugs found by Bishop (1981) in fens and freshwater marshes in East Anglia were *A. intermedius* and *D. laeve*. In Scottish wetland habitats, more species were found to be associated with *Juncus effusus–Sphagnum* mires than with *Molinia–Myrica* mires, the most frequent species being *A. intermedius*, *D. laeve* and *A. ater*. *Deroceras laeve* appears to be the most characteristic and widespread of wetland slugs (e.g. Bruijns et al., 1959; Cameron, 1978) while other common slugs are generally widespread species such as *A. intermedius* and *A. ater*. *Deroceras reticulatum* is less widespread and abundant (Table 10.4) than on drier grasslands.

Deroceras agreste occurs in small numbers at several wetland sites in the British Isles (eg. Cameron, 1978; Bishop, 1977b, 1981) where it is mainly a northern species. Bishop (1977b) described this species as widespread though local in the Central Highlands of Scotland, where it was also found on richer grasslands. Further north, Sølhoy (1981) found that *D. agreste* was widespread in the Faroe Islands although nowhere abundant, occurring mainly in fields and human settlements. Sølhoy also recorded this species from an attitude of 800 m in western Norway. According to Kerney and Cameron (1979), *D. agreste* has a mainly northern and eastern distribution and this was confirmed by Likharev and Rammel'meier (1952), who described its distribution as the European USSR and northern and central Europe. In the USSR its ecological distribution is broader than in Britain and Likharev and Rammel'meier recorded it from open country and woodland habitats and described it as a serious pest species. *Deroceras agreste* appears to replace *D. reticulatum* in eastern Europe and Shikov (1979) described how, in the flood plain of the Volga, *D. reticulatum* and *D. sturanyi* were essentially synanthropic species while *D. agreste* and *D. laeve* were widespread species. Although *D. laeve* was common in damp places, *D. agreste* preferred less moist areas and tolerated drought. Shikov suggested that both *D. reticulatum* and *D. sturanyi* were warmth loving species which could not survive in moist cold northern latitudes unless associated with human activities. *Deroceras reticulatum* was able to colonize fields in the western provinces where it was milder and moister. Von Proschwitz (1988b) concluded that *D. reticulatum* was a long-established synanthropic species in south-west Sweden, although it was widely distributed.

Gardens present a more favourable environment to slugs than arable ground because they represent a mosaic of habitats and offer more shelter. Barnes and Weil (1944, 1945) carried out an extensive survey of about 50 gardens in Hertfordshire over three years. The most abundant slugs were those species characteristic of cultivated ground, i.e. *A. hortensis*, *D. reticulatum* and *T. budapestensis*. In some instances *D. reticulatum* was less abundant than the other species and Barnes and Weil suggested that this species was more a slug of fields than gardens. However, *D. reticulatum* and *A. hortensis* were found in every garden and on virtually every sampling date. Other slugs present in smaller numbers were *A. ater*, *A. subfuscus*, *T. sowerbyi*, *A. circumscriptus* and *L. maximus* while ten individuals of *L. flavus* were also found. Neighbouring gardens were also found to have very different slug faunas in some instances, reflecting the poor dispersive capacity of slugs (see Section 10.2.2). Bruijns *et al.* (1959) recorded seven species from gardens and arable ground in the Netherlands, including *D. reticulatum*, *A. hortensis*, *A. ater*, *A. circumscriptus s.s.*, *L. maximus*. *Milax gagates* and *T. sowerbyi* occurred locally while *T. budapestensis* does not occur in that region (Kerney and Cameron, 1979). *Limax flavus* is also associated with human settlements. Thus the fauna in these gardens was similar to that described by Barnes and Weil (1944, 1945) and Bruijns *et al.* (1959) concluded that garden slugs were normally invaders from other habitats.

The spatial distribution of slugs in a garden in Kent was investigated by Lyth (1972). *Deroceras reticulatum* and *T. budapestensis* were the most abundant slugs, each representing about one-third of the total number of slugs collected. *Arion hortensis* was also common (21%) and there were smaller numbers of *A. fasciatus* (10%), *A. subfuscus* and *T. sowerbyi*. *Arion ater* was also present in the garden. This fauna was also similar to that described by Barnes and Weil (1944, 1945) except that no *Limax* were present. Lyth found significantly fewer slugs on a lawn than on cultivated ground although all the species present except *T. sowerbyi* were found on the lawn. This suggests that although there was little shelter on the lawn, slugs dispersed there from adjacent shelter during the time they were active. *Arion intermedius* is notable for its absence from gardens, particularly as it is one of the most widespread species in other habitats (Boycott, 1934). Taylor (1902–1907) described how this slug was not a species of gardens or arable land, while McCracken and Selander (1980) found that *A. intermedius* occurred most frequently in relatively undisturbed habitats. This observation is supported by the fauna lists in Tables 10.3–10.5. However, the reason for this is uncertain.

10.4 PRODUCTION AND ENERGY FLOW

Some earlier authors, for example Frömming (1954), suggested that gastropods might play an important role in the functioning of terrestrial ecosystems, particularly woodlands. Mason (1970) was probably the first to investigate the role of snails in litter decomposition in woodland. He calculated that the total snail population ingested 0.35–0.43% of the annual litter input in a beechwood and that 49% of this energy was assimilated. Slugs were not included in this investigation. Pallant (1974) calculated regressions of food eaten on faeces produced, using data from laboratory experiments (Chapter 3). Measurements were then made of the quantities of faeces produced by *D. reticulatum* under field conditions in order to estimate the rates of assimilation for *D. reticulatum* on grassland and in woodland using the regression equations. Jennings (1975) also found a significant positive correlation between the dry weight of food consumed and the dry weight of faeces produced for seven slug species, including *D. reticulatum*. Pallant (1974) estimated the rates of assimilation to be 161.3 J $(100 \, mg)^{-1}$ dry weight in 24 hours on grassland and 141.5 J $(100 \, mg)^{-1}$ dry weight in woodland. The corresponding assimilation value for slugs fed in the laboratory of 134.9 J was slightly lower and Pallant suggested that this might reflect additional activities such as reproduction and seeking shelter. Wallwork (1975) recorded mean calorific values of 20.3 kJ and 22.0 kJ g^{-1} dry weight for *A. hortensis* and *L. maximus* respectively from beech forest floor. These specific differences were common to other invertebrates and Wallwork concluded that the calorific value for a species was directly related to its mobility.

In a review of the literature on soil fauna populations and their role in decomposition processes, Petersen and Luxton (1982) commented on the scarcity of data for gastropods and listed some biomass estimates (dry weight) for these animals. These included estimates of 883 and 822 mg m^{-2} in

Danish deciduous forests where the majority of the gastropod biomass consisted of *A. subfuscus*. Only the biomasses of diplopods and earthworms exceeded these values. Other biomass estimates included 270 mg m^{-2} for *A. ater* and 220 mg m^{-2} for total slugs, both for woodlands. By the extrapolation of laboratory measurements of faecal production to field data for faecal production, Jennings (1975) estimated that mature *A. ater* living in mixed deciduous woodland consumed a mean weight of 9.34 mg in 24 hours and for immature slugs the corresponding consumption rate was estimated at 20.15 mg. These values are measured as milligrams of dry weight per gram of live weight of slug. Corresponding values for consumption by other species from the wood were: *D. reticulatum* 29.06 mg, *A. fasciatus* 26.96 mg, *A. hortensis* 20.94 mg, *A. intermedius* 29.66 mg, *A. subfuscus* 28.09 mg and *L. marginatus* 20.90 mg. Data for *A. ater* agreed with the general principle that specific metabolic rate decreases with age and with increasing weight and also that consumption per unit weight is usually greater for smaller animals. Differences between laboratory and field values for consumption rates, with the laboratory rate being generally lower, were taken as an indication of the higher activity rates of animals under natural conditions. In the same mixed woodland, Jennings and Barkham (1976) estimated the mean biomass of slugs to be 788 mg dry wt m^{-2}. *Arion ater*, which made up about 29% of the total slug biomass with a mean live weight of 1.52 g m^{-2}, ingested about 8.17 g dry wt m^{-2} yr^{-1}, representing 1.54% of the total leaf litter input. They considered that the effect on plant decomposition processes of the other slug species (71% of the total live slug biomass) might be proportionately much greater because of their smaller size. This is a higher proportion than that recorded previously by Mason for snails in beechwood.

Lutman (1978) investigated the role of slugs in montane *Agrostis–Festuca* grassland. *Deroceras reticulatum* and *A. intermedius* were the most abundant species (12 m^{-2} and 14 m^{-2} respectively) although *D. reticulatum* accounted for 56% of the mean annual biomass while the corresponding figure for *A. intermedius* was only 13%. The least abundant slug in the area, *A. ater*, accounted for 19% of the mean annual biomass because of its much larger size. There were considerable seasonal variations in these proportions. The other three species present (Table 10.4) made up the remaining 12% of biomass. The annual production for a generation of *A. intermedius* was estimated at 0.41 g m^{-2} and egg production by this generation was 0.04 g m^{-2}. The total production by *D. reticulatum* in a year was estimated at 0.85 g m^{-2} and egg production was equivalent to 0.06 g m^{-2}. While the production rates for three other less common species were relatively insignificant, an appreciable rate of annual production of 0.58 g m^{-2} was recorded for *A. ater* on account of its larger size. Lutman estimated that slugs consumed 16.3 g dry wt m^{-2} yr^{-1}, equivalent to 304.54 kJ m^{-2} yr^{-1}, more than half of this being consumed by *D. reticulatum* which ingested live plant material more frequently than other slug species. Net primary production above ground of the grassland was 1145 g dry wt m^{-2} yr^{-1} and slug consumption represented about 1.4% of primary production and about 6% of total herbivore consumption, the major part of which was by sheep.

The mean annual population density of slugs in a beech wood was estimated at 34 m^{-2} by Phillipson (1983) and over 90% of these were *A. hortensis*. The equivalent mean biomass was 266 mg dry wt m^{-2}, a figure comparable with those given by Petersen and Luxton (1982) for woodland but only a third of the biomass in the mixed woodland examined by Jennings and Barkham (1976). The estimated biomass was less than two-thirds of the biomass given by Lutman (1978) for slug populations on grassland. Annual food consumption was estimated at between 3 and 4 g dry wt m^{-2} yr^{-1}, which represented 1.28–1.7% of the rate at which ground litter disappeared in a year. Although the estimated proportion (1–2%) of plant material consumed by slugs in woodland and on grassland appears to be low, slugs probably have a more important role in these ecosystems because they feed on a wide range of both live and dead plant material and have a broad spectrum of enzymes in the gut including cellulases and chitinases. Thus they assist in the fragmentation and breakdown of plant material and their faeces provide a substrate for the activity of decomposers in the soil. Richter (1979) suggested that the major role of the slug *A. columbianus* lay in the cycling of inorganic nutrients in a forest ecosystem. The high specific productivity and lack of specific environmental requirements of *D. reticulatum*, *D. laeve* and *A. circumscriptus*, together with their omnivorous diet and potential food value for man, led to an evaluation of the suitability of these slugs as components of life-support systems of the regenerative type by Burenkov and Agre (1975).

10.5 CONSERVATION

There is no evidence that any indigenous land mollusc has become extinct in Britain since the climatic optimum about 7000 B.P., when the native fauna of land and freshwater Mollusca was probably complete (Kerney, 1968). The range of some species has, however, continued to decline over the past century, a process that began after the climatic optimum with the change in temperature. The majority of British species of slug are widespread and do not appear to be threatened while species such as *D. reticulatum*, *A. hortensis* and *T. budapestensis* appear to have benefited from agriculture and horticulture. Several other species have shown a considerable increase in range as the result of their association with man. These synanthropic species include milacid slugs, particularly *B. pallens* which has spread rapidly in north-western Europe in recent years, *Testacella* spp., *L. flavus* and, possibly, *D. caruanae* in localities away from south-west England (South, 1974; Kerney and Cameron, 1979).

It is clear, however, from Tables 10.3–10.5 that species diversity is reduced by human activities and the increased rate of habitat disturbance and destruction in recent years, particularly to woodlands and wetlands, may have affected some species. Human activities likely to affect slug and snail populations were reviewed by Kerney and Stubbs (1980). Species which appear to be intolerant of human disturbance, the anthropophobic species of Boycott (1934), may be eliminated by the disturbance of their ancient woodland habitat. Slugs such as *L. cinereoniger* and *L. tenellus* may be threatened in this way by clearance and replanting although there is little evidence to suggest that they have declined over the past century (South, 1974). Although local, these species are more widely distributed in Europe. The distribution of *A. lusitanicus* and *D. agreste* is unclear because of earlier problems with identification. As both species are probably near the edge of their geographical range in Britain, they might be vulnerable to disturbance, although there is evidence that *A. lusitanicus* is a synanthropic species away from the west of Britain.

In the British Isles, *Geomalacus maculosus* is restricted to Co. Kerry and adjacent parts of Co. Cork in south-west Ireland, although it also occurs in northern Spain and northern Portugal. The status of this slug as a threatened species has been discussed by Platts and Speight (1988). It is locally common in the south-west of Ireland with populations occurring at several protected sites although the ecological requirements of this slug are not properly understood. On the Iberian peninsula its status is less clear and forestry operations appear to have eradicated it from many sites. In 1987 *G. maculosus* was proposed for addition to the lists of European fauna to be given special protection under the provisions of the Berne Convention and it was added to Appendix II which lists 'strictly protected fauna' requiring strict habitat protection for this species. The EC Habitat Directive incorporating this listing has now been formally adopted.

10.6 POLLUTION

10.6.1 Atmospheric pollution

Atmospheric pollution appears to have caused the impoverishment of snail faunas in some urban and industrial environments (Kerney and Stubbs, 1980). Holyoak (1978) found that atmospheric pollution had caused a considerable reduction in the distribution of the clausilid snail *Balea perversa* (Linnaeus) and, to a lesser extent, *Clausilia bidentata* (Ström). It was uncertain whether this reduction was due to the increased acidity of the tree bark or a reduction in the abundance of epiphytic lichens, or some combination of these effects. However, Holyoak found no evidence that the slug *L. marginata* had decreased in polluted areas to the same extent as *B.*

perversa. There appears to be little other information on the acidifying effects of atmospheric sulphur dioxide and oxides of nitrogen on slugs although Greven (1985) showed that the number of gland cells in the inner mantle epithelium of the slug *L. flavus* increased significantly when the slug was continuously exposed to 2 ppm sulphur dioxide, and that this was probably a response to a stress situation. Opalinski (1981) measured the rates of respiration for individuals of *Limax* sp. taken from a site subjected to atmospheric pollution from a mine and from a relatively unpolluted area. Oxygen consumption was higher at the polluted site in September although there was no appreciable difference in July or August. There appears to be little other information on the effect of sulphur dioxide on slugs.

10.6.2 Pollution by metals

Soils normally contain at least trace quantities of the 'heavy metals', including copper, lead, zinc, nickel, cadmium and mercury. In some areas levels of these metals have been substantially increased from mining waste tips, from industrial fallout (e.g. from smelting works), by lead derived from car exhausts and by the use of some pesticides and herbicides. The concentrations of these metals may reach levels that would render them toxic to plants and animals. While molluscs have been used to monitor environmental heavy metal pollution in aquatic environments, little work has been carried out in terrestrial environments although Cavalloro and Ravera (1966) suggested that *A. ater* might be used as a biological indicator of ^{54}Mn contamination in terrestrial environments. Schoettli and Seiler (1970) suggested that zinc might become linked with calcium spherules in the digestive gland of *A. ater* after absorption. Coughtrey and Martin (1976) showed that the digestive gland of the snail *H. aspersa* stored lead, zinc and cadmium while copper was evenly distributed throughout the body tissues. They suggested that analysis of the digestive gland might be useful in studying levels of cadmium at different sites. However, Coughtrey and Martin (1977) showed that trace metal concentrations in *H. aspersa* were a function of the size of the snail and that the uptake of trace elements was affected by the concentrations of trace metals already present in the mollusc.

A study of the concentrations of nine metals in *A. ater* from locations close to and far away from highways in Canada, led Popham and D'Auria (1980) to suggest that *A. ater* might be useful as an indicator to assess environmental quality. This slug was particularly suitable as it was widely distributed in both urban and rural environments and tended to occupy a specific home range. The distributions of metals in *A. ater* collected from a relatively unpolluted site and from a site near a disused lead and zinc mine, where manganese levels were also high, were compared by Ireland (1979). All metals analysed, except magnesium, were higher in concentration at the polluted site. Tissue analyses showed high concentrations of magnesium, zinc, cadmium and also phosphate in the digestive gland and high concentrations of calcium in both digestive gland and mantle. Lead and copper were not concentrated to any extent in specific tissues. Manganese was located mainly in the mantle, skin and foot, which were all mucus-producing tissues, and Ireland suggested that mucus might be the site for excretion of manganese since high concentrations were found in mucus and manganese was not stored in association with digestive gland spherules. Further studies on the uptake and distribution of cadmium derived from food showed that uptake was greater in July compared with September (Ireland, 1981). Although most cadmium was located in the digestive gland it was, unlike zinc, not associated with any subcellular organelles. A protein isolated from cadmium-treated slugs showed characteristics similar to metallothioneins, substances usually synthesized in response to sublethal heavy metal doses. Similar results were obtained by Dallinger *et al.* (1989) for cadmium and zinc after feeding *A. lusitanicus* on metal-contaminated substrate from a lead mining area and they also concluded that the cadmium binding component was a metallothionein-like protein, probably involved in the detoxification of cadmium. Carter (1983) showed that cadmium was concentrated by slugs beyond the level recorded in their food and noted that caution was needed when using these animals to monitor heavy metal concentrations in ecosystems.

The effect of chronic and acute lead treatment on the slug *A. ater* was investigated by Ireland (1984). Accumulation of lead in the tissues was greater after acute exposure, with most of the metal being deposited in the intestine and least in the foot. This was because many food materials and metals are not readily transported across the gut wall although it was not clear whether the high deposition of lead in the gut was due to precipitation within the gut or to adsorption by the tissues of the gut. After chronic exposure to lead, the calcium content of the intestine decreased and both chronic and acute exposure to lead resulted in a decrease in the calcium content of the digestive gland and foot. The activity of delta-aminolaevulinic acid (ALAD), an enzyme involved in the biosynthetic pathway to porphyrin formation (see Chapter 3) was reduced after acute treatment with lead, due to increased lead concentration in the digestive gland. Ireland (1988) compared the uptake and distribution of silver in the slug *A. ater* with that in the snail *A. fulica*. *Achatina fulica* accumulated more silver than *A. ater* although this was not associated with the presence of a shell in that snail. The main site of silver deposition was the digestive gland where this metal was associated with calcium spherules, while the concentration of faecal silver was higher in *A. ater* than in *A. fulica*. Silver was also excreted in the mucus of *A. ater* and mucus from silver-treated animals also contained twice as much protein.

Molluscs can accumulate higher concentrations of metal ions than other groups of invertebrates although these levels have not generally increased mortality except where the exposure was very high (Marigomez *et al.*, 1986). However, little is known about the effect of abnormal environmental concentrations of heavy metals on the physiology and metabolism of terrestrial slugs and snails. Marigomez *et al.* (1986) studied the effect of different levels (0–1000 ppm) of copper, zinc, mercury and lead in the diet on feeding activity in *A. ater*. *Arion ater* was tolerant of high levels of zinc and copper although high levels (1000 ppm) of lead and mercury resulted in a slightly higher mortality in cultures. Marigomez *et al.* concluded that normal environmental levels of these metals would not cause a significant mortality in *A. ater* populations. Feeding activity was not affected by low doses of copper and zinc but higher doses (100–1000 ppm for copper and 300–1000 ppm for zinc) decreased food uptake until acclimation occurred. There was a close parallel between copper and cadmium levels in copper-exposed slugs. Feeding activity in mercury-treated slugs was closely related to dosage with food consumption decreasing as mercury dosage increased. Lead treatment did not affect feeding activity at the metal levels studied.

Recio *et al.* (1988a, b) found that *A. ater* tolerated high doses of zinc (0–1000 ppm) and that the metal accumulated in the digestive gland. It was located mainly in lipofuscin granules of the excretory cells and in the perinuclear cytoplasm and spherules of the calcium cells. Schoettli and Seiler (1970) had concluded that zinc was restricted to the calcium spherules of the digestive gland but Recio *et al.* (1988a) considered this to be true only at low zinc concentrations and after short-time exposures to higher concentrations. Excretion of the lipofuscin material was one of the main mechanisms for zinc release into the lumen of the acini of the digestive gland for subsequent elimination in the faeces. Pathological effects, such as a reduction in polysaccharide reserves, were only evident at exposure to the highest zinc level. At this level there was also a trend in reduction of mean epithelial thickness of digestive gland cells. The concentrations of five metals were measured in six species of slug from deciduous woodland near a disused lead and zinc mine site (Greville and Morgan, 1989a). They found marked interspecific differences in rates of metal accumulation, with species belonging to the same genus being widely separated although the rank scores of the six species varied with the different metals. *Arion subfuscus* was the most efficient metal accumulator and the order of metal accumulation capability was:

A. hortensis < *D. reticulatum* < *T. budapestensis* < *A. ater* < *D. caruanae* < *A. subfuscus*

Greville and Morgan suggested that the higher rate of accumulation of metals by *A. subfuscus* might be explained by the fungivorous diet of this species since fungi are known to accumulate metals from the soil. They discussed the implications of these interspecific differences in the biological monitoring of metal pollution and stressed the importance of using the same species for measurements from different sites. The seasonal levels of five metals in *D. reticulatum* collected from a disused lead and zinc mine site were recorded by Greville and Morgan (1989b). Although some significant monthly differences in metal levels were found, there were considerable variation in the concentrations of the five metals investigated. Greville and Morgan concluded that the presence of these irregular variations would restrict the potential value of slugs as biomonitors of heavy metal contamination. The heavy metal component may be expressed either as the metal content (body burden) of the slug or as a function of body weight (i.e. as metal concentration). Greville and Morgan (1990) showed that both interspecific and intermetal differences in metal accumulation could be described more effectively by comparing the metal content : dry body weight relationship rather than the metal concentration : dry body weight relationship. However, they concluded that the intrinsic variability in metal levels reduced the likelihood of slugs being used as biological indicators of metal contamination in terrestrial environments.

10.6.3 Pesticide pollution

The effects of molluscicides such as metaldehyde and methiocarb are discussed under slug control (Chapter 12). However, pollution by other pesticides may affect slug populations indirectly. Davis and Harrison (1966) reported residues of 0.3 ppm of the organochlorine insecticide, dieldrin, from a single sample of slugs taken from arable ground and concentrations of 5.3–23.8 ppm *pp'*-DDT, 0.4–2.6 ppm *op'*-DDT, 1.2–9.9 ppm *pp'*-TDE and 0.4–3.4 ppm *pp'*-DDE from three samples of slugs taken from apple orchards where the organochlorine compounds were largely concentrated at the top of the soil profile. The residues from slugs and earthworms were among the highest recorded for invertebrates at that time. All the animals were active and apparently healthy at the time of collection. Davis (1966) described how both slugs and earthworms were able to accumulate organochlorine residues at concentrations higher than those in their environment without being seriously affected and discussed the implications of this to predatory birds. Earthworms and slugs acquired much higher levels of DDT in sprayed fields than carabid beetles and these residues were more persistent. The breakdown product in slugs was TDE while that in earthworms was DDE (Davis and French, 1969). Gish (1970) reported organochlorine insecticide residues of 89 ppm in slugs from arable land in the USA. In another study in the USA, Kuhr *et al.* (1974) found residues of DDT, DDE and DDD in slugs from apple orchards which had not been sprayed with DDT for 12 years. Individual slugs of the species *L. valentiana* were exposed to turf treated four years previously with ^{36}Cl-DDT (Forsyth and Peterle, 1982). Slugs accumulated equilibrium levels of DDT residues in 55–80 days under glasshouse conditions. However, they lost DDT residues faster than the rate of uptake when transferred to clean turf. In outside enclosures, residues in slugs did not exceed 5.0 ppm, although here uptake was hindered by lack of rainfall and low temperatures. The minimal uptake of residues by *L. valentiana* from contaminated vegetation and detritus, in the absence of soil, showed that soil was a more important source of residues than the food eaten by the slug.

The fate of a 1 kg ha^{-1} application of ^{36}Cl-DDT on an old-field study area was recorded by Forsyth *et al.* (1983) over the next five years. During this period DDT residues (DDT plus metabolites) declined in the tissues of the slug, *D. laeve*, from 9.3 to 2.1 mg kg^{-1}, and in earthworms. There was a corresponding decline in residues in the tissues of two predators of slugs and earthworms. Ebling *et al.* (1984) concluded that slugs (*D. reticulatum*) were unsuitable for monitoring residues of selected organic compounds, including organochlorine compounds and carbamates, in the soil.

11

Slugs as pests

11.1 SLUGS AND MAN

Few species of slug appear to have been adversely affected by human activities, apart from the localized effects of habitat destruction, while several species have extended their geographical distribution. Several have become considerably more abundant as the result of agricultural and horticultural activities (see Sections 1.2 and 10.5). Arable farming has tended to reduce species diversity and few slug species are found on cultivated land, although several, including *D. reticulatum, A. hortensis* and *T. budapestensis* have successfully adapted to these conditions and have become established in many areas as pests of field and horticultural crops (Section 10.3.3). Slugs may also hinder the regeneration of deciduous woodlands and Elliott (1985) showed that *A. ater* and *A. hortensis* damaged first-season tree seedlings while Urquart (1952) also reported slugs eating tree seedlings. In addition to their importance as crop pests, slugs have been shown to act as vectors of helminth parasites of domestic and wild mammals and birds, including the nematodes, *A. cantonensis*, which can also cause eosinophilic meningo-encephalitis in man, and *A. costaricensis*, which causes human abdominal angiostrongyliasis. The economic loss due to abdominal angiostrongyliasis in Central America was estimated by Andrews (1989) to be US\$ 5 million per year. The role of slugs as vectors of disease has been discussed in Section 9.2. Slugs are considered to be unpleasant by many people and have been the subject of at least one horror story (Hutson, 1982). It is not surprising, therefore, that an example of long-standing phobia of slugs was described by Gustavson and Weight (1981).

A number of anecdotal reports exist of the medicinal use of slugs. Taylor (1902–1907) described how *D. reticulatum* or *A. ater* could be used as a cure for tuberculosis and other diseases of the lungs and chest. They were used in the form of a broth or eaten alive. The consumption of raw slugs in Japan as a remedy for lumbago was recorded by Kojima *et al.* (1979). A poultice of *D. reticulatum* or *A. ater* was also considered to be effective against throat and chest complaints. The shell granules of *A. ater* were also considered to have medicinal properties, including a remedy for dysentery. *Arion ater* could be rubbed on warts to remove them and also used for the purpose of making an ointment. Snails, particularly *Helix* spp., have been used as sources of specialized chemicals, including lectins, and one lectin, *Limax flavus* agglutinin, has been extracted from slugs (Section 9.2.4). This is one of the few known lectins which bind to sialic acid and it is now available commercially. It is possible that further lectins may be isolated from slugs. Mead (1961, 1979) discussed the use of snails, particularly A. *fulica*, as a food source and in recent years there has been a considerable increase in the number of snails, particularly *Helix* spp. and species of achatinids, raised for food by snail farming (e.g. Daguzan, 1989; Runham, 1989). Slugs are not generally used as a source of human food although, like snails, they represent a means of turning waste plant material into potentially usable biomass. However, it is clear that raw slugs are sometimes consumed as medicine or food – for example, in Hawaii (Kuberski and Wallace, 1979; Sharma *et al.*, 1981). Harper (1988) described how the Yurok Indians of California used the banana slug, A. *columbianus*, for food when other food sources became scarce. She also recorded that A. *columbianus* was eaten by German immigrant families in the 1800s and early 1900s and, more recently, how it was eaten as a novelty in California. An unusual use for slugs was recorded from the Berlin region by Stein (in Janssen, 1985), who described how A. *ater* had been formerly used to grease cart wheels because of the high natural fat content of the slug.

Slugs have been shown to be capable of acting as vectors of plant diseases. Fourman (1938) found that viable fungal spores were present in the gut and faeces of slugs while Gregg (1957) showed that passage through the gut of the snail, H. *aspersa*, stimulated germination of the spores of *Phytophthora* spp. The transmission of the downy mildew (*Phytophthora phaseoli* Thaxt.) by slugs on lima beans was reported by Webster *et al.* (1964) and transmission of the brassica dark leaf spot (*Alternaria brassicicola* Schw.) in cabbages by A. *ater* was described by Hasan and Vago (1966). Sano (1977) recorded a slug feeding on the white chrysanthemum rust (*Puccinia horiana*). Damage to grapevines by D. *reticulatum* and A. *hortensis* in vineyards where *Botrytis* was present, increased the extent of the damage by the fungus (Hering, 1969). *Lehmannia marginata* was observed feeding on chestnut blight cankers (*Endothia parasitica* (Murr.)) on chestnut trees by Turchetti and Chelazzi (1984) and they concluded from experiments that L. *marginata*, and possibly two other *Limax* spp., could be important vectors of this blight. Dawkins *et al.* (1985) demonstrated that A. *hortensis*, D. *reticulatum* and T. *budapestensis* were capable of transmitting the carrot liquorice rot (*Mycocentrospora acerina* (Hartig)) in the laboratory but concluded that slugs were unlikely to be important vectors in the field.

Arion hortensis, D. *reticulatum*, T. *budapestensis* and L. *maculatus* were capable of transmitting the bacterial soft rot (*Erwinia carotovora* (Jones)), a cause of spoilage of potatoes during storage, in the laboratory and slugs remained infected for up to 42 days (Dawkins *et al.*, 1986). However, Dawkins *et al.* considered that other means of dispersal of the pathogen were more important than these slugs. Although of no economic importance, the dispersal of lichen propagules by slugs was described by McCarthy and Healy (1978). Transmission of the tobacco mosaic virus by slugs was considered to be unlikely by Purdy (1928), although Heinze (1958) showed that the virus was still present in the gut of *D. reticulatum* and *L. marginata* two days after feeding and could be transferred to plants in this state. Cook *et al.* (1989) showed that when individuals of *D. reticulatum* were fed on white clover plants infected with white clover mosaic potexvirus (WCMV), they acquired the virus and subsequently transmitted it to healthy white clover plants. Slug faeces were also found to contain infective WCMV. The transmission rate was probably low and transmission probably occurred by incidental contamination during feeding and via contaminated faeces. Slugs were also found to be capable of distributing the adults, juveniles and eggs of the stem nematode, *Ditylenchus dipsaci*, after feeding on *D. dipsaci*-infested white clover (Cook *et al.*, 1989). Although of no economic importance, the dispersal of lichen propagules by slugs was described by McCarthy and Healy (1978).

11.2 SLUG DAMAGE IN GREAT BRITAIN

11.2.1 Introduction

The slug problem is a long-established one and, in an advisory leaflet, Anon. (1905) stated that over the past four years there had been a considerable increase in the number of slugs all over Britain, to the extent that whole fields of cabbage, wheat and other plants had been destroyed. The extent of slug damage to many agricultural and horticultural crops in Great Britain has become increasingly apparent over the past 30 or 40 years. Press reports of severe slug damage were common after the warm wet summers and autumns experienced during several years in the 1980s (e.g. *Farmers Weekly* 13/11/1981; 27/11/1981 and 28/10/1988), although slug populations were reduced as the result of dry conditions in 1989 and 1990 (*Farmers Weekly* 16/11/1990). However, accurate information on the economic importance of slugs is difficult to obtain and tends to be more unreliable than that for many important insect pests. Reports of slug damage tend to be confused at times with damage due to other pests including wireworms, cutworms, leatherjackets and even rabbits and birds. Examples of slug damage are shown in Figs 11.1–11.3.

Hunter (1969a) showed that the proportion of crops damaged was not evenly distributed throughout England and Wales but was generally greater in the eastern counties. This probably reflected the concentration of wheat and potatoes, the two most heavily damaged crops, in these areas. Hunter described how the distribution of severe damage was localized and closely described how the distribution of heavy soils (clay and loam). This agrees

Fig. 11.1 Damage to potatoes by *A. hortensis*.

Fig. 11.2 Wheat grains hollowed by *Deroceras reticulatum*.

Fig. 11.3 *Arion hortensis* damaging strawberry fruit.

with the conclusions of Gould (1961). The effect of soil type on the distribution of slugs is discussed further in Section 10.2.2.

11.2.2 Pest species of slug

Theobald (1895) listed the three most destructive species of slug as *D. reticulatum* (as *Limax agrestis*), *A. ater* and *L. maximus*, without any justification for the inclusion of the latter two. However, Taylor (1902–1907) did not cite *A. ater* or *L. maximus* as pests although he stated that *L. maximus* would eat young garden plants but preferred fungi. Taylor considered *D. reticulatum* and *A. hortensis* to be very destructive slugs and described how the milacid slugs *M. gagates* and *T. sowerbyi* readily destroyed garden plants. *Tandonia budapestensis* was either overlooked by Taylor or is a more recent introduction (South, 1974). Miles *et al.* (1931) listed the slug species of economic importance as *D. reticulatum*, *T. sowerbyi*, *A. hortensis*, *A. ater* and *A. subfuscus*. In Britain at the present time, *D. reticulatum*, *A. hortensis* agg. and *T. budapestensis* are the most important slug pests although the *A. circumscriptus* agg. and *T. sowerbyi* are also pests. This is reflected in the distribution of these five species on arable land (Table 10.3). *Arion ater* and *A. subfuscus* have been cited as pest species on several occasions but appear to be relatively unimportant. Relatively large numbers of *A. intermedius* have also been reported from fields of winter wheat by Glen *et al.* (1984) although the pest status of this species remains uncertain. *Deroceras caruanae* has been cited as a pest (e.g. Runham and Hunter, 1970) and this species may be responsible for some of the damage attributed to *D. reticulatum*.

Although the *A. circumscriptus agg.* have only recently been recorded as separate species, *A. fasciatus s.s.* seems less likely to be a pest than *A. circumscriptus s.s.* and *A. silvaticus* because it has a more restricted distribution (Kerney, 1976; Kerney and Cameron, 1979) in Britain. *Arion silvaticus* has been recorded from winter wheat fields (Glen *et al.*, 1984).

Little information is available about the pest status of individual species in the *A. hortensis agg.* but since *A. owenii* appears to be less widely distributed than *A. hortensis s.s.* and the widely distributed *A. distinctus* (see Section 1.4.4), *A. owenii* seems less likely to be a pest than the other two species. *Arion distinctus* has been reported as a pest species by Glen *et al.* (1989b). It has been suggested that different species of slug attack different crops. For example, both Thomas (1947) and Rayner (1975) concluded that *T. budapestensis* and *A. hortensis* were mainly responsible for damage to potatoes and that *D. reticulatum* was a secondary feeder. However, this now seems unlikely and the major pest species *D. reticulatum*, *A. hortensis agg.* and *T. budapestensis* are all capable of attacking undamaged potato tubers in the field and in the laboratory (Pinder, 1969; Storey, 1985; Airey, 1986a). A wide range of other slug species will also feed on potato in the laboratory (Section 7.1.5). Several species have been recorded feeding on wheat grains but *D. reticulatum* is the species most frequently reported damaging wheat crops (Martin and Kelly, 1986). The reason for this is probably that *D. reticulatum* is more widely distributed (e.g. Table 10.3) and tends to feed at the surface of the soil, nearer to the germinating grain.

11.2.3 Horticultural crops

There are few reliable estimates of the extent of slug damage to horticultural crops but there seems little doubt that slugs are responsible for considerable losses in this area. The environment offered by horticultural activities is particularly suitable for slugs, smaller market gardens and allotments especially offering a diversity of habitats within a relatively small area. The organic matter content of the soil is often high and additional shelter is provided by straw and rotting organic debris lying on the soil surface and by cloches, frames and plant containers. In a survey of pests in allotment sites around Leeds (West Yorkshire), Atkinson *et al.* (1979) found that slugs were twice as frequent as any other pest and that they were the main pests reported for brassicas and potatoes. Estimates showed that large numbers of slugs were present (Table 10.4). The most common species were *T. budapestensis*, *D. reticulatum* and *A. hortensis*.

The following summary is based on information from unpublished damage reports and from Anon. (1979). Two crops extensively damaged are lettuce and Brussels sprouts, particularly the sprouts or 'buttons'. Crops frequently damaged include other brassicas (cauliflower, broccoli, spring cabbage, etc.), carrots, celery and runner, broad and French beans. Damage is occasionally reported to many other crops, e.g. cucumbers and chicory. In many instances maximum damage occurs at the seedling stage. Increasing damage has been reported to ripening strawberries and this may be partly due to the greater use of polythene cloches with this crop. Onions and parsnips are rarely damaged. Many flowers and ornamentals are damaged, with underground parts (bulbs, corms and tubers) being holed and emerging foliage shredded or young shoots emerging from stools of herbaceous perennials may be grazed. Mature foliage and flowers are also sometimes spoilt by slugs. Eaton and Tompsett (1976) reported damage by *A. fasciatus* to lily bulbs. Slug damage has been reported to raspberry canes (Pitcher, 1950). Severe damage to a cucumber crop in a polythene-clad house in Ayrshire was reported by Foster (1977). The most abundant slugs were *D. caruanae* (35%), *A. subfuscus* (32%) and *D. reticulatum* (27%). Stem and bud damage were associated with attack by *A. subfuscus* and the less abundant *M. gagates* while leaf damage was caused by *Deroceras* spp. Foster (1977) found that *D. caruanae* was common in polythene-clad houses in Lanarkshire and on field crops in Ayrshire and suggested that this species had a pest status in south-west Scotland similar to that of *D. reticulatum* in Britain. A survey of organic vegetable growers by Peacock and Norton (1990) showed that slugs and birds followed weeds as the most important pests of organic crops.

11.2.4 Agricultural crops – general

Slug damage has been reported for a wide range of agricultural crops but the most serious losses occur to winter cereals and potatoes. Only occasional damage has been recorded to root crops, including turnip, swede and fodder beet. Sugar beet seedlings have been damaged but it is estimated that, up to the stage of singling, 30% of plants can be removed without significant effect on the crop (Johnson, 1968). Strickland (1965) also considered that the mean annual slug damage to sugar beet was equivalent to less than 0.1% of the total acreage planted while Dunning and Davies (1975) calculated that the value of the crop loss for 1975 due to slug damage was £10 000. However, Runham and Hunter (1970) suggested that slug damage would become more important as traditional methods of sowing sugar beet were abandoned in favour of lower seed rates using monogerm seeds. Field beans (*Vicia faba*), especially winter varieties, are often damaged by slugs in the early stages of growth. There are relatively few records of significant damage to pea crops but Wharton and Ensor (1969) reported that slugs could form a troublesome source of contamination when peas were harvested for freezing. Significant damage, mainly to young seedlings, has been recorded for several forage crops including clover, lucerne, kale, cabbage and maize. Slugs may damage newly established grass leys by grazing soon after shoots emerge from the soil. Ong *et al.* (1978) showed that although physiological causes accounted for most tiller deaths in a grass sward, grazing by slugs and rodents was more important than the damage caused by stem-boring insect larvae. There is evidence that cultivated grasses are preferred by slugs and, in one instance, S53 meadow fescue undersown in barley was seriously damaged while weed grasses *Poa annua* and *Agrostis* were left alone. Cottam (1986) showed that grazing by *D. reticulatum* on mixed swards of *T. repens* and *D. glomerata* could reduce the yield per plant of *T. repens* by as much as 60%. Clements *et al.* (1982) found that slugs were a problem when grass was established by minimal cultivation techniques and slug damage was described by Lewis (1980) as being more serious after direct drilling than in ploughed soils. Oilseed rape crops are also at risk from slug attack at the seedling stage (Port and Port, 1986). Glen *et al.* (1989a, 1990) showed that slug damage to oilseed rape seedlings was inversely related to their total glucosinolate content and concluded that glucosinolates in oilseed rape seedlings protect them from slug attack. They suggested that slug damage to the oilseed rape crop was likely to increase in severity as the trend towards the use of double-low (low glucosinolate, low erucic acid) cultivars continued.

11.2.5 Agricultural crops – potatoes

Economic damage to the potato crop is caused by slugs excavating holes in tubers of maincrop cultivars from late summer until the crop is harvested. Later harvested crops are particularly at risk, especially during wet autumns. Early and early maincrop potato cultivars tend to be less susceptible to slug damage. Slugs do not generally cause economic damage to seed potatoes after planting, although Hunter *et al.* (1968) showed that slugs may damage the leaves and stems of the growing potato crop. Slug damage can usually be distinguished from other damage, e.g. cutworms and wireworms, as it generally consists of a small hole at the surface leading to a larger chamber hollowed out underneath. Stephenson (1965) found that tubers of the less susceptible cv. Majestic had only one hole while those of the more susceptible cv. Redskin tended to have two or more holes. This damage usually occurs in the field although tubers may also be damaged during storage. Damaged tubers are more accessible to other pests and diseases, especially bacterial rots, and Webley (in Runham and Hunter, 1970) suggested that slugs might be involved in the transmission of potato blight (*Phytophthora infestans* (Mont.)). Although slug damage does not reduce the actual yield of the potato crop by any significant amount, damaged tubers are unsuitable for human consumption and are usually sold as stock feed at about 25% of the ware price (Baker and Waines, 1957). Only a small amount of slug damage (5% or less) can be tolerated in a ware crop and although a small amount of damage can be dressed out to ware standard, this sorting adds to the cost of the crop. When potato sets were planted more deeply, Jarvis (1973) found that the yield of slug damaged tubers was larger. Airey (1984) found that, even in a uniformly irrigated potato crop, slug damage was not uniform across the site owing to the aggregated distribution of the slugs. Wareing and Bailey (1989) obtained a similar result with distinct 'hot-spots' of high damage. The location of slugs corresponded with areas of most crop damage. The reasons for this type of aggregated distribution are discussed in Section 10.2.2.

The extent of slug damage to potatoes is more easily assessed than for most other crops. Baker and Waines (1957) estimated that in England and Wales over the years 1954–56, the quantity of ware potatoes unsaleable because of slug damage varied from 11 800 tonnes (0.29% of production) in 1954 to 5100 tonnes (0.12%) in 1956. In Scotland the loss was negligible in 1955 but over 2600 tonnes (0.26%) in 1954. More slug damage was recorded in established arable fields than in freshly ploughed grassland. Bardner and Waines (1964) found that the proportion of potato crops damaged by wireworms and slugs in Scotland between the years 1954 and 1960 was approximately the same, with a mean of between 3 to 9% being damaged by either pest. Tuber damage rarely fell into the severe category and Bardner and Waines concluded that wireworm and slug damage in Scotland were unimportant. Figures given by Strickland (1965) based on Potato Marketing Board Reports for the previous five years, showed that slugs were probably the third most important pest of potatoes in England and Wales. He found that on the average 23% of the maincrop acreage received some slug damage each year although this represented only a relatively small net loss of about 36 000 tonnes per year (less than 0.75%). This was only about a tenth of the loss due to aphids and a third of the loss caused by potato root eelworm but twice that caused by wireworm. Hunter (1969a) estimated that of about 168 000 hectares of second early and maincrop potatoes grown during the 1966/67 season, about 38 900 tonnes (0.81%) were damaged although only 800 hectares were chemically treated. The cost to farmers of slug damage to the potato crop was estimated at £645 700. While this net loss was relatively low, the cost of dressing-out damaged crops was estimated to be about £1.6 million per year although this was only partly due to slug damage. Church *et al.* (1970) estimated from samples of stored potatoes that 2.1% (1965) and 2.6% (1966) had damage making them unacceptable for ware. Stephenson and Bardner (1976) concluded from published data that estimates of net damage varied between 0.12 and 2.6% of the total crop, an amount equivalent at 1974 values to between £0.14 and £3.17 million. They suggested that losses would be slightly less than net damage because slug damage could be confused with cutworm damage and also because some slug damaged tubers would probably have injuries from other causes.

Port and Port (1986) estimated that these losses would be equivalent to between £0.3 and £7 million at 1984 and 1985 values. They described how the potato area had declined by 11% since 1974–76 although production had increased by 27%. Although the potential value of this loss was considered to be larger than ten years previously, this could be modified by factors such as the wide variation in slug damage between seasons and by fluctuations in the price of potatoes. The potential slug damage had increased because larger areas of susceptible cultivars were being grown but the actual loss might have been reduced by the increased use of molluscicides. In 1984, Maris Piper was the most widely grown cultivar and it is also one of the most susceptible to slug attack (Airey, 1986a). Damage by slugs is causing increased problems in potatoes grown primarily for the washed, prepack market resulting in additional costs to the grower (Beer, 1989). Beer described how relatively high proportions of batches of potatoes grown by a cooperative of 15 growers in eastern England for this market were affected by slug damage (1987, 34.3%; 1988, 67.1%) and this damage was very difficult to remove during grading. The main cultivars grown tended to include the more susceptible Maris Piper and Estima and growers were turning to cultivars which were less susceptible to slug damage although they were less preferred by the supermarkets.

The differential susceptibility of potato cultivars to slug attack is an important aspect of slug damage to the potato crop. Warley (1970) showed that there could be a considerable difference in financial loss for ware potatoes between, for example, Majestic (up to £4.50 per hectare) and King Edward (£10–£62.50) for an average crop at 1970 prices, while the loss to Pentland Dell was negligible. Table 11.1 summarizes information obtained from field trials on the differential susceptibility of potato cultivars. This confirms that substantial differences in slug damage are found between different cultivars when grown under the same conditions. The use of a susceptibility index for comparing cultivars and the possible reasons for this variation in susceptibility between cultivars to slug attack is discussed further in Section 7.1.5. It has been suggested that first early cultivars are more resistant to slug attack because they are lifted before damage begins and before tubers mature. While this may be true for most cultivars, the first early cv. Epicure was the most susceptible grown in a field trial of 15 potato cultivars (South, 1973a). Damage to maincrop potatoes is generally light during the early summer but increases rapidly from August onwards and in eastern England may increase tenfold between the first of September and the first of October (Runham and Hunter, 1970). This increase is the result of interaction between several factors.

Firstly, slug activity is often restricted by a lack of moisture during the earlier part of the summer. Stephenson (1967) suggested that activity was restricted to furrows until the haulm died off in the autumn, exposing the ridges to rainfall. However, Warley (1970) showed that from early August,

Table 11.1 Differential susceptibility of potato cultivars under field conditions.

Source	Cultivar	% slug damage		
Thomas (1947)	Arran Banner	29		Lifting
	Majestic	7		date
	Arran Peak	5		7 September
	Arran Banner	50		Lifting
	Majestic	26		date
	Arran Peak	18		2 November
Stephenson (1965)	Majestic	5		
	Redskin	11		
Gould (1965)		*1961*	*1963*	
	Pentland Falcon	–	4	
	Pentland Dell	–	4	
	Desireé	–	5	
	Majestic	12	7	
	Pentland Crown	20	15	
	King Edward	22	13	
	Ulster Glen	25	–	
	Ulster Ranger	41	–	
Winfield *et al.* (1967)	Pentland Falcon	12		
	Pentland Dell	18		
	Majestic	19		
	Pentland Crown	20		
	Record	25		
	King Edward	29		
	Maris Piper	43		
South (1973a)	Sharpes Express (FE)	3		
	Home Guard (FE)	6		
	Ulster Supreme (M)	8		
	Majestic (M)	11		
	Arran Banner (EM)	12		
	King Edward (EM)	19		
	Pentland Crown (EM)	20		
	Epicure (FE)	20		
Pinder (1974)	Dr MacIntosh	15		
	Record	17		Approximate
	Majestic	28		values
	King Edward	37		
	Redskin	66		
Stephenson and Bardner (1976)	Majestic	5.7		Mean for years
	King Edward	8.5		1961–74
Rogers-Lewis (1977a)	Pentland Crown	19		
	Pentland Ivory	12		

Table 11.1 contd

Source	Cultivar	% slug damage	
Airey et al. (1984)	Desireé	14 ⎫	
	Romano	16 ⎪	
	Pentland Hawk	18 ⎬	Approximate values
	King Edward	18 ⎪	
	Maris Piper	53 ⎭	

FE, first early; EM, early maincrop; M, maincrop.

although there was a full canopy, heavy rain showers wet the ridge to almost the same extent as the furrows. The dispersal of slugs in the potato crop is discussed further in Section 10.2.2. Secondly, most of the available information, e.g. Bett (1960), Runham and Hunter (1970) and Warley (1970), has shown that numbers and size of slugs tended to increase in autumn when damage is greatest, although there are considerable regional differences from year to year, in the timing of egg laying and maximum numbers of slugs. Thirdly, Hunter et al. (1968) suggested that tuber damage might increase when the haulm died off because an alternative source of food had been lost. Both Pinder (1969) and Warley (1970) showed that this was not true and that early removal of haulm did not affect the extent of tuber damage. Finally, the autumn increase in damage may be related to tuber maturation after growth has ceased (Thomas, 1947). This view has been supported by several authors, including Pinder (1969), Warley (1970) and Wareing (1982). Pinder (1969) showed that tubers of earlier planted cv. Majestic received significantly more damage than tubers from seed planted a month later and thus maturing later. However, Airey (1986a) found that tuber damage began early in the summer although slugs attacked larger (older) tubers in preference to smaller (newer) tubers. Atkin (1979) showed that differences were present between cultivars regardless of their state of development or the lifting date. Stephenson (1965) found that although irrigation of a potato crop of two cultivars, cv. Majestic and cv. Redskin, increased the yield of tubers, it also increased the damage by slugs, particularly to the more susceptible cv. Redskin.

11.2.6 Agricultural crops – cereals

Slugs damage cereal crops in two ways. The first and most serious is grain hollowing resulting from, first, the germ and then the endosperm being eaten before the grain can germinate. Hogan (1985) showed that *D. reticulatum* followed wheat seeds along drill lines, feeding on the grain. The other type of damage is caused by grazing on the young shoots as they appear above the ground. Some of this damage, especially at the base of the stems, may be confused with wireworm damage. The resulting shredding of the leaves may retard the growth of the plant and even kill it, especially when temperatures are low. Kemp and Newell (1987) showed that severe damage to the flag leaves of winter wheat was correlated with a reduction in grain yield although leaf damage is generally considered to be less important than grain hollowing (Anon., 1979). It is difficult to assess the final effect of slug damage to the winter wheat crop. Jessop (1969) removed winter wheat plants by hand in December to simulate the loss caused by different levels of slug attack. He found that loss was compensated for by increased tillering of the remaining plants and that a mean loss in yield of 19%, following a 75% simulated loss of seedlings in December (actual loss between 56 and 72%), did not generally justify the ploughing-in of the crop and redrilling with spring corn. However, Jessop's conclusions applied to randomly thinned plants. Under normal conditions, damage would be more patchy and redrilling might be justified in those parts of a crop where tillering could not compensate for loss in yield. There is evidence that winter wheat cultivars differ in their susceptibility to slug attack. The cv. Avalon is more susceptible than cv. Parade and this difference may be related to the total water-soluble sugar content of the seeds during germination, with Avalon having the greatest total water-soluble sugar content (Spaull and Eldon, 1990).

Damage has been recorded to other cereals including barley, oats and even rye. This damage has been generally on a small scale until recently and there are several reasons for this. In the case of rye this is probably because of the relatively small acreage grown. Traditionally oats and barley were spring sown, and spring cereals, including wheat, are not so susceptible because they germinate and become established more quickly, giving slugs less opportunity to damage them. However, damage has been reported to all spring cereals at some time. There is evidence that oat seeds are less palatable than other cereals. Duthoit (1964) showed, in laboratory tests, that *D. reticulatum*, the species most frequently responsible for cereal damage, had a preference for wheat grains when given a choice between oats, barley and wheat. *Arion hortensis* and *T. budapestensis* showed as much, if not more, preference for barley. Oat grains were left virtually alone by all three species and Duthoit suggested that oats might be grown in fields known to be infested with slugs. Port and Port (1986) described how the removal of the barley husk made the seed more palatable to *D. reticulatum*. The increased growing of winter barley over the past decade has resulted in increased slug damage to the barley crop.

Duthoit (1961) described how winter wheat seed loss was heavy before germination during the mild wet winter of 1960/61. Gould (1961) showed that slugs were more likely to be a serious pest on winter wheat after a wet summer on heavy land, especially when the seedbed was very rough and cloddy. He also found that damage was more frequent when winter wheat followed dry harvesting peas, leys, cereals (if stubble was ploughed without burning) and brassica seed crops. The large amount of plant debris left after harvesting initially provided shelter and, after being ploughed back into the soil, acted as a source of food and increased the water-holding capacity of the soil. Damage was less likely to occur after potato and sugar beet crops and when the land was left fallow. These crops and fallow land receive more extensive cultivation and Hunter (1967a) showed that cultivations may significantly reduce numbers of slugs on arable ground. Slug numbers were also reduced after vining peas, probably because slugs were removed when the crop was harvested. Martin and Kelly (1986) found that the risk of slug damage to winter wheat was greatest when the crop followed oilseed rape although previous crops of legumes, cereals and grass also carried an increased risk of damage.

Slug damage to winter wheat has increased with the development of minimal cultivation techniques. In an investigation into the effects of direct drilling on the soil fauna, Edwards (1975) showed that slug populations were many times greater on direct-drilled plots on sandy loam than on ploughed plots and that slug attacks were more severe, particularly in mild wet winters and when oilseed rape occurred in the rotation.

Stephenson and Bardner (1976) described how in these trials there were 6.4 times as many slugs in direct-drilled plots as in ploughed plots. They also confirmed that the dense cover afforded by oilseed rape crops and the debris after harvest provided favourable conditions for slugs. Stephenson and Bardner (1976) concluded that, with the expected increase in direct drilling and the increasing acreage of oilseed rape, losses due to slug damage could be expected to rise. Glen et al. (1984, 1988) showed that slug numbers and damage were considerably higher in winter wheat grown on heavy clay soil when straw was not burnt, than after straw burning. Shallow cultivation to incorporate straw had little effect on slug numbers but it did reduce damage in comparison with direct drilling into straw or stubble. Glen et al. (1988) suggested that, in fields where straw residues were incorporated rather than burnt, the presence of more slugs increased the risk of damage only if seed bed conditions enabled slugs to move though the soil in search of seeds. It was not clear whether the effect of straw burning on slugs was due to the heat generated, to the removal of food and shelter or to other factors such as ash residues.

The effect of soil type and structure on the distribution of slugs has been described in Section 10.2.2. A survey of slug problems in winter wheat fields showed that slugs were more important pests on medium and heavy soils and calcareous soils (Martin and Kelly, 1986). Stephenson (1975b) found that D. reticulatum could enter the spaces between soil aggregates and reach wheat seed at normal planting depths even where the aggregates were as small as 6.2 mm. However, when aggregates of fairly dry, medium–coarse soil were broken down and firmed over the seed by moderate or heavy pressure, slug damage to winter wheat was reduced. In practice, seed could be protected in this way by rolling immediately after drilling provided that soil and weather conditions allowed this. Davis (1989) demonstrated in the laboratory that *A. hortensis, D. reticulatum* and *T. budapestensis* slugs were less able to enter soil and damage buried wheat grains at high levels of consolidation than at low levels. Davis concluded that increased consolidation, at levels within the range found in arable fields, could reduce slug mobility and thus the damage to crops. In a field trial, Glen *et al.* (1989b) investigated the effects of tilth (soil particle size) and consolidation on slug numbers and biomass in the top 10 cm of soil and on damage to winter wheat seeds and seedlings in a clay soil. Depending on the type of seed bed, between 3 and 33% of seeds and seedlings were killed by slugs. Unlike Davis (1989), they found that most damage occurred on consolidated seed beds of fine or medium tilth, while least damage occurred on loose seed beds with the same texture. Damage at an intermediate level occurred on cloddy seed beds where consolidation had little effect. The level of damage found in different types of seed bed was directly related to slug biomass and inversely related to the depth of sowing and the percentage of fine soil present, and these three factors accounted for over 90% of the variance in slug damage. Glen *et al.* (1989b) concluded that increased slug damage was associated with consolidation because slug biomass was increased, seed was located at shallower depths and consolidation failed to break down clods of soil into finer particles. Moens (1983) concluded that the extent of slug damage to winter wheat was determined by three factors: (i) slug activity, (ii) lack of seed covering and (iii) delayed germination which extended the vulnerable stages of the young plant. When the soil was broken down to a fine tilth, this provided an effective protection for seeds against slug attack. Seedlings were especially vulnerable between the imbibition stage and the emergence of the coleoptile as this provided protection for the foliar tissues.

It is difficult to estimate the extent of slug damage to wheat in Great Britain since this varies from year to year and from one region to another, depending on climate, soil type and other factors. In 1970, slugs were considered to be one of the three most important pests of winter wheat in Britain (Runham and Hunter, 1970). Strickland (1965) stated that, in England and Wales, slugs were responsible for an average loss of about 17 000 hectare-equivalents out of about 745 000 hectares, i.e. about 2.2% of the wheat grown. Hunter (1969a), who considered this estimate rather high as it was based on localized surveys by Gould (1961) and Duthoit (1964), made an extensive survey of slug damage to winter wheat for England and Wales during the period autumn 1966 to spring 1967 and found that, out of 881 000 hectares sown with winter wheat, about 10 000 hectares (1.1%) were redrilled and between 4000 and 5000 hectares (about 0.5%) had to be treated with chemicals. The total financial loss was estimated at £191 700 (1969), i.e.

about 0.2% of the total value of the crop. Port and Port (1986) estimated that this loss was equivalent to £2.69 million in 1985. Although the proportion of the total crop lost was small, the economic consequences of slug damage may be serious for individual farmers because of the patchy distribution of slug damage. The effect of changing agriculture on slugs as pests of cereals was reviewed by Martin and Kelly (1986). They described how a combination of factors had led to an increased risk of slug damage to winter wheat over the previous decade as agricultural practices had changed and continued to change. There had been a substantial rise in the area devoted to wheat, which by 1985 was virtually all autumn drilled. There had also been a significant increase in the growing of oilseed rape and of peas for harvesting dry. The weather in the early 1980s had been particularly favourable and there was a tendency to reduce cultivations on medium and heavy land. Glen (1989) described how slugs were the pests causing greatest concern to wheat growers in Britain, especially in crops following oilseed rape. Furthermore, slugs had become second only to aphids as a pest problem in barley.

11.2.7 Damage forecasting

A number of attempts have been made to predict the level of slug damage likely to occur in a crop. Duthoit (1961, 1964) devised a method in which wheat grains were placed on the soil surface and the numbers of these damaged by slugs within one week were used as a measure of population density. Birds and mice could interfere with these baits, and various ways have been suggested to overcome this problem (Runham and Hunter, 1970). Hunter (1969a) described how a similar method had been used for the potato crop. This involved putting small samples of the main crop cultivar under the soil in the field about a month before the crop was planted. The extent of slug damage to these sample tubers was then assessed after a week. Symonds (1973) described the use of this method to measure slug activity before planting potato crops in fields in central and southern England between 1968 and 1970. Results suggested that there was a relationship between this activity and subsequent slug damage to the ware crop. Glen (1989) described how slug damage to winter wheat was directly related to slug biomass in the top 10 cm of soil at the time of sowing, and inversely related to the depth of sowing and the percentage of fine soil in the seed bed. Although the size distribution of soil aggregates in the seed bed and the depth of sowing could be assessed or adjusted at the time of sowing, methods for obtaining a reliable estimate of the slug biomass tended to be too slow and time-consuming for a practical forecasting system. Further investigations were needed into the relationship between trap catches and the numbers of slugs present in the soil.

11.3 SLUG DAMAGE IN EUROPE

Ricou (1961) reported large-scale slug damage in Normandy and other regions of France, particularly during wet years and especially on winter wheat. She described *D. reticulatum* and *A. hortensis* as the most important pest species of slug. Chevallier (1973b) considered that *D. reticulatum* was the most important pest species of slug in France. He also listed six other species as pests, of which *D. laeve* and *A. hortensis* were most widespread, although *D. laeve* required damp soils. Damage by *A. circumscriptus* to cereals and potatoes was mainly in the north of the country while damage by *A. ater, T. sowerbyi* and *A. lusitanicus* was localized. *Tandonia budapenstensis* was probably absent from France and Chevallier emphasized the risk that this species might be introduced into the country. The restricted distribution of the milacid slugs in Europe has been discussed in Sections 1.4.2 and 10.3.3. Mallet (1973) confirmed that *D. reticulatum, A. hortensis* and *A. circumscriptus* were important pests of agriculture. *Arion ater* was not considered a pest species while damage by *D. laeve* was restricted to moist areas. Several possible occurrences of *T. budapestensis* in France had been recorded, including damage in a sunflower field, but this species was not an established crop pest there. The distribution of economic damage by slugs in France was assessed by Hommay and Briard (1989) in terms of the areas treated by molluscicide baits. Most baits were used in Lorraine, central France and Burgundy while treatments were reduced in northern France, the Champagne and Aquitaine. In the north-east, rape and cereals were the main crops damaged while in the south-west spring crops, such as maize and sunflowers, were damaged. Hommay and Briard considered that the slug problem had increased over the past few years and this was related to changes in agriculture, including increased acreages of oilseed plants and winter cereals. Mallet (1973) estimated that a slug population of about $100 \, m^{-2}$ could consume about $250 \, kg \, ha^{-1}$ of plant material each day.

In France a wide range of horticultural and agricultural crops were damaged (Mallet, 1973). Damage to cereal crops is similar to that in Britain, with early sown crops being more susceptible to slug damage. Damage is more severe following oilseed rape. Maize seeds may be badly damaged, particularly in the south-west of the country. Dussart (1989) described how, in 1986, nearly a half of the maize crop in that region was damaged by slugs. According to Mallet (1973), the maize seeds were preferred to the plants themselves. Sugar beet damage was higher in years when the spring was moist and growth was slow. Mallet (1973) described how, because oilseed rape was sown when slug activity was highest in September and October, the young plants were especially susceptible to slug damage. Leguminous forage crops (clover, lucerne) were susceptible particularly in the early stages of growth in spring. Other crops damaged by slugs in France included potatoes, tobacco, brassicas, peas and medicinal herbs. Quantitative samples taken from arable land in 12 *départements* by Hommay and Lavanceau (1986) showed that two species, *D. reticulatum* and *A. hortensis*, were most abundant, with *D. reticulatum* making up between 48.1 and 99.4% of the total slugs. Other species were in the minority and individually rarely formed more than 5% of the total slugs, although species diversity of arable land was generally higher in the west and south-west of the country. Milacid slugs were generally uncommon. Maurin *et al.* (1989) also confirmed that *D. reticulatum* and *A. hortensis* were the main slug pests in France.

In Belgium, Moens and van den Bruel (1958a) reported damage by *D. reticulatum* to rye grown in the Polder region, and Mallet (1973) described how this damage occurred when the rye crop followed clover. Moens (1980) described damage by *D. reticulatum* to autumn sown cereal crops, which was aggravated by plant debris remaining from the previous crop. In Belgium the extent of the damage was partly dependent on the soil type, with the greatest damage occurring on the Polders and on wet soils. Slugs also showed preference for coarser and clay soils. On some clays and silts, *A. hortensis* was more common than *D. reticulatum*, and the slug population was more similar to that of horticultural soils. Cereals were often severely damaged and in some years severe slug damage was also found in sowings of legumes (clover), forage crops (rye-grass and maize) and in sugar beet. The use of monogerm sugar beet seed had aggravated the slug problem. *Tandonia budapestensis* was common in market and vegetable gardens in central and northern Belgium (van den Bruel and Moens, 1958a) and Moens (1980) described how the association of *T. budapestensis* and *A. hortensis* in these gardens resulted in severe damage to potatoes, carrots, celeriac and to bulbs and roots of various ornamentals, although less damage was done to early potatoes. A loss of about 30% of the crop could occur with maincrop potatoes and this could reach as high as 70% of the yield in wet summers. In Belgium this scale of damage tended to be limited to the vegetable gardens of amateur gardeners and the impact was, therefore, limited. Spring sowings of carrots, grass beans and other plants were regularly damaged by *A. hortensis* and *D. reticulatum*; these slugs also reduced the value of cabbages and lettuces at harvest and damaged strawberries. In glasshouses, *L. valentiana* damaged the flowers of orchids and other ornamental plants. A small proportion of this damage was also attributed to *D. laeve* and *A. hortensis*.

Godan (1983) has given an account of slug damage in western Germany. She considered that the main pest species of slugs were *D. reticulatum* and *D. agreste* and that damage by *Arion* spp. was more localized. This was also true for central Europe. *Arion* spp. and *M. gagates* were described as pests of rape and rape seed crops in Germany. *Boettgerilla pallens*, which was first recorded from Germany in 1962, was common in gardens and greenhouses where it had been reported as a pest. Godan (1973) summarized the damage reports over the ten-year period from 1962 to 1971. Slug damage to the cabbage crop represented the largest proportion (40%) of the total reported damage to crops. Damage to winter cereals (wheat, rye and barley) formed only 24.6% of the reports. Other crops affected, in order of relative damage, were lettuce, green vegetables, ornamentals and flowers, oilseed rape, forage plants, tobacco, root vegetables, strawberries and grassland. While this represents the economic damage to the crops, it is partly determined by the area of each crop grown. The extent of damage to a crop was often greater than that to the cabbage crop and Godan (1983) showed that lettuce, green vegetables, forage plants, rape, tobacco and grassland were all more heavily damaged than the cabbage crop. Potatoes were also damaged and the cv. Isola was particularly prone to attack. Among the oilseed plants, young plants of winter rape were damaged in the autumn. However, linseed and mustard suffered very little damage. *Deroceras reticulatum* and *D. agreste* also damaged the leaves and shoots of grape vines. Both metaldehyde and methiocarb baits are registered in western Germany for use in cereals, rape, vegetable crops, strawberries and ornamental plants (Büchs *et al.*, 1989). Detailed lists of agricultural and horticultural crops, fruit and forest trees and their slug and snail pests were included in the book by Godan (1983). She also described how *L. flavus* was a pest on agricultural products during storage, feeding on apples, pears, potatoes, carrots, pumpkin and other produce. It was also possible that fungal and bacterial diseases of stored potatoes and other produce were spread by the movements of these slugs. Moens (1980) reported that *L. valentiana* was a pest of glasshouse crops in Germany.

Most pest species of slugs are widely distributed throughout Europe (Chapter 1), although there is relatively little information about slug pests in southern Europe. *Deroceras reticulatum* is the main slug pest in Austria, Czechoslovakia and Poland, while in the latter country *D. agreste* may also be a pest species (Godan, 1983). Damage by other slugs has also been reported from these countries. Tan (1987) described how a moss cover sheltering seeds and seedlings of the fir *Abies alba* protected them against damage by *A. ater* and *A. circumscriptus* in the Jura mountains. In Romania, *D. agreste* was reported as a pest of glasshouse crops by Perju *et al.* (1981) and Lacatusu *et al.* (1984). Several slugs, including *D. agreste*, *D. reticulatum*, *D. laeve*, *A. subfuscus* and *A. circumscriptus*, were described as pests in the European USSR by Likharev and Rammel'meier (1952). *Deroceras agreste* and *D. reticulatum* were particularly destructive to fields and gardens. It was possible to grade the affected areas of the European USSR into heavily afflicted and moderately afflicted areas. Slug damage occurred in other parts of the USSR and *Parmacella olivieri* caused considerable damage to citrus crops in the subtropical parts of Talysh while *Parmacella* also damaged tea plantations and vegetable crops in Tadzhikistan (Likharev and Rammel'meier, 1952). Damage to young potatoes, cabbages and tomato plants by *D. caucasicum* was reported by Likharev and Izzatullaev (1972) from Tadzhikistan. Slugs were not considered to be harmful to field crops in the south-eastern USSR by Kosoglazov (1972), although *D. laeve, L. valentiana* and *L. flavus* were regarded as glasshouse pests. Dmitrieva (1969) developed an equation for the prediction of the degree of infestation by *D. reticulatum* in fields and Shapiro (1981) also developed equations for the prediction of numbers of *Deroceras* on arable land. In a study of the effect of soil erosion on invertebrates, Atlavinyte *et al.* (1974) found that arionid slugs were carried away by rainwater on fallow and arable land. In northern Europe, the tendency for *L. valentiana* and other limacid slugs to extend their distribution in glasshouses has been described in Section 1.4.3. In some countries, e.g. Sweden (Waldén, 1960), *L. valentiana* has caused considerable damage to glasshouse crops.

11.4 SLUG DAMAGE IN NORTH AMERICA

With the exception of the genus *Prophysaon* and *D. laeve*, all the slugs of economic importance in North America are introduced species, mostly from northern Europe. The recent spread of introduced veronicellid slugs in the southern USA, described by Dundee (1977), represents a potential pest problem as these slugs are pests in Central America (Andrews, 1989). The spread of European slugs in north-eastern North America has been discussed by Chichester and Getz (1968, 1969) and is described in Sections 1.2, 1.4 and 10.3.2. About 15 species, including the members of the *A. fasciatus agg.*, have been introduced and Chichester and Getz considered that the most likely means of introduction was with imported plant materials, with slugs subsequently being dispersed within the USA by similar passive means and also by active dispersal. Only one milacid slug, *M. gagates*, had been introduced and this was scarce and restricted to glasshouses and gardens. Chichester and Getz (1973) considered that native slugs were seldom abundant enough under natural conditions to be considered serious pests although *D. laeve* was often abundant in glasshouses where it could do considerable damage. Several species are widespread, including *D. reticulatum* which Lovatt and Black (1920) described as occurring from California in the south to British Columbia and through to Quebec in the north. Early records showed that *D. reticulatum* was well established in Massachusetts by 1843, reached as far west as Colorado in 1890, and was first mentioned from California and Oregon by 1891. Lovatt and Black (1920) considered *D. reticulatum* to be a serious pest of field and horticultural crops and described how *M. gagates* had become established as a pest in glasshouses while *P. andersoni* was a garden pest. Crowell (1967) considered that *A. ater* was becoming more economically important than *D. reticulatum* in Oregon. A survey of private gardens and market gardens in southern Manitoba by Howe and Findlay (1972) showed that three species of slugs were present.

The native species, *D. laeve*, and the introduced species, *D. reticulatum*, were widely distributed while *L. valentiana* was restricted to glasshouses. Only *D. reticulatum* was considered to be a pest slug and this species appeared to be extending its range.

Slugs, particularly *D. reticulatum*, are important pests of forage crops in the USA and are favoured by conservation tillage systems. *Deroceras reticulatum* was associated with the poor establishment of ladino clover and alfalfa during an investigation into glyphosate timing effects on the establishment of sod-seeded legumes and grasses in Montana (Welty *et al.*, 1981). A less favourable slug environment was produced when the interval between spraying glyphosate and seeding was extended to 28 days, as this allowed sufficient time for grass desiccation and drying out of the drill furrow. In a comparison of complete tillage with no-tillage (slit seeding followed by paraquat) techniques for the establishment of alfalfa, Grant *et al.* (1982) found that *D. reticulatum* caused more defoliation to alfalfa seedlings in the no-tillage plots than in the complete tillage plots and that slug densities were higher on no-tillage plots. Ladino clover stand reductions by slugs, particularly *D. laeve*, were also found by Rogers *et al.* (1985) to be much greater in no-tillage planting than in conventional plantings. Alfalfa damage associated with conservation tillage methods was also reported by Byers and Bierlein (1984). Most serious slug damage in conservation tillage systems occurred in the first month after planting (Byers *et al.*, 1983). Slugs preferred the youngest seedlings at the cotyledon stage and as plants matured, the slugs stopped cutting the stems and fed exclusively on leaflets. In Pennsylvania, *D. reticulatum* destroyed most of the alfalfa seedlings in oat stubble during the first weeks after planting. Byers *et al.* concluded that once past the susceptible period, legume seedlings could withstand grazing by slugs.

In 1968, slugs, especially *D. reticulatum*, caused widespread damage in no-tillage or minimum tillage maize fields after heavier than normal spring rainfall (Barry, 1969). Another example of the association of slug damage with conservation tillage systems was described by Hammond (1985). During 1983 in Ohio, *D. reticulatum* was a minor but widespread problem on soybean crops and the majority of the infested fields were planted as no-tillage or minimum tillage, having corn as their previous crop. Hammond concluded that slugs were likely to remain a minor or secondary pest of soybeans and only really occurred as pests in situations where heavy plant residues remained from the previous crop. Three species of slug, *D. reticulatum*, *D. laeve* and *A. fasciatus*, were associated with damage to legumes planted with a minimum tillage drill into grass fields in Pennsylvania (Byers *et al.*, 1985) while Dowling and Linscott (1985) concluded that insects and slugs, particularly the latter, could seriously limit the establishment of sod-seeded lucerne in the north-eastern USA. Hammond and Stinner (1987) described how slugs were the most serious non-insect pest of reduced tillage corn (maize). They found that slug numbers were highest in no-tillage systems where crop residue cover was greatest and lowest where no residue was present and there was a trend for numbers to be greater when the previous crop had been soybeans. In experimental plots of alfalfa, soybeans and maize sown by conservation tillage methods, plant losses did not increase in proportion to the density of slugs (*D. reticulatum*). Barratt *et al.* (1989) suggested that even low slug densities could cause significant crop losses and that these losses were particularly important in dicotyledonous crops, such as soybean and alfalfa. Plant mortality was considerably lower in maize seedlings. Reports of other crops damaged in North America include tobacco (Mistic and Morrison, 1979) and strawberries in Québec (Duval and Banville, 1989), western USA (Howitt and Cole, 1962) and Manitoba (*D. laeve*) (Prystupa *et al.*, 1987).

11.5 SLUG DAMAGE IN OTHER COUNTRIES

In addition to North America, some European pest species of slug have been introduced to other temperate countries, including Australia, New Zealand and South Africa (Sections 1.2 and 1.4). A list of slugs introduced into Australia included the pest species *M. gagates*, *D. reticulatum* and *D. caruanae* (Altena and Smith, 1975) and it was suggested that since *T. budapestensis* and *T. sowerbyi* had been recorded from New Zealand, these species might occur as introductions in Australia although they had not been confirmed by 1975. Yamashita *et al.* (1979) reported another introduced species, *L. nyctelia*, which occurred on Kenya white clover in subtropical pastures in Queensland, although most of the slug damage to the clover was due to a *Deroceras* sp. Both species consumed the clover but avoided 14 species of tropical grasses in feeding trials. Martin (1978) listed the pest species *A. hortensis*, *M. gagates*, *T. budapestensis*, *D. reticulatum* and *D. laeve* among slugs that had been introduced into New Zealand. The two most common species on hill pastures were described as *D. reticulatum* and *A. hortensis* by Charlton (1978). Additional introduced pest species from New Zealand pastures which were listed by Barker (1989a) included *A. intermedius* and *Deroceras panormitanum* (Lessonera and Pollonera). Reports of slug damage from New Zealand have been mainly concerned with damage to forage crops and to conservation tillage systems, recalling the slug problems encountered in the USA. Charlton (1978) listed several previous reports which attributed legume seedling loss in hill pastures to slug attack. Charlton demonstrated experimentally that when slugs were present, less than 10% of oversown legumes survived at six weeks. Seedlings in the early stages of development were severely depleted although once seedlings reached the 1–5 true leaf stage, there was little seedling mortality.

In experiments which measured the seed loss from different pasture seed types laid in lines on pasture, the slugs *D. reticulatum* and *D. panormitanum* were found to have consumed low levels of legume seed during the winter and early spring, although they did not consume grass seed (Blank and Bell, 1982). Hughes and Gaynor (1984) described how observations made in a long-term forage cropping rotation suggested that maize direct drilled into pasture stubble was being damaged by slugs. Slugs ate either the seed or the emerging coleoptile, causing the death of pre-emergent seedlings and also grazed the leaves of established plants although this rarely caused the death of the plant. In field trials, Hughes and Gaynor (1984) direct drilled maize into pasture and when wet weather followed sowing, slugs killed up to 44% of the pre-emergent maize seedlings. The slugs mainly responsible for this damage were *M. gagates* and *D. reticulatum*. However, slug damage was reduced when a winter oat forage crop was grown before the summer maize crop, despite oats being a host crop for slugs. The later grazing of maize leaves still occurred in a wetter year but this did not result in plant death. Hughes and Gaynor suggested that the reduction in slug damage might be due to increased protection of germinating maize seeds by better closure of the drill slits in the more friable soil under the oat crop. Ferguson *et al.* (1988) investigated the possible control of *D. reticulatum* by stock management prior to direct-drilled pasture establishment. They found that, at stocking rates of 1500 or more sheep days per hectare, slug control of more than 90% could be achieved through mortality of slugs due to physical injury sustained by treading of the pastures. There was the added flexibility of employing various combinations of stock numbers and length of grazing period. This treatment gave better slug control than methiocarb slug baits.

Barker (1989a) reviewed slug problems associated with pastoral agriculture. He described the New Zealand climate as only marginally suitable for slugs, as the summers in many regions were too hot and dry. In some regions slugs were able to survive over the summer, because there was adequate summer rainfall, the pasture type or grazing management provided adequate cover to insulate the soil surface, and surface cracking of soil provided refuges during the summer. Under these conditions severe slug damage occurred during pasture establishment by no-tillage methods. Slug damage during pasture establishment consisted of hollowing seeds, severing young shoots and shredding leaves on young seedlings and in this way they reduced plant populations. Slugs also defoliated clovers in established swards thus reducing seedling vigour. Barker suggested that slug problems were accentuated by poor farming practices and emphasized the need for effective stock management such as that described by Ferguson *et al.* (1988). The relative importance of different pest species varied according to the agroclimatic zones (Barker, 1989a). In the dry arable zone, where a pasture component was included within a rotation, only *D. reticulatum* was common and slug populations were generally low unless the land was under irrigation. *Deroceras reticulatum* and *A. intermedius* occurred together in low numbers in the cold inter-montane and high altitude zone and in variable but higher numbers in the low to moderate fertility hill zone. However, on high fertility, moderate to high rainfall lowlands in the milder climate of the North Island, *D. reticulatum* was the most common slug with *D. panormitanum* being equally abundant locally, especially on wet soils. Other slugs present included *A. intermedius*, *A. hortensis* and *M. gagates*. In the same zone, but in the cooler conditions of the South Island, the slug fauna consisted mainly of *D. reticulatum*, *D. panormitanum* and *M. gagates*.

While the pest status of slugs in temperate regions is well documented, less information is available about slug damage in the tropics. In their review of the slug problem, Hunter and Runham (1971) described how slug damage to crops had been reported from East Africa, Australia, South Africa, tropical America and Asia. The most common forms of slug damage were grazing of the fruits and leaves of coffee or brassicas and by hollowing out the seed of newly planted grain crops. Several reports of slug damage from tropical countries were listed by Godan (1983). The arionid slug, *Anadenus altivagus* (Theobald) is a pest of the potato crop in India (Agrawal, 1977) while slug damage to young rubber trees was reported by Bertrand (1941). *Deroceras reticulatum* was described as a pest of field and horticultural crops in the cooler regions of Colombia by Ruppel (1959). The Veronicellidae are known pests in tropical America, parts of Africa and Asia. Crops damaged by these slugs in Central America include coffee, bananas, tobacco and dry beans (*Phaseolus vulgaris* (Linnaeus)) (Andrews, 1989). Although slug pest problems due to veronicellids were reported from Central America in the late nineteenth century, Andrews (1989) described how severe attacks only began in the late 1960s and early 1970s, which coincided with the introduction of the veronicellid, *S. plebeia*, to Central America in the mid 1960s. *Sarasinula plebeia* was now a serious pest of dry beans and the slugs severed seedlings at soil level as well as consuming leaves, growing shoots and young pods. Regionally, losses had been estimated at US$ 27–45 million per year, with severe effects on agriculture in the region (Andrews, 1989). Veronicellid slugs, including *S. plebeia*, acted as intermediate hosts of the nematode, *A. costaricensis*, the cause of abdominal angiostrongyliasis in humans, and this has been discussed in Section 9.2.2.

12

Control of slug pests

12.1 HISTORY OF CHEMICAL CONTROL

12.1.1 Early methods

Early methods of slug control depended largely on the use of dry dressings which irritated slugs so that they eventually became desiccated. Konrad von Megenberg suggested in 1349 that slugs could be destroyed by the use of salt (Weiss, 1973). Theobald (1895) suggested the use of dressings of salt and of lime to control slugs in field crops. Anon. (1905) recommended the use of (a) soot and lime; (b) salt and lime; (c) lime and caustic soda and (d) powdered coke. Taylor (1902–1907) also recommended sawdust, chaff and sand for this purpose. Anon. (1905) suggested the use of barriers, such as trenches filled with lime or tar, and the penning of ducks and poultry in gardens. Slugs could also be trapped using baits of bran mash, moist oatmeal or other attractive substances and then collected and destroyed. In the USA, Lovatt and Black (1920) described a series of slug control investigations carried out at the Oregon Agricultural Experiment Station between 1912 and 1919, using mainly *D. reticulatum* and the greenhouse slug, *M. gagates*. They concluded that irritants were of little use because slugs often recovered; and also the effectiveness of the substances used was spoilt by moisture. Contact sprays were also of little use, although sprays of 1 and 10% copper sulphate solutions killed slugs under laboratory conditions. However, Bordeaux mixture was less effective although it acted as a slug repellent. Lovatt and Black concluded that calcium arsenate baits were the most effective slug poisons available at that time.

A number of compounds, mainly inorganic, were tested by Miles *et al.* (1931) against slugs. They included the use of copper sulphate for the control of slugs and noted that aluminium sulphate killed slugs. Hall (1932) suggested the use of zinc collars for the protection of plants. In an advisory leaflet, Anon. (1932) advised the use of multiple applications of quicklime or salt at night when slugs were active on the soil surface. Broadcasting a copper sulphate–kainit mixture (see below) had also given good results in growing crops such as young oats, peas and sugar beet. The use of barriers was still recommended for market gardens and on allotments (Anon., 1932). Lange and MacLeod (1941) described the use of calcium arsenate and metaldehyde in slug baits. Stapley (1949) described two substances which had been used for slug control. Powdered copper sulphate mixed with kainit (potassium chloride/sodium chloride) or salt as a carrier could be broadcast when slugs were active although it had the disadvantage of scorching foliage. This mixture was first described by Anderson and Taylor in 1925 (Miles *et al.*, 1931). The mixture was still in use in 1949 and Stapley considered copper sulphate to be more reliable than metaldehyde bait under wet conditions. In an advisory leaflet, Anon. (1968) suggested that copper or aluminium sulphate might be more effective for the control of slugs under glass where the high humidity allowed recovery of metaldehyde-poisoned slugs. The other substance described by Stapley (1949), which had been superseded by metaldehyde, was a bait based on bran mixed with Paris Green (copper aceto-arsenite) and had been used mainly against cutworms and leatherjackets. The use of iron sulphate for slug control was described by Palombi (1949) and by van den Bruel and Moens (1957).

12.1.2 Metaldehyde

Metaldehyde was formerly marketed as a solid fuel (Meta-fuel) for picnic stoves and is effective against both slugs and snails. These are immobilized by metaldehyde and death then follows by desiccation. It is unique in its specificity for slugs and snails, and baits containing metaldehyde are ineffective against soil insects such as wireworms and leatherjackets and also against woodlice (Gimingham, 1940). The way in which the slug-killing properties of Meta-fuel were discovered remains uncertain. Although these properties were first described in the gardening press in Britain in 1936, Gimingham described how the use of meta bait had been known in South Africa since 1934 and related two stories of the way in which discarded Meta-fuel had been found to attract and kill slugs. By 1940, metaldehyde–bran baits were in widespread use for slug control in field crops and in gardens and allotments (Gimingham, 1940). Metaldehyde pellets continue to be in use today and are approved for use for the control of slugs and snails in field crops, vegetables, fruit crops, ornamentals and glasshouse crops (Ivens, 1989).

12.1.3 Carbamates including methiocarb

The carbamates, a group of residual contact insecticides, were introduced in the 1950s. Ruppel (1959) compared the effectiveness of carbaryl (Sevin) as a bran bait (1.3 and 2.5% carbaryl) and a contact spray against *D. reticulatum*. He also used a bait based on a mixture of 1.3% carbaryl with 1.3% metaldehyde. Calcium arsenate (7.5%) and metaldehyde (2.5%) baits and a spray of 0.3% metaldehyde were used as standards. Carbaryl sprays were ineffective although the metaldehyde spray was very effective within 24 hours. The carbaryl–metaldehyde bait gave a similar slug mortality to the calcium arsenate–metaldehyde bait. The baits of carbaryl alone also caused high mortalities but took longer to act. Ruppel (1959) concluded that carbaryl was effective against slugs as a bait poison but not as a spray. Webley (1962, 1965) also showed that the addition of carbaryl to meta bran baits improved the slug catch. The potential synergistic effect of metaldehyde when added to carbamate baits was investigated by Crowell (1977). He concluded that the addition of metaldehyde to baits containing carbaryl, methiocarb and another carbamate, UC-30045, would have little or no value. However, Mallet and Bougaran (1971) found that the combination of a carbamate PE-9683 with metaldehyde in baits gave effective control at low temperatures, was fast acting and also irreversible.

Riemschneider and Hecker (1979) reported that dimethylglyoxime, although not a molluscicide, showed synergistic activity in combination with metaldehyde in baits used against snails. Getzin and Cole (1964) screened 60 chemicals as contact poisons in the laboratory, using the slug, *P. andersoni*, as a test species. They found that only one substance, the carbamate Mesurol (Bayer 37344) possessed molluscicidal properties equivalent to metaldehyde when tested in the field and that these substances were more effective when used in 4% bran bait formulations than when applied as sprays. Getzin (1965) demonstrated that Mesurol (methiocarb) had a toxic action against *D. reticulatum* equal to that of metaldehyde. Getzin and Cole (1964) found in tests that carbaryl 2 and 3% baits showed little activity against slugs. Crowell (1967) described a laboratory screening technique for presenting experimental baits to molluscs and found that five experimental carbamate materials, including methiocarb, showed considerable molluscicidal activity against *A. ater* and the snail, *H. aspersa*. Crowell (1967) emphasized the need to maintain a balance between too low a concentration of active ingredient, which would produce a sublethal (immobilizing) dose; and too high a concentration, which would tend to repel slugs.

Barry (1969) described a field trial to evaluate the carbamates, carbaryl and Zectran, the organophosphorous insecticide phorate, the soil insecticide Zinophos (thionazin), the aquatic molluscicide Frescon, the herbicide dinoseb and a commercial calcium arsenate–metaldehyde bait; for the control of *D. reticulatum* and other slugs. Phorate was the only chemical tested that produced appreciable reductions in slug numbers within 48 hours after application. This confirmed some previous reports that carbaryl was not very effective as a molluscicide against slugs. Out of 74 materials, including over 20 carbamates, initially screened by Judge (1969) as candidate molluscicides, only 23 showed molluscicidal properties. In a second stage, only four were effective when applied as sprays and two others were effective when applied in granular form. These six compounds included the carbamates methiocarb, UC-30045, Hercules 18526 and the carbamate-type insecticide aldicarb together with metaldehyde and phorate. Hunter and Johnston (1970) used two methods to screen seven carbamates for toxicity against *D. reticulatum*. They presented baits to slugs confined in Petri dishes out of doors, enabling them to compare the toxicity of the different carbamates to slugs. The second method involved the introduction of known concentrations of chemical mixed in a 1.5% agar solution directly into the gut, using a micrometer syringe. This technique was described by Henderson (1969). Hunter and Johnston (1970) were able to obtain a more accurate estimate for the LD_{50} of each substance in this way and found that several carbamates, including methiocarb, were more toxic to slugs than metaldehyde. A number of carbamates have been shown to be effective molluscicides and suitable for use against slugs and methiocarb is one of the most effective of these. Methiocarb pellets are currently in use today and are approved for use for the control of slugs and snails in field crops, vegetables, blackcurrants and strawberries (Ivens, 1989).

12.1.4 Other substances

Van den Bruel and Moens (1958a) found that slugs were killed when crossing a band of soil treated with DNOC, a dinitro insecticide and herbicide. Van den Bruel and Moens (1957, 1958b) then showed that DNOC was effective in field trials against slugs. However, the rate of application required was higher than when this compound was used as a herbicide in cereals. Calcium cyanamide (Cyanamid), a herbicide and defoliant, has also been used effectively as a dust and an aqueous suspension to prevent slug damage (van den Bruel and Moens, 1956b, 1958b).

The effect of organochlorine insecticide residues, particularly DDT and its derivatives, on slugs has been described in Section 10.6.3. Although residues were found in slugs, these appeared to cause no harmful effects. However, Buckhurst (1947) reported limited slug kills when DDT was sprayed as a dust and an emulsion. Stephenson (1959) found that treatment with 1.25% aldrin dust at the time of planting, significantly reduced slug damage to a potato crop. Ricou (1961) described how slugs that had been fed the organochlorine insecticide heptachlor were initially immobilized but subsequently recovered. Ricou also reported that organophosphorous compounds had little effect on slugs. Neither endrin (an organochlorine insecticide) nor dimethoate (an organophosphorous insecticide) were found by Webley (1962) to increase the slug catch when tested as synergists in meta-baits and instead they acted as slug repellents. Endrin was also ineffective in a field trial against the slug *L. alte* (Brar and Simwat, 1973). However, Manna and Ghose (1972) showed that endrin was an effective gut poison for the snail, *A. fulica*, bringing about histopathological changes in the gut. El-Okda (1981) showed experimentally that DDT and another organochlorine insecticide, lindane, and also phoxime, an organophosphorous insecticide, were not effective against *L. maximus*.

Judge (1969) found that granules of the organophosphorous insecticide, phorate, were very effective against *D. reticulatum* when broadcast on trays of growing peas, while Barry (1969) found that phorate applied as an emulsion and in the form of granules was effective against *D. reticulatum* in field trials with no-tillage and minimum tillage maize crops. Musick (1972) also showed that phorate greatly reduced slug numbers when applied to the surface of the soil of a no-tillage maize field and suggested that maximumn reductions could be achieved by careful timing of the spray application and by a second application. Phorate also killed *L. alte* in both laboratory and field trials (Brar and Simwat, 1973). Despite the success of phorate granules in the USA, Symonds (1975) found that this compound did not give significant slug kills even at an application rate of 60 kg ha^{-1}. However, when Edwards (1976) examined the uptake of two organophosphorous insecticides, phorate and diazinon, by slugs in field experiments in spring wheat, he confirmed the toxicity of phorate to slugs. Diazinon, a non-systemic insecticide, had no direct effect on slugs. Edwards suggested that phorate, a systemic insecticide, was taken up from the soil into the wheat plants and then acted as a stomach poison to the slug. The phorate was applied as granules and the diazinon as a spray, and they were applied in these experiments at a higher rate than that normally recommended for them. Linzell and Madge (1986) found that slug populations on grass and wheat plots were not immediately affected when phorate and aldicarb were applied, although numbers were subsequently reduced by these compounds.

Webley (1965) showed that addition of the dithiocarbamates dazomet (a soil sterilant) and zineb (a fungicide) to meta-bran baits repelled slugs. Gorokhov and Aliev (1971) obtained kills of from 42 to 100% of *Deroceras* sp. in field trials by using daily sprays of an organic acid amide, N-(3-chloro-4-methyl phenyl) butyric acid amide applied when slugs were becoming active in June. Daxl (1970, 1971) demonstrated that *L. flavus, D. reticulatum, D. laeve* and *A. hortensis* were susceptible to the herbicide ioxynil applied as a spray and in baits, although metaldehyde and the carbamate insecticide isolan were more effective than ioxynil. All three substances acted as ovicides against the eggs of *L. flavus*. The compound 2,6-dibromo-4-(4-nitrophenylazo) phenol (Yurimin) irritated and produced strong contractions in slugs when applied as a dust to cabbage and the skin of exposed slugs showed degenerative and destructive tissue changes when examined (Inoue and Hayashi, 1983). Chemical control methods used for slugs tend to be less effective than those used for other crop pests. The problems involved in developing chemical control methods for slugs were discussed by Henderson and Parker (1986), based on laboratory studies on the effect of bait and contact poisons on the pest slug, *D. reticulatum*. Metaldehyde and methiocarb reduced the acceptability of baits to slugs and *D. reticulatum* was able to detect metaldehyde in sweetened agar gel at concentrations as low as 10 ppm. The copious mucus produced by the slug in response to stimulation tended to reduce the effect of contact poisons and the latter needed to be sufficiently hydrophilic to enter the mucous layer rapidly. Henderson and Parker concluded that, while there was no lack of poisonous substances, the problem with baits and contact poisons lay in getting the poison into the slug. The feeding behaviour of slugs on poison baits is discussed further in Section 12.5.2.

12.2 RECENT DEVELOPMENTS IN SLUG CONTROL BY CHEMICALS

12.2.1 Plant derivatives

Thompson (1928) suggested that the insecticide rotenone, derived from the roots of *Derris* and *Lonchocarpus* spp., might be used in poison baits for slugs. Considerable interest has been shown in the past in molluscicides of plant origin, particularly those that might be used for the control of aquatic snail vectors of schistosomiasis (Mozley, 1952; Marston and Hostettmann, 1985). Mozley described how finely powdered *Derris* root mixed with water would kill one of the snail hosts of schistosomiasis. He also described how saponins (steroidal glycosides), such as those from the fruit of the trees *Balanites aegyptiaca*, *Sapindus saponaria* and *Swartzia madagascariensis* would also kill these snails. Krochmal and Lequesne (1970) reported the molluscicidal properties of the American pokeweed, *Phytolacca americana*, while the insecticidal and molluscicidal properties of the related species, *Phytolacca dodecandra* (endod) were described by Spielman and Lemma (1973). Water and *n*-butanol extracts of dried, ground berries of *P. americana* were considered by Nilsson and Sorensson (1989) to be worth further investigation for slug and snail control. Webbe and Lambert (1983) found that saponins with an oleonic acid derivative seemed to offer the greatest potential as molluscicides against aquatic snails. They described other compounds in addition to saponins, including flavenols, alkaloids and terpenoids which might be molluscicidal. Airey *et al.* (1989) reviewed work on identifying natural compounds, mainly from plants, which might be used in slug control. They showed that the saponin fraction from the common starfish, *Asterias rubens* Linnaeus, was highly toxic to *D. reticulatum* when absorbed from a glass surface. Piperine, a lipophilic amide from the black pepper, *Piper nigrum*, showed an LD_{50} against *D. reticulatum* similar to that for methiocarb. Airey *et al.* (1989) also attempted to identify compounds that would interfere with biochemical processes, such as calcium metabolism and mucus production, which are important to slugs. No suitable compounds were found which would disrupt calcium metabolism or mucus production although a number of irritants, including capsaicin from the chilli pepper, *Capsicum annuum*, and synthetic tetronic acids, were tested. Although the slug reacted to these compounds and they were sloughed off the skin by increased mucus production, slug mortality from desiccation only approached 50% at very high levels of these compounds. Various behaviour-controlling chemicals (semiochemicals) were investigated and some attractant and phagostimulant compounds were identified. Airey *et al.* (1989) concluded that compounds which interfered with feeding, e.g. the monoterpenoid ketone, (+)-fenchone, appeared to offer the most promising leads for new crop protection agents.

12.2.2 Aluminium and other metal compounds

Durham (1920) recommended the use of an aluminium sulphate spray as a snail poison and Miles *et al.* (1931) reported that this chemical killed slugs. In the 1960s, a wettable powder containing aluminium sulphate, Fertosan, was marketed as a contact poison for the control of slug damage in gardens. It was applied as a powder or a spray. Fertosan was compared with metaldehyde and methiocarb in field trials on Chinese cabbage and in laboratory trials (Glen and Orsman, 1986). They found that all treatments were fairly successful in protecting the Chinese cabbage plants for the first two days after planting in the field, which was up to four days after the application of the molluscicides. Aluminium sulphate did not affect the survival or food consumption of *D. reticulatum* 24 hours after exposure in the laboratory and Glen and Orsman (1986) suggested that prolonged exposure of slugs in the field to the chemical might be required and that it might also act as a repellent. Glen *et al.* (1986) also described a wettable powder containing aluminium sulphate, copper sulphate and borax which was sold under the trade name Nobble. This was intended to be sprayed on to the soil surface either before, during or soon after planting and was sold for the control of slug damage in a wide range of crops, including potatoes. Nobble remains effective for up to six months and claims to kill slug eggs and recently hatched slugs. In a field trial designed to test the effects of Nobble on slug damage to potatoes, in a situation where neonate slugs were hatching over a period of six weeks after application, there was no significant difference in damage between untreated plots and those treated with Nobble (Glen *et al.*, 1986). The slugs added were *D. reticulatum*, *A. distinctus* and *A. silvaticus*, while *D. reticulatum* was the predominant slug found on the plot at harvest. However, more recent unpublished reports suggest that the timing of application is important to more effective slug control by Nobble.

Many metals are known to be toxic to slugs and the effect of some metals on slugs has been discussed in Section 10.6.2. Copper salts have been used with limited success against slugs in the past. Interest in the use of metal-containing compounds has been revived recently, particularly as it has been demonstrated that copper can be absorbed through the epithelium of the foot, by a paracellular route and by pinocytosis (Ryder and Bowen, 1977c). Copper can also be taken up by the slug egg and initially retained in the perivitelline membranes (Ryder and Bowen, 1977d). Henderson *et al.* (1989) investigated the toxicity of inorganic salts of ten metals to *D. reticulatum*, using dry glass plates coated with the chemical. Salts of sodium, magnesium and potassium had little effect. Copper(II), zinc, iron(II) and aluminium sulphates all killed slugs by contact. They also showed that the toxic effects of these compounds were less severe when dusted on a layer of wet soil. Henderson *et al.* (1989) showed that when iron and aluminium were applied as organic complexes, chelation delayed the inactivation which occurred when metal salts were broadcast as contact-acting molluscicides on to wet soil. However, even the most active chelates only killed significant numbers of slugs when applied at impracticably high rates in the field although Henderson and Martin (1990) subsequently showed that slugs could be killed by a broadcast application of the chelate iron(III) 2,4-pentanedionate applied at a rate of 40 kg a.i. ha^{-1}. Trials of an aluminium chelate slug bait, based on aluminium acetylacetonate, were described by Henderson *et al.* (1990). Henderson *et al.* (1989) concluded that the most effective method for delivery of the metal complex was in the form of a bait. In field trials, milled wheat baits containing these compounds killed slugs as effectively as baits based on metaldehyde and methiocarb and were shown to be more effective under wet conditions since slugs did not recover from metal poisoning.

12.3 METALDEHYDE

The name metaldehyde is accepted by BSI and ISO in lieu of a common name for 2,4,6,8-tetramethyl-1,3,5,7-tetroxocane. It is also known as metacetaldehyde and by the trade name Meta, particularly when applied to the solid fuel. Metaldehyde is produced by the polymerization of acetaldehyde and consists mainly of the tetramer. It is an effective molluscicide against slugs, although neither acetaldehyde nor paraldehyde (a trimer) were considered to share these molluscicidal properties by Cragg and Vincent (1952). However, Henderson (1970) showed that this was incorrect (see below). The pure tetramer of metaldehyde forms colourless crystals, while the commercial grade is a white crystalline powder and has a characteristic sweet odour, due partly to the presence of acetaldehyde. It is sparingly soluble in water (200 ppm) and is soluble in benzene and chloroform. Although metaldehyde has a high m.p. of 246°C in a sealed tube, it sublimes before this point at about 112°C. Metaldehyde also tends to depolymerize, especially under acid conditions. Further details of the structure and properties of metaldehyde can be found in Worthing (1983) and in Raw (1970).

Metaldehyde is currently approved for use in Britain as a bait in pellets generally containing 6% a.i. metaldehyde together with bran or another cereal product (Ivens, 1989). Paper-based tapes and pads containing a thin layer of metaldehyde mixed with a small quantity of cereal are also available and contain about 7% a.i. These are applied as a barrier around plants or buried with seed potatoes. Metaldehyde baits containing 3, 4 or 5% a.i. were formerly available but have now been discontinued since the introduction of smaller (mini) slug pellets. Trials reported by Runham and Hunter (1970) showed that metaldehyde pellets containing 6% a.i. had given significantly better slug control than those containing 5% a.i. or less. Metaldehyde sprays and dusts have been used in the past to control slugs (Jefferson, 1952; Lange and Sciaroni, 1952; Runham and Hunter, 1970) and Cragg and Vincent (1952) suggested that these were the best methods of application. Webley (1965) recommended that metaldehyde sprays should be used for slug epidemics and baits only applied as long-term measures to keep down slug populations which were already partly controlled. However, sprays do not persist on wet soils or on foliage when the weather is wet, which is the time when slugs are most active.

Thomas (1948) described how metaldehyde poisoning had both an anaesthetic effect and an irritant effect and caused the gut wall to become transparent. He also found that high humidity encouraged the recovery of slugs from metaldehyde poisoning. The action of metaldehyde on *D. reticulatum* was investigated by Cragg and Vincent (1952). They showed that metaldehyde, both in powder form and in aqueous solution (0.02%), acted on slugs either as a contact or stomach poison. The toxic effect appeared more rapidly with contact treatment because metaldehyde was rapidly absorbed through the foot. The characteristic effects of metaldehyde poisoning were immobilization broken by outbursts of uncoordinated muscular activity and excessive mucus production, resulting in a considerable loss of water. Feeding was also inhibited. Toxicity increased with temperature and slugs could recover from moderate doses of metaldehyde if kept in a saturated atmosphere. No gut lesions were found in slugs poisoned with metaldehyde and the transparent effect noted by Thomas (1948) was due to a sloughing off of the thin mucous membrane of the crop, which occurred after death from any cause.

Unlike Cragg and Vincent (1952), Martin (in Sparks, 1972) described extensive histological effects of metaldehyde poisoning in *A. ater*. Maximum tissue damage occurred at the site where metaldehyde was absorbed (i.e. the stomach and anterior loop of the intestine), although the posterior loop was not affected. Tissue damage also occurred in the kidney, the site of excretion. Other histological changes were related to excessive mucus discharge, desiccation, physical trauma in the violent reaction to poisoning and to the onset of autolysis. Cragg and Vincent (1952) found no evidence that metaldehyde had any fumigant action and suggested that broadcasting and spraying were the best methods of application. They compared the effects of metaldehyde with acetaldehyde and paraldehyde by contact action and by injection into the gut and into the haemocoele. Acetaldehyde and paraldehyde appeared to have no toxic effects and Cragg and Vincent suggested that the toxic effects of metaldehyde were not due to the formation of acetaldehyde. Contact with concentrations of metaldehyde equivalent to 0.0063 mg cm^{-2} produced lethal effects while 0.06 mg of solid metaldehyde taken through the mouth killed average-sized slugs.

In contrast to the conclusions of Cragg and Vincent, Henderson (1970) showed that metaldehyde had a fumigant action, probably caused by the acetaldehyde present in impure metaldehyde. He showed that acetaldehyde was toxic to slugs generally and was slightly more toxic to *D. reticulatum* than to *A. hortensis*. Henderson suggested that this might be due to the thicker skin and less watery mucus of *A. hortensis*. However, the fumigant action was unlikely to be important in the field. Henderson (1966, 1969, 1970) also compared the toxicity of several chemicals, including metaldehyde, as contact and stomach poisons to *D. reticulatum* and *A. hortensis*. Acetaldehyde was considerably less toxic than the other substances tested in aqueous solution while metaldehyde was less toxic than copper sulphate when they were applied as dry powders, the LD_{50}'s being 43 370 ppm and 2027 ppm respectively. Metaldehyde was more toxic than copper sulphate when used as a stomach poison, with the median LD_{50}'s being 85 μg slug^{-1} and 129–132 μg slug^{-1} respectively. While the forced ingestion method used allowed the effect of chemicals to be assessed as stomach poisons free from effects due to contact or fumigant action, it was still possible that an effective poison might have a repellent action if incorporated into a slug bait (Henderson, 1969, 1970). Henderson (1966) concluded that the slug poisons in use at that time were not the most toxic compounds to slugs and that their effectiveness depended on other factors such as palatability. New substances were needed with a higher intrinsic toxicity.

Bourne *et al.* (1988) estimated the LD_{50} value of metaldehyde for *D. reticulatum* to be in the range of 70–100 μg per adult slug, similar to that obtained by Henderson. Wedgwood and Bailey (1988) estimated the lethal dose of metaldehyde to be about 0.2 mg g^{-1} body weight of slug in *Deroceras* spp., while *A. hortensis* was less susceptible. Wedgwood and Bailey described how after feeding on a pellet containing metaldehyde, slugs developed writhing activity, undirected mouthing movements and moved with an awkward gait, followed by a period of immobility. This suggested that metaldehyde had a toxic effect on the nervous system. In mammals there is evidence that metaldehyde is decomposed to acetaldehyde in the gut (see Section 12.7.1) and acetaldehyde has been detected in the blood of slugs after a meal on a metaldehyde pellet (Mills *et al.*, 1989). Mills *et al.* used the isolated central nervous system of the aquatic snail, *L. stagnalis*, to examine the effects of metaldehyde and acetaldehyde on the neural circuitry controlling feeding. Application of metaldehyde and acetaldehyde increased firing activity in feeding motoneurons in the buccal ganglia and these motoneurons subsequently developed paroxysmal depolarizing shifts. Mills *et al.* suggested that this type of activity could explain some of the symptoms of metaldehyde poisoning seen in slugs after ingestion of baits, such as writhing activity and uncoordinated movement. Young and Wilkins (1989a) found that metaldehyde failed to inhibit slug acetylcholinesterase activity and concluded that metaldehyde was ineffective as a nerve poison. They did not, however, examine the effect of acetaldehyde on acetylcholinesterase activity.

There is evidence that different species of slug vary in their susceptibility to metaldehyde, reflecting the different susceptibilities of *D. reticulatum* and *A. hortensis* to acetaldehyde described by Henderson (1970). *Milax gagates* was found to be less susceptible to metaldehyde than *D. reticulatum* by Ruppel (1959) although baits and sprays containing carbaryl mixed with metaldehyde were equally effective against both species. Webley (1962) found that *A. hortensis* was not attracted to metaldehyde to the same extent as *D. reticulatum* and *T. budapestensis*.

12.4 METHIOCARB

The common name methiocarb is accepted by BSI and ISO for 4-(methylthio)-3, 5-xylyl methylcarbamate. In Germany it has also been known as mercaptodimethur. Methiocarb was introduced in 1965 by Bayer AG under the trade name Mesurol and is a white odourless crystalline powder (m.p. 121.5°C), insoluble in water and soluble in many organic solvents. It is stable under normal conditions up to about 120–150°C. Further details of the structure and properties of methiocarb can be found in Worthing (1987) and Martin and Forrest (1969). Methiocarb is currently approved for use in Britain as a bait in pellets generally containing 4% a.i. together with cereal (Ivens, 1989). Methiocarb was first marketed in Britain by Bayer under the tradename Draza and the subsequent development of Draza for slug control was fully described by Martin and Forrest (1969) and Kelly and Martin (1989), including wildlife studies on the molluscicide. Getzin (1965) showed that methiocarb baits containing 3 and 4% a.i. were more effective than 1 and 2% baits and they provided economic control for *D. reticulatum*. However, Byers and Bierlein (1984) obtained increased yields of alfalfa using 2% methiocarb bait. Methiocarb has been applied in the form of a spray for slug control (e.g. Charlton, 1978). Methiocarb pellets are generally used for surface overall application or for admixture with seed at drilling. The latter use applies mainly to cereals, direct-drilled brassicas including oilseed rape and ryegrass and clover leys.

In slugs, methiocarb produces marked paralysis and/or loss of muscle tone but does not cause the excessive mucus production that is characteristic of metaldehyde poisoning. The pathway of a ^{14}C-labelled carbamate through the digestive tract of *D. reticulatum* was followed by Triebskorn *et al.* (1990), using autoradiography and scintillation counting. The carbamate was absorbed by the cells of the oesophagus and crop and was rapidly transported by the blood to the periphery of the body. It then re-entered the cells of the digestive tract and digestive gland from the haemocoele. Triebskorn *et al.* suggested that digestive cells of the digestive gland might be involved in detoxification and that connective tissue cells might be major storage sites for the carbamate. Secretion in mucus and excretion in faeces were likely routes for ^{14}C elimination.

Methiocarb acts as a stomach poison and, in common with other carbamate insecticides, inhibits insect cholinesterase. The cholinesterases of *L. maximus* were described by Pessah and Sokolove (1983a). They then investigated the interaction of cholinesterases from the foot muscle and blood with several organophosphates and carbamates, including methiocarb (Pessah and Sokolove, 1983b). Two types of cholinesterase were found, one an acetylcholinesterase type enzyme in the foot and circumoesophageal ganglion and the second a propionylcholinesterase type in the blood. Pessah and Sokolove concluded that the molluscicidal activity of carbamates in *L. maximus* could be attributed, at least in part, to the differential inhibition of cholinesterases found in tissues of the animal. The antiesterase activity of carbaryl was also demonstrated by Bowen and Lloyd (1971).

Young and Wilkins (1989a) described the effect of methiocarb on the acetylcholinesterase of three slug species *D. reticulatum*, *T. budapestensis* and *A. distinctus*. All three slug species had acetylcholinesterase of relatively low sensitivity to methiocarb when compared with insects and the earthworm, *L. terrestris*. There was evidence that five isoenzymes were included in the *D. reticulatum* enzyme and that the sensitivity of individual isoenzymes to inhibition by methiocarb varied. Slightly less susceptible acetylcholinesterase was present in those individuals of *D. reticulatum* that survived methiocarb poisoning. Young and Wilkins (1989a) concluded that, if the acetylcholinesterase isoenzymes in slugs were genetically based, then acetylcholinesterase sensitivity could be a limiting factor in the control of slug pests because of the possibility that resistant strains might develop. Although the selection pressure exerted by current slug control methods was not high enough to lead to the development of resistance in the field, the use of more effective methods could result in resistance developing. Young and Wilkins (1989b) described a technique for comparing the contact toxicity of molluscicides, including methiocarb, to slugs and estimated the LD_{50} of methiocarb for *D. reticulatum* as 0.32 mg g^{-1}. Bourne *et al.* (1988) estimated the LD_{50} value of methiocarb to be approximately 130–140 μg per adult *D. reticulatum*, when injested as a bait.

12.5 APPLICATION OF MOLLUSCICIDES

12.5.1 Seed dressings

Gould (1962b) carried out laboratory and field trials using seed dressings of copper salts and metaldehyde at a rate of about 1% a.i. wt^{-1} of seed. Thirty-one compounds were tested against *D. reticulatum* and ten of these were also tested against *A. hortensis* as seed treatments on winter wheat as an alternative to slug pellets for the control of slugs in cereals (Scott *et al.*, 1977). The compounds were formulated as dusts with talc and stuck to the seeds with 3% methyl cellulose to give 0.2% a.i. wt^{-1} of seed. Thiocarboxime and methiocarb reduced feeding and killed more than half of the slugs. However, thiocarboxime was considered to be too hazardous for use as a seed treatment. Metaldehyde was considerably less effective than methiocarb as a seed treatment. These compounds had no serious effect on germination. Scott *et al.* (1977) also found that three copper salts tested were initially effective in preventing seed damage, supporting Gould's (1962b) suggestion that some copper seed dressings were mild repellents but were unlikely to kill slugs. In further laboratory tests of new compounds as seed treatments, Scott *et al.* (1984) found that methiocarb, at 0.1% a.i. wt^{-1} of seed was still the best seed treatment, increasing the yield of winter wheat by 50% compared with untreated controls. Several seed treatments, including methiocarb, were more effective than methiocarb pellets mixed with the seed. Charlton (1978) showed that a seed coating of methiocarb (methiocarb 75% a.i. with 40% gum arabic) on legume seeds increased seedling survival in oversown legumes when exposed to *D. reticulatum* and *A. hortensis*. Moens (1983) showed that damage by *D. reticulatum* to winter wheat could be reduced by dressing the grain with a 50% methiocarb dust (0.7% a.i. wt^{-1} of seed). A reduction could also be achieved by distributing methiocarb pellets in the rows with the seed. These techniques were only effective when grains or pellets were exposed on the surface or in soil spaces. Glen *et al.* (1984) showed that while slug pellets had little effect on high populations of slugs present on plots with straw residues, and did not reduce damage to seeds and seedlings, slug damage was reduced when methiocarb dressed seed was used. However, Barker *et al.* (1984) found that a seed coating of methiocarb (0.3% a.i. wt^{-1} of seed) was phytotoxic to maize seed in a direct-drilled crop. Although methiocarb seed dressings are not used to protect seeds from slug attack, methiocarb pellets are recommended for use on cereals, oilseed rape and ryegrass for admixture with seed at drilling.

12.5.2 Feeding behaviour

The control of slugs by poison baits requires pellets which are attractive enough to initiate feeding and palatable enough to ensure that a lethal amount is consumed. For this reason the feeding behaviour of slugs with respect to slug pellets has been investigated by several authors. Crowell (1967) found in a laboratory screening test that, for some poisons, a bait with a concentration of 2% a.i. killed more slugs than a 4% a.i. bait. He concluded that the effectiveness of a given carbamate bait was determined largely by the amount eaten before the slug became paralysed. There was a balance between the concentration that would produce a sublethal dose and that which was sufficiently high to be a repellent. The consumption of bait by *D. reticulatum* and *T. budapestensis* was measured by Wright and Williams (1980), using baits containing chromic oxide as a marker. The amount of chromium in a slug and in any faeces produced by the slug after feeding on the bait provided a measure of the amount of bait consumed. Wright and Williams (1980) found the main effect of adding metaldehyde, methiocarb or another molluscicide to the bait was to reduce the amount of bait eaten at all concentrations tested, i.e. 1–6%. Increasing the amount of poison from 1 to 6% resulted in a progressive reduction in the amount of bait eaten, with *D. reticulatum* being more sensitive than *T. budapestensis*. A higher concentration of molluscicide meant that slugs eating the bait received a higher dose of poison. Wright and Williams suggested that, as bait was eaten, the poison was absorbed and exerted its paralytic effect on the gut wall, stopping the animal from feeding. Thus, if a molluscicide could be presented in a form that was not so rapidly absorbed, it might be more effective.

These observations were confirmed by Wedgwood and Bailey (1988), who analysed single meals on maize flour pellets containing six concentrations of metaldehyde from 0.5 to 8% a.i. together with a control with no metaldehyde. Three species of slug, *D. reticulatum*, *D. caruanae* and *A. hortensis*, were used in the experiments. The presence of metaldehyde in the bait reduced bite rates, while meal length and number of bites per meal were markedly inhibited by metaldehyde. Bite size was reduced by about 30% in meals on 6% pellets compared with 0% (control). They suggested that at higher concentrations of metaldehyde, aversion might be more important than toxic effects in ending the meal and this could result in sublethal amounts of poison being ingested. The elimination of faeces was also delayed after metaldehyde meals and this would contribute to mortality. The effect of sublethal doses of methiocarb and metaldehyde on fecundity and egg viability of *D. reticulatum* was investigated by Kemp and Newell (1989). Neither molluscicide caused a loss of fecundity and some slugs laid eggs within two weeks of the toxic meal. The size of egg batches from slugs which had received sublethal doses of metaldehyde were smaller than normal but more batches may have been laid by these slugs. Kemp (1987) found that methiocarb-poisoned slugs were less fertile than those that were poisoned by metaldehyde and that methiocarb gave better control of neonate slugs.

Frain and Newell (1982) suggested that improvements in bait consumption might be achieved by the incorporation of various attractants in the food and that variation in the flavour of the attractant within a commercial bait might improve numbers of slugs killed. Pellets containing methiocarb were more effective against *D. reticulatum* and *A. hortensis* than metaldehyde pellets in laboratory trials, in the absence of an alternative food (Airey, 1986b). However, when leaf discs of potato were present methiocarb was no more effective than metaldehyde against *D. reticulatum* although leaf discs did not interfere with the effectiveness of methiocarb against *A. hortensis*. Airey concluded that *A. hortensis* was more susceptible to methiocarb pellets than *D. reticulatum* but that there was no difference between the species with respect to metaldehyde. Bailey *et al.* (1989) suggested that both metaldehyde and methiocarb shorten meals by interfering with the neural control of feeding, while metaldehyde also probably disabled the muscles involved in feeding. While slugs would accept a non-toxic pellet after a methiocarb meal, they rarely accepted a pellet after a metaldehyde meal.

12.5.3 Attractants for baits

The previous section described how some molluscicides were repellent to slugs and several investigations have attempted to identify attractant substances which might be added to baits to improve the catch. Gimingham and Newton (1937) and Jary and Austin (1937) showed that tablets of Meta-fuel were attractive to slugs without any additive such as bran. Barnes and Weil (1942) reviewed earlier attempts to find an alternative carrier to bran for metaldehyde-based slug baits. They also tested a number of substances for this purpose and concluded that baits containing foodstuffs were better than baits of metaldehyde alone or with non-foodstuffs such as soil or sand. Barnes and Weil confirmed previous observations that the majority of slugs were caught at baits during the first two or three nights, with most slugs being caught on the first night's exposure of the baits. Numbers trapped were usually insignificant after the fourth or fifth night. However, Webley (1964) described the effective life of metaldehyde baits as being three weeks for *D. reticulatum* and two weeks for other species. The effectiveness of commercial and laboratory prepared metaldehyde and methiocarb baits on wheat bran or on plastic foam was increased when these were moistened with beer (Smith and Boswell, 1970). Since pressed yeast in beer was found to be an attractant for slugs (Section 7.1.4), Frain (1981) compared several metaldehyde baits, including one with added yeast, in a field trial to see whether metaldehyde-based baits could be made more attractive to slugs. The results were, however, inconclusive.

12.5.4 Behaviour with respect to baits in the field

The size of slug pellets will affect their distribution. Webley (1970) investigated the effects of distribution and numbers of slug pellets on the catch of slugs. He found that ten pellets per $0.75 \, m^{-2}$ gave the greatest catch, although in economic terms the best treatment was 2–5 pellets for *D. reticulatum*. The original pellets were relatively large but Hunter and Symonds (1970) showed that a given weight of small pellets presented more killing points per unit area to a slug than larger pellets. They found that the optimum density of pellets was between 25 and $100 \, m^{-2}$, i.e. between 10 and 20 cm apart. Most baits are now presented in the form of small (mini) pellets to ensure a more effective distribution. The behaviour of slugs with respect to baits was recorded by Howling and Port (1989) using time-lapse video techniques. Commercial metaldehyde and methiocarb formulations were used at manufacturers' recommended rates. Howling and Port showed that the time spent foraging before a pellet was contacted did not differ between metaldehyde and methiocarb although the time spent feeding on pellets was reduced in formulations containing metaldehyde. This recalls the observations of several authors on the feeding behaviour of slugs at baits reported in Section 12.5.2. However, this reduction in feeding was related to the presence or absence of metaldehyde rather than the concentration. Symptoms of poisoning appeared more rapidly after slugs ceased feeding on pellets containing metaldehyde than on methiocarb pellets. Thus slugs travelled greater distances from methiocarb baits than from metaldehyde baits. Commercial baits contain various additives including animal repellents, colouring, antifungal agents and a binder to hold the pellet together. Early formulations tended to disintegrate rapidly under moist conditions. Modern pellet formulations tend to be more resistant to weathering and Hogan (1985) found that commercial methiocarb baits remained attractive to slugs after 30 days' exposure in the field. Stephenson (1972) described how gelatin could replace the fibrous fraction of wheat bran in the formulation of slug baits although an aqueous bran extract was still necessary to act as a phagostimulant. When this gelatin was hardened with formaldehyde, the durability of the bait in wet weather was considerably increased although the toxicant remained effective.

12.5.5 Timing of control measures

The timing of slug control treatments is important as slugs need to be actively feeding at the soil surface for pellets to be effective. The factors regulating slug locomotor and feeding activity have already been discussed (Section 7.4) and it is clear that activity is determined by the interaction of endogenous rhythms with a complex of physical factors (Rollo, 1982; Cook and Ford, 1989). Damage forecasting (see Section 11.2.6) has been used to predict the level of slug damage likely to occur in a crop. Webley (1964) showed that the two most important factors affecting the numbers of slugs attracted by metaldehyde baits were the length of time the baits had been exposed and the average night air temperature. The relative humidity, when considered alone, had little effect on numbers trapped. Rogers-Lewis (1977a) suggested that slug pellets should be broadcast at the first opportunity after about mid-July as otherwise there might not be another chance before damage started in the autumn. This was also late enough to catch the spring-hatched slugs. Although slug damage to potatoes could be reduced by applying baits in the spring, before or soon after planting potatoes, Rayner (1975) described how better control could be achieved if applications were made between July and September, although this was not always adequate. Rayner et al. (1978) confirmed by further field trials that a useful, although not wholly adequate, degree of control of slugs in maincrop potatoes could be achieved with one application of methiocarb pellets (5.6 kg ha^{-1}) to the growing crop from mid-July. Slightly improved control was achieved by increasing the rate of application to 11.2 kg ha^{-1} and by a second application later in the season. Rayner et al. (1978) also found that measurements of rainfall and slug activity, estimated from bait trapping, did not provide a basis for the timing of control measures, although bait trapping did give an indication of the level of crop damage that might be expected. In some instances, for example on winter wheat when slug activity is at a high level, it may be necessary to make three applications of pellets: the first, pre-drilling to reduce slug numbers, the second, using bait mixed with seed at drilling to avoid grain hollowing and, the third, post-drilling to prevent grazing of the emerging seedlings.

12.6 EFFICIENCY OF METALDEHYDE AND METHIOCARB

12.6.1 Factors involved in testing baits

The effect of surface slug movement on the results of trials on slug control in potatoes was investigated by Rogers-Lewis (1977b), who concluded that surface slug movement was unlikely to affect the results of such trials. Frain and Newell (1983) suggested that counting dead slugs in field trials as a means of assessing the effectiveness of different molluscicides was likely to give misleading results. Slugs poisoned by metaldehyde tended to crawl away and die elsewhere while methiocarb-killed slugs tended to remain in the vicinity of the baits. Frain and Newell suggested that, in field trials, instead of counting dead slugs, the residual population of slugs remaining after molluscicide treatment should be estimated.

12.6.2 Recovery from metaldehyde poisoning

Webley (1964) reported that in tests with metaldehyde baits, some slugs that had been immobilized, recovered later and moved away. Ricou (1964) also found, in wet weather, that some slugs poisoned by metaldehyde were able to rehydrate in contact with moisture on plants and so recovered. Since slugs poisoned by metaldehyde tend to crawl away and die elsewhere (Frain and Newell, 1983), it seems likely that some of these may find shelter and recover from the poison. In laboratory experiments, a number of slugs initially poisoned by metaldehyde were found to recover during the following 48-hour period (Crawford-Sidebotham, 1970). Bourne *et al.* (1988) also confirmed that, under conditions of high humidity and constant temperature, more metaldehyde-poisoned slugs recovered compared to those poisoned with methiocarb. Briggs and Henderson (1987) described recovery rates of 80% with metaldehyde-poisoned *D. reticulatum* when these were removed from the field to more favourable conditions. However, 50% of methiocarb-poisoned slugs also recovered under these conditions.

The requirements of a slug poison were discussed by Briggs and Henderson (1987). The choice of route of entry into the slug, i.e. via the gut or through the skin, imposed different chemical requirements. In the case of contact poisons, the chemical needed initially to dissolve in the mucous layer before becoming available for uptake through the integument and the foot of the animal, and this depended upon the lipophilicity of the compound. Since lipophilicity was inversely related to water solubility, there were conflicting requirements. In wet soils, aqueous concentrations of chemicals would be highest for more soluble, less lipophilic materials although such materials would show a greater tendency for leaching under wet conditions, resulting in poor persistence and increased risk of pollution. In the case of stomach poisons, the effective life of baits was usually limited by the degradation of the carrier, usually bran, and persistence was less important. The main requirement for the slug poison incorporated in a bait was a lack of repellency. Both metaldehyde and methiocarb showed comparatively low toxicity and took several days to kill slugs. Increasing the proportion of active ingredient in baits decreased its consumption (Section 12.5.2). Briggs and Henderson (1987) considered that the effectiveness of metaldehyde and methiocarb as bait poisons was due more to their lack of repellency and slow speed of action than to their molluscicidal activity, which was low compared with many other pesticides. Since relatively high recovery rates had been recorded for both compounds, their success depended partly on their causing death from desiccation while the slug was immobilized on the soil surface.

12.6.3 Laboratory experiments

A number of laboratory comparisons have been made of the efficiencies of metaldehyde and methiocarb as slug poisons. Results of a number of these experiments are summarized in Table 12.1. Laboratory comparisons tend to support the idea that methiocarb is a more effective molluscicide than metaldehyde. However, Kemp and Newell (1985) and Kemp (1987) showed that there was no significant difference in the total mortality when poisoned

Table 12.1 Comparison of metaldehyde and methiocarb as slug poisons: laboratory experiments.

Method	Mode of assessment	Species	a.i. (%)	Mortality % (metaldehyde/ methiocarb)	Reference
Metaldehyde or methiocarb mixed with loose bran. Soil-filled wooden boxes	Mortality over 14-day period	A. ater	4 2	20/85 24/69	Crowell (1967)
Discs of carrot soaked in suspension of metaldehyde or methiocarb in water	Mortality after 7-day period	D. reticulatum D. laeve	–	100/80 */80	Judge (1969)
Metaldehyde or methiocarb spray applied to young peas in trays of compost	Mortality after 7-day period	D. reticulatum	1	65/68	Judge (1969)

Metaldehyde/ loose bran (10% a.i.) or methiocarb pellets (4% a.i.) in soil-filled cage	Mortality after 48-h recovery period	*T. souerbyi*	27/58	Crawford-Sidebotham (1970)
		T. budapestensis –	26/32	
		D. reticulatum	22/49	
		D. caruanae	45/55	
		A. hortensis	12/40	
		A. subfuscus	15/60	
		A. fasciatus	28/55	
		A. ater	18/49	
Metaldehyde (6% a.i.) or methiocarb (4% a.i.)	(a) Mortality after first meal + 7-day recovery period	*D. reticulatum*	18/46	Kemp and Newell (1985)
	(b) Mortality of survivors after second meal + 7-day recovery period	*D. reticulatum*	52/31	
	(c) Total mortality	*D. reticulatum*	61/62	

*Not tested

slugs which had recovered were offered the bait second time (Table 12.1). In most instances slugs poisoned by either metaldehyde or methiocarb accepted a second batch of the same bait, in contrast to the findings of Bailey *et al.* (1989). Kemp and Newell (1985) also found that sexually immature *D. reticulatum* appeared less susceptible to poisoning than sexually mature adults. Bourne *et al.* (1988) described how the symptoms of poisoning in *D. reticulatum* were different. Methiocarb-poisoned slugs gained weight and appeared bloated, while metaldehyde-poisoned slugs lost weight and adopted a darker, shrunken appearance. They attributed these differences to the different modes of action of the two poisons, with metaldehyde-poisoned slugs becoming dehydrated and shrinking while slugs poisoned by methiocarb appeared to absorb moisture and appeared bloated. A trend towards recovery over a 14-day period was recorded with both metaldehyde- and methiocarb-poisoned slugs and Bourne *et al.* concluded from the laboratory comparison that there seemed to be little difference in the efficacy of the two compounds against *D. reticulatum*. However, Bourne *et al.* (1990) suggested that the differing modes of action of the two molluscicides provided some advantages for metaldehyde during warm weather and for methiocarb during cool, moist weather. These different modes of action were exploited by combining the two active ingredients in baits when it was found that the total concentration of active ingredients could be reduced to 3% without adversely affecting the efficacy of the bait formulation.

The effects of carbamate and of metaldehyde poisoning on mucous cells in the skin and digestive tract of *D. reticulatum* were described by Triebskorn and Ebert (1989) and Triebskorn (1989). After carbamate application, the damage to mucous cells was less severe than after metaldehyde although carbamates had a greater impact generally on the epithelia of the skin and digestive tract, and this was primarily attributed to nuclear damage. Carbamates caused a deformation of the mucous cells while metaldehyde induced the total destruction of these cells, particularly the immature mucous cells, resulting in a fatal mucus deficiency. Metaldehyde caused an increased secretion of mucus immediately after application, which might serve to dilute the toxin and pass it through the digestive tract, but the loss of fluid through intensified mucus production in the gut and on the skin surface resulted in desiccation which could only be compensated for in a humid environment. Mucus production in the digestive tract was increased after carbamate poisoning, with increased secretion of acid mucopolysaccharides which might contribute to the detoxification of the carbamate. Triebskorn and Ebert (1989) emphasized the importance of mucous cells in the reaction of slugs to molluscicides and suggested that these cells might be regarded as potential targets in the development of new specific molluscicides. In a study of the effects on the crop of *D. reticulatum* after ingestion of baits containing 4% w/w metaldehyde or 4% w/w methiocarb, Bourne *et al.* (1991) found that damage to the surface of the cells became apparent only 24 hours after ingestion. Damage to the surface membrane of the crop epithelial cells by metaldehyde was more severe than by methiocarb. However, methiocarb appeared to have an effect on the osmotic balance of the slug resulting in the formation of large intracellular spaces within the cytoplasm of the cells and swelling between the basal muscle layers of the crop.

12.6.4 Field trials

Laboratory tests have tended to measure the direct poisoning effect under controlled conditions while field trials are exposed to a range of rather variable factors, including weather, methods of application and availability of alternative food. A number of comparisons of metaldehyde and methiocarb were made by Martin and Forrest (1969) during development work with Draza (methiocarb) pellets. They concluded that Draza trapped and killed significantly more slugs than metaldehyde. Port and Port (1986) summarized the results of a number of field trials comparing methiocarb with metaldehyde and concluded that results were rarely decisive. Where differences occurred, methiocarb was generally rated as more effective than metaldehyde. However, problems were frequently encountered in assessing the mortality due to the treatment, such as those discussed in Section 12.6.1.

12.7 ENVIRONMENTAL EFFECTS OF MOLLUSCICIDES

12.7.1 Metaldehyde

Several writers, e.g. Gimingham (1940), have described how metaldehyde baits were ineffective against soil insects. Büchs *et al.* (1989) showed that when metaldehyde baits were tested, they had little effect on non-target carabid beetles. Metaldehyde is toxic to vertebrates and, in the presence of gastric hydrochloric acid in the mammalian gut, it is decomposed to acetaldehyde. Since the rate of production of acetaldehyde exceeds its rate of oxidation, acetaldehyde accumulates to produce toxic blood levels. In mice, metaldehyde administered orally at a dose of 1 g kg^{-1} produced convulsions and death within two hours (Homeida and Cooke, 1982). Metaldehyde was originally incorporated with bran into fairly large pellets which were palatable to most domesticated animals, and instances of poisoning have been reported in several species. Homeida and Cooke (1982) listed several examples including dogs and cats, sheep and poultry. Longbottom and Gordon (1979) described an incident of metaldehyde poisoning in a dairy herd where a broken bag of slug pellets had been left in a field. Changes in the size of pellets and the use of repellents have reduced this danger.

Kitchen *et al.* (1978) described how there had been a dramatic decrease in instances of poisoning to dogs in California after criteria were established to make pellets less attractive to animals. There was evidence that baits were attractive to dogs, since they deliberately searched for slug bait, even after having been previously poisoned by bait (Studdert, 1985). While bait can be made less attractive to domestic animals and to wildlife, concern has been expressed about the effect on predators, particularly hedgehogs, of consuming slugs poisoned by metaldehyde or methiocarb. Although there is a great deal of anecdotal information, few systematic studies have been made of this problem. Published data for toxic doses of metaldehyde and methiocarb for slugs and for residues of these chemicals in poisoned slugs (see below) suggest that predators would need to consume relatively large numbers of poisoned slugs to ingest a toxic dose of metaldehyde or methiocarb. However, the accumulative effect of these chemicals in a predator may be important, although metaldehyde would be depolymerized fairly rapidly and the acetaldehyde formed would be metabolized. Levels of residues were estimated in citrus groves after the broadcast application of a 7.5% a.i. granular formulation of metaldehyde to control the snail, *H. aspersa*. After three days, rind samples showed a maximum residue level of 0.02 ppm while after ten days this had fallen to < 0.01 ppm, and similar levels were found in edible pulp samples.

12.7.2 Methiocarb

Tests with Draza (methiocarb pellets) have shown that small numbers of earthworms (< 5 %) were affected when pellets were applied for slug control, probably through accidental contact with pellets when earthworms came to the surface (Martin and Forrest, 1969). Symonds (1975) also showed that earthworms were at risk from methiocarb pellets. Bieri *et al.* (1989a) demonstrated by laboratory experiments, using slug pellets spread on the soil surface, that methiocarb caused a significantly higher mortality in *Lumbricus terrestris* Linnaeus than metaldehyde. None of the earthworms exposed to metaldehyde baits died during the experiment. Bieri *et al.* (1989b) discussed the possible long-term effects of this mortality on the structure of agricultural soils. However, the probability of an earthworm encountering a slug pellet in the field is considerably less than in a laboratory experiment and it is difficult to assess the importance of this effect on arable land. The application of Sevin to spruce forest in the USSR was found to reduce earthworm numbers and this indirectly affected mole populations. Sevin residues were reported from moles (*Talpa europea* Linnaeus) and the numbers of moles declined as the result of lack of food (Shilova *et al.*, 1971). Methiocarb is also an insecticide and Büchs *et al.* (1989) described how methiocarb baits were responsible for high mortality among non-target carabid beetles, which are among the most important groups of invertebrate predators on arable land. However, the effects of baits on beetles could be expected to be less in autumn, when the insects were less active. The later that baits were applied in spring, the greater would be the numbers of species affected.

Kelly and Curry (1985) examined the effect of methiocarb on the arthropods of a winter wheat crop. Methiocarb was applied for the control of slugs in winter wheat in the following ways:

(i)   as a seed dressing (1% a.i. wt^{-1} grain);
(ii) 4% methiocarb granules broadcast to the seed bed pre-sowing;
(iii) 4% methiocarb granules combine drilled with the seed.

Kelly and Curry concluded that single applications of methiocarb granules would not adversely affect the predatory and decomposer fauna. The effect of metaldehyde and methiocarb slug pellets on surface-living organisms in grassland was compared by Bieri *et al.* (1989). No adverse effects were recorded with metaldehyde pellets but methiocarb reduced numbers of staphilinids and smaller carabid beetles that were generally considered to be beneficial. Kelly and Curry (1985) described the half-life of methiocarb in the soil as from 4 to 56 days, depending on soil type. Martin and Forrest (1969) found that birds responded least to short wavelength colours (blue and violet) and feeding tests with caged birds and in gardens showed that birds were able to distinguish between blue-coloured Draza pellets and food pellets, and rejected the blue pellets. Eight-week-old pullets were reluctant to eat slugs poisoned with Draza. No adverse effects on wildlife, including birds, were reported from field trials where Draza was applied to wheat and barley fields, and Martin and Forrest (1969) concluded that Draza should not present a significant hazard to wildlife. Both methiocarb and metaldehyde pellets are now coloured blue to discourage birds from eating them. Nelson (in Martin and Forrest, 1969) found that Draza did not appear to attract dogs in the same way that metaldehyde baits attracted them (see Studdert, 1985). However, Jarvis (1988) found evidence that populations of the wood mouse, *Apodemus sylvaticus* Linnaeus, were reduced when methiocarb pellets were applied to the soil surface on arable land for slug control. Domestic and farm animals have been poisoned less frequently by methiocarb pellets than by metaldehyde pellets. Giles *et al.* (1984) described methiocarb poisoning in a sheep while Studdert (1985) reported that, in Australia, 43% of poisonings due to the consumption of slug baits by cats and dogs were due to methiocarb compared with 57% due to metaldehyde. Although mammal deaths may be reduced by the addition of repellents during the formulation of slug baits, such repellents may also make the pellets less attractive to slugs.

12.8 OTHER METHODS OF CONTROL

The introduction and subsequent spread of several European pest species of slug to other temperate countries and also of tropical veronicellid slugs in the southern USA (see Sections 11.4 and 11.5) demonstrate the capacity which these animals have for dispersal. Lists of pest slugs and snails introduced to the USA and to Canada (Godan, 1983) showed how most of these introductions occurred passively via imported plant materials. Godan also stressed that there was a general lack of regulations in Europe relating to plant inspection and quarantine, which had a specific bearing on gastropod pests although several species of snail were quarantine pests in the USA and other countries. The accidental importation of live arionid slugs in pre-packaged carrots from the USA (South, 1974), and the rapid spread of apparently introduced species, such as *Boettgerilla pallens*, emphasizes the need for efficient plant inspection and quarantine regulations relating specifically to slugs.

Biological control techniques, using natural enemies to control pest populations, have been used successfully against a number of invertebrate pests, although usually in conjunction with other compatible control methods (Samways, 1981). Known predators of slugs include several vertebrates and a larger number of invertebrates, particularly sciomyzid flies and carabid beetles (Section 9.2). The majority of these predators take a wide range of prey and so would be unsuitable for the control of slugs, although a few predators do feed almost exclusively on snails and slugs. Slugs are also intermediate hosts of several helminth and nematode parasites of mammals and birds (Section 9.3). It is difficult to assess the effect of predators or parasites on slug numbers but it seems unlikely that they exert a significant effect on natural slug populations (Section 10.2.3).

The reduced spatial heterogeneity of arable habitats may improve the prey-searching ability of predators. Dempster (1968) found that when Brussels sprouts were grown on the same plots for three successive years, significantly more *D. reticulatum* were found at the end of three years on plots that had received several applications of DDT (2 per year) compared with unsprayed plots. He suggested that this might be due to a reduction in numbers of natural invertebrate predators of the slug by DDT. Symondson (1989) showed that a significant reduction in both damage to lettuces and slug numbers could be achieved by using the predatory carabid beetle *A. parallelepipedus*, when the crop was grown in a polytunnel on plots delimited by plastic and other barriers. While this form of biological control might be applicable in a protected horticultural environment, it is difficult to envisage how carabid beetles or other insect predators could be used for the biological control of slugs on open arable sites where the use of barriers to prevent dispersal of predators would not be practicable. These insect predators would also be exposed to predation by birds and other predators. Although little is known about the diseases of slugs, both the microsporidian parasite *Microsporidian novacastriensis* and the ciliate *Tetrahymena rostrata* (Section 9.3.3) represent potential biological control agents.

Slug populations may be reduced by cultural control. Additional cultivations on heavy soils to give a finer tilth and firmer seed bed reduce slug damage, particularly in winter wheat (Section 11.2.6). Some crops, such as oilseed rape and cereals, encourage slugs and crop rotation may be used to give some degree of control whilst the destruction of crop residues also reduces slug activity. The use of resistant crop cultivars offers another means of slug control. Resistant cultivars of potatoes and winter wheat have been described (Sections 11.2.5 and 11.2.6) and some cultivars of oilseed rape are more susceptible to slug attack than others (Section 11.2.4). However, resistance to slug damage is not always compatible with market requirements (e.g. Beer, 1989; Glen *et al.*, 1989a). It may be concluded that, while biological, cultural and varietal control methods alone are probably unable to provide adequate control of slug pests, these techniques may make a useful contribution to a pest management programme when integrated with chemical methods of control.

Appendix

A.1 CULTURE METHODS

Many authors of papers dealing with the biology and life cycles of slugs under laboratory conditions have described techniques for the rearing of slugs (see especially Chapter 8). In addition, several papers have been written specifically on this subject. Sivik (1954) kept slugs in wooden trays covered with gauze, using soil as a substrate. Stephenson (1962) used glass crystallizing dishes half filled with a 5 parts loam/3 parts peat/1 part sharp sand mixture and covered by terylene net to rear several species of slug to maturity. A high humidity was maintained by stacking the dishes in a screw-topped jar containing a small amount of water. A technique was also described for rearing slugs individually. Since few *T. budapestensis* penetrated into the loam/peat/sand substrate, Stephenson (1966) tested media with a more open texture and found that coarser soils provided an acceptable substrate. Fine gravel or moistened filter paper were used as susbstrates for slugs and snails by Kingston (1966), with additional calcium provided in the form of blackboard chalk, while the use of vermiculite as a substrate was described by Gray *et al.* (1985). The diets used for other slug species in laboratory cultures have tended to consist mainly of natural or seminatural foods, although synthetic diets have been developed for slugs (e.g. Wright, 1973; Rueda, 1989). The use of both natural and synthetic diets is described in Section 3.1.8. The carnivorous slugs *T. scutulum* and *T. haliotidea* have been successfully reared in captivity using corked glass tubes of soil containing living earthworms as food (Barnes and Stokes, 1951; Stokes, 1958).

A.2 ANAESTHETIZING, NARCOTIZING AND PRESERVING SLUGS

Techniques for the anaesthetization of slugs, so that various experimental procedures can be carried out, have been described by several authors, particularly in experimental work associated with organ transplantation (Section 5.5.2) and the feeding motor programme (Section 6.7). In addition, several papers have been written specifically on this subject. Runham *et al.* (1965) evaluated several techniques that had been used for narcotizing and anaesthetizing gastropods. They found that anaesthesia could be induced in

A. ater by immersing the slug in a 0.5–1.0% propylene phenoxetol solution in water. Survival rates ranged from 70% for anaesthesia alone to between 8 and 50% after various internal operations. Successful anaesthesia was also obtained for mature *A. ater* by injecting them with a 10% solution of magnesium chloride. Survival rates ranged between 32 and 38% although very young *A. ater*, which recovered from anaesthesia, all died within 24 hours. A recovery rate of 100% was obtained when snails (*H. aspersa*) were anaesthetized by injection with 1 ml of a Ringer solution saturated with ether. However, the snails produced large quantities of mucus and retracted into their shells. A technique for the use of carbon dioxide as an anaesthetic for *D. reticulatum* was described by Bailey (1969). The recovery rate was found to be 100% provided that the exposure to carbon dioxide was not in excess of one hour. Gelperin *et al.* (1978) immersed slugs in ice for 15 minutes or longer, prior to dissecting out the brain, lip area and buccal mass for experimental work on the feeding motor programme in *L. maximus*.

Slugs may be preserved in 70–80% industrial methylated spirit or in a 4–10% formaldehyde solution. However, they generally contract when fixed in this way and their colours change. Several methods have been described for narcotizing slugs so that they become relaxed and do not contract when fixed. Slugs may be immersed in solutions such as Nembutal (sodium pentabarbitone), but they generally take many hours to become completely relaxed. Runham *et al.* (1965) described how a solution containing 0.08% Nembutal and 1% propylene phenoxetol relaxed slugs within a few minutes, even the larger species requiring no more than about 10 minutes. The animals were then fixed in formalin. It was necessary, however, to squeeze the head to express the retracted tentacles before fixation. De Winter (1985) found that immersion in dilute (25%) ethanol or methylated spirit provided a rapid method for the relaxation of slugs before fixation. Preservation of slugs in alcohol or formalin removes much of their colour and may distort their form. Freeze-dried specimens retain much of their form and colour and Crowell (1973) described a technique for freeze-drying slugs, which could then be coated with clear lacquer and mounted on entomological pins.

References

The bibliography by Lindquist *et al.* (1977) includes many of the earlier references to slugs. The book by Godan (1983) also contains an extensive bibliography of terrestrial slugs and snails.

Abeloos, M. (1942) Les étapes de la croissance chez la limace rouge (*Arion rufus*). *C.R. Acad. Sci., Paris*, **215**, 38–39.

Abeloos, M. (1943) Forme de la croissance des espèces du genre *Limax*. *C.R. Acad. Sci., Paris*, **217**, 159–161.

Abeloos, M. (1944) Recherches expérimentales sur la croissance. La croissance des mollusques Arionidés. *Bull. Biol. France et Belg.*, **78**, 215–256.

Adiyodi, K. G. and Adiyodi, R. G. (1983a) *Reproductive Biology of Invertebrates*. Vol. 1. *Oogenesis, Oviposition and Oosorption*, Wiley, Chichester.

Adiyodi, K. G. and Adiyodi, R. G. (1983b) *Reproductive Biology of Invertebrates*. Vol. 2. *Spermatogenesis and Sperm Function*, Wiley, Chichester.

Agrawal, H. P. (1977) Control of the slug *Anadenus altivagus* (Theobald). Damage to the potato crop by certain pesticides. *Food, Farm. Agric.*, **8**, 16–18.

Airey, W. J. (1984) The distribution of slug damage in a potato crop. *J. Mollusc. Stud.*, **50**, 239–240.

Airey, W. J. (1986a) Some aspects of the feeding behaviour of slugs in the potato crop. PhD thesis, Univ. of London.

Airey, W. J. (1986b) The influence of an alternative food on the effectiveness of proprietary molluscicidal pellets against two species of slugs. *J. Mollusc. Stud.*, **52**, 206–213.

Airey, W. J. (1987a) Laboratory studies on damage to potato tubers by slugs. *J. Mollusc. Stud.*, **53**, 97–104.

Airey, W. J. (1987b) The influence of food deprivation on the locomotor activity of slugs. *J. Mollusc. Stud.*, **53**, 37–45.

Airey, W. J. (1988a) The response of the grey field slug, *Deroceras reticulatum* (Müller), to potato root and tuber diffusates. *J. Mollusc. Stud.*, **54**, 132–135.

Airey, W. J. (1988b) The influence of an alternative food on tuber damage by slugs. *J. Mollusc. Stud.*, **54**, 131–132.

Airey, W. J., Henderson, I., Jones, A. and Selman, B. (1984) Slug damage and control. In *Potato Progress* (ed. M. Storey), Potato Marketing Board, London, pp. 16–17.

Airey, W. J., Henderson, I., Pickett, J. A., Scott, G. C., Stephenson, J. W. and Woodcock, C. M. (1989) In *Slugs and Snails in World Agriculture* (ed. I. Henderson), British Crop Protection Council, Thornton Heath, pp. 301–307.

Alonso, M. R. (1975) Terrestrial malacological fauna of the Granada depression (Spain): I. Slugs. *Cuad. Cien. Biol., Univ. Granada*, **4**, 71–88.

Alonso, M. R. and Ibanez, M. (1984) Slugs (Mollusca, Gastropoda) from Pla de Manlleu (Spain, Tarragona). *Misc. Zool.*, **8**, 35–40.

Altena, C. O. van R. and Smith, B. J. (1975) Notes on introduced slugs of the families Limacidae and Milacidae in Australia, with two new records. *J. Malacol. Soc. Australia*, **3**, 63–80.

Altieri, M. A., Hagen, K. S., Trujillo, J. and Caltagirone, L. E. (1982) Biological control of *Limax maximus* and *Helix aspersa* by indigenous predators in a daisy field in central coastal California. *Acta Oecol.*, **3**, 387–390.

Ancel, P. (1902) Histogénèse et structure de la gonade hermaphrodite d'*Helix pomatia*. *Arch. Biol. (Paris)*, **19**, 389–652.

Anderson, J. B. and McCracken, G. F. (1986) Breeding system and population genetic structure in philomycid slugs (Mollusca: Pulmonata). *Biol. J. Linn. Soc.*, **29**, 317–329.

Anderson, R. C. (1960) On the development and transmission of *Cosmocercoides dukae* of terrestrial molluscs in Ontario. *Canad. J. Zool.*, **38**, 801–825.

Anderson, R. and Norris, A. (1976) *Boettgerilla pallens* Simroth new to Ireland. *J. Conchol.*, **28**, 207–208.

Anderson, W. A. and Personne, P. (1970) The cytochemical localization of glycolytic and oxidative enzymes within mitochondria of spermatozoa of some pulmonate gastropods. *J. Histochem. Cytochem.*, **18**, 783–793.

Andrewartha, H. G. and Birch, L. C. (1954) *The Distribution and Abundance of Animals*, Univ. of Chicago Press, Chicago.

Andrews, K. L. (1989) Slug pests of dry beans in Central America. In *Slugs and Snails in World Agriculture* (ed. I. Henderson), British Crop Protection Council, Thornton Heath, pp. 85–89.

Angseesing, J. P. A. (1974) Selective eating of the acyanogenic form of *Trifolium repens. Heredity*, **32**, 73–83.

Angseesing, J. P. A. and Angseesing, W. J. (1973) Field observations on the cyanogenic polymorphism in *Trifolium repens. Heredity*, **31**, 276–282.

Angulo, E. and Moya, J. (1985) Subsurface cisterns in the nephrocytes of the slug, *Arion ater* (Mollusca, Pulmonata). *Arch. Anat. Microsc. Morphol. Exp.*, **74**, 107–110.

Angulo, E. and Moya, J. (1986) A basement membrane substructured in the kidney sac of arionid slugs (Mollusca: Pulmonata). *Cytologia (Tokyo)*, **51**, 749–752.

Angulo, E. and Moya, J. (1987) Tubular structures in the rough endoplasmic reticulum cisterns of the pedal gland cells of the slug *Arion ater* Linnaeus 1758 (Mollusca: Pulmonata). *Zool. Anzeiger*, **218**, 185–192.

Angulo, E. and Moya, J. (1989a) Electron microscopical studies on the renal concretions of the slug *Arion ater* (Linnaeus, 1758) (Mollusca, Pulmonata). *Zool. Anzeiger*, **222**, 273–280.

Angulo, E. and Moya, J. (1989b) A rough endoplasmic reticulum-rich (RER) cell in the postreproductive ovotestis of the slug, *Arion ater* (L). (Mollusca, Pulmonata). *Z. Mikrosk.-Anat. Forsch. (Leipzig)*, **103**, 1–7.

Angulo, E. and Moya, J. (1989c) The ultrastructure of the epicardium epithelium of the slug, *Arion ater* (Linnaeus, 1758) (Pulmonata, Stylommatophora). *Folia Histochem. Cytobiol.*, **27**, 113–119.

Anon. (1905) *Slugs and Snails*. Board of Agriculture and Fisheries Leaflet No. 132, London.

Anon. (1920a) *J. Conchol.*, **16**, 125.

Anon. (1920b) *J. Conchol.*, **16**, 125.

Anon. (1932) *Slugs and Snails*, Ministry of Agriculture, Fisheries and Food Advisory Leaflet No. 115, London.

Anon. (1968) *Slugs and Snails*, Ministry of Agriculture, Fisheries and Food Advisory Leaflet No. 115, London.

Anon. (1979) *Slugs and Snails*, Ministry of Agriculture, Fisheries and Food Advisory Leaflet No. 115, London.

Anon. (1986) The hit list of public enemies. *Farmers Weekly*, 23 May, p. 17.

Anon. (1988) Slugs go on rampage in record bad season. *Farmers Weekly*, 28 October, p. 46.

Anon. (1990) Feedback. *New Scientist*, 6 October, p. 64.

Arcadi, J. A. (1963) Some mucus producing cells of the garden slug, (*Lehmannia poirieri*). *Ann. N.Y. Acad. Sci.*, **106**, 451–457.

Arcadi, J. A. (1968) Tissue response to the injection of charcoal into the pulmonate gastropod, *Lehmannia poirieri*. *J. Invert. Pathol.*, **11**, 59–62.

Arias, R. O. and Crowell, H. H. (1963) A contribution to the biology of the gray garden slug. *Bull. S. California Acad. Sci.*, **62**, 83–97.

Arni, P. (1973) Vergleichende Untersuchungen an Schluepfstadien von neun Pulmonaten-Arten (Mollusca, Gastropoda). *Revue Suisse Zool.*, **80**, 323–402.

Arnold, S. J. (1981) Behavioural variation in natural populations: 2. The inheritance of a feeding response in crosses between geographic races of the garter snake, *Thamnophis elegans*. *Evolution*, **35**, 510–515.

Arvanitaki, A. and Cardot, H. (1932) Sur les variations de la concentration du milieu interior chez les mollusques térrestres. *J. Physiol.*, **30**, 577–592.

Ash, L. R. (1976) Observations on the role of mollusks and planarians in the transmission of *Angiostrongylus cantonensis* infection to man in New Caledonia. *Revista Biol. Tropical*, **24**, 163–174.

Asitinskaya, S. E. and Shulepova, T. M. (1977) [Ovostatic and ovicidal effects of some invertebrates on ascarid eggs]. *Med. Parazitol. Parazitar. Bolezni*, **46**, 231–233.

Atkin, J. C. (1979) Varietal susceptibility of potatoes to slug attack. PhD thesis. Univ. of Newcastle-upon-Tyne.

Atkinson, H. J., Gibson, N. H. E. and Evans, H. (1979) A study of common crop pests in allotment gardens around Leeds, Yorkshire, England, UK. *Plant Pathol.*, **28**, 169–177.

Atlavinyte, O., Kuginyte, Z. and Pileckis, S. (1974) Erosion effect on soil fauna under different crops. *Pedobiologia*, **14**, 35–40.

Babor, J. F. (1894) Über den Cyclus der Geschlechtentwicklung der Stylommato-phoren. *Verhandl. deut. Zool. Gessell.*, **4**, 55–61.

Babrakzai, N. and Miller, W. B. (1974) A colchicine hypotonic squash technique for chromosome spreads of pulmonate land snails. *Malacol. Rev.*, **7**, 37.

Babu, A. and Skowronska-Wendland, D. (1988) Histological and histochemical studies of the digestive system of the slug, *Deroceras reticulatum* (Müller) (Pulmonata). *Bull. Soc. Amis Sci. Lettres de Poznan, Ser. D, Sci. Biol.*, **26**, 65–72.

Babu, K. S., Venkatachari, S. A. T. and Muralikrishna Das, P. (1977)

Multipolar heteropolar giant neurons in the central nervous system of the slug, *Laevicaulis alte. J. Animal Morphol. Physiol.*, **24**, 292–301.

Babula, A. and Wielinska, L. (1988) Ultrastructural studies of the digestive system of the slug, *Deroceras reticulatum* (Müller) (Pulmonata). *Bull. Soc. Amis Sci. Lettres de Poznan, Ser. D, Sci. Biol.*, **26**, 73–78.

Bachrach, E. and Cardot, H. (1924) Développement des Limaces et des Limnées, à différent températures. *C.R. Soc. Biol. (Paris)*, **91**, 260.

Backeljau, T. (1985a) *Arion intermedius* Normand 1852 in southern Italy. *J. Conchol.*, **32**, 69–70.

Backeljau, T. (1985b) Estimation of genic similarity within and between *Arion hortensis sensu lato* and *Arion intermedius* by means of isoelectric focused esterase patterns in hepatopancreas homogenates (Mollusca, Pulmonata: Arionidae). *Z. Zool. System. Evolutionsforsch.*, **23**, 38–49.

Backeljau, T., Davies, S. M. and de Bruyn, L. (1988) An albumen gland protein polymorphism in the terrestrial slug *Arion owenii. Biochem. System. Ecol.*, **16**, 425–430.

Backeljau, T. and Marquet, R. (1985) An advantageous use of multivariate statistics in a biometrical study on the *Arion hortensis* complex (Pulmonata: Arionidae) in Belgium. *Malacol. Rev.*, **18**, 57–72.

Bailey, S. E. R. (1981) Circannual and circadian rhythms in the snail, *Helix aspersa* Müller and the photoperiodic control of annual activity and reproduction. *J. Comp. Physiol. A. Sensory, Neural Behav. Physiol.*, **142**, 89–94.

Bailey, S. E. R. (1989) Foraging behaviour of terrestrial gastropods: integrating field and laboratory studies. *J. Mollusc. Stud.*, **55**, 263–272.

Bailey, S. E. R., Cordon, S. and Hutchinson, S. (1989) Why don't slugs eat more bait? A behavioural study of early meal termination produced by methiocarb and metaldehyde baits in *Deroceras reticulatum*. In *Slugs and Snails in World Agriculture* (ed. I. Henderson), British Crop Protection Council, Thornton Heath, pp. 385–390.

Bailey, T. G. (1969) A new anaesthetic technique for slugs. *Experientia*, **25**, 1225.

Bailey, T. G. (1970) Studies on organ cultures of slug reproductive tracts. PhD thesis, Univ. of Wales.

Bailey, T. G. (1971a) Osmotic pressure and pH of slug haemolymph. *Comp. Biochem. Physiol. A. Comp. Physiol.*, **40**, 83–88.

Bailey, T. G. (1971b) Haemolymph proteins of the slug, *Agriolimax reticulatus. Comp. Biochem. Physiol. B. Comp. Biochem.*, **40**, 699–707.

Bailey, T. G. (1973) The *in vitro* culture of reproductive organs of the slug, *Agriolimax reticulatus. Netherlands J. Zool.*, **23**, 72–85.

Baker, C. B. M. and Waines, R. A. (1957) Wireworm and slug damage to ware potatoes 1954–1956. *Plant Pathol.*, **6**, 115–122.

Baker, R. A. (1970a) The food of *Riccardoella limacum* (Schrank) (Acari – Trombidiformes) and its relationship with pulmonate molluscs. *J. Nat. Hist.*, **4**, 521–530.

Baker, R. A. (1970b) Studies on the life-history of *Riccardoella limacum* (Schrank) (Acari – Trombidiformes). *J. Nat. Hist.*, **4**, 511–519.

Banta, P. A., Welsford, I. G. and Prior, D. J. (1990) Water-orientation behaviour in the terrestrial gastropod, *Limax maximus*: The effect of dehydration and arginine vasotocin. *Physiol. Zool.*, **63**, 683–696.

Barker, G. M. (1979) The introduced slugs of New Zealand (Gastropoda: Pulmonata). *N.Z. J. Zool.*, **6**, 411–438.

Barker, G. M. (1989a) Slug problems in New Zealand pastoral agriculture. In *Slugs and Snails in World Agriculture* (ed. I. Henderson), British Crop Protection Council, Thornton Heath, pp. 59–68.

Barker, G. M. (1989b) Flatworm predation of terrestrial molluscs in New Zealand and a brief review of previous records. *N.Z. Entomol.*, **12**, 75–79.

Barker, G. M. and McGhie, R. A. (1984) The biology of introduced slugs in New Zealand – *Limax maximus*. *N.Z. Entomol.*, **8**, 106–111.

Barker, G. M., Pottinger, R. P., McGhie, R. A. and Addison, P. J. (1984) Slug control in direct-drilled maize. *N.Z. J. Exp. Agric.*, **12**, 155–160.

Barnes, H. F. (1952) The absence of slugs in a garden and an experiment in re-stocking. *Proc. Zool. Soc. London*, **123**, 49–58.

Barnes, H. F. and Stokes, B. M. (1951) Marking and breeding *Testacella* slugs. *Ann. Appl. Biol.*, **38**, 540–545.

Barnes, H. F. and Weil, J. W. (1942) Baiting slugs using metaldehyde mixed with various substances. *Ann. Appl. Biol.*, **29**, 56–68.

Barnes, H. F. and Weil, J. W. (1944) Slugs in gardens: their numbers, activities and distribution. Part I. *J. Animal Ecol.*, **13**, 140–175.

Barnes, H. F. and Weil, J. W. (1945) Slugs in gardens: their numbers, activities and distribution. Part II. *J. Animal Ecol.*, **14**, 71–105.

Baronio, P. (1974) The insect enemies of gastropod mollusks. *Boll. Inst. Entomol. Univ. Studi Bologna*, **32**, 169–187.

Barr, R. A. (1926) Some observations on the pedal gland of *Milax*. *Qtly J. Microsc. Sci.*, **70**, 647–667.

Barr, R. A. (1928) Some notes on the mucous and skin glands of *Arion ater*. *Qtly J. Microsc. Sci.*, **71**, 503–526.

Barrantes, F. J. (1970) The neuromuscular junctions of a pulmonate mollusc: I. Ultrastructural study. *Z. Zellforsch. Mikrosk. Anat. Abteil. Histochem.*, **104**, 205–212.

Barratt, B. I. P., Byers, R. A. and Bierlein, D. L. (1989) Conservation tillage crop establishment in relation to density of the field slug (*Deroceras reticulatum* (Müller)). In *Slugs and Snails in World Agriculture* (ed. I. Henderson), British Crop Protection Council, Thornton Heath, pp. 93–100.

Barry, B. D. (1969) Evaluation of chemicals for control of slugs on field corn in Ohio. *J. Econ. Entomol.*, **62**, 1277–1279.

Barry, S. A. and Gelperin, A. (1982a) Exogenous choline augments transmission at an identified cholinergic synapse in the terrestrial mollusc, *Limax maximus*. *J. Neurobiol.*, **48**, 439–448.

Barry, S. A. and Gelperin, A. (1982b) Dietary choline augments blood choline and cholinergic transmission in the terrestrial mollusc, *Limax maximus*. *J. Neurobiol.*, **48**, 451–457.

Baunacke, W. (1913) Studien zur Frage nach der Statocystenfunktion. *Biol. Zentral.*, **33**, 427–452.

Bayford, E. G. (1906) *Drilus flavescens* Rossi, female and its larva. *Entomol. Monthly Mag.*, **42**, 267–268.

Bayne, C. J. (1966) Observations on the composition of the layers of the egg of *Agriolimax reticulatus*, the grey field slug (Pulmonata, Stylommatophora). *Comp. Biochem. Physiol.*, **19**, 317–338.

Bayne, C. J. (1967a) A study of the layers investing the gastropod egg. PhD thesis, Univ. of Wales.

Bayne, C. J. (1967b) Studies on the composition of extracts of the reproductive glands of *Agriolimax reticulatus*, the grey field slug (Pulmonata, Stylommatophora). *Comp. Biochem. Physiol.*, **23**, 761–773.

Bayne, C. J. (1968a) A study of the desiccation of egg capsules of eight gastropod species. *J. Zool. (London)*, **155**, 401–411.

Bayne, C. J. (1968b) Molluscan organ culture. *Malacol. Rev.*, **1**, 125–135.

Bayne, C. J. (1968c) Histochemical studies on the egg capsules of eight gastropod molluscs. *Proc. Malacol. Soc. London*, **38**, 199–212.

Bayne, C. J. (1969) Survival of the embryos of the grey field slug *Agriolimax reticulatus*, following desiccation of the egg. *Malacologia*, **9**, 391–401.

Bayne, C. J. (1970) Organization of the spermatozoon of *Agriolimax reticulatus*, the grey field slug (Pulmonata, Stylommatophora). *Z. Zellforsch. Mikrosk. Anat.*, **103**, 75–89.

Beaver, R. A. (1972) Ecological studies on Diptera breeding in dead snails. *The Entomologist*, **105**, 41–52.

Beer, G. J. (1989) Levels and economics of slug damage in potato crops 1987 and 1988. In *Slugs and Snails in World Agriculture* (ed. I. Henderson), British Crop Protection Council, Thornton Heath, pp. 101–106.

Beeson, G. E. (1960) Chromosome numbers of slugs. *Nature (London)*, **186**, 257–258.

Beiswanger, C. M., Sokolove, P. G. and Prior, D. J. (1981) Extraocular photo-entrainment of the circadian locomotor rhythm of the garden slug *Limax*. *J. Exp. Zool.*, **216**, 13–24.

Beltz, B. and Gelperin, A. (1979) An ultrastructural analysis of the salivary system of the terrestrial mollusc. *Tissue & Cell*, **11**, 31–50.

Beltz, B. and Gelperin, A. (1980a) Mechanisms of peripheral modulation of salivary burster in *Limax maximus*: A presumptive sensorimotor neuron. *J. Neurophysiol. (Bethesda)*, **44**, 675–686.

Beltz, B. and Gelperin, A. (1980b) Mechanosensory input modulates activity of an autoactive bursting neuron in *Limax maximus*. *J. Neurophysiol. (Bethesda)*, **44**, 665–674.

Beresford-Jones, W. P. (1966) Observations on *Muellerius capillaris* (Müller, 1889), Comeron, 1927. II. Experimental infection of mice, guinea-pigs and rabbits with third stage larvae. *Res. Vet. Sci.*, **7**, 287–291.

Berg, C. O. (1964) Snail control in trematode diseases: the possible value of Sciomyzid larvae snail-killing Diptera. *Adv. Parasitol.*, **2**, 259–307.

Bertrand, H.W.R. (1941) Prevention of damage to young rubber by snails and slugs. *Trop. Agric. (Colombo)*, **97**, 327.

Bett, J. A. (1960) The breeding seasons of slugs in gardens. *Proc. Zool. Soc. London*, **135**, 559–568.

Beyer, W. N. and Saari, D. M. (1977) Effect of tree species on the distribution of slugs. *J. Animal Ecol.*, **46**, 697–702.

Beyer, W. N. and Saari, D. M. (1978) Activity and ecological distribution of the slug *Arion subfuscus* (Stylommatophora: Arionidae). *Amer. Mid. Nat.*, **100**, 359–367.

Biedermann, W. (1906) Studien zur vergleichenden Physiologie der peristaltischen Bewegungfen. III. Die Innervation der Schneckensohle. *Pfluegers Arch. Gesamte Physiol.*, **111**, 251–297.

Bieri, M., Schweizer, H., Christensen, K. and Daniel, O. (1989a) The effect of metaldehyde and methiocarb slug pellets on *Lumbricus terrestris* Linne. In *Slugs and Snails in World Agriculture* (ed. I. Henderson), British Crop Protection Council, Thornton Heath, pp. 237–244.

Bieri, M., Schweizer, H., Christensen, K. and Daniel, O. (1989b) The effect of metaldehyde and methiocarb slug pellets on surface dwelling organisms in grassland. In *Slugs and Snails in World Agriculture* (ed. I. Henderson), British Crop Protection Council, Thornton Heath, pp. 391–392.

Binot, D. and Chétail, M. (1968a) Les mucocytes du manteau d'*Arion rufus* (L.) et leur développement: Histologie et histochimie. *Ann. Histochim.*, **13**, 73–82.

Binot, D. and Chétail, M. (1968b) Histogénèse des mucocytes de la glande et de la sole pédieuses d'*Arion rufus* (Stylommatophora: Arionidae). *Malacol. Rev.*, **1**, 91–102.

Bishop, J. A. and Korn, M. E. (1969) Natural selection and cyanogenesis in white clover, *Trifolium repens. Heredity*, **24**, 423–430.

Bishop, M. J. (1976) The terrestrial molluscs of the Province of Novara. *Atti Soc. Italiana Sci. Nat. Museo Civico Storia Nat. Milano*, **117**, 265–299.

Bishop, M. J. (1977a) The mollusca of acid woodland in West Cork and Kerry. *Proc. Roy. Irish Acad., Sect. B – Biol., Geol. Chem. Sci.*, **77**, 227–244.

Bishop, M. J. (1977b) The habitats of Mollusca in the Central Highlands of Scotland. *J. Conchol.*, **29**, 189–197.

Bishop, M. J. (1980) The Mollusca of acid woodland in an Italian province of Novara. *J. Conchol.*, **30**, 181–188.

Bishop, M. J. (1981) Quantitative studies on some living British wetland mollusc faunas. *Biol. J. Linn. Soc.*, **15**, 299–326.

Blackman, G. E. (1942) Statistical and ecological studies in the distribution of species in plant communities I. Dispersion as a factor in the study of changes in plant populations. *Ann. Bot. (London)*, **6**, 35–70.

Blanchard, D. and Getz, L. L. (1979) *Arion subfuscus* in southeastern Michigan. *Nautilus*, **93**, 36–37.

Blank, R. H. and Bell, D. S. (1982) Seasonal patterns of pasture seed loss to black field crickets (*Teleogryllus commodus*) and other invertebrates. *N.Z. J. Agric. Res.*, **25**, 123–130.

Blinn, W. C. (1964) Water in the mantle cavity of land snails. *Physiol. Zool.*, **37**, 329–337.

Bliss, C. I. and Fisher, R. A. (1953) Fitting the negative binomial distribution to biological data. *Biometrics*, **9**, 176–200.

Boag, D. A. and Wishart, W. D. (1982) Distribution and abundance of terrestrial gastropods on a winter range of bighorn sheep (*Ovis canadensis*) in southwestern Alberta (Canada). *Canad. J. Zool.*, **60**, 2633–2640.

Boer, H. H. and Joosse, J. (1975) Endocrinology. In *Pulmonates*, Vol. 1 (eds V. Fretter and J. Peake), Academic Press, New York and London, pp. 245–308.

Boettger, C. R. (1956) Über einen Fall von pathologischer Geestaltreranderung dei einer Wegschnecke der Art *Arion ater* (L). *Biol. Zentral.*, **75**, 257–267.

Bonaventura, C. and Bonaventura, J. (1983) Respiratory pigments: Structure and function. In *The Mollusca*, Vol. 2 (ed. P. W. Hochachka), Academic Press, New York and London, pp. 1–50.

Borden, V. C. (1948) Ciliates from *Deroceras agrestis*. Master's thesis, Univ. of Virginia.

Bourne, N. B., Jones, G. W. and Bowen, I. D. (1988) Slug feeding behaviour in relation to control with molluscicidal baits. *J. Mollusc. Stud.*, **54**, 327–338.

Bourne, N. B., Jones, G. W. and Bowen, I. D. (1990) Feeding behaviour and mortality of the slug *Deroceras reticulatum* in relation to control with molluscicidal baits containing various combinations of metaldehyde with methiocarb. *Ann. Appl. Biol.*, **117**, 455–468.

Bourne, N. B., Jones, G. W. and Bowen, I. D. (1991) Endocytosis in the crop of the slug *Deroceras reticulatum* (Müller) and the effect of the ingested molluscicides, metaldehyde and methiocarb. *J. Mollusc. Stud.*, **57**, 71–80.

Bowen, I. D. (1970) The fine structural localization of acid phosphate in the gut epithelial cells of the slug, *Arion ater* (L.). *Protoplasma*, **70**, 247–260.

Bowen, I. D. and Davies, P. (1971) The fine structural distribution of acid phosphatase in the digestive gland of *Arion hortensis* (Fer.). *Protoplasma*, **73**, 73–81.

Bowen, I. D. and Lloyd, D. C. (1971) A technique for the electron cytochemical localization of the site of carbaryl metabolism. *J. Invert. Pathol.*, **18**, 183–190.

Boycott, A. E. (1934) The habitats of land mollusca in Britain. *J. Animal Ecol.*, **22**, 1–38.

Branson, B. A. (1976) *Testacella haliotidea* (Draparnaud) free living in Pennsylvania. *Sterkiana*, **62**, 20.

Brar, H. S. and Simwat, G. S. (1973) Control of the common slug, *Laevicaulis alte* Ferussac (Gastropoda) with certain chemicals. *J. Res. Punjab Agric. Univ.*, **10**, 99–101.

Bridgeford, H. B. and Pelluet, D. (1952) Induced changes in the cells of the ovotestis of the slug *Deroceras reticulatum* (Müller) with special reference to the nucleolus. *Canad. J. Zool.*, **30**, 323–337.

Briggs, G. G. and Henderson, I. F. (1987) Some factors affecting the toxicity of poisons to the slug *Deroceras reticulatum* Müller (Pulmonata: Limacidae). *Crop Protect.*, **6**, 341–346.

Bright, S. W. J., Lea, P. J., Kueh, J. S. H., Woodcock, C., Hollomon, D. W. and Scott, G. C. (1982) Proline content does not influence pest and disease susceptibility of barley. *Nature (London)*, **295**, 592–593.

Broadley, D. G. and Cock, E. V. (1975) *Snakes of Zimbabwe*, Longman Zimbabwe, Harare.

Brockie, R. E. (1959) Observations on the food of the hedgehog (*Erinaceus europaeus* L.) in New Zealand. *N. Z. J. Sci.*, **2**, 121–136.

Brooks, W. M. (1968) Tetrahymenid ciliates as parasites of the gray garden slug. *Hilgardia*, **39**, 205–276.

Brown, F. J. (1933) Life history of the fowl tapeworm *Davainea proglottina*. *Nature (London)*, **131**, 276–277.

Broyles, J. L. and Sokolove, P. G. (1978) Pedal wave recovery following transection of pedal nerves in the slug, *Limas maximus*. *J. Exp. Zool.*, **206**, 371–380.

Bruijns, M. F. M., van Altena, R. and Butot, L. J. M. (1959) The Netherlands as an enviroment for land Mollusca. *Basteria*, **23**, supplement.

Büchs, W., Heimbach, U. and Czarnecki, E. (1989). Effects of snail baits on non-target carabid beetles. In *Slugs and Snails in World Agriculture* (ed. I. Henderson), British Crop Protection Council, Thornton Heath, pp. 245–252.

Buckhurst, A. S. (1947) Slug control with DDT. *Fruit Grower (London)*, **103**, 146–149.

Bullock, T. H. and Horridge, G. A. (1965). *Structure and Function of the Nervous Systems of Invertebrates*, Vol. II, Freeman, San Francisco and London.

Burenkov, M. S. (1977) [Population age structure of three slug species (Order Stylommatophora, subclass Pulmonata)]. *Zh. Obsh. Biol.*, **38**, 296–304.

Burenkov, M. S. and Agre, A. L. (1975) [Preliminary evaluation of the possibility of using land molluscs as a component of closed ecological systems]. *Kosmich. Biol. Aviakosmich. Med.*, **9**, 90–92.

Burgess, R. S. L. and Emos, R. A. (1987) Selective grazing of acyanogenic white clover: Variation in behaviour among populations of the slug *Deroceras reticulatum. Oecologia (Berlin)*, **73**, 432–435.

Burghardt, G. M. (1968) Chemical preference studies on newborn snakes of three sympatric species of *Natrix. Copeia*, **968**, 732–737.

Burnet, B. (1972) Enzyme protein polymorphism in the slug *Arion ater. Genet. Res.*, **20**, 161–173.

Burton, D. W. (1978) Anatomy, histology and function of the reproductive system of the tracheopulmonate slug, *Athoracophorus bitentaculatus. Zool. Pub. Victoria Univ. Wellington*, **68**, 1–18.

Burton, R. F. (1964) Variations in the volume and concentration of the blood of the snail *Helix pomatia* L., in relation to the water content of the body. *Canad. J. Zool.*, **42**, 1085–1097.

Burton, R. F. (1968) Ionic regulation in the blood of Pulmonata. *Comp. Biochem. Physiol.*, **25**, 509–516.

Burton, R. F. (1969) Buffers in the blood of the snail, *Helix pomatia* L. *Comp. Biochem. Physiol.*, **29**, 919–930.

Burton, R. F. (1972) The storage of calcium and magnesium phosphates and of calcite in the digestive glands of the Pulmonata (Gastropoda). *Comp. Biochem. Physiol. A. Comp. Physiol.*, **43**, 655–663.

Byers, R. A. and Bierlein, D. L. (1982) Feeding preferences of 3 slug species in the laboratory. *Melsheimer Entomol. Ser.*, **32**, 5–11.

Byers, R. A. and Bierlein, D. L. (1984) Continuous alfalfa: Invertebrate pests during establishment. *J. Econ. Entomol.*, **77**, 1500–1503.

Byers, R. A., Barratt, B. I. P. and Calvin, D. (1989) Comparison between defined-area traps and refuge traps for sampling slugs in conservation-tillage crop enviroments. In *Slugs and Snails in World Agriculture* (ed. I. Henderson), British Crop Protection Council, Thornton Heath, pp. 187–192.·

Byers, R. A., Mangan, R. L. and Templeton, W. C. (1983) Insect and slug pests in forage legume seedlings. *J. Soil Water Conserv.*, **38**, 224–226.

Byers, R. A., Templeton, W. C., Mangan, R. L., Bierlein, D. L., Campbell, W. F. and Donley, H. J. (1985) Establishment of legumes in grass swards: Effects of pesticides on slugs, insects, legume seedling numbers and forage yield and quality. *Grass Forage Sci.*, **40**, 41–48.

Cabaret, J. (1982a) The estimation of the infection of molluscs by protostrongylids: Parameters used and their relationship. *Ann. Parasitol. Humaine Comp.*, **57**, 367–374.

Cabaret, J. (1982b) L'infestation des chèvres par *Muellerius capillaris* au paturage. *Ann. Parasitol. Humaine Group.*, **57**, 637–638.

Cabaret, J., Dakkak, A. and Bahaida, B. (1980) Etude de l'infestation des mollusques terrestres de la région de Rabbat (Maroc) par les larves de protostrongylids dans les conditions naturelles. *Rev. Elev. Méd. Vet. Pays Trop.*, **33**, 159–165.

Cabaret, J. and Pandey, V. S. (1986) The use of tracer lambs for monitoring protostrongylidid infection on extensive pastures of Morocco. *Ann. Recherches Vet.*, **17**, 69–74.

Cabaret, J. and Vendroux, P. (1986) The response of four terrestrial molluscs to the presence of herbivore faeces: its influence on infection by protostrongylids. *Canad. J. Zool.*, **64** 850–854.

Cabaret, J. and Weber, H. (1987) The infection of the slug *Deroceras reticulatum* by the protostrongylid nematode *Muellerius capillaris*: aspects of early reactions. *Ann. Parasitol. Humaine Comp.*, **62**, 37–46.

Cambridge, P. (1981) *Parmacella*, a slug new to the Pleistocene of Britain. *J. Conchol.*, **30**, 329–330.

Cameron, A. E. (1939) The control of potato slugs. *J. Min. Agric.*, **46**, 454–462.

Cameron, R. A. D. (1978) Terrestrial snail faunas of the Malham Area. *Field Stud.*, **4**, 715–728.

Cameron, R. A. D. and Carter, M. A. (1979) Intra and interspecific effects of population density on growth and activity in some helicid land snails (Gastropoda. Pulmonata). *J. Animal Ecol.*, **48**, 237–246.

Cameron, R. A. D., Jackson, N. and Eversham, B. (1983) A field key to the slugs of the British Isles. *Field Stud.*, **5**, 807–824.

Cameron, R. A. D. and Redfern, M. (1972) The terrestrial Mollusca of the Malham Area. *Field Stud.*, **3**, 589–602.

Campion, H. M. (1957) The structure and function of some of the cutaneous glands in some molluscs. PhD thesis, Univ. of London.

Carmichael, E. B. and Rivers, T. D. (1932) The effect of dehydration upon hatchability of *Limax flavus* eggs. *Ecology*, **13**, 375–380.

Carpenter, A., Cherrett, J. M., Ford, J. B., Thomas, M. and Evans, E. (1985) An inexpensive rhizotron for research on soil and litter-living organisms. In *Ecological Interactions in Soil: Plants, Microbes and Animals* (eds A. H. Fitter, D. Atkinson, D. J. Read and M. B. Usher), Blackwell Scientific Publications, Oxford.

Carrick, R. (1938) The life history and development of *Agriolimax agrestis* L. the grey field slug. *Trans. Roy. Soc. Edinb.*, **59**, 563–597.

Carrick, R. (1942) The grey field slug *Agriolimax agrestis* L. and its environment. *Ann. Appl. Biol.*, **29**, 43–55.

Carriker, M. R. (1946) Morphology of the alimentary system of the snail, *Lymnaea stagnalis appressa* Say. *Trans. Wisconsin Acad. Sci., Arts Lett.*, **38**, 1–88.

Carter, A. (1983) Cadmium, copper and zinc in soil animals and their food in a red clover system. *Canad. J. Zool.*, **61**, 2751–2757.

Catalan, R. E., Castillon, M. P. and Rallo, A. (1977) Lipid metabolism during development of the mollusc *Arion empiricorum*, distribution of lipids in midgut gland, genitalia and foot muscle. *Comp. Biochem. Physiol. B. Comp. Biochem.*, **57**, 73–79.

Cates, R. G. (1975) The interface between slugs and wild ginger: some evolutionary aspects. *Ecology*, **56**, 391–400.

Cates, R. G. and Orians, G. H. (1975) Successional status and the palatability of plants to generalized herbivores. *Ecology*, 56, 410–418.

Cather, J. N. and Tompa, A. S. (1972) The podocyst in pulmonate evolution. *Malacol. Rev.*, 5, 1–3.

Catling, P. M. and Freedman, B. (1980a) Variation in distribution and abundance of 4 sympatric species of snakes at Amherstburg, Ontario, Canada. *Canad. Field-Nat.*, 94, 19–27.

Catling, P. M. and Freedman, B. (1980b) Food and feeding behaviour of sympatric snakes at Amherstburg, Ontario, Canada. *Canad. Field-Nat.*, 94, 28–33.

Cavalloro, R. and Ravera, O. (1966) Biological indicator of manganese-54 contamination in terrestrial environments. *Nature (London)*, 209, 1259.

Chan, C. Y. and Moffett, S. (1982a) Cerebral motoneurons mediating tentacle retraction in the land slug, *Ariolimax columbianus*. *J. Neurobiol.*, 13, 163–172.

Chan, C. Y. and Moffett, S. (1982b) Neurophysiological mechanisms underlying habituation of the tentacle retraction reflex in the land slug *Ariolimax columbianus*. *J. Neurobiol.*, 13, 173–184.

Chang, J. J. and Gelperin, A. (1976) Molluscan feeding motor programme: Response to lip chemostimulation and modulation by identified serotonergic interneurons. *Soc. Neurosci. Abstr.*, 2, 322.

Chang, J. J. and Gelperin, A. (1980) Rapid taste-aversion learning by an isolated molluscan central nervous system. *Proc. Nat. Acad. Sci. USA*, 77, 6204–6206.

Chang, N.-S. and Lim, Y. S. (1989) Ultrastructural and histochemical studies on the epithelial cell of Korean terrestrial slug *Incilaria fruhstorferi*. *Korean J. Zool.*, 32, 93–106.

Charlton, J. F. L. (1978) Slugs as a possible cause of establishment failure in pasture legumes oversown in boxes. *N.Z. J. Exp. Agric.*, 6, 313–318.

Charrier, M. (1990) Evolution, during digestion, of the bacterial flora in the alimentary system of *Helix aspersa* (Gastropoda: Pulmonata): A scanning electron microscope study. *J. Mollusc. Stud.*, 56, 425–433.

Chase, R. and Croll, R. P. (1981) Tentacular function in snail (*Achatina fulica*) olfactory orientation. *J. Comp. Physiol. A. Sensory, Neural Behav. Physiol.*, 143, 357–362.

Chatfield, J. E. (1976a) *Limax grossui* Lupu 1970, a slug new to the British Isles. *J. Conchol.*, 29, 1–4.

Chatfield, J. E. (1976b) Studies on food and feeding in some European Land Molluscs. *J. Conchol.*, 29, 5–20.

Chatwin, A. L. (1973) Some aspects of the nervous control of locomotion and posture in *Agriolimax reticulatus*. PhD thesis, Univ. of Bath.

Chelazzi, G., Le Voci, G. and Parpagnoli, D. (1988) Relative importance of airborne odors and trails in the group homing of *Limacus flavus* (Linnaeus) (Gastropoda, Pulmonata). *J. Mollusc. Stud.*, 54, 173–180.

Chen, D., Jiaxiang, G. and Jicheng, S. (1984) Observations on the breeding of a single slug, *Agriolimax agrestis*. *Acta Zool. Sinica*, 30, 362–367.

Chétail, M. (1963) Étude de la régénération du tentacle oculaire chez un Arionidae (*Arion rufus* L.) et un Limacidae (*Agriolimax agrestis* L.). *Arch. Anat. Microsc. Morphol. Exp.*, 52, 129–203.

Chétail, M. and Binot, D. (1967) Particularites histochimiques de la glande et de la sole pedieuses d'*Arion rufus* (Stylommatophora: Arionidae). *Malacologia*, 5, 269–284.

Chevallier, H. (1969) Taxonomie et Biologie des Grands *Arion* de France (Pulmonata: Arionidae). *Malacologia*, 9, 73–78.

Chevallier, H. (1971) Cycle Biologique des Grands *Arion* en France. *Atti Soc. Italiana Sci. Nat. Museo Civico Storia Nat. Milano*, 112, 316–320.

Chevallier, H. (1973a) Répartition en France de *Deroceras caruanae* (Pollonera 1891). *Haliotis*, 3, 205–207.

Chevallier, H. (1973b) Les mollusques continentaux de France d'intérêt économique. *Haliotis*, 3, 9–18.

Chevallier, H. (1976) Observations sur le polymorphisme des limaces rouges (*Arion rufus* Linne et *Arion lusitanicus* Mabille) et de l'escargot petit-gris (*Helix aspersa* Müller). *Haliotis*, 6, 41–48.

Chevallier, H. (1979) *Les Escargots: Un Elevage d'Avenir*, Dargaud Editeur, Paris.

Chevallier, H. (1982) Facteurs de croissance chez des Gastéropodes Pulmonés terrestres paléarctiques en élevage. *Haliotis*, 12, 29–46.

Chichester, L. F. and Getz, L. L. (1968) Terrestrial slugs (Ecology, biology, vector studies, laboratory culture). *Biologist (Phi Sigma Soc.)*, 50, 148–166.

Chichester, L. F. and Getz, L. L. (1969) The zoogeography and ecology of arionid and limacid slugs introduced into northeastern North America. *Malacologia*, 7, 313–346.

Chichester, L. F. and Getz, L. L. (1973) The terrestrial slugs of northeastern North America. *Sterkiana*, 51, 11–42.

Church, B. M., Hampson, C. P. and Fox, W. R. (1970) The quality of stored maincrop potatoes in Great Britain. *Potato Res.*, 13, 41–58.

Clements, R. O., French, N., Guile, C. T., Golightly, W. H., Lewis, S. and Savage, M. J. (1982) *Ann. Appl. Biol.*, 101, 305–314.

Coe, M. J. (1971) The activity of the slug *Trichotoxon copleyi copleyi* Verdcourt. *East African Wildlife J.*, 3, 170–172.

Coifmann, I. (1934) Sul sistema nervoso dello *Vaginula solea*. *Boll. Zool.*, 5, 109–122.

Collinge, W. E. (1921) The starling: is it injurious to agriculture? *J. Min. Agric.*, 27, 1114–1121.

Colville, B., Lloyd-Evans, L. and Norris, A. (1974) *Boettgerilla pallens*: a new British species. *J. Conchol.*, 28, 203–208.

Comfort, A. (1957) The duration of life in Molluscs. *Proc. Malacol. Soc.*, 32, 219–241.

Conroy, B. A. (1980) Coexistence of two closely related species of *Arion* in natural habitats. *J. Conchol.*, 30, 189–200.

Cook, A. (1977) Mucus trail following by the slug *Limax grossui* Lupu. *Animal Behav.*, 25, 774–781.

Cook, A. (1979a) Homing by the slug *Limax pseudoflavus*. *Animal Behav.*, 27, 545–552.

Cook, A. (1979b) Homing in the Gastropoda. *Malacologia*, 18, 315–318.

Cook, A. (1980) Field studies of homing in the pulmonate slug, *Limax pseudoflavus*. *J. Mollusc. Stud.*, 46, 100–105.

Cook, A. (1981a) Huddling and the control of water loss by the slug. *Limax pseudoflavus*. *Animal Behav.*, 29, 289–298.

Cook, A. (1981b) A comparative study of aggregation in pulmonate slugs (Genus *Limax*). *J. Animal Ecol.*, **50**, 703–714.

Cook, A. (1985) Tentacular function in trail following *Limax pseudoflavus. J. Mollusc. Stud.*, **51**, 240–247.

Cook, A. (1987) Functional aspects of the mucus producing glands of the systellommatophoran slug *Veronicella floridana. J. Zool. (London)*, **211**, 291–306.

Cook, A. and Ford, D. J. G. (1989) The control of activity of the pulmonate slug, *Limax pseudoflavus*, by weather. In *Slugs and Snails in World Agriculture* (ed. I. Henderson), British Crop Protection Council, Thornton Heath, pp. 337–342.

Cook, A. and Gelperin, A. (1988a) Distribution of FMRF amide-like immuno-reactivity in the nervous system of the slug *Limax maximus. Cell Tissue Res.*, **253**, 69–76.

Cook, A. and Gelperin, A. (1988b) Distribution of GABA-like immunoreactive neurons in the slug, *Limax maximus. Cell Tissue Res.*, **253**, 77–81.

Cook, A. and Radford, D. J. (1988) The comparative ecology of four sym-patric limacid slug species in Northern Ireland UK. *Malacologia*, **28**, 131–146.

Cook, A. and Shirbhate, R. (1983) The mucus producing glands and the distribution of cilia of the pulmonate slug, *Limax pseudoflavus. J. Zool. (London)*, **201**, 97–116.

Cook, R., Thomas, B. J. and Mizen, K. A. (1989) Dissemination of white clover mosaic virus and stem nematode, *Ditylenchus dipsaci*, by the slug, *Deroceras reticulatum*. In *Slugs and Snails in World Agriculture* (ed. I. Henderson), British Crop Protection Council, Thornton Heath, pp. 107–112.

Copeland, J. (1981) Effects of larval firefly extracts on molluscan cardiac activity. *Experientia (Basel)*, **37**, 1271–1272.

Copeland, J. and Gelperin, A. (1983) Feeding and a serotonergic interneuron activate an identified autoactive salivary neuron in *Limax maximus. Comp. Biochem. Physiol. A. Comp. Physiol.*, **76**, 21–30.

Corbet, G. B. and Southern, H. N. (eds) (1977) *The Handbook of British Mammals*, Blackwell Scientific Publications, Oxford.

Corliss, J. O., Smith, A. C. and Foulkes, J. (1962) A species of *Tetrahymena* from the British garden slug *Milax budapestensis. Nature (London)*, **196**, 1008–1009.

Cornic, J. F. (1973) Étude du régime alimentaire de trois espèces de Carabiques et de ses variations en verger de pommiers. *Ann. Soc. Entomol. France*, **9**, 69–87.

Cottam, D. A. (1985) Frequency-dependent grazing by slugs and grasshoppers. *J. Ecol.*, **73**, 925–934.

Cottam, D. A. (1986) The effects of slug-grazing on *Trifolium repens* and *Dactylis glomerata* in monoculture and mixed sward. *Oikos*, **47**, 275–279.

Cottrell, G. A. and Osborne, N. N. (1970) *Nature (London)*, **225**, 470–472.

Coughtrey, P. J. and Martin, M. H. (1976) The distribution of Pb, Zn, Cd and Cu within the pulmonate mollusc *Helix aspersa* Müller. *Oecologia (Berlin)*, **23**, 315–322.

Coughtrey, P. J. and Martin, M. H. (1977) The uptake of lead, zinc, cadmium and copper within the pulmonate mollusc *Helix aspersa* Müller, and its relevance to the monitoring of heavy metal contamination of the environment. *Oecologia (Berlin)*, **27**, 65–74.

Cragg, J. B., Forster, R. and Vincent, M. (1957) Larval trematodes (Brachylaemidae) from the slugs *Milax sowerbyi* (Fér.), *Agriolimax reticulatus* (Müll.) and *Arion lusitanicus* Mab. *Parasitology*, **47**, 396–404.

Cragg, J. B. and Vincent, M. H. (1952) The action of metaldehyde on the slug *Agriolimax reticulatus* (Müller). *Ann. Appl. Biol.*, **39**, 392–406.

Crampton, D. M. (1975) The anatomy and method of functioning of the buccal mass of *Testacella maugei* Férussac. *Proc. Malacol. Soc. London*, **41**, 549–570.

Crawford-Sidebotham, T. J. (1970) Differential susceptibility of species of slugs to metaldehyde/bran and to methiocarb baits. *Oecologia (Berlin)*, **7**, 303–324.

Crawford-Sidebotham, T. J. (1971) Studies on aspects of slug behaviour and the relation between molluscs and cyanogenic plants. PhD thesis, Univ. of Birmingham.

Crawford-Sidebotham, T. J. (1972a) The role of slugs and snails in the maintenance of the cyanogenesis polymorphisms of *Lotus corniculatus* and *Trifolium repens. Heredity*, **28**, 405–411.

Crawford-Sidebotham, T. J. (1972b) The influence of weather upon the activity of slugs. *Oecologia (Berlin)*, **9**, 141–154.

Crawshay, L. R. (1903) On the life-history of *Drilus flavescens* Rossi. *Trans. Roy. Entomol. Soc. London*, 39–51.

Crowell, H. H. (1967) Slug and snail control with experimental poison baits. *J. Econ. Entomol.*, **60**, 1048–1050.

Crowell, H. H. (1977) Chemical control of terrestrial slugs and snails. *Oregon Agric. Exp. Station. Bull. No. 628*, 70 pp.

Crowley, T. E. (1978) Island life: St. Helena. *J. Conchol.*, **29**, 233–237.

Crowson, R. A. (1981) *The Biology of the Coleoptera*, Academic Press, New York and London.

Crozier, W. J. and Federighi, H. (1925a) Phototropic circus movement of *Limax* as affected by temperature. *J. Gen. Physiol.*, **7**, 151–169.

Crozier, W. J. and Federighi, H. (1925b) The locomotion of *Limax*. II. Vertical ascension with added loads. *J. Gen. Physiol.*, **7** 415–419.

Crozier, W. J. and Pilz, G. F. (1924) The locomotion of *Limax*. I. Temperature coefficient of pedal activity. *J. Gen. Physiol*, **6**, 711–721.

Cuendet, G. (1983) Predation on earthworms by the black-headed gull. In *Earthworm Ecology* (ed. J. E. Satchell), Chapman and Hall, London, pp. 415–424.

Culligan, N. and Gelperin, A. (1983) 1-trial associative learning by an isolated molluscan central nervous system: Use of different chemoreceptors for training and testing. *Brain Res.*, **266**, 319–328.

Curtis, J. (1860) *Farm Insects*, Blackie & Son, London.

Curtis, S. and Cowden, R. R. (1977) Ultrastructure and histochemistry of the supportive structures associated with the radula of the slug *Limax maximus. J. Morphol.*, **151**, 187–212.

Curtis, S. and Cowden, R. R. (1978) Responsiveness of the slug *Limax maximus* to injections of fluorescein- and rhodamine-conjugated immunogens. *Devel. Comp. Immunol.*, **2**, 727–733.

Curtis, S. and Cowden, R. R. (1979) Histochemical and ultrastructural features of the aorta of the slug (*Limax maximus*). *J. Morphol.*, **161**, 1–22.

Czarna, Z., Krasowska, E. and Turska, R. (1985) The tubular structures in the cells of the digestive gland of *Arion rufus* (Mollusca Gastropoda). *Zool. Poloniae*, **32**, 23–28.

Czeczuga, B. (1984) Occurrence of α-doradexanthin (3,3'-dihydroxy-4-keto-α-carotene) in some species of Mollusca. *Zool. Poloniae*, **31**, 13–20.

Daguzan, J. (1989) Snail rearing or heliciculture of *Helix aspersa* Müller. In *Slugs and Snails in World Agriculture* (ed. I. Henderson), British Crop Protection Council, Thornton Heath, pp. 3–10.

Dainton, B. H. (1943) Effect of air currents, light, humidity and temperature on slugs. *Nature (London)*, **151**, 25.

Dainton, B. H. (1954a) The activity of slugs. I. The induction of activity by changing temperatures. *J. Exp. Biol.*, **31**, 165–187.

Dainton, B. H. (1954b) The activity of slugs. II. The effect of light and air currents. *J. Exp. Biol.*, **31**, 188–197.

Dainton, B. H. (1989) Field and laboratory observations on slug and snail behaviour. In *Slugs and Snails in World Agriculture* (ed. I. Henderson), British Crop Protection Council, Thornton Heath, pp. 201–208.

Dainton, B. H. and Wright, J. (1985) Falling temperature stimulates activity in the slug *Arion ater. J. Exp. Biol.*, **118**, 439–443.

Dalal, Y. M. and Pandya, G. T. (1976) A histochemical and structural study of suprapedal gland of the garden slug, *Laevicaulis alte* (Fér.). *J. Animal Morphol. Physiol.*, **23**, 46–51.

Dallinger, R., Janssen, H. H., Bauer-Hilty, A. and Berger, B. (1989) Characterization of an inducible cadmium-binding protein from hepatopancreas of metal-exposed slugs (Arionidae, Mollusca). *Comp. Biochem. Physiol. C. Comp. Pharmacol. Toxicol.*, **92**, 355–360.

Dalton, L. M. and Widdowson, P. S. (1989) The involvement of opioid peptides in stress-induced analgesia in the slug *Arion ater. Peptides (Elmsford)*, **10**, 9–14.

Dance, S. P. (1972) Vertical range of molluscs on Ben Lawers, Scotland. *J. Conchol.*, **27**, 509–515.

David, H. and Götze, J. (1963) Elektronenmikroskopische Befunde an der Mitteldarmdrüse von Schnecken. *Z. Mikrosk.-Anat. Forsch. (Leipzig)*, **70**, 252–272.

David, W. A. L., Taylor, C. E. and Atkey, P. T. (1977) Nonoccluded virus-like particles in the mollusc *Agriolimax reticulatus* (Stylommatophora: Limacinae). *J. Invert. Pathol.*, **29**, 242–243.

Davidson, A. J. and Longton, R. E. (1987). Acceptability of mosses as food for a herbivore, the slug *Arion hortensis. Symp. Biol. Hung.*, **35**, 707–720.

Davidson, A. J., Harborne, J. B. and Longton, R. E. (1989) Identification of hydroxycinnamic and phenolic acids in *Mnium hornum* and *Brachythecium rutabulum* and their possible role in protection against herbivory. *J. Hattori Bot. Lab.*, **0**(67), 415–422.

Davidson, D. H. (1976) Assimilation efficiencies of slugs on different food materials. *Oecologia (Berlin)*, **26**, 267–273.

Davies, M. J. (1953) The contents of the crops of some British carabid beetles. *Entomol. Monthly Mag.*, **89**, 18–23.

Davies, R. C. (1974) Some aspects of the anatomy, fine structure and neurophysiology of the pneumostome of the garden slug *Limax flavus* (L). MSc thesis, Univ. of St Andrews.

Davies, S. M. (1977) The *Arion hortensis* complex with notes on *A. intermedius* Normand (Pulmonata: Arionidae). *J. Conchol.*, **29**, 173–187.

Davies, S. M. (1979) Segregates of the *Arion hortensis* complex (Pulmonata: Arionidae), with the description of a new species, *Arion owenii. J. Conchol.*, **30**, 123–128.

Davies, S. M. (1987) *Arion flagellus* Collinge and *A. lusitanicus* Mabille in the British Isles. A morphological, biological and taxonomical investigation. *J. Conchol.*, **32**, 339–354.

Davis, A. J. (1989) Effects of soil compaction on damage to wheat seeds by three pest species of slug. *Crop Protect.*, **8**, 118–121.

Davis, B. N. K. (1966) Soil animals as vectors for organochlorine insecticides. In *Pesticides in the Environment and their Effects on Wildlife; J. Appl. Ecol.*, (Suppt), June 1966.

Davis, B. N. K. and French, M. C. (1969) The accumulation and loss of organochlorine insecticide residues by beetles, worms and slugs in sprayed fields. *Soil Biol. Biochem.*, **1**, 45–55.

Davis, B. N. K. and Harrison, R. B. (1966) Organochlorine insecticide residues in soil invertebrates. *Nature (London)*, **211**, 1424–1425.

Dawkins, G., Hislop, J., Luxton, M. and Bishop, C. (1986) Transmission of bacterial soft rot of potatoes by slugs. *J. Mollusc. Stud.*, **52**, 25–29.

Dawkins, G., Luxton, M. and Bishop, C. (1985) Transmission of liquorice rot of carrots by slugs. *J. Mollusc. Stud.*, **51**, 83–85.

Dawson, R. B. (1932) Leatherjackets. *J. Board of Greenkeeping Res.*, **2**, 183.

Daxl, R. (1969) Beobachtungen zur diurnalen und saisonellen Aktivität einiger Nacktschneckenarten. *Z. Angew. Zool.*, **56**, 357–370.

Daxl, R. (1970) Der Einfluss von Temperatur und relativer Luftfeuchte auf die molluskizide Wirkung des Metaldehyd, Isolan und Ioxynil auf *Limax flavus* L. und dessen Eier. *Z. Angew. Zool.*, **67**, 57–87.

Daxl, R. (1971) Das Verhalten der molluskiziden Verbindungen Metaldehyd, Isolan und Ioxynil gegen Nacktschnecken unter Freilandbedingungen. *Z. Angew. Zool.*, **58**, 203–241.

De Wilde, J. J. A. (1983) Notes on the *Arion hortensis* complex in Belgium (Mollusca, Pulmonata: Arionidae). *Ann. Soc. Roy. Zool. Belg.*, **113**, 87–96.

De Wilde, J. J. (1986) Further notes on the species of the *Arion hortensis* complex in Belgium (Mollusca, Pulmonata: Arionidae). *Ann. Soc. Roy. Zool. Belg.*, **116**, 71–74.

Delaney, K. and Gelperin, A. (1986) Post-ingestive food-aversion learning to amino acid-deficient diets by the terrestrial slug, *Limax maximus. J. Comp. Physiol. A. Sensory, Neural Behav. Physiol.*, **159**, 281–296.

Delaney, K. and Gelperin, A. (1987) Central correlates of taste discrimination in the terrestrial slug *Limax maximus. Soc. Neurosci. Abstr.*, **13**, 388.

Delaney, K. and Gelperin, A. (1990a) Cerebral interneurons controlling fictive feeding in *Limax maximus*: I. Anatomy and criteria for re-identificationi. *J. Comp. Physiol. A. Sensory, Neural Behav. Physiol.*, **166**, 297–310.

Delaney, K. and Gelperin, A. (1990b) Cerebral interneurons controlling fictive feeding in *Limax maximus*: II. Initiation and modulation of fictive feeding. *J. Comp. Physiol. A. Sensory, Neural Behav. Physiol.*, **166**, 311–326.

Delaney, K. and Gelperin, A. (1990c) Cerebral interneurons controlling fictive feeding in *Limax maximus*: III. Integration of sensory inputs. *J. Comp. Physiol. A. Sensory, Neural Behav. Physiol.*, **166**, 327–344.
Delaunay, H. (1931) L'excrétion azotée des Invertébrés. *Biol. Rev.*, **6**, 265–302.
Dell, R. K. (1964) Land snails from sub-Antarctic islands. *Trans. Roy. Soc. N.Z. (Zoology)*, **4**, 167–173.
Dempster, J. P. (1968) The control of *Pieris rapae* with DDT. III. Some changes in the crop fauna. *J. Appl. Ecol.*, **5**, 463–475.
Denny, M. W. (1980) The role of gastropod pedal mucus in locomotion. *Nature (London)*, **285**, 160.
Denny, M. W. (1981) A quantitative model for the adhesive locomotion of the terrestrial slug *Ariolimax columbianus*. *J. Exp. Biology*, **91**, 195–217.
Denny, M. W. and Gosline, J. M. (1980) The physical properties of the pedal mucus of the terrestrial slug *Ariolimax columbianus*. *J. Exp. Biol.*, **88**, 375–393.
Deyrup-Olsen, I., Luchtel, D. and Martin, A. W. (1989) Secretory cells of the body wall of *Ariolimax columbianus*: structure, function and control. In *Slugs and Snails in World Agriculture* (ed. I. Henderson), British Crop Protection Council, Thornton Heath, pp. 407–412.
Deyrup-Olsen, I. and Martin, A. W. (1982a) Roles of calcium in fluid responses of the body wall of terrestrial slugs. *Amer. Zool.*, **22**, 978.
Deyrup-Olsen, I. and Martin, A. W. (1982b) Surface exudation in terrestrial slugs. *Comp. Biochem. Physiol. C. Comp. Pharmacol.*, **72**, 45–52.
Deyrup-Olsen, I. and Martin, A. W. (1987) Osmolyte processing in the gut and an important role of the rectum in the land slug *Ariolimax columbianus* (Pulmonata: Arionidae). *J. Exp. Zool.*, **243**, 33–38.
Deyrup-Olsen, I., Martin, A. W. and Paine, R. T. (1986) The autotomy escape response of the terrestrial slug *Prophysaon foliolatum* (Pulmonata: Arionidae). *Malacologia*, **27**, 307–312.
Dickinson, P. S., Prior, D. J. and Avery, C. (1988) The pneumostome rhythm in slugs. A response to dehydration controlled by haemolymph osmolality and peptide hormones. *Comp. Biochem. Physiol. A. Comp. Physiol.*, **89**, 579–586.
Dimelow, E. J. (1963) Observations on the feeding of the hedgehog (*Erinaceus europaeus*). *Proc. Zool. Soc. London*, **141**, 291–309.
Dirzo, R. (1980) Experimental studies on slug–plant interactions: 1. The acceptability of thirty plant species to the slug *Agriolimax caruanae*. *J. Ecol.*, **68**, 981–998.
Dirzo, R. and Harper, J. L. (1980) Experimental studies on slug–plant interactions 2. The effect of grazing by slugs on high density monocultures of *Capsella bursa-pastoris* and *Poa annua*. *J. Ecol.*, **68**, 999–1011.
Dirzo, R. and Harper, J. L. (1982a) Experimental studies on slug–plant interactions: 3. Differences in the acceptability of individual plants of *Trifolium repens* to slugs and snails. *J. Ecol.*, **70**, 101–118.
Dirzo, R. and Harper, J. L. (1982b) Experimental studies on slug–plant interactions: 4. The performance of cyanogenic and acyanogenic morphs of *Trifolium repens* in the field. *J. Ecol.*, **70**, 119–138.
Disney, R. H. L. (1976) A further case of a scuttle fly (Diptera: Phoridae) whose larva attacks slug eggs. *Entomol. Monthly Mag.*, **112**, 174.
Dmitrieva, E. F. (1969) [Population dynamics, growth, feeding and reproduction

of field slug (*Deroceras reticulatum*) in Leningrad Oblast]. *Zool. Zh.*, **48**, 802–810.

Dobson, D. and Bailey, S. (1982) Duration of feeding and crop fullness in *Deroceras reticulatum*. *J. Mollusc. Stud.*, **48**, 371–372.

Dobson, R. M. (1963) Observations on the occurrence of keeled slugs (*Milax*, Gray) in the Glasgow area. *The Glasgow Naturalist*, **18**, 246–248.

Dobzhansky, T. and Wright, S. (1943) Genetics of natural populations. X. Dispersion rates in *Drosophila pseudoobscura*. *Genetics*, **28**, 304–340.

Dochita, L. (1971) Étude sur le développement de *Deroceras padisii* Lupu, 1969 et son cycle biologique. *Trav. Mus. Hist. Nat. 'Grigore Antipa'*, **11**, 139–147.

Dochita, L. (1972) Étude anatomique comparée de quelques espèces du genre *Deroceras* Rafinesque 1820. *Trav. Mus. Hist. Nat. 'Grigore Antipa'*, **12**, 85–105.

Dohmen, M. R. (1983) Gametogenesis. In *The Mollusca*, Vol. 3. *Development* (eds N. H. Verdonk, J. A. M. van den Biggelaar, A. S. Tompa), Academic Press, New York and London, pp. 1–48.

Dolan, S. and Fleming, C. C. (1988) Isoenzymes in the identification and systematics of terrestrial slugs of the *Arion hortensis* complex. *Biochem. System. Ecol.*, **16**, 195–198.

Dowling, P. M. and Linscott, D. L. (1985) Slugs as a primary limitation to establishment of sod-seeded lucerne. *Crop Protect.*, **4**, 394–402.

Dubnitskii, A. A. (1956) [Studies on the development cycle of the nematode *Skrjabingylus nasicola*, a parasite affecting the frontal sinus of fur bearers of the marten family]. *Karakulev Zverov*, **9**, 59–61.

Duncan, C. J. (1975) Reproduction. In *Pulmonates*, Vol. 1 (eds V. Fretter, J. Peake), Academic Press, New York and London, pp. 309–366.

Dundee, D. S. (1977) Observations on the veronicellid slugs of the southern United States. *Nautilus*, **91**, 108–114.

Dundee, D. S., Hermann, H. R. and Hermann, P. W. (1968) New records for introduced mollusks. *Nautilus*, **82**, 43–45.

Dundee, D. S., Tizzard, M. and Traub, M. (1975) Aggregative behaviour in veronicellid slugs. *Nautilus*, **89**, 69–72.

Dunnett, G. M. (1956) The autumn and winter mortality of starlings, *Sturnus vulgaris*, in relation to their food supply. *Ibis*, **98**, 220–230.

Dunning, R. A. and Davies, N. B. (1975) Sugar beet and its pest and virus problems in England. *Proc. 8th Brit. Insect. Fung. Conf.*, pp. 453–463.

Durham, H. E. (1920) Of slugs. *Gardeners Chronicle*, **68**, 85–86.

Dussart, G. (1989) Slugs and snails and scientist's tales. *New Scientist*, No. 1674, 22 July 1989.

Dussourd, D. E. and Eisner, T. (1987) Vein-cutting behaviour: Insect counterploy to the latex defense of plants. *Science*, **237**, 898–901.

Duthoit, C. M. (1961) Assessing the activity of the field slug in cereals. *Plant Pathol.*, **10**, 165.

Duthoit, C. M. (1964) Slugs and food preferences. *Plant Pathol.*, **13**, 73–78.

Dutton, G. J. (1966) Uridine diphosphate glucose and the synthesis of phenolic glucosides by mollusks. *Arch. Biochem. Biophys.*, **116**, 399–405.

Duval, A. (1982) Le système circulatoire des limaces. *Malacologia*, **22**, 627–630.

Duval, A. (1983) Heartbeat and blood pressure in terrestrial slugs. *Canad. J. Zool.*, **61**, 987–992.

Duval, A. and Banville, G. (1989) Ecology of *Deroceras reticulatum* (Müll.)

(Stylommatophora, Limacidae) in Quebec strawberry fields. In *Slugs and Snails in World Agriculture* (ed. I. Henderson) British Crop Protection Council, Thornton Heath, pp. 147–160.

Duval, A. and Runham, N. W. (1981) The arterial system of 6 species of terrestrial slug. *J. Mollusc. Stud.*, **47**, 43–52.

Duval, D. M. (1970a) Some aspects of the behaviour of pest species of slugs. *J. Conchol.*, **27**, 163–170.

Duval, D. M. (1970b) A note on the body wall of three species of slugs, *Agriolimax reticulatum*, *Arion hortensis* and *Milax budapestensis*. *Proc. Malacol. Soc. London*, **39**, 227–229.

Duval, D. M. (1971) A note on the acceptability of various weeds as food for *Agriolimax reticulatus* (Müller). *J. Conchol.*, **27**, 249–251.

Duval, D. M. (1972) A record of slug movements in late summer. *J. Conchol.*, **27**, 505–508.

Duval, D. M. (1973) A note on the acceptability of various weeds as food for *Arion hortensis* Férrusac. *J. Conchol.*, **28**, 37–39.

Dyson, M. (1964) An experimental study of wound healing in *Arion*. PhD thesis, Univ. of London.

Eakin, R. M. and Brandenburger, J. L. (1975) Retinal differences between light-tolerant and light-avoiding slugs (Mollusca: Pulmonata). *J. Ultrastruc. Res.*, **53**, 382–394.

Eakin, R. M., Brandenburger, J. L. and Barker, G. M. (1980) Fine structure of the eye of the New Zealand slug *Athoracophorus bitentaculatus*. *Zoomorphologie*, **94**, 225–240.

Eaton, H. J. and Tompsett, A. A. (1976) Avoiding slug damage to lily bulbs at Rosewarne Experimental Horticulture Station, Camborne, Cornwall. In *Lilies 1976 and other Liliaceae*, Royal Horticultural Society, London, pp. 63–65.

Ebling, W., Pflugmacher, J. and Haque, A. (1984) Der regenwurm als schlüsselorganismus zur messung der bodenbelastung mit organischen fremdchemikalien. *Ber. Landwirtschaft*, **62**, 222–255.

Eckert, J. and Laemler, G. (1972) Angiostrongylosis in man and animals. *Z. Parasitenkunde*, **39**, 303–322.

Edelstram, C. and Palmer, C. (1950) Homing behaviour in gastropods. *Oikos*, **2**, 259–270.

Edwards, C. A. (1975) Effects of direct drilling on the soil fauna. *Outlook Agric.*, **8**, 243–244.

Edwards, C. A. (1976) The uptake of two organophosphorus insecticides by slugs. *Bull. Env. Contam. Toxicol.*, **16**, 406–410.

Edwards, C. A. and Lofty, J. R. (1977) *Biology of Earthworms.*, 2nd edn, Chapman and Hall, London.

Edwards, P. J. and Wratten, S. D. (1980) *Ecology of Insect–Plant Interactions.* Edward Arnold, London.

Egan, M. E. and Gelperin, A. (1981) Olfactory inputs to a bursting serotonergic interneuron in a terrestrial mollusk. *J. Mollusc. Stud.*, **47**, 80–88.

El-Okda, M. M. K. (1981) Response of 2 land Mollusca to certain insecticides. *Bull. Entomol. Soc. Egypt Econ. Ser.*, **12**, 53–58.

Elliott, J. M. (1967) The food of trout (*Salmo trutta*) in a Dartmoor stream. *J. Appl. Ecol.*, **4**, 59–71.

Elliott, L. P. (1969) Certain bacteria, some of medical interest, associated with the slug *Limax maximus*. *J. Invert. Pathol.*, **15**, 306–312.

Elliott, M. M. (1985) Some effects of invertebrate herbivores on deciduous woodland regeneration. PhD thesis, Univ. of East Anglia.

Ellis, A. E. (ed.) (1951) Census of the distribution of British non-marine Mollusca. *J. Conchol.*, **23**, 171–244.

Ellis, A. E. (1964a) *Arion lusitanicus* Mabille in Cornwall. *J. Conchol.*, **25**, 285–287.

Ellis, A. E. (1964b) *Milax budapestensis* (Hazay) in woodland. *J. Conchol.*, **25**, 298.

Ellis, A. E. (1969) *British Snails*, Clarendon Press, Oxford.

Evans, J. G. (1972) *Land Snails in Archaeology*, Seminar Press, London and New York.

Evans, J. R. (1971). Further observations on the biology of *Prosternon tessellatum* (L.) (Col., Elateridae). *Entomol. Monthly Mag.*, **107**, 73–78.

Evans, N. J. (1977) Studies on the variation and taxonomy in two species aggregates of terrestrial slugs: *Limax flavus*, L. agg. and *Arion ater* L. agg. PhD thesis, Univ. of Liverpool.

Evans, N. J. (1978) *Limax pseudoflavus* Evans. A critical description and comparison with related species. *Irish Nat. J.*, **19**, 231–236.

Evans, N. J. (1982) Observations on variation in body coloration and patterning in *Limax flavus* and *Limax pseudoflavus*. *J. Nat. Hist.*, **16**, 847–858.

Evans, N. J. (1985) The use of electrophoresis in the separation of two closely related species of terrestrial slugs. *Biochem. System. Ecol.*, **13**, 325–328.

Evans, N. J. (1986) The status of *Limax maculatus* (Kaleniczenko 1851), *Limax grossui* Lupu 1970, and *Limax pseudoflavus* Evans 1978 (Gastropoda, Limacidae). *Proc. Acad. Nat. Sci. Philadelphia*, **138**, 576–588.

Evans, W. A. L. and Jones, E. G. (1962a) Carbohydrases in the alimentary tract of the slug *Arion ater* L. *Comp. Biochem. Physiol.*, **5.**, 149–160.

Evans, W. A. L. and Jones, E. G. (1962b) A note on the proteinase activity in the alimentary canal of the slug *Arion ater* L. *Comp. Biochem. Physiol.*, **5.**, 223–225.

Eversham, B. (1989) Mammals as vectors of slugs. *The Conchologists' Newsletter*, No. 109, June 1989, p. 197.

Feare, C. J., Dunnet, G. M. and Patterson, I. J. (1974) Ecological studies of the rook (*Corvus frugilegus* L.) in north-east Scotland: Food intake and feeding. *J. Appl. Ecol.*, **11**, 867–896.

Ferguson, C. M., Barratt, B. I. P. and Jones, P. A. (1988) Control of the grey field slug (*Deroceras reticulatum* (Müller)) by stock management prior to direct-drilled pasture establishment. *J. Agric. Sci.*, **111**, 443–449.

Ferguson, C. M., Barratt, B. I. P. and Jones, P. A. (1989) A new technique for estimating density of the field slug, *Deroceras reticulatum* (Müller). In *Slugs and Snails in World Agriculture* (ed. I. Henderson), British Crop Protection Council, Thornton Heath, pp. 331–336.

Ferguson, C. M. and Hanks, C. B. (1990) Evaluation of defined-area trapping for estimating the density of the field slug, *Deroceras reticulatum* (Müller). *Ann. Appl. Biol.*, **117**, 451–454.

Fincher, F. (1947) *Lampyris noctiluca* L. (Col. Lampyridae) attacking slug and feeding on it. *Entomol. Monthly Mag.*, **83**, 149.

Fleming, C. C. (1989). Population structure of *Deroceras reticulatum*. In *Slugs and Snails in World Agriculture* (ed. I. Henderson), British Crop Protection Council, Thornton Heath, pp. 413–416.

Florkin, M. and Scheer, B. T. (eds) (1972) *Chemical Zoology*, Vol. VII. *Mollusca*, Academic Press, New York and London.

Focardi, S. and Quattrini, D. (1972) Structure of the reproductive apparatus and life cycle of *Milax gagates* (Draparnaud): Mollusca Gastropoda Pulmonata. *Boll. Zool.*, 39, 9–27.

Foltz, D. W., Ochman, H., Jones, J. S., Evangelisti, M. and Selander, R. K. (1982a) Genetic population structure and breeding systems in arionid slugs (Mollusca: Pulmonata). *Biol. J. Linn. Soc.*, 17, 225–242.

Foltz, D. W., Ochman, H. and Selander, R. K. (1984) Genetic diversity and breeding systems in terrestrial slugs, *Malacologia*, 25, 593–606.

Foltz, D. W., Schaitkin, B. M. and Selander, R. K. (1982b) Gametic disequilibrium in the self-fertilizing slug *Deroceras laeve*. *Evolution*, 36, 80–85.

Foote, B. A. (1963) Biology of slug-killing *Tetanocera* (Diptera: Sciomyzidae). *Proc. North Central Branch Entomol. Soc. America*, 18, 97.

Forcart, L. (1960) Die systematische von *Milax barypa* Bourguignant. *Arch. Mollusk.*, 89, 111–112.

Ford, D. J. G. and Cook, A. (1987) The effects of temperature and light on the circadian activity of the pulmonate slug *Limax pseudoflavus* Evans. *Animal Behav.*, 35, 1754–1765.

Ford, D. J. G. and Cook, A. (1988) Responses to pulsed light in a pulmonate slug *Limax pseudoflavus*. *J. Zool. (London)*, 214, 663–672.

Forsyth, D. J. and Peterle, T. J. (1982) Uptake of chlorine-36-labelled DDT residues by slugs and isopods in the laboratory and field. *Env. Pollution Ser. A. Ecol. Biol.*, 29, 135–144.

Forsyth, D. J., Peterle, T. J. and Bandy, L. W. (1983) Persistence and transfer of chlorine-36-labelled DDT in the soil and biota of an old-field ecosystem: A 6-year balance study. *Ecology*, 64, 1620–1636.

Fosshagen, M. S., Palmgren, P. and Valovirta, I. (1972) The invertebrate fauna of the Kilpisjarvi area, Finnish Lapland: 2. Terrestrial gastropods. *Acta Soc. Fauna Flora Fennica*, 80, 37–39.

Foster, G. N. (1977) Problems in cucumber crops caused by slugs, cuckoo-spit insect, mushroom cecid, hairy fungus beetle and the house mouse. *Plant Pathol.*, 26, 100–101.

Foster, R. (1954) Studies on the ecology of the trematode *Brachylaemus*. PhD thesis, Univ. of Durham.

Foster, R. (1958a) Infestation of the slugs *Milax sowerbii* Fér. and *Agriolimax reticulatus* Müll. by trematode metacercariae (Brachylaemidae). *Parasitology*, 48, 303–311.

Foster, R. (1958b) The effects of trematode metacercariae (Brachylaemidae) on the slugs *Milax sowerbii* Férr. and *Agriolimax reticulatus* Müll. *Parasitology*, 48, 261–268.

Fourman, K. L. (1938) Untersuchungen über die Bedeutung der Bodenfauna bei der biologischen Umwandlung des Bestandesabfalles forstlichen Standorte. *Mitt. Forstwirt. Forstwissen.*, 9, 144–169.

Fournié, J. (1979a) Formation de la coquille des mollusques: les problems posés par la prèsence et le comportement de cellules libres dans la coquille normale et regénérée chez *Agriolimax* reticulatum (Gastéropode, Pulmoné). *Malacologia*, **18**, 543–548.

Fournié, J. (1979b) Étude des cellules libres présentes à la surface interne de la coquille d' *Agriolimax reticulatus* (Müller): Origine et rôle dans la mise en place de l'hypostracum. *Ann. Sci. Nat. Zool. (Paris)*, **1**, 169–185.

Fournié, J. and Zylberberg, L. (1987) Ultrastructural organisation ontogenesis and regeneration of the periostracum in the slug *Deroceras reticulatum* (Mollusca). *Canad. J. Zool.*, **65**, 1935–1941.

Fournié, J. and Chétail, M. (1982a) Accumulation calcique au niveau cellulaire chez les mollusques. *Malacologia*, **22**, 265–284.

Fournié, J. and Chétail, M. (1982b) Evidence for a mobilization of calcium reserves for reproduction requirements in *Deroceras reticulatum*. *Malacologia*, **22**, 285–291.

Fox, C. J. S. (1962) First record of the keeled slug *Milax gagates* (Drap.) in Nova Scotia. *Canad. Field Nat.*, **76**, 122–123.

Fox, D. L. (1983) Biochromy of the Mollusca. In *The Mollusca*, Vol. 2. *Environmental Biochemistry and Physiology* (ed. P. W. Hochachka), Academic Press, New York and London, pp. 281–304.

Frain, M. J. (1981) Chemoreception and feeding in the grey field slug *Deroceras reticulatum* (Müller) with reference to molluscicide formation. PhD thesis, Univ. of London.

Frain, M. J. and Newell, P. F. (1982) Meal size and a feeding assay for *Deroceras reticulatum*. *J. Mollusc. Stud.*, **48**, 98–99.

Frain, M. J. and Newell, P. F. (1983) Testing molluscicides against slugs: The importance of assessing the residual population. *J. Mollusc. Stud.*, **49**, 164–173.

François, E., Riga, A. and Moens, R. (1965) Labelling the grey field slug *Agriolimax reticulatus* by means of radionuclides. *Parasitica*, XXI, **4**, 139–151.

François, E., Moens, R. and Riga, A. (1966) Les radioisotopes en écologie animale. De placements dans un milieu de *Agriolimax reticulatus* Müller. *Med. Rijks. Landbouw. Gent*, **31**, 1032–1042.

François E., Riga, A. and Moens, R. (1968) Estimation des populations de *Agriolimax reticulatus* Müller au moyen de la technique de marquage au radiophosphore ^{32}P, et recapture. *Parasitica*, **24**, 63–78.

Frandsen, D. (1901) Studies on the reaction of *Limax maximus* to directive stimuli. *Proc. Amer. Acad. Arts Sci.*, **37**, 185–228.

Fretter, V. (1952) Experiments with ^{32}P and ^{131}I on species of *Helix*, *Arion* and *Agriolimax*. *Qtly J. Microsc. Sci.*, **93**, 133–146.

Frömming, E. (1954) *Biologie der mittel-europaischen Landgastropoden*, Dunkler und Humblot, Berlin.

Frömming, E., Peter, H. and Reichmuth, W. (1961) Beitrag zur Frange der pathologische Gestaltveränderung und der Geschwülste bei unserer Nacktschnecken. *Zool. Anzeiger*, **166**, 139–147.

Furbish, D. R. and Furbish, W. J. (1984) Structure, crystallography and morphogenesis of the cryptic shell of the terrestrial slug *Limax maximus* (Mollusca, Gastropoda). *J. Morphol.*, **180**, 195–212.

Furuta, E., Yamaguchi, K.-I. and Shimozawa, A. (1988) Morphological and functional studies on hemolymph cells of land slug *Incilaria* spp. *Devel. Comp. Immunol.*, **12**, 413.

Furuta, E. and Shimozawa, A. (1983) Primary culture of cells from the foot and mantle of the slug, *Incilaria fruhstorferi*. *Zool. Mag.* (*Tokyo*), **92**, 290–296.

Gahan, A. B. (1907) Greenhouse pests of Maryland. *Bull. Maryland Agric. Exp. Station*, **119**, 36.

Galangau, V. (1964) Le cycle sexuel annual de *Milax gagates* Drap. (Gastéropodes pulmonés) et ses deux pontes. *Bull. Soc. Zool. France*, **89**, 510–513.

Galangau, V. and Tuzet, O. (1966) Cou et présence de formations de nature ergastoplasmique dans les spermatides de *Milax gagates* Drap. (Gastéropodes pulmonés). *C.R. Hebdom. Séances Acad. Sci.*, **262**, 2364–2366.

Galtsoff, P. S., Lutz, F. E., Welch, P. S. and Needham, J. G. (1937) *Culture Methods for Invertebrate Animals*, Comstock Publishing Co., New York.

Garner, J. H. (1974) A study of excretion in the slug, *Agriolimax reticulatum* (Müller). PhD thesis, Univ. of Wales.

Gelperin, A. (1974) Olfactory basis of homing behaviour in the giant garden slug, *Limax maximus*. *Proc. Nat. Acad. Sci. USA*, **71**, 966–970.

Gelperin, A. (1975) Rapid food-aversion learning by a terrestrial mollusk. *Science*, **189**, 567–570.

Gelperin, A. (1986) Complex associative learning in small neural networks. *Trends Neurosci.*, **9**, 323–328.

Gelperin, A., Chang, J. J. and Reingold, S. C. (1978) Feeding motor program in Limax: I. Neuromuscular correlates and control by chemosensory input. *J. Neurobiol.*, **9**, 285–300.

Gelperin, A. and Culligan, N. (1984) *In vitro* expression of *in vivo* learning by an isolated molluscan central nervous system. *Brain Res.*, **304**, 207–214.

Gelperin, A. and Reingold, S. C. (1981) Plasticity of feeding response emitted by isolated brain of a terrestrial mollusc. In *Advances in Physiological Sciences*, Vol. 23, *Neurobiology of Invertebrates* (ed. J. Salanki), Pergamon Press, Oxford, pp. 249–266.

Gerhardt, U. (1933) Zur Kopulation der Limaciden. *Z. Morphol. Oekol. Tiere*, **27**, 401–450.

Gerhardt, U. (1935) Weitere Untersuchungen zur Kopulation der Nacktschnecken. *Z. Morphol. Oekol. Tiere*, **30**, 297–332.

Germain, L. (1930) *Faune de France. 21 Mollusques Terrestres et Fluviatiles* (première partie), P. Lechevalier, Paris.

Getz, L. L. (1959) Notes on the ecology of slugs: *Arion circumscriptus*, *Deroceras reticulatum* and *D. laeve*. *Amer. Mid. Nat.*, **61**, 485–498.

Getz, L. L. (1962) Color forms of *Arion subfuscus* in New Hampshire. *Nautilus*, **76**, 70–71.

Getz, L. L. (1974) Species diversity of terrestrial snails in the Great Smoky Mountains. *Nautilus*, **88**, 6.

Getzin, L. W. (1965) Control of the gray garden slug with bait formulations of a carbamate molluscicide. *J. Econ. Entomol.*, **58**, 158–159.

Getzin, L. W. and Cole, S. G. (1964) Evaluation of potential molluscicides for slug control. *Washington Agric. Exp. Sta. Tech. Bull. 658*. 9pp.

Ghiretti, F. and Ghiretti-Magaldi, A. (1975) Respiration. In *Pulmonates*, Vol. 1 (eds V. Fretter and J. Peake), Academic Press, New York and London, pp. 33–52.

Gietzen, D. W., Harris, A. S., Caprile, A., Larson, J. M., Leung, P. M. B. and

Rogers, O. R. (1987) Feeding responses to a tryptophan devoid diet correlation with behaviour but not plasma tryptophan or brain serotonin levels. *Soc. Neurosci. Abstr.*, **13**, 463.

Giles, C. J., Pycock, J. F., Humphreys, D. J. and Stodulski, J. B. J. (1984) Methiocarb poisoning in a sheep. *Vet. Record*, **114**, 642.

Gimingham, C. T. (1940) Pests of vegetable crops. *Ann. Appl. Biol.*, **27**, 167–168.

Gimingham, C. T. and Newton, H. C. F. (1937) A poison bait for slugs. *J. Min. Agric.*, **44**, 242–246.

Gish, C. D. (1970) Organochlorine insecticide residues in soils and soil invertebrates from agricultural lands. *Pest. Monit. J.*, **3**, 241–246.

Gittenberger, E. (1980) *Limax flavus*, new record living on the island of St. Helena. *Basteria*, **44**, 2.

Gittenberger, E. and de Winter, A. J. (1980) New data about Dutch slugs. *Basteria*, **44**, 71–76.

Gleich, J. G. and Gilbert, F. F. (1976) A survey of terrestrial gastropods from central Maine. *Canad. J. Zool.*, **54**, 620–627.

Glen, D. M. (1989) Understanding and predicting slug problems in cereals. In *Slugs and Snails in World Agriculture* (ed. I. Henderson), British Crop Protection Council, Thornton Heath, pp. 253–262.

Glen, D. M., Jones, H. and Fieldsend, J. K. (1989) Effect of glucosinolates on slug damage to oilseed rape. Production and protection of oilseed rape and other brassica crops. *Aspects Appl. Biol.*, **23**, 377–381.

Glen, D. M., Jones, H. and Fieldsend, J. K. (1990) Damage to oilseed rape seedlings by the field slug *Deroceras reticulatum* in relation to glucosinolate concentration. *Ann. Appl, Biol.*, **117**, 197–207.

Glen, D. M., Milsom, N. F. and Wiltshire, C. W. (1986) Evaluation of a mixture containing copper sulphate, aluminium sulphate and borax for control of slug damage to potatoes. *Tests of Agrochemicals and Cultivars*, No. 7 (*Ann. Appl. Biol.*, **108** Suppl.), 26–27.

Glen, D. M., Milsom, N. F. and Wiltshire, C. W. (1989) Effects of seed bed conditions on slug numbers and damage to winter wheat in clay soil. *Ann. Appl. Biol.*, **115**, 177–190.

Glen, D. M. and Orsman, I. A. (1986) Comparison of molluscicides based on metaldehyde, methiocarb or aluminium sulphate. *Crop Protect.*, **5**, 371–375.

Glen, D. M. and Wiltshire, C. W. (1986) Estimating slug populations from bait-trap catches. *1986 British Crop Protection Conf. – Pests and Diseases*, **3**, 1151–1158.

Glen, D. M., Wiltshire, C. W. and Milsom, N. F. (1984) Slugs and straw disposal in winter wheat. *1984 British Crop Protection Conf. – Pests and Diseases*, pp. 139–144.

Glen, D. M., Wiltshire, C. W. and Milsom, N. F. (1988) Effects of straw disposal on slug problems in cereals. *Env. Aspects Appl. Biol.*, Part 2. *Aspects Appl. Biol.*, **17**, 173–179.

Godan, D. (1973) Les dégâts des Limacidés et des Arionidés et leur importance économique en République Fédérale d'Allemagne. *Haliotis*, **3**, 27–32.

Godan, D. (1983) *Pest Slugs and Snails – Biology and Control* (translated by S. Gruber), Springer-Verlag, Berlin.

Golding, D. W. (1974) A survey of neuroendocrine phenomena in non-arthropod invertebrates. *Biol. Rev.*, **49**, 161–224.

Goldring, J. M., Kater, J. W. and Kater, S. B. (1983) Electrophysiological and morphological identification of action potential generating secretory cell types isolated from the salivary gland of *Ariolimax*. *J. Exp. Biol.*, **102**, 13–24.

Gorokhov, V. V. and Aliev, N. A. (1971) [The use of amides of organic acids in the control of terrestrial molluscs]. *Byul. Vsesoyuznogo Inst. Gel'mintologii*, No. 5, 15–19.

Gottfried, H., Dorfman, R. I. and Wall, P. E. (1967) Steroids of invertebrates: Production of oestrogens by an accessory reproductive tissue of the slug *Arion ater rufus* (Linn.). *Nature (London)*, **215**, 409–410.

Gottfried, H. and Dorfman, R. I. (1970a) Steroids of invertebrates: IV. On the optic tentacle-gonadal axis in the control of the male-phase ovotestis in the slug *Ariolimax californicus*. *Gen. Comp. Endocrinol.*, **15**, 101–119.

Gottfried, H. and Dorfman, R. I. (1970b) Steroids of invertebrates: VI. Effect of tentacular homogenates *in vitro* upon post-androstenedione metabolism in the male phase of *Ariolimax californicus* ovotestis. *Gen. Comp. Endocrinol.*, **15**, 139–142.

Gottfried, H. and Dorfman, R. I. (1970c) Steroids of invertebrates: V. The *in vitro* biosynthesis of steroids by the male-phase ovotestis of the slug *Ariolimax californicus*. *Gen. Comp. Endocrinol.*, **15**, 120–138.

Gottfried, H. and Lusis, O. (1966) Steroids of invertebrates: the *in vitro* production of 11-ketotestosterone and other steroids by the eggs of the slug *Arion ater rufus* (Linn.). *Nature (London)*, **212**, 1488–1489.

Goudsmit, E. M. (1972) Carbohydrates and carbohydrate metabolism in Mollusca. In *Chemical Zoology*, Vol. VII, *Mollusca* (eds M. Florkin and B. T. Scheer), Academic Press, New York and London, pp. 219–244.

Gould, H. J. (1961) Observations on slug damage to winter wheat in East Anglia. *Plant Pathol.*, **10**, 142–146.

Gould, H. J. (1962a) Trials on the control of slugs on arable fields in autumn. *Plant Pathol.*, **11**, 125–130.

Gould, H. J. (1962b) Tests with seed dressings to control grain hollowing of winter wheat by slugs. *Plant Pathol.*, **11**, 147–152.

Gould, H. J. (1965) Observations on the susceptibility of maincrop potato varieties to slug damage. *Plant Pathol.*, **14**, 109–111.

Gould, H. J. and Webley, D. (1972) Field trials for the control of slugs on winter wheat. *Plant Pathol.*, **21**, 77–82.

Gouyon, P. H., Port, P. and Caraux, G. (1983) Selection of seedlings of *Thymus vulgaris* by grazing slugs. *J. Ecol.*, **71**, 299–306.

Grant, J. F., Yeargan, K. V., Pass, B. C. and Parr, J. C. (1982) Invertebrate organisms associated with alfalfa seedling loss in complete-tillage and no-tillage plantings. *J. Econ. Entomol.*, **75**, 822–826.

Grassé, P.-P. (ed.) (1968) *Traite de Zoologie: Anatomie, Systématique, Biologie*, Vol. 5, Part III, *Mollusques Gastéropodes et Scaphopodes*, Masson et Cie, Editeurs, Paris.

Gray, J. B., Kralka, R. A. and Samuel, W. M. (1985) Rearing of eight species of terrestrial gastropods (order Stylommatophora) under laboratory conditions. *Canad. J. Zool.*, **63**, 2474–2476.

Gray, J. E. (1840) *A Manual of the Land and Freshwater Shells of the British Islands by William Turton MD*, New Edition, London.

Grega, D. S. and Prior, D. J. (1985) The effects of feeding on heart activity in the terrestrial slug, *Limax maximus*: Central and peripheral control. *J. Comp. Physiol. A. Sensory, Neural Behav. Physiol.*, **156**, 539–546.

Grega, D. S. and Prior, D. J. (1986) Modification of cardiac activity in response to dehydration in the terrestrial slug, *Limax maximus*. *J. Exp. Zool.*, **237**, 185–190.

Gregg, M. (1957) Germination of oospores of *P. erythroseptica. Nature (London)*, **180**, 150.

Gregory, P. T. (1978) Feeding habits and diet overlap of three species of garter snakes (*Thamnophis*) on Vancouver Island. *Canad. J. Zool.*, **56**, 1967–1974.

Greven, H. (1985) Vermehrung epidermaler Schleimzellen als Antwort von Lumbriciden und Gastropoden auf Stresssituationen. *Ver. Gesell. Okologie*, **15**, 321–325.

Greville, R. W. and Morgan, A. J. (1989a) Concentration of metals (Cu, Pb, Cd, Zn, Ca) in six species of British terrestrial gastropods near a disused Pb/Zn mine. *J. Mollusc. Stud.*, **55**, 31–36.

Greville, R. W. and Morgan, A. J. (1989b) Seasonal changes in metal levels (copper, lead, cadmium, zinc and calcium) within the gray field slug, *Deroceras reticulatum*, living in a highly polluted habitat. *Env. Pollution*, **59**, 287–304.

Greville, R. W. and Morgan, A. J. (1990) The influence of size on the accumulated amounts of metals (Cu, Pb, Cd, Zn and Ca) in six species of slug sampled from a contaminated woodland site. *J. Mollusc. Stud.*, **56**, 355–362.

Grime, J. P., MacPherson Stewart, S. F. and Dearman, R. S. (1968) An investigation of leaf palatability using the snail, *Cepaea nemoralis* L. *J. Ecol.*, **56**, 405–420.

Gronvold, J. and Nansen, P. (1984) The possible role of slugs in the transmission of infective larvae of *Cooperia onocophora. J. Helminthol.*, **58**, 239–240.

Grossu, A. V. (1970). Comparative study of the species of *Lytopelte* Boettger of Roumania (Fam. Limacidae, Gastropoda, Pulmonata) and a description of a new species *L. lotrensis* n.sp. *Proc. Malacol. Soc. London*, **39**, 105–110.

Grossu, A. V. (1977) Le polymorphisme des gasteropodes et la possibilité d'identifier et de limiter l'espèce avec des methodes biochimiques. *Malacologia*, **16**, 15–19.

Grossu, A. V. and Lupu, D. (1961) Die Gattung *Lytopelte* (Limacidae) in den Karpathen. *Arch. Mollusk.*, **90**, 27–31.

Grossu, A. V. and Tesio, C. (1971) Études biochimiques pour la taxonomie et la systématique du genre *Limax* (Gastropoda, Limacomorpha). *Lavori della Soc. Malacol. Italiana*, **8**, 289–300.

Grossu, A. V. and Tesio, C. (1975) Suggestions for species grouping within the family Limacidae (Gastropoda, Pulmonata) by biochemical methods. *Proc. Malacol. Soc. London*, **41**, 321–329.

Guerrier, P. (1971) Nouvelles données expérimentales sur la segmentation et l'organogenèse chez *Limax maximus* (Gastéropode Pulmone). *Ann. Embryol. Morph.*, **3**, 283–294.

Guilhon, J. (1963) Recherche sur le cycle évolutif de strongle des raisseaux du chien. *Bull. Acad. Vet. France*, **36**, 431–442.

Guilhon, J. and Cens, B. (1973) *Angiostrongylus vasorum* (Baillet, 1986). Étude biologique et morphologique. *Ann. Parasitol. Humaine Comparée*, **48**, 567–596.

Guraya, S. S. (1969) Histochemical observations on the ooplasmic components in the developing slug oocyte. *Acta Embryol. Exp.*, **2/3**, 197–209.

Gustavson, J. L. and Weight, D. G. (1981) Hypnotherapy for a phobia of slugs: A case report. *Amer. J. Clin. Hypnosis*, **23**, 258–262.

Habets, L., Vieth, U. C. and Hermann, G. (1979) Isolation and new properties of *Arion empiricorum* lectin. *Biochim. Biophys. Acta*, **582**, 154–163.

Hall, A. H. (1932) Eradication of slugs and snails. *Nature (London)*, **130**, 170.

Hamilton, P. V. (1977) The use of mucous trails in gastropod orientation studies. *Malacol. Rev.*, **10**, 73–76.

Hamilton, P. A. and Wellington, W. G. (1981a) The effects of food and density on the movement of *Arion ater* and *Ariolimax columbianus* (Pulmonata: Stylommatophora) between habitats. *Res. Pop. Ecol. (Kyoto)*, **23**, 299–308.

Hamilton, P. A. and Wellington, W. G. (1981b) The effects of food supply and density on the nocturnal behaviour of *Arion ater* and *Ariolimax columbianus* (Pulmonata: Stylommatophora). *Res. Pop. Ecol. (Kyoto)*, **23**, 309–317.

Hammond, R. B. (1985) Slugs as a new pest of soybeans. *J. Kansas Entomol. Soc.*, **58**, 364–366.

Hammond, R. B. and Stinner, B. R. (1987) Seedcorn maggots (Diptera, Anthomyiidae) and slugs in conservation tillage systems in Ohio, USA. *J. Econ. Entomol.*, **80**, 680–684.

Hansson, I. (1967) Transmission of the parasitic nematode *Skrjabingylus nasicola* to species of *Mustela* (Mammalia). *Oikos*, **18**, 247–252.

Harbourne, J. B. (1977) *Introduction to Ecological Biochemistry*, Academic Press, New York and London.

Harper, A. B. (1988) *The Banana Slug*, Bay Leaves Press, California.

Hartenstein, R. (1982) Soil macroinvertebrates, aldehyde oxidase (EC 1.2.3.1.), catalase (EC 1.11.1.6.), cellulase (EC 3.2.1.4.) and peroxidase (EC 1.11.1.7.). *Soil Biol. Biochem.*, **14**, 387–392.

Hasan, S. and Vago, C. (1966) Transmission of *Alternaria brassicicola* by slugs. *Plant Dis. Rep.*, **50**, 764–767.

Hayward, J. F. (1954) *Agriolimax caruanae* Pollonera as a Holocene fossil. *J. Conchol.*, **23**, 403–404.

Heinze, K. (1958) Können Schnecken pflanzliche Virosen übertragen? *Z. Pflanzenkrank. Pflanzenschutz*, **65**, 193–198.

Henderson, I. F. (1966) Studies on the laboratory assessment of the toxicity of chemicals to slugs, especially *Agriolimax reticulatus* (Müller). PhD thesis, Univ. of London.

Henderson, I. F. (1968) Laboratory methods for assessing the toxicity of contact poisons to slugs. *Ann. Appl. Biol.*, **62**, 363–369.

Henderson, I. F. (1969) A laboratory method for assessing the toxicity of stomach poisons to slugs. *Ann. Appl. Biol.*, **63**, 167–171.

Henderson, I. F. (1970) The fumigant effects of metaldehyde. *Ann. Appl. Biol.*, **65**, 507–510.

Henderson, I. F., Briggs, G. G., Bullock, J. I., Coward, N. P., Dawson, G. W., Larkworthy, L. F. and Pickett, J. A. (1989) A new group of molluscicidal compounds. In *Slugs and Snails in World Agriculture* (ed. I. Henderson), British Crop Protection Council, Thornton Heath, pp. 289–294.

Henderson, I. F. and Martin, A. P. (1990) Control of slugs with contact-action molluscicides. *Ann. Appl. Biol.*, **116**, 273–278.

Henderson, I. F., Martin, A. P. and Parker, K. A. (1990) Laboratory and field assessment of a new aluminium chelate slug poison. *Crop Protect.*, **9**, 131–134.

Henderson, I. F. and Parker, K. A. (1986) Problems in developing chemical control of slugs. Crop protection of sugar beet and crop protection and quality of potato, Part II. *Aspects Appl. Biol.*, **13**, 341–347.

Henderson, N. E. and Pelluet, D. (1960) The effect of visible light on the ovotestis of the slug, *Deroceras reticulatum* (Müller). *Canad. J. Zool.*, **38**, 173–178.

Hering, M. (1969) Nacktschneckenfrass an Trauben. *Weinberg und Keller*, **16**, 201–204.

Herter, K. (1938) Die Biologie der Europäischen Igel. *Monog. Wildsäuget*, **5**.

Hess, S. D. and Prior, D. J. (1985) Locomotor activity of the terrestrial slug, *Limax maximus*: Response to progressive dehydration. *J. Exp. Biol.*, **116**, 323–330.

Hess, S. D. and Prior, D. J. (1989) Small cardioactive peptide B modulates *Limax* feeding motoneurones. *J. Exp. Biol.*, **142**, 473–478.

Hill, R. S. (1977) Studies on the ovotestis of the slug, *Agriolimax reticulatus* (Mueller): 2. The epithelia. *Cell Tissue Res.*, **183**, 131–142.

Hill, R. S. (1978) Oogenesis in the slug, *Deroceras reticulatum* (Müller). PhD thesis, Univ. of Wales.

Hill, R. S. and Bowen, I. D. (1976) Studies on the ovotestis of the slug *Agriolimax reticulatus* (Müller): 1. The Oocyte. *Cell Tissue Res.*, **173**, 465–482.

Hill, R. S., Bryant, J. A. and Greenway, S. C. (1978) Characterization of ribosomal RNA in the slug, *Deroceras reticulatum*. *Comp. Biochem. Physiol. B. Comp. Biochem.*, **61**, 203–206.

Hobmaier, M. (1941) Extramammalian phase of *Skrjabingylus chitwoodorum* (Nematoda). *J. Parasitol.*, **27**, 237–239.

Hoffmann, R. J. (1983) The mating system of the terrestrial slug, *Deroceras laeve*. *Evolution*, **37**, 423–425.

Hogan, J. M. (1985) The behaviour of the grey field slug (*Deroceras reticulatum* Müller), with particular reference to control in winter wheat. PhD thesis, Univ. of Newcastle-upon-Tyne.

Hogan, J. M. and Steele, G. R. (1986) Dye-marking slugs. *J. Mollusc. Stud.*, **52**, 138–143.

Hogg, N. A. S. and Wijdenes, J. (1979) Gonadal organogenesis, and factors influencing regeneration following surgical castration in *Deroceras reticulatum* (Pulmonata: Limacidae). *Cell Tissue Res.*, **198**, 295–308.

Holyoak, D. J. (1968) A comparative study of the food of some British Corvidae. *Bird Study*, **15**, 147–153.

Holyoak, D. J. (1978) Effects of atmospheric pollution on the distribution of *Balea perversa* (Linnaeus) (Pulmonata: Clausiliidae) in Southern Britain. *J. Conchol.*, **29**, 319–323.

Holyoak, D. J. and Seddon, M. B. (1983) Land Mollusca from Norway, Finland and Sweden. *J. Conchol.*, **31**, 190.

Homeida, A. M. and Cooke, R. G. (1982) Pharmacological aspects of metaldehyde poisoning in mice. *J. Vet. Pharmacol. Therapeut.*, **5**, 77–82.

Hommay, G. and Briard, P. (1989) A few aspects of slug damage in France. In *Slugs and Snails in World Agriculture* (ed. I. Henderson), British Crop Protection Council, Thornton Heath, pp. 379–384.

Hommay, G. and Lavanceau, P. (1986) Les différentes espèces de limaces présentes en grandes cultures. *Phytoma – Défense des Cultures*, February 1986, pp. 19–22.

Hopfield, J. F. and Gelperin, A. (1989) Differential conditioning to a compound stimulus and its components in the terrestrial mollusc, *Limax maximus*. *Behav. Neurosci.*, **103**, 329–333.

Hora, S. L. (1928) Hibernation and aestivation in gastropod molluscs. *Rec. Indian Mus.*, **30**, 357–373.

Hori, E., Kano, R. and Ishigaki, Y. (1976) Experimental intermediate hosts of *Angiostrongylus cantonensis*: Studies on snails and a slug. *Japan. J. Parasitol.*, **25**, 434–440.

Hori, E., Yamaguchi, K., Fujimoto, K., Nishina, M. and Takahashi, M. (1985) Experimental studies on the development of *Angiostrongylus cantonensis* larvae in mollusks: Development under low temperatures. *Japan. J. Parasitol.*, **34**, 273–284.

Horne, F. R. (1977a) Ureotelism in the slug, *Limax flavus* Linne. *J. Exp. Zool.*, **199**, 227–232.

Horne, F. R. (1977b) Regulation of urea biosynthesis in the slug, *Limax flavus* Linne. *Comp. Biochem. Physiol. B. Comp. Biochem.* **56**, 63–69.

Horne, F. R. (1979) Comparative aspects of an aestivating metabolism in the gastropod, *Marisa cornaurietis*. *Comp. Biochem. Physiol., A. Comp. Physiol.*, **64**, 309–312.

Horne, F. R. and Barnes, G. (1970) Re-evaluation of urea biosynthesis in prosobranch and pulmonate snails. *Z. Vergleich. Physiol.*, **69**, 452–457.

Horne, F. R. and Beck, S. (1979) Purine production during fasting in the slug, *Limax flavus*. *J. Exp. Zool.*, **209**, 309–316.

Horne, F. R. and Boonkoom, V. (1970) The distribution of the ornithine cycle enzymes in twelve gastropods. *Comp. Biochem. Physiol.*, **32**, 141–153.

Horrill, J. C. and Richards, A. J. (1986) Differential grazing by the mollusk *Arion hortensis* on cyanogenic and acyanogenic seedlings of the white clover, *Trifolium repens*. *Heredity*, **56**, 277–281.

Hostettman, K. and Marston, A. (1987) Antifungal, molluscicidal and cytotoxic compounds from plants used in traditional medicine. *Ann. Proc. Phytochem. Soc. Europe*, 65–83.

Howe, G. J. and Findlay, G. M. (1972) Distribution of terrestrial slug species in Manitoba. *Manitoba Entomol.*, **6**, 46–48.

Howes, N. H. and Wells, G. P. (1934a) The water relations of snails and slugs. I. Weight rhythms in *Helix pomatia* L. *J. Exp. Biol.*, **11**, 328–343.

Howes, N. H. and Wells, G. P. (1934b) The water relations of snails and slugs. II. Weight rhythms in *Arion ater* L. and *Limax flavus* L. *J. Exp. Biol.*, **11**, 344–351.

Howitt, A. J. (1961) Chemical control of slugs in orchard grass-Ladino white clover pastures in the Pacific Northwest. *J. Econ. Entomol.*, **54**, 778–781.

Howitt, A. J. and Cole, S. G. (1962) Chemical control of slugs affecting vegetables and strawberries in the Pacific Northwest. *J. Econ. Entomol.*, **55**, 320–325.

Howling, G. G. and Port, G. R. (1989) Time-lapse video assessment of molluscicide baits. In *Slugs and Snails in World Agriculture* (ed. I. Henderson), British Crop Protection Council, Thornton Heath, pp. 161–166.

Hsiao, T. H. (1972) Chemical feeding requirements of oligophagous insects. In *Insect and Mite Nutrition* (ed. J. G. Rodrigez) North Holland Publishing Co., Amsterdam, pp. 225–240.

Hubendick, B. (1957) The eating function in *Lymnaea stagnalis*. *Ark. Zoologi*, **10**, 511–521.

Hughes, G. M. and Kerkut, G. A. (1956) Electrical activity in a slug ganglion in relation to the concentration of Locke solution. *J. Exp. Biol.*, **33**, 282–294.

Hughes, K. A. and Gaynor, D. L. (1984) Comparison of Argentine stem weevil (*Listronotus bonariensis*) and slug damage in maize direct-drilled into pasture or following winter oats. *N.Z. J. Exp. Agric.*, **12**, 47–54.

Humphreys, J. (1982) *Testacella maugei* in a Cornish garden. *The Conchologist's Newsletter*, No. 83, December 1982, p. 311.

Hunter, P. J. (1966) The distribution and abundance of slugs on an arable plot in Northumberland. *J. Animal Ecol.*, **35**, 543–557.

Hunter, P. J. (1967a) The effect of cultivations on slugs of arable land. *Plant Pathol.*, **16**, 153–156.

Hunter, P. J. (1967b) A note on nematodes in the alimentary canal of slugs. *Proc. Malacol. Soc. London*, **37**, 385.

Hunter, P. J. (1968a) Studies on slugs of arable ground: I. Sampling methods. *Malacologia*, **6**, 369–377.

Hunter, P. J. (1968b) Studies on slugs of arable ground: II. Life cycles. *Malacologia*, **6**, 379–389.

Hunter, P. J. (1968c) Studies on slugs of arable ground: III. Feeding habits. *Malacologia*, **6**, 391–399.

Hunter, P. J. (1969a) An estimate of the extent of slug damage to wheat and potatoes in England and Wales. *NAAS Qtly Rev.*, **85**, 31–36.

Hunter, P. J. (1969b) Slugs and their control. *Proc. 5th British Insecticide and Fungicide Conf.* (1969), pp. 715–719.

Hunter, P. J. and Johnston, D. L. (1970) Screening carbamates for toxicity against slugs. *J. Econ. Entomol.*, **63**, 305–306.

Hunter, P. J. and Runham, N. W. (1971) Slugs: A world problem. *Tropical Sci.*, **13**, 191–197.

Hunter, P. J. and Symonds, B. V. (1970) The distribution of bait pellets for slug control. *Ann. Appl. Biol.*, **65**, 1–7.

Hunter, P. J. and Symonds, B. V. (1971) The leap-frogging slug. *Nature (London)*, **229**, 349.

Hunter, P. J., Symonds, B. V. and Newell, P. F. (1968) Potato leaf and stem damage by slugs. *Plant Pathol.*, **17**, 161–164.

Hutson, S. (1982) *Slugs*, Allen, London.

Ikeda, K. (1937) Cytogenetic studies on the self-fertilization of *Philomycus bilineatus*. (Studies on hermaphroditism in Pulmonata II.) *J. Sci. Hiroshima Univ., Ser. B, Div. 5*, 66–123.

Ingram, W. M. (1946) Mollusk food of the beetle, *Scaphinotus interruptus* (Men.). *Bull. S. Carolina Acad. Sci.*, **45**, 34–46.

Ingram, W. M. and Adolph, H. M. (1943) Habitat and observations of *Ariolimax columbianus*. *Nautilus*, **56**.

Inoue, T. and Hayashi, E. (1983) Molluscicidal properties of 4-(Phenylazo) phenol derivatives. *Japan. J. Appl. Entomol. Zool.*, **27**, 84–91.

Intermill, R. W., Palmer, C. P., Fredrick, R. M. and Tamashiro, H. (1972)

Angiostrongylus cantonensis on Okinawa. *Japan. J. Exp. Med.*, **42**, 355–359.

Ireland, M. P. (1979) Distribution of essential and toxic metals in the terrestrial gastropod *Arion ater. Env. Pollution*, **20**, 271–278.

Ireland, M . P. (1981) Uptake and distribution of cadmium in the terrestrial slug, *Arion ater. Comp. Biochem. Physiol. A. Comp. Physiol.*, **68**, 37–42.

Ireland, M. P. (1982) Sites of water, zinc and calcium uptake and distribution of these metals after cadmium administration in *Arion ater* (Gastropoda, Pulmonata). *Comp. Biochem. Physiol. A. Comp. Physiol.*, **73**, 217–222.

Ireland, M. P. (1984) Effect of chronic and acute lead treatment in the slug *Arion ater* on calcium and Δ-aminolevulinic acid dehydratase activity. *Comp. Biochem. Physiol. C. Comp. Pharmacol. Toxicol.*, **79**, 287–290.

Ireland, M. P. (1988) A comparative study of the uptake and distribution on silver in a slug *Arion ater* and a snail *Achatina fulica. Comp. Biochem. Physiol. C. Comp. Pharmacol. Toxicol.*, **90**, 189–194.

Isarankura, K. and Runham, N. W. (1968) Studies on the replacement of the gastropod radula. *Malacologia*, **7**, 71–91.

Ivens, G. W. (ed.) (1989) *The UK Pesticide Guide*, British Crop Protection Council, Thornton Heath.

Janssen, H. H. (1985) Some histophysiological findings on the mid-gut gland of the common garden snail, *Arion rufus*, Gastropoda: Stylommatophora. *Zool. Anzeiger*, **215**, 33–51.

Jarvis, R. H. (1973) Effect of depth of planting on the yield of maincrop potatoes. *Exp. Husbandry*, **24**, 37–40.

Jarvis, R. H. (1988) The Boxworth Project. In *Britain since 'Silent Spring'* (ed. D. J. L. Harding), Institute of Biology, London, pp. 46–55.

Jary, S. G. and Austin, M. D. (1937) 'Meta-fuel' and slug control. *J. South East Agric. College, Wye*, **40**, 183–186.

Jefferson, R. N. (1952) The control of slugs: varying degrees of success achieved with metaldehyde sprays and dusts. *Florists Rev.*, **111**, 29–30, 123–127.

Jennings, T. J. (1975) The role of slugs in woodland ecosystems processes. PhD thesis, Univ. of East Anglia.

Jennings, T. J. and Barkham, J. P. (1975a) Slug populations in mixed deciduous woodland. *Oecologia (Berlin)*, **20**, 279–286.

Jennings, T. J. and Barkham, J. P. (1975b) Food of slugs in mixed deciduous woodland. *Oikos*, **26**, 211–221.

Jennings, T. J. and Barkham, J. P. (1976) Quantitative study of feeding in woodland by the slug *Arion ater. Oikos*, **27**, 168–173.

Jennings, T. J. and Barkham, J. P. (1979) Niche separation in woodland slugs. *Oikos*, **33**, 127–131.

Jensen, P. and Corbin, K. W. (1966) Some factors affecting aggregation of *Isotoma viridis* Bourlet and *Arion fasciatus* Nilsson. *Ecology*, **47**, 332–334.

Jessop, N. H. (1969) The effects of simulated slug damage on the yield of winter wheat. *Plant Pathol.*, **18**, 172–175.

Jeuniaux, C. (1954) Sur la chitinase et la flore bactérienne intestinale des mollusques gastéropodes. *Mem. Acad. Roy. Belg. Class Sci.*, **80**, **28**, 1–45.

Jezewska, M. M. (1969) The nephridial excretion of guanine, xanthine and uric acid in slugs (Limacidae) and snails (Helicidae). *Acta Biochim. Polonica*, **16**, 313–320.

Jezewska, M. M. (1971) The output of purines as related with the nitrogen intake in *Limax maximus* L. (Gastropoda). *Bull. Acad. Polonaise Sci., Serie Sci. Biol.*, **19**, 23–26.

Jezewska, M. M. (1972) Purinotelism in slugs Limacidae and Arionidae. *Bull. Acad. Polonaise Sci., Serie Sci. Biol.*, **20**, 365–368.

Johnson, G. (1965) *Entomology Department Annual Report*, Rothamsted Experimental Station Report for 1964, pp. 146–160.

Johnson, G. (1968) *Entomology Department Annual Report*, Rothamsted Experimental Station Report for 1967, pp. 197–198.

Johnson, G. (1969) *Entomology Department Annual Report*, Rothamsted Experimental Station Report for 1968, pp. 199–222.

Johnson, G. (1978) *Entomology Department Annual Report*, Rothamsted Experimental Station Report for 1977, pp. 155–156.

Johnston, K. A., Kershaw, W. J. S. and Pearce, R. S. (1989) Biochemical mechanisms of resistance of potato cultivars to slug attack. In *Slugs and Snails in World Agriculture* (ed. I. Henderson), British Crop Protection Council, Thornton Heath, pp. 281–288.

Jones, A. A. (1985) Evaluation of a microsporidian parasite of the grey field slug, *Deroceras reticulatum (Müller)*. PhD thesis, Univ. of Newcastle-upon-Tyne.

Jones, A. A. and Selman, B. J. (1984) A possible biological control agent of the grey field slug (*Deroceras reticulatum*). *1984 British Crop Protection Conf. – Pests and Diseases*, pp. 261–266.

Jones, A. A. and Selman, B. J. (1985) *Microsporidium novacastriensis*, new species, a microsporidian parasite of the gray field slug, *Deroceras reticulatum. J. Protozool.*, **32**, 581–586.

Jones, D. A. (1962) Selective eating of the acyanogenic form of the plant *Lotus corniculatus* L. by various animals. *Nature (London)*, **193**, 1109–1110.

Jones, D. A. (1973a) Co-evolution and cyanogenesis. In *Taxonomy and Ecology* (ed. V. H. Heywood), Academic Press, New York and London, pp. 213–242.

Jones, H. D. (1973b) The mechanism of locomotion of *Agriolimax reticulatus* (Mollusca: Gastropoda). *J. Zool. (London)*, **171**, 489–498.

Jones, H. D. (1975) Locomotion. In *Pulmonates*, Vol. 1 (eds V. Fretter and J. Peake), Academic Press, New York and London, pp. 1–32.

Jones, H. D. (1983) Circulatory systems of gastropods and bivalves. In *The Mollusca*, Vol. 5, *Physiology*, Part 2 (eds A. S. M. Saleuddin and K. M. Wilbur), Academic Press, New York and London, pp. 189–239.

Jong-Brink, M. de, Boer, H. H. and Joosse, J. (1983) Mollusca. In *Reproductive Biology of Invertebrates*, Vol. 1. *Oogenesis, Oviposition and Oosorption* (eds K. G. Adiyodi abd R. G. Adiyodi), Wiley, Chichester, pp. 297–355.

Joosse, J. and Reitz, D. (1969) Functional anatomical aspects of the ovotestis of *Lymnaea stagnalis. Malacologia*, **9**, 101–109.

Jourdane, J. (1972) Étude expérimentale du cycle biologique de deux espèces de *Choanotaenia intestinaux* des Soricidae. *Z. Parasitenkunde*, **38**, 333–343.

Jourdane, J. (1977) Ecology of the development and transmission of the platyhelminth parasites of soricids of the Pyrenees. *Mem. Mus. Nat. Hist. Nat. A. Zool.*, **103**, 1–174.

Judge, F. D. (1969) Preliminary screening of candidate molluscicides. *J. Econ. Entomol.*, **62**, 1393–1397.

Judge, F. D. (1972) Aspects of the biology of the grey garden slug (*Deroceras reticulatum* Müller). *Search Agric. (Geneva New York)*, **2**, 1–18.

Jungbluth, J. H., Likharev, I. M. and Wiktor, A. (1980) Comparative morphologic studies on the radula of the land slug: 1. Limacoidea and Zonitoidea (Gastropoda: Pulmonata). *Arch. Mollusk.*, **111**, 15–36.

Karlin, E. J. (1961) Temperature and light as factors affecting the locomotor activity of slugs. *Nautilus*, **74**, 125–130.

Karlin, E. J. and Bacon, C. (1961) Courtship, mating and egg laying behaviour in the Limacidae (Mollusca). *Trans. Amer. Microsc. Soc.*, **80**, 399–406.

Karlin, E. J. and Naegele, J. A. (1958) *Slugs and Snails in New York Greenhouses*. New York State College of Agriculture, Ext. Bulletin 1004, 1–16.

Karlin, E. J. and Naegele, J. A. (1960) *Biology of the Mollusca of Greenhouses in New York State*. Cornell Univ., Agriculture Experiment Station Memoir 372, 1–35.

Kasinath, B. S. and Singh, A. K. (1987) Slug lectin SL binding detects sialic acid on surface of cultured glomerular epithelial cells GEC. *Fed. Proc.*, **46**, 369.

Kassanis, B., Woods, R. D. and MacFarlane, I. (1984) Galactogen, a virus-like particle from slugs. *Ann. Appl. Biol.*, **105**, 587–589.

Kataoka, S. (1975) Fine structure of the retina of a slug, *Limax flavus* L. *Vision Res.*, **15**, 681–686.

Kataoka, S. (1976) Fine structure of the epidermis of the optic tentacle in a slug, *Limax flavus* L. *Tissue & Cell*, **8**, 47–60.

Kataoka, S. (1977) Ultrastructure of the cornea and accessory retina in a slug, *Limax flavus* L. *J. Ultrastruct. Res.*, **60**, 296–305.

Kavaliers, M. and Hirst, M. (1985) Naloxone-reversible stress-induced feeding and analgesia in the slug *Limax maximus*. *Life Sci.*, **38**, 203–210.

Kavaliers, M., Hirst, M. and Teskey, G. C. (1984) Opioid-induced feeding in the slug, *Limax maximus*. *Physiol. Behav.*, **33**, 765–768.

Kavaliers, M., Ossenkopp, K.-P. and Mathers, A. (1985) Magnetic fields inhibit opioid-induced feeding in the slug, *Limax maximus*. *Pharmacol. Biochem. Behav.*, **23**, 727–730.

Kavaliers, M., Rangeley, R. W., Hirst, M. and Teskey, G. C. (1986) Mu- and kappa-opiate agonists modulate ingestive behaviours in the slug, *Limax maximus*. *Pharmacol. Biochem. Behav.*, **24**, 561–566.

Keilin, D. (1919) On the life-history and larval anatomy of *Melinda cognata* (Diptera, Calliphorinae) parasitic in the snail *Helicella virgata* Da Costa, with an account of the other Diptera living upon molluscs. *Parasitology*, **11**, 430–455.

Kelly, M. T. and Curry, J. P. (1985) Studies on the arthropod fauna of a winter wheat crop and its response to the pesticide methiocarb. *Pedobiologia*, **28**, 413–421.

Kelly, J. R. and Martin, T. J. (1989) Twenty-one years experience with methiocarb bait. In *Slugs and Snails in World Agriculture* (ed. I. Henderson), British Crop Protection Council, Thornton Heath, pp. 131–145.

Kemp, N. J. (1987) Field estimation and the chemical control of the grey field slug, *Deroceras reticulatum* (Müller). PhD thesis, Univ. of London.

Kemp, N. J. and Newell, P. F. (1985) Laboratory observations on the effectiveness of methiocarb and metaldehyde baits against the slug *Deroceras reticulatum* (Müller). *J. Mollusc. Stud.*, **51**, 228–230.

Kemp, N. J. and Newell, P. F. (1987) Slug damage to the flag leaves of winter wheat. *J. Mollusc. Stud.*, **53**, 109–111.

Kemp, N. J. and Newell, P. F. (1989) Temporal changes in the lipid and glycogen content of adult *Deroceras reticulatum* (Müller) and *Arion hortensis agg.* In *Slugs and Snails in World Agriculture* (ed. I Henderson), British Crop Protection Council, Thornton Heath, pp. 209–216.

Kendall, C. E. T. (1921) The Mollusca of Oundle. *J. Conchol.*, **16**, 240–251.

Kennard, A. S. (1923) The Holocene non-marine Mollusca of England. *Proc. Malacol. Soc. London*, **16**, 241–259.

Kennedy, G. Y. (1959) A porphyrin pigment in the integument of *Arion ater* (L.). *J. Mar. Biol. Ass. UK*, **38**, 27–32.

Kerkut, G. A. and Taylor, B. J. R. (1956) The sensitivity of the pedal ganglion of the slug to osmotic pressure changes. *J. Exp. Biol.*, **33**, 493–501.

Kerkut, G. A. and Walker, R. J. (1975) Nervous system, eye and statocyst. In *Pulmonates*, Vol. 1 (eds V. Fretter and J. Peake), Academic Press, New York and London, pp. 165–244.

Kerney, M. P. (1966) Snails and man in Britain. *J. Conchol.*, **26**, 3–14.

Kerney, M. P. (1968) Britain's fauna of land Mollusca and its relation to the post-glacial thermal optimum. *Symp. Zool. Soc. London*, No. 22, 273–291.

Kerney, M. P. (ed.) (1976) *Atlas of the Non-marine Mollusca of the British Isles*, Institute of Terrestrial Ecology, Cambridge.

Kerney, M. P. (1982) Vice-comital census of the non-marine Mollusca of the British Isles (8th edn). *J. Conchol.*, **31**, 63–71.

Kerney, M. P. (1986) A 19th century record of *Limax maculatus* in the British Isles. *The Conchologist's Newsletter*, No. 97, June 1986, p. 361.

Kerney, M. P. (1987) Recorder's Report: Non-marine Mollusca. *J. Conchol.*, **32**, 380–383.

Kerney, M. P. (1989) Recorder's Report: Non-marine Mollusca. *J. Conchol.*, **33**, 264–265.

Kerney, M. P. and Cameron, R. A. D. (1979) *A Field Guide to the Land Snails of Britain and North-west Europe*, Collins, London.

Kerney, M. P. and Stubbs, A. (1980) *The Conservation of Snails, Slugs and Freshwater Mussels*, Nature Conservancy Council, London.

Kerney, M. P., Cameron, R. A. D., and Jungbluth, J. H. (1983) *Die Landschnecken Nord- und Mitteleuropas*, Verlag Paul Parey, Hamburg and Berlin.

Kerth, L. (1973) Radulaersatz und Zellproliferation der roentgenbestrahlten Radulascheide der Nacktschnecke *Limax flavus* L. Ergebnisse zur Arbetisteilung der Scheidengewebe. *Wilhelm Roux' Arch. Entwicklungs. Organismen*, **172**, 317–348.

Kerth, K. (1976) Light and electron microscopical studies of the radula transport of the pulmonate *Limax flavus* L. (Gastropoda, Stylommatophora). *Zoomorphologie*, **83**, 271–281.

Kerth, K. (1979) Electron microscopic studies on radular tooth formation in the snails *Helix pomatia* and *Limax flavus* (Pulmonata, Stylommatophora). *Cell Tissue Res.*, **203**, 283–290.

Kerth, K. and Krause, G. (1969) Untersuchungen mittels Roentgenbestrahlung ueber den Radula-Ersatz der Nacktschnecke *Limax flavus* L. *Wilhelm Roux' Arch. Entwicklungs. Organismen*, **164**, 48–82.

Kim, Y. K. and Chang, J. J. (1987) A study on the identified neurons related to the visceral nerve in the terrestrial slug, *Incilaria fruhstorferi-daiseniana*. *Korean J. Zool.*, **30**, 29–43.

Kingston, N. (1966) Observations on the laboratory rearing of terrestrial molluscs. *Amer. Mid. Nat.*, **76**, 528–532.

Kitchen, R. L., Schubert, T. A., Mull, R. L. and Knaak, J. B. (1978) Palatability studies of snail and slug poisons using dogs. *J. Amer. Vet. Med. Ass.*, **173**, 85–90.

Kittel, R. (1956) Untersuchungen über den Geruchs und Geschmackssin bei den Gattungen *Arion* und *Limax* (Mollusca: Pulmonata). *Zool. Anzeiger*, **157**, 185–195.

Kneidel, K. A. (1983) Fugitive species and priority during colonization in carrion-breeding Diptera communities. *Ecol. Entomol.*, **8**, 163–170.

Kneidel, K. A. (1984) Competition and disturbance in communities of carrion-breeding Diptera. *J. Animal Ecol.*, **53**, 849–866.

Knight, G. H. (1964) Some factors affecting the distribution of *Endymion nonscriptus* (L). Garcke in Warwickshire woods. *J. Ecol.*, **52**, 405–421.

Knight, W. E., Gibson, P. B., Cope, W. A., Miller, J. D. and Barnett, O. W. (1978) Comparison of cyanogenesis in four sources of white clover plants. *Crop Sci.*, **18**, 996–998.

Kniprath, E. (1980) Shell sac formation by cell delamination? *Haliotis*, **10**, 81.

Kniprath, E. (1981) Ontogeny of the molluscan shell field: A review. *Zool. Scripta*, **10**, 61–79.

Knutson, L. V., Stephenson, J. W. and Berg, C. O. (1965) Biology of a snail-killing fly, *Tetanocera elata* (Diptera, Sciomyzidae). *Proc. Malacol. Soc. London*, **36**, 213–220.

Knutson, L. V., Stephenson, J. W. and Berg, C. O. (1970) Biosystematic studies of *Salticella fasciata* (Meigen), a snail killing fly (Diptera: Sciomyzidae). *Trans. Roy. Entomol. Soc. London*, **122**, 81–100.

Ko, R. (1978) Occurrence of *Angiostrongylus cantonensis* in the heart of a spider monkey. *J. Helminthol.*, **52**, 229–230.

Kojima, S., Hata, H., Kobayashi, M., Yokogawa, M., Takahashi, N., Takaso, T. and Kaneda, J. (1979) Eosinophilic meningitis: a suspected case of angiostrongylosis found in Shizuoka Prefecture, Honshu, Japan. *Amer. J. Trop. Med. Hygiene*, **28**, 36–41.

Kosoglazov, A. A. (1972) [Malacofauna (Gastropoda, Pulmonata) of greenhouses in the southeast part of the European RSFSR]. *Zool. Zh.*, **51**, 289–290.

Kothbauer, H. (1970) Die Bedeutung von Anti-A_{HP}, einem Agglutinin aus der Eiweibdruese der Wein-bergschnecke *Helix pomatia* L. *Oecologia (Berlin)*, **6**, 48–57.

Kothbauer, H. and Schenkelbrunner, H. (1971) Haemagglutinine aus Schnecken: Zur Frage ihrer biologischen Fuhktion. *Z. Naturforsch. B.*, **26**, 1082–1084.

Kotov, A. A. (1978) Data on the ecology and behaviour of the rock dove in the southern Urals and western Siberia. *Byul. Moskovs. Obsh. Ispyt. Prirody Otdel Biol.*, **83**, 71–80.

Koval, N. F. (1976) Data on the ecology of the wryneck in the gardens of the middle Dnieper. *Vestnik Zool.*, **4**, 87–90.

Kozloff, E. N. (1956a) Experimental infection of the gray garden slug, *Deroceras reticulatum* (Müller), by the holotrichous ciliate *Tetrahymena pyriformis* (Ehrenberg). *J. Protozool.*, **3**, 17–19.

Kozloff, E. N. (1956b) *Tetrahymena limacis* from the terrestrial pulmonate gastropods *Monadenia fidelis* and *Prophysaon andersoni*. *J. Protozool.*, **3**, 204–208.

Kozloff, E. N. (1957) A species of *Tetrahymena* parasitic in the renal organ of the slug *Deroceras reticulatum*. *J. Protozool.*, **4**, 75–79.

Krajniak, K. G., Greenberg, M. J., Price, D. A., Doble, K. E. and Lee, T. D. (1989). The identification, localization and pharmacology of FMRF amide-related peptides and SCP_B in the penis and crop of the terrestrial slug *L. maximus*. *Comp. Biochem. Physiol. C. Comp. Pharmacol.*, **94**, 485–492.

Krissinger, W. A. (1984) The life history of *Lutztrema monenteron* (Trematoda: Dicrocoeliidae). *Proc. Helminthol. Soc. Washington*, **51**, 275–281.

Krochmal, A. and Lequesne, P. W. (1970) Pokeweed (*Phytolacca americana*): Possible source of a molluscicide. *US Forest Service Res. Paper N.E.*, **177**, 1–8.

Krüpe, M. and Pieper, H. (1966) Hämagglutinine von Anti-A- und Anti-B-Charakter bei einigen Landlungenschnecken. *Z. Immunitaetsforsch.*, **130**, 296–300.

Kuberski, T. and Wallace, G. D. (1979) Clinical manifestations of eosinophilic meningitis due to *Angiostrongylus cantonensis*. *Neurology*, **29**, 1566–1570.

Kugler, O. E. (1965) A morphological and histochemical study of the reproductive system of the slug *Philomycus carolinianus* (Bosc.). *J. Morphol.*, **116**, 117–132.

Kühlhorn, F. (1986) Diptera occurrence on carcasses of *Arion rufus* (Gastropoda) and its possible sanitary importance. *Angew. Parasitol.*, **27**, 123–130.

Kuhr, R. J., Davis, A. C. and Bourke, J. B. (1974) DDT residues in soil, water and fauna from New York apple orchards. *Pest. Monit. J.*, **7**, 200–204.

Kulkarni, A. B. (1971) Studies on the embryology and the early development of the land slug, *Laevicaulis alte*. *Marathwada Univ. J. Sci. Sect. B. Biol. Sci.*, **10**, 165–172.

Kulkarni, A. B. (1972) Some observations on the anatomy and histology of the digestive system of the land slug, *Laevicaulis alte*. *Marathwada Univ. J. Sci. Sect. B. Biol. Sci.*, **11**, 183–191.

Kulkarni, A. B. (1978) Effect of optic tentacle on the growth of albumin gland in the land slug *Laevicaulis alte* (Stylommatophora: Pulmonata). *Ann. Zool. (Agra)*, **14**, 171–180.

Kulkarni, A. B., Choudhari, M. V. and Vankhede, G. N. (1983) Thermal relations of the slug *Semperula maculata*. II. Changes in the neurosecretory cells. *Comp. Physiol. Ecol.*, **8**, 331–334.

Kulkarni, A. B., Jawalikar, D. P. and Bhatekawande, S. B. (1988) Environmental control over endocrine regulation of gametogenesis in a land slug, *Laevicaulis alte*. *Indian J. Comp. Animal Physiol.*, **6**, 112–119.

Kumar, T. P. and Babu, K. S. (1979) Locomotor rhythms in the slug *Laevicaulis alte* (Ferussac, 1821). *Indian J. Exp. Biol.*, **17**, 421–423.

Kumar, T. P., Mohan, T. P. M. and Babu, K. S. (1982) Rhythm in acetylcholine-acetylcholinesterase system in the slug *Laevicaulis alte*. *Acta Physiol. Polonica*, **33**, 495–502.

Kumar, T. P., Ramamurthi, R. and Babu, K. S. (1981) Circadian fluctuations in total protein and carbohydrate contents in the slug *Laevicaulis alte*. *Biol. Bull. (Woods Hole)*, **160**, 114–121.

Künkel, K. (1903) Zur Locomotion unserer Nacktschnecken. *Zool. Anzeiger*, **26**, 560–568.

Künkel, K. (1916) *Zur Biologie der Lungenschnecken*, Carl Winter, Heidelberg.

Lacatusu, M., Tudor, C., Teodorescu, I., Suciu, M., Nastasescu, M., Popescu, G. and Dinescu, I. (1984) Studies of fauna harmful and useful to successive greenhouse cultivations. *Trav. Mus. Hist. Nat. 'Grigore Antipa'*, **25**, 249–256.

Lainé, H. A. (1971) Some observations on the structure of the skin of *Agriolimax reticulatus*. MSc thesis, Univ. of Keele.

Lamotte, M. and Stern, G. (1987) Les bilans énergétiques chez les mollusques pulmonés. *Haliotis*, **16**, 103–128.

Landauer, M. R. and Cardullo, L. (1983) Feeding aversion in terrestrial slugs. *Experientia*, **39**, 1165–1166.

Lane, N. J. (1963) Microvilli on the external surfaces of gastropod tentacles and body walls. *Qtly J. Microsc. Sci.*, **104**, 495–504.

Lane, N. J. (1964a) Semper's organ, a cephalic gland in certain gastropods. *Qtly J. Microsc. Sci.*, **105**, 331–342.

Lane, N. J. (1964b) The fine structure of certain secretory cells in the optic tentacles of the snail, *Helix aspersa*. *Qtly J. Microsc. Sci.*, **105**, 35–47.

Lange, W. H. and MacLeod, G. F. (1941) Metaldehyde and calcium arsenate in slug and snail baits. *J. Econ. Entomol.*, **34**, 321–322.

Lange, W. H. and Sciaroni, R. H. (1952) Metaldehyde dusts for control of slugs affecting brussel sprouts in central California. *J. Econ. Entomol.*, **45**, 896–897.

Lankester, M. W. and Anderson, R. L. (1968) Gastropods as intermediate hosts of *Pneumostrongylus tenuis* Dougherty of white-tailed deer. *Canad. J. Zool.*, **46**, 373–383.

Lanza, B. and Quattrini, D. (1964) Richerche sulla biologia dei Veronicellidae (Gastropoda, Soleolifera). I. La reproduzione in isolamento individuale di *Vaginulus borellianus* (Colosi) e di *Laevicaulis alte* (Férussac). *Monit. Zool. Italiano*, **72**, 93–141.

Larochelle, A. (1972) Notes on the food of *Cychrini* (Coleoptera: Carabidae). *Great Lakes Entomologist*, **5**, 81–83.

Laryea, A. A. (1969) The arterial gland of *Agriolimax reticulatus* (Pulmonata: Limacidae). *Malacologia*, **9**, 273 (Abstract).

Laviolette, P. (1950a) Rôle de la gonade dans la morphogénèse du tractus génital chez quelques mollusques Limacidae et Arionidae. *C. R. Acad. Sci.*, **231**, 1567–1569.

Laviolette, P. (1950b) Régénération de la gonade après castration chez le Mollusque Gastropode *Arion rufus* L. *C. R. Acad. Sci.*, **231**, 1168–1170.

Laviolette, P. (1950c) Sur un retard de la maturité génitale observé chez *Arion rufus* L. *Bull. Mensuel Soc. Linn. Lyon*, **19**, 52–56.

Laviolette, P. (1954) Étude cytologique et expérimentale de la régénération germinale après castration chez *Arion rufus* L. *Ann. Sci. Nat. Zool. Biol. Animale*, **16**, 427–533.

Lawrence, T. C. (1939) Notes on the feeding habits of *Scolopendra subspinipes* Leach (Myriapoda). *Proc. Hawaiian Entomol. Soc.*, **8**, 497–498.

Lawrey, J. D. (1983) Lichen herbivore preference: a test of two hypotheses. *Am. J. Bot.*, **70**, 1188–1194.

Leatherdale, D. (1955) Galls of *Rondaniola bursaria* (Diptera, Cecidomyidae) eaten by slugs. *Entomol. Monthly Mag.*, **91**, 63.

Lee, J. H. and Chang, J. J. (1986) Learning of post-ingestive food-aversion by the land slug *Incilaria fruhstorferi daiseniana* to amino acid deficient diets. *Soc. Neurosci. Abstr.*, **12**, 39.

Lewis, R. D. (1969a) Studies on the locomotor activity of the slug *Arion ater* (Linnaeus): I. Humidity, temperature and light reactions. *Malacologia*, **7**, 295–306.

Lewis, R. D. (1969b) Studies on the locomotor activity of the slug *Arion ater* (Linnaeus): II. Locomotor activity rhythms. *Malacologia*, **7**, 307–312.

Lewis, T. (1980) Faunal changes and potential pests associated with direct drilling. *Bull. OEPP (Organisation Européene et Mediterranéenne pour la Protection des Plantes)*, **10**, 187–194.

Liang, C.-K. and Rosenberg, H. (1968) On the distribution and biosynthesis of 2-aminoethylphosphonate in two terrestrial molluscs. *Comp. Biochem. Physiol.*, **25**, 673–681.

Liat, L. B. (1966) Land molluscs as food of Malayan rodents and insectivores. *J. Zool. (London)*, **148**, 554–560.

Likharev, I. M. and Izzatullaev, Z. (1972) Transported species of slugs new for the fauna of Tadzhikistan. *Dokl. Akad. Nauk Tadzhikskoi SSR*, **15**, 66–68.

Likharev, I. M. and Rammel'meier, E. S. (1952) [*Terrestrial Mollusks of the Fauna of the USSR*]. Akademiya Nauk SSSR Zoologicheskii Institut, Moskva (translated from Russian by Israel Program for Scientific Translations, Jerusalem, 1962).

Likharev, I. M. and Wiktor, A. (1980) [The fauna of the slugs of the USSR and adjacent countries]. In *Fauna SSSR, Molljuski*, Vol III, Part 5, 'Nauka', Leningrad.

Limaye, L. S., Pradham, V. R., Bhopale, M. K., Renapurkar, D. M. and Sharma, K. D. (1988) *Angiostrongylus cantonensis*. Study of intermediate paratenic and definitive hosts in Greater Bombay, India. *Helminthol. (Bratislava)*, **25**, 31–39.

Linaweaver, P. G. (1966) Eosinophilic meningitis first reported cases on Guam, M. I. *Military Med.* **131**, 579–587.

Lindquist, R. K., Rollo, C. D., Ellis, C. R. L., Johnson, B. A. and Krueger, H. R. (1977) A bibliography of terrestrial slugs (Gastropoda: Stylommatophora and Systellommatophora) for agricultural researchers in North America. *Ohio Agric. Res. Devel. Center Res. Circ.* 232.

Linzell, B. S. and Madge, D. S. (1986) Effects of pesticides and fertilizer on invertebrate populations of grass and what plots in Kent (UK) in relation to productivity and yield. *Grass Forage Sci.*, **41**, 159–174.

Lisicky, M. and Ponec, J. (1979) Supplements to the work of J. Ponec: Molluscs of the lesser Carpathians, Czechoslovakia. *Zb. Slov. Narodneho Muzea Prirodne Vedy*, **25**, 105–108.

Lissmann, H. W. (1945) The mechanism of locomotion in gastropod molluscs. I. Kinematics. *J. Exp. Biol.*, **21**, 58–69.

Lissmann, H. W. (1946) The mechanism of locomotion in gastropod molluscs. II. Kinetics. *J. Exp. Biol.*, **22**, 37–50.

Lloyd, P. E. (1978) Distribution and molecular characteristics of cardioactive peptides in the snail, *Helix aspersa. J. Comp. Physiol.*, **128**, 269–276.

Lloyd, M. (1963) Numerical observations on movements of animals between beech litter and fallen branches. *J. Animal Ecol.*, **32**, 157–163.

Lockie, J. D. (1956) The food and feeding behaviour of the jackdaw, rook and carrion crows. *J. Animal Ecol.*, **25**, 421–428.

Loest, R. A. (1979) Ammonia volatilization and absorption by terrestrial gastropods: A comparison between shelled and shell-less species. *Physiol. Zool.*, **52**, 461–469.

Lohmander, H. (1937) Über die nordischen Formen von *Arion circumscriptus* Johnson. *Acta Soc. Fauna Flora Fennica*, **60**, 90–112.

Longbottom, G. M. and Gordon, A. S. M. (1979) Metaldehyde poisoning in a dairy herd. *The Vet. Rec.*, **104**, 454–455.

Lovatt, A. L. and Black, A. B. (1920) The gray garden slug. *Oregon Agric. Exp. Sta. Bull.*, **170**, 1–43.

Luchtel, D. L. (1972a) Gonadal development and sex determination in pulmonate molluscs: I. *Arion circumscriptus. Z. Zellforsch. Mikrosk. Anat.*, **130**, 279–301.

Luchtel, D. L. (1972b) Gonadal development and sex determination in pulmonate molluscs: II. *Arion ater rufus* and *Deroceras reticulatum. Z. Zellforsch. Mikrosk. Anat.*, **130**, 302–311.

Luchtel, D. L., Martin, A. W. and Deyrup-Olsen, I. (1984) The channel cell of the terrestrial slug *Ariolimax columbianus* (Stylommatophora, Arionidae). *Cell Tissue Res.*, **235**, 143–152.

Luff, M. L. (1965) The morphology and microclimate of *Dactylis glomerata* tussocks. *J. Ecol.*, **53**, 771–787.

Lupu, D. (1970) Contributions à l étude des Limacides de Roumanie. *Trav. Mus. Hist. Nat. 'Grigore Antipa'*, **10**, 61–71.

Lupu, D. (1977) Le polymorphism chez quelques éspece appartenant aux famille Limacidae et Arionidae de Roumanie. *Malacologia*, **16**, 21–33.

Lusis, O. (1961) Post embryonic changes in the reproductive system of the slug *Arion rufus* L. *Proc. Zool. Soc. London*, **137**, 433–468.

Lusis, O. (1962) Pigment of the hermaphrodite gland of *Arion ater rufus* L. *Nature (London)*, **194**, 1191–1192.

Lusis, O. (1966) Changes induced in the reproductive system of *Arion ater rufus* L. *Proc. Malacol. Soc. London*, **37**, 19–26.

Luther, A. (1915) Zuchtversuche an Ackerschnecken (*Agriolimax reticulatus* Müll. und *A. agrestis* L.). *Acta Soc. Pro Fauna Flora Fennica*, **40**, 1–42.

Lutman, J. (1978) The role of slugs in an Agrostis-Festuca grassland. In *Ecological Studies: Analysis and Synthesis*, Vol. 27, *Production Ecology of British Moors and Montane Grasslands* (eds O. W. Heal and D. F. Perkins), Springer-Verlag, New York and Berlin, pp. 332–347.

Lyth, M. (1972) Aspects of the spatial distribution of slugs (Gastropoda: Pulmonata) with some reference to their water relations. PhD thesis, Univ. of London.

Lyth, M. (1982) Water contents of slugs (Gastropoda: Pulmonata) maintained in standardized culture conditions. *J. Mollusc. Stud.*, **48**, 214–218.

Lyth, M. (1983) Water contents of slugs (Gastropoda: Pulmonata) in natural habitats and the influence of culture conditions on water content stability in *Arion ater* (L). *J. Mollusc. Stud.*, **49**, 179–184.

Machin, J. (1964a) The evaporation of water from *Helix aspersa*. I. The nature of the evaporating surface. *J. Exp. Biol.*, **41**, 759–769.

Machin, J. (1964b) The evaporation of water from *Helix aspersa*. II. Measurement of air flow and the diffusion of water vapour. *J. Exp. Biol.*, **41**, 771–781.

Machin, J. (1964c) The evaporation of water from *Helix aspersa*. III. The application of evaporation formulae. *J. Exp. Biol.*, **41**, 783–792.

Machin, J. (1966) The evaporation of water from *Helix aspersa*. IV. Loss from the mantle of the inactive snail. *J. Exp. Biol.*, **45**, 269–278.

Machin, J. (1972) Water exchange in the mantle of a terrestrial snail during periods of reduced evaporative loss. *J. Exp. Biol.*, **57**, 103–111.

Machin, J. (1974) Osmotic gradients across snail epidermis: Evidence for a water barrier. *Science (Washington DC)*, **183**, 759–760.

Machin, J. (1975) Water relationships. In *Pulmonates*, Vol. 1 (eds V. Fretter and J. Peake), Academic Press, New York and London, pp. 105–164.

Maina, J. N. (1989) The morphology of the lung of a tropical terrestrial slug, *Trichotoxon copleyi* (Mollusca: Gastropoda: Pulmonata). A scanning and transmission electron microscopic study. *J. Zool. (London)*, **217**, 355–366.

Maiorana, V. C. (1981) Herbivory in sun and shade. *Biol. J. Linn. Soc.*, **15**, 151–156.

Makings, P. (1959) *Agriolimax carunae* Poll. new to Ireland. *J. Conchol.*, **24**, 354–355.

Makra, M. E. and Prior, D. J. (1985) Angiotensin II can initiate contact-rehydration in terrestrial slugs. *J. Exp. Biol.*, **119**, 385–388.

Mallet, C. (1973) Les limaces, ennemies des jardins mais aussi des grandes cultures. *Phytoma*, No. 250, July–August 1973.

Mallet, C. and Bougaran, H. (1971) Action molluscicide de divers carbamates. *Med. Fac. Landbouw. Rijks. Gent*, **36**, 207–215.

Manaka, K.-I., Furuta, E. and Shimozawa, A. (1980) Primary tissue culture of cells in the land slugs *Incilaria bilineata*. *Zool. Mag. (Tokyo)*, **89**, 252–263.

Manna, B. and Ghose, K. C. (1972) Histopathological changes in the gut of *Achatina fulica* Bowditch caused by endrin: A molluscicide. *Indian J. Exp. Biol.*, **10**, 461–463.

Marban-Mendoza, N., Jeyaprakash, A., Jansson, H.-B., Damon, R. A. and Zuckerman, B. M. (1987) Control of root-knot nematodes on tomato by lectins. *J. Nematol.*, **19**, 331–335.

Marchand, C-R., Sokolove, P. G. and Dubois, M. P. (1984) Immunocytological localization of a somatostatin-like substance in the brain of the giant slug, *Limax maximus*. *Cell Tissue Res.*, **238**, 349–354.

Marcuzzi, G. and Lafisca, M. T. (1975) Observations on the digestive enzymes of some litter-feeding animals. in *Progress in Soil Zoology* (ed. J. Vanek), W. Junk, pp. 593–598.

Marigomez, J. A., Angulo, E. and Saez, V. (1986) Feeding and growth responses to copper, zinc, mercury and lead in the terrestrial gastropod *Arion ater* (Linne). *J. Mollusc. Stud.*, **52**, 68–78.

Marston, A. and Hostettmann, K. (1985) Plant molluscicides. *Phytochem.*, **24**, 639–652.

Martin, A. W. (1983) Excretion. In *The Mollusca*, Vol. 5, *Physiology*, Part 2 (eds A. S. Saleuddin and K. M. Wilbur), Academic Press, New York and London, pp. 353–405.

Martin, A. W. and Deyrup-Olsen, I. (1982) Blood venting through the pneumostome in terrestrial slugs. *Comp. Biochem. Physiol. C. Comp. Pharmacol.*, **72**, 53–58.

Martin, A. W. and Deyrup-Olsen, I. (1986) Function of the epithelial channel cells of the body wall of a terrestrial slug, *Ariolimax columbianus*. *J. Exp. Biol.*, **121**, 301–314.

Martin, A. W., Deyrup-Olsen, I. and Stewart, D. M. (1990) Regulation of body volume by the peripheral nervous system of the terrestrial slug *Ariolimax columbianus*. *J. Exp. Zool.*, **253**, 121–131.

Martin, A. W., Harrison, F. M., Hustar, M. J. and Stewart, D. M. (1958) The blood volume of some representative molluscs. *J. Exp. Biol.*, **35**, 260.

Martin, N. A. (1978) Introduced slugs and snails that cause damage to plants. *Proc. New Zealand Weed and Pest Control Conf. 1978*, pp. 124–126.

Martin, T. J. and Forrest, J. P. (1969) Development of DRAZA in Great Britain. *Pflanz.-Nachr. Beyer*, **22**, 205–243.

Martin, T. J. and Kelly, J. R. (1986) The effect of changing agriculture on slugs as pests of cereals. *1986 British Crop Protection Conf. – Pests and Diseases*, pp. 411–424.

Martoja, M. (1972) Endocrinology of Mollusca. In *Chemical Zoology*, Vol. VII, *Mollusca* (eds M. Florkin and B. T. Scheer), Academic Press, New York and London, pp. 349–392.

Mason, C. F. (1970) Snail populations, beech litter production, and the role of snails in litter decomposition. *Oecologia (Berlin)*, **5**, 215–239.

Mason, J. and Copeland, J. (1988) The incidence and variety of *Lehmannia valentiana* conjoined twins: related breeding experiments (Gastropoda: Pulmonata). *Malacologia*, **28**, 17–27.

Mason, J. and Copeland, J. (1989) A mechanism for the creation of conjoined twinning in *Lehmannia valentiana* present in the primary oocyte (Gastropoda: Pulmonata). *Malacologia*, **30**, 325–340.

Matekin, P. V. and Pakhorukova, L. V. (1980) Genotypic interpretation of some ecological parameters of population dynamics. *Zool. Zh.*, **59**, 163–174.

Matzke, M. (1987) Gastropods at succession sites of brown coal mining beside Halle (Saale) (East Germany). *Malakol. Abh. (Dresden)*, **12**, 39–48.

Maurin, G., Lavanceau, P. and Fougeroux, A. (1989) A method for the study of molluscicides. In *Slugs and Snails in World Agriculture* (ed. I. Henderson), British Crop Protection Council, Thornton Heath, pp. 167–170.

Maury, M. F. and Reygrobellet, D. (1963) Sur les distinctions spécifique chez les mollusques limacidés du genre *Deroceras*. *C.R. Hebdom. Séances Acad. Sci.*, **288**, 2902–2903.

Maxwell, W. L. (1977) Freeze-etching studies of pulmonate spermatozoa. *Veliger*, **20**, 71–74.

Maxwell, W. L. (1980) Distribution of glycogen deposits in 2 euthyneuran sperm tails. *Internat. J. Invert. Reproduct. Devel.*, **2**, 245–250.

Maxwell, W. L. (1983) Mollusca. In *Reproductive Biology of Invertebrates*, Vol. II, *Spermatogenesis and Sperm Function* (eds K. G. Adiyodi and R. G. Adiyodi), Wiley, Chichester and New York, pp. 275–319.

Maze, R. J. and Johnstone, C. (1986) Gastropod intermediate hosts of the meningeal worm *Parelophostrongylus tenuis* in Pennsylvania (USA): Observations on their ecology. *Can. J. Zool.*, **64**, 185–188.

Mead, A. R. (1943) Revision of the giant west coast land slugs of the genus *Ariolimax* Mörch (Pulmonata: Arionidae). *Amer. Mid. Nat.*, **30**, 675–717.

Mead, A. R. (1961) *The Giant African Snail: A Problem in Economic Malacology*, Univ. of Chicago Press, Chicago.

Mead, A. R. (1979) Economic malacology, with particular reference to *Achatina fulica*. In *Pulmonates*, Vol. 2B (eds V. Fretter and J. Peake), Academic Press, New York and London, pp. 1–144.

Meenakshi, V. R. and Scheer, B. T. (1968) Studies on the carbohydrates of the slug *Ariolimax columbianus* with special reference to their distribution in the reproductive system. *Comp. Biochem. Physiol.*, **26**, 1091–1097.

Meenakshi, V. R. and Scheer, B. T. (1969) Regulation of galactogen synthesis in the slug *Ariolimax columbianus*. *Comp. Biochem. Physiol.*, **29**, 841–845.

Meenakshi, V. R. and Scheer, B. T. (1970) Chemical studies of the internal shell of the slug, *Ariolimax columbianus* (Gould) with special reference to the organic matrix. *Comp. Biochem. Physiol.*, **34**, 953–957.

Mellanby, K. (1961) Slugs at low temperatures. *Nature (London)*, **189**, 944.

Melrose, G. R., O'Neil, M. C. and Sokolove, P. G. (1983) Male gonadotropic factor in brain and blood of photoperiodically stimulated slugs (*Limax maximus*). *Gen. Comp. Endocrinol.*, **52**, 319–328.

Michelson, E. H. (1971) Distribution and pathogenicity of *Tetrahymena limacis* in the slug *Deroceras reticulatum*. *Parasitology*, **62**, 125–131.

Mienis, H. K. (1988) Slugs in Jerusalem seem to be fond of milk too. *The Conchologist's Newsletter*, June 1988, 104–105.

Mienis, H. K. (1989) The marsh slug *Deroceras laeve* (Mollusca, Gastropoda) feeding on the Florida wax scale *Ceroplastes floridens* (Insecta, Rhynchota) in Israel. *Z. Angew. Zool.*, **76**, 377–378.

Miles, H. W. (1924) Observations on the hatching of the field slug *Agriolimax agrestis* Linn. *Scottish Naturalist*, No. 149, 131–134.

Miles, H. W., Wood, J. and Thomas, I. (1931) On the ecology and control of slugs. *Ann. Appl. Biol.*, **18**, 370–400.

Miles, P. M. (1969) Technique of recording slug movement and some results of tests with metaldehyde and bran, and bran baits only. *Lab. Pract.*, **18**, 437–440.

Miller, R. L. (1982) A sialic acid-specific lectin from the slug, *Limax flavus*. *J. Invert. Pathol.*, **39**, 210–214.

Miller, R. L., Collawn, J. F. and Fish, W. W. (1982) Purification and macromolecular properties of a sialic acid-specific lectin from the slug, *Limax flavus*. *J. Biol. Chem.*, **257**, 7574–7580.

Mills, J. D., Bailey, S. E. R., Wedgwood, M. A. and McCrohan, C. R. (1989) Effects of molluscicides on feeding behaviour and neuronal activity. In *Slugs and Snails in World Agriculture* (ed. I. Henderson), British Crop Protection Council, Thornton Heath, pp. 77–84.

Milne, A. (1957) The natural control of insect populations. *Canad. Entomol.*, **89**, 193–213.

Milne, A. (1962) On a theory of natural control of insect populations. *J. Theor. Biol.*, **3**, 19–50.

Milne, A., Coggins, R. E. and Laughlin, R. (1958) The determination of numbers of leatherjackets in sample turves. *J. Animal Ecol.*, **27**, 125.

Mistic, W. J. and Morrison, D. W. (1979) Control of slugs in Burley tobacco fields in the Appalachian Mountains of North Carolina. *Internat. Tobacco*, **181**, 60–61.

Moens, R. (1971) Test of the effectiveness of some new molluscicidal preparations in the control of slugs. *Med. Fac. Landbouw. Rijks. Gent*, **36**, 216–223.

Moens, R. (1980) Le problème des limaces dans la protection des végétaux. *Rev. Agric. (Bruxelles)*, **33**, 117–132.

Moens, R. (1983) Essais sur le protection des grains de froment contre l'attaque des limaces. *Rev. Agric. (Bruxelles)*, **36**, 1303–1317.

Moens, R., François, E. and Riga, A. (1966) Les radioisotopes en écologie animale: Deplacements dans un milieu naturel de *Agriolimax reticulatus* Müller. *Med. Fac. Landbouw. Rijks. Gent*, **31**, 1032–1042.

Moens, R., François, E., Riga, A. and van den Bruel, W. E. (1965) Les radioisotopes en écologie animale. Premières informations sur le comportement de *Agriolimax reticulatus* Müller. *Med. Landbouw. Opzoekings. Staat Gent*, **3**, 1810–1823.

Moens, R., François, E., Riga, A. and van den Bruel, W. E. (1967) A mechanical barrier against terrestrial gastropods. *Parasitica*, **23**, 22–27.

Moens, R. and Rassel, A. (1985) Étude comparée de radulas de quelques stylommatophores (Mollusques, Gastropodes) par microscopie électronique à balayage. *Ann. Soc. Roy. Zool. Belg.*, **115**, 45–60.

Mølgaard, P. (1986a) Food plant preferences by slugs and snails: A simple method to evaluate the relative palatability of the food plants. *Biochem. System. Ecol.*, **14**, 113–1222.

Mølgaard, P. (1986b) Population genetics and geographical distribution of caffeic acid esters in leaves of *Plantago major* in Denmark. *J. Ecol.*, **74**, 1127–1138.

Moore, C. H. (1934) Slug and beetle. *J. Conchol.*, **20**, 85.

Mordan, P. B. (1973) Aspects of the ecology of terrestrial gastropods at Monks Wood NNR and other woodlands, with special reference to the Zonitidae. PhD thesis, Univ. of London.

Morris, P. (1983) *Hedgehogs*, Whittet Books, Weybridge.

Morton, B. (1979) The diurnal rhythm and the cycle of feeding and digestion in the slug *Deroceras caruanae*. *J. Zool. (London)*, **187**, 135–152.

Moya, J. and Rallo, A. M. (1975) Intracisternal polycylinders: A cytoplasmic structure in cells of the terrestrial slug *Arion empiricorum* Férussac (Pulmonata, Stylommatophora). *Cell Tissue Res.*, **159**, 423–433.

Mozley, A. (1952) *Molluscicides*, Lewis, London.

Müller, S. and Ohnesorge, B. (1985) The use of marked slugs in estimating population densities and studying migrations of *Arion* sp. *Anz. Schaed. Pflanz. Umwelt.*, **58**, 123–126.

Murthy, M. S. and Ramamurthi, R. (1978) Body component indices and amino acid pools in the slug *Laevicaulis alte* (Ferr. 1881). *Comp. Physiol. Ecol.*, **3**, 7–12.

Musick, G. J. (1972) Efficacy of phorate for control of slugs in field corn. *J. Econ. Entomol.*, **65**, 220–222.

McCarthy, P. M. and Healy, J. A. (1978) Dispersal of lichen propagules by slugs. *The Lichenologist*, **10**, 131.

McCracken, G. F. and Selander, K. S. (1980) Self fertilization and monogenic strains in natural populations of terrestrial slugs. *Proc. Nat. Acad. Sci. USA*, **77**, 684–688.

McCrone, E. J. and Sokolove, P. G. (1979) Brain-gonad axis and photoperiodically-stimulated sexual maturation in the slug, *Limax maximus. J. Comp. Physiol. A. Sensory, Neural Behav. Physiol.*, **133**, 117–124.

McCrone, E. J. and Sokolove, P. G. (1986) Photoperiodic activation of brains in castrates and the role of the gonad in reproductive maturation of *Limax maximus. J. Comp. Physiol. A. Sensory, Neural Behav. Physiol.*, **158**, 151–158.

McCrone, E. J., van Minnen, J. and Sokolove, P. G. (1981) Slug (*Limax maximus*) reproductive maturation hormone: *In vivo* evidence for long-day stimulation of secretion from brains and cerebral ganglia. *J. Comp. Physiol. A. Sensory, Neural Behav. Physiol.*, **143**, 311–316.

MacDougall, R. S. (1942) The mole. Its life-history, habits and economic importance. *Trans. Roy. Highland Agric. Soc., Scotland*, **54**, 80–107.

MacKay, A. R. and Gelperin, A. (1972) Pharmacology and reflex responsiveness of the heart in the giant garden slug, *Limax maximus. Comp. Biochem. Physiol. A. Comp. Physiol.*, **43**, 877–896.

McMillan, N. F. (1954) The Mollusca of Patrick's Wood, Bromborough, Cheshire. *J. Conch.*, **24**, 13–16.

McMillan, N. F. (1969) *Agriolimax caruanae* Pollonera in Ireland. *Irish Nat.*, **16**, 178.

Nagabhushanam, R. and Kulkarni, A. B. (1971a) Histochemical observations on the egg capsules of the land slug, *Laevicaulis alte. Marathwada Univ. J. Sci., Sect. B. Biol. Sci.*, **10**, 121–122.

Nagabhushanam, R. and Kulkarni, A. B. (1971b) Reproductive biology of the land slug, *Laevicaulis alte. Riv. Biol. (Perugia)*, **64**, 15–44.

Nagabhushanam, R. and Kulkarni, A. B. (1971c) Neurosecretion in the slug, *Laevicaulis alte. Proc. Indian Acad. Sci. Sect. B*, **73**, 290–302.

Neck, R. W. (1984) Living terrestrial gastropods from the Eastern Caprock Escarpment, Texas (USA). *Nautilus*, **98**, 68–74.

Neijzing, M. G. and Zeven, A. C. (1976) Anther eating by snails and slugs in *Streptocarpus. Acta Bot. Neerlandica*, **25**, 337–339.

Nelson, J. M. (1971) The invertebrates of an area of Pennine moorland within the Moor House Nature Reserve in Northern England. *Trans. Soc. Brit. Entomol.*, **19**, 173–235.

Newell, P. F. (1965) Recent methods of marking invertebrate animals for behavioural studies. *Animal Behav.*, **13**, 579 (Abstract).

Newell, P. F. (1966a) The analysis of the nocturnal behaviour of slugs on the surface of the soil. *Med. Biol. Illus.*, **16**, 146–159.

Newell, P. F. (1966b) Time lapse cine recording the soil surface activity of slugs. *Animal Behav.*, **13**, 583.

Newell, P. F. (1967) Molluscs: methods for estimating production and energy flow. In *Methods of Study in Soil Ecology* (ed. J. Phillipson), UNESCO Paris, pp. 285–291.

Newell, P. F. (1971). Molluscs. In *Methods of Study in Quantitative Soil Ecology: Population, Production and Energy Flow* (ed. J. Phillipson), IBP Handbook No. 18, Blackwell, Oxford, pp. 128–149.

Newell, P. F. (1973) Étude de l'ultrastructure de l'épithelium dorsal et pédieux des Limaces *Arion hortensis* Férussac et *Agriolimax reticulatus* Müller. *Haliotis*, **3**, 131–142.

Newell, P. F. (1977) The structure and enzyme histochemistry of slug skin. *Malacologia*, 16, 183–195.

Newell, P. F. and Appleton, T. C. (1979) Aestivating snails – the physiology of water regulation in the mantle of the terrestrial pulmonate *Otala lactea*. *Malacologia*, 18, 575–581.

Newell, P. F. and Machin, J. (1976) Water regulation in aestivating snails: Ultrastructural and analytical evidence for an unusual cellular phenomenon. *Cell Tissue Res.*, 173, 417–421.

Newell, P. F. and Newell, G. E. (1968) The eye of the slug *Agriolimax reticulatus*. *Symp. Zool. Soc. London*, 23, 97–111.

Nicholas, J. (1984) The biology of reproduction in two British pulmonate slugs. PhD thesis, Univ. of Wales.

Nicklas, N. L. and Hoffmann, R. J. (1981) Apomictic parthenogenesis in a hermaphroditic terrestrial slug, *Deroceras laeve*. 160, 123–135.

Niemelä, P., Tuomi, J. and Molarius, A. (1988) Feeding aversion to conspecific material in a terrestrial slug *Deroceras agreste*. *Agric. Eco. Env.*, 20, 175–180.

Nilsson, C. and Sorensson, A. (1989) Molluscicidae effects of extracts of endod *Phytolacca* from Sweden. *Vaxtskyddsnotiser*, 53, 104–105.

Noda, S., Sato, A., Nojima, H., Watanabe, Y., Kawabata, N. and Matayoshi, S. (1982) A survey of *Angiostrongylus cantonensis* in the Amami Islands (Japan). 2. The occurrence of *Angiostrongylus cantonensis* in snails, slugs and rodents in Okierabu-jima. *Japan. J. Parasitol.*, 31, 329–338.

Noel, P. (1891) Destruction des limacons par le crapaud et la genouille. *Rev. Sci. Nat. Ouest (Paris)*, 1, 261–262.

Norris, A. (1987) The generic classification of British Milacidae. *J. Conchol.*, 32, 387–388.

North, M. C. and Bailey, S. E. R. (1989) Distribution of *Boettgerilla pallens* in North-West England. In *Slugs and Snails in World Agriculture* (ed. I. Henderson), British Crop Protection Council, Thornton Heath, pp. 327–330.

Ogren, R. E. (1959a) The nematode *Cosmocercoides dukae* as a parasite of the slug. *J. Parasitol.*, 45, suppt 45.

Ogren, R. E. (1959b) The nematode *Cosmocercoides dukae* as a parasite of the slug. *Proc. Pennsylvania Acad. Sci.*, 33, 236–241.

Ojanova, N. (1964) Über die Biologie und Ökologie von Zwei für die Fauna Bulgariens neuen arten der Familie Arionidae. *Izv. Zool. Inst. Sofya*, 15, 203–214.

Oldham, C. (1915) *Testacella scutulum* in Hertfordshire. *Trans. Hertfordshire Nat. Hist. Soc.*, 15, 193–194.

Oldham, C. (1922) *Limax tenellus* in Gloucester West, Hereford and Montgomery. *J. Conchol.*, 16, 276.

Oldham, C. (1942) Auto-fecundation and duration of life in *Limax cinereoniger*. *Proc. Malacol. Soc. London*, 25, 9–10.

Ong, C. K., Marshall, C. and Sagar, G. R. (1978) The physiology of tiller death in a grass sward. *J. Brit. Grassland Soc.*, 33, 205–212.

Opalinski, K. W. (1981) Respiration of a slug, *Limax* sp. in a degraded area. *Polish Ecol. Stud.*, 7, 29–36.

Orive, E., Berjon, A. and Otero, M. P. (1979) A comparative study of intestinal absorption in *Arion empiricorum* and *Helix pomatia*. *Comp. Biochem. Physiol. A. Comp. Physiol.*, 64, 557–564.

Orive, E., Berjon, A. and Otero, M. P. (1980) Metabolism of nutrients during intestinal adsorption in *Helix pomatia* and *Arion empiricorum* (Gastropoda: Pulmonata). *Comp. Biochem. Physiol. B. Comp. Biochem.*, **66**, 155–158.

Orth, R. E., Moore, I., Fisher, J. W. and Legner, E. F. (1975) A rove beetle *Ocypus olens* with potential for biological control of the brown garden snail *Helix aspersa* in California, including a key to the nearctic species of *Ocypus*. *Canad. Entomol.*, **107**, 1111–1116.

Osanova, N. (1970) Die Nacktschnecken im westlichen Teil des Balkan-Gebirges (Bulgarien). *Malakol. Abh. (Dresden)*, **3**, 71–79.

Osborne, N. N. (1971) Distribution of biogenic amines in the slug, *Limax maximus*. *Z. Zellforsch. Mikrosk. Anat.*, **112**, 15–30.

Pakhorukova, L. V. (1976). A quantitative study of the feeding of the gray field (*Deroceras agrestis*) and field (*Deroceras reticulatum*) slugs. *Zool. Zh.*, **55**, 29–33.

Pakhorukova, L. V. and Matekin, P. V. (1977) Interspecific differences in the effect of temperature on the duration of embryonic development in slugs. *Zh. Obsh. Biol.*, **38**, 116–122.

Pallant, D. (1967) Studies on the feeding of the grey field slug Agriolimax Reticulatus (Müller) in the laboratory and in woodland. MSc thesis, Univ. of Durham.

Pallant, D. (1969) The food of the grey field slug (*Agriolimax reticulatus*) (Müller) in woodland. *J. Animal Ecol.*, **38**, 391–397.

Pallant, D. (1970) A quantitative study of feeding in woodland by the grey field slug (*Agriolimax reticulatus* (Müller)). *Proc. Malacol. Soc. London*, **39**, 83–87.

Pallant, D. (1972) The food of the grey field slug (*Agriolimax reticulatus*) (Müller) on grassland. *J. Animal Ecol.*, **41**, 761–769.

Pallant, D. (1974) Assimilation in the grey field slug (*Agriolimax reticulatus*) (Müller). *Proc. Malacol. Soc. London*, **41**, 99.

Palombi, A. (1949) I molluschi dannosi all'agricoltura. *Rivista Fitosani.*, **2**, 7–13.

Parivar, K. (1978) A histological survey of gonadal development in *Arion ater* L. (Mollusca, Pulmonata). *J. Mollusc. Stud.*, **44**, 250–264.

Parivar, K. (1980) Differentiation of Sertoli cells and postreproductive epithelial cells in the hermaphrodite gland of *Arion ater* (Mollusca, Pulmonata). *J. Mollusc. Stud.*, **46**, 139–147.

Parivar, K. (1981) Spermatogenesis and sperm distribution in the land slug *Arion ater*. *Z. Mikrosk.-Anat. Forsch. (Leipzig)*, **95**, 81–92.

Parivar, K. (1982) Organ culture studies on cell differentiation in the hermaphrodite gland of *Arion ater* L. (Mollusca, Pulmonata). *J. Mollusc. Stud.*, **48**, 355–361.

Patterson, C. M. and Burch, J. B. (1978) Chromosomes of pulmonate molluscs. In *Pulmonates*, Vol. 2A (eds V. Fretter and J. Peake), Academic Press, New York and London, pp. 171–213.

Paul, C. R. C. (1975) The ecology of Mollusca in ancient woodland. 1. The fauna of Hayley Wood, Cambridgeshire. *J. Conchol.*, **28**, 301–327.

Paul, C. R. C. (1978a) The ecology of Mollusca in ancient woodland. 2. Analysis of distribution and experiments in Hayley Wood, Cambridgeshire. *J. Conchol.*, **29**, 281–294.

Paul, C. R. C. (1978b) The ecology of Mollusca in ancient woodland. 3

Frequency of occurrence in West Cambridgeshire Woods. *J. Conchol.*, **29**, 295–300.

Pavan, K. T. and Babu, K. S. (1979) Locomotor rhythms in the slug *Laevicaulis alte* (Ferunsac, 1921). *Ind. J. Exp. Biol.*, **17**, 421–423.

Pavan, K. T., Mohan, P. M. and Babu, K. S. (1982) Rhythm in acetylcholine-acetylcholinesterase system in the slug *Laevicaulis alte*. *Act. Physiol. Pol.*, **33**, 495–502.

Peacock, L. and Norton, G. A. (1990) A critical analysis of organic vegetable crop protection in the UK. *Agric. Ecosys. Env.*, **31**, 187–198.

Peake, J. (1978) Distribution and ecology of the Stylommatophora. In *Pulmonates*, Vol. 2A (eds V. Fretter and J. Peake), Academic Press, New York and London, pp. 429–526.

Pearl, R. (1928) *The Rate of Living*, Knopf, New York.

Pearl, R. and Miner, J. R. (1935) Experimental studies in the duration of life. XIV. The comparative mortality of certain lower organisms. *Qtly Rev. Biol.*, **10**, 60–79.

Pellerdy, L. P. (1965) *Coccidia and Coccidiosis*, Akademia Kiado, Budapest.

Pelluet, D. (1964) On the hormonal control of cell differentiation in the ovotestis of slugs (Gastropoda: Pulmonata). *Canad. J. Zool.*, **42**, 195–199.

Pelluet, D. and Lane, N. J. (1961) The relation between neurosecretion and cell differentiation in the ovotestis of slugs (Gastropoda: Pulmonata). *Canad. J. Zool.*, **39**, 789–805.

Pelseener, P. (1928) Les parasites des mollusques et les mollusques parasites. *Bull. Soc. Zool. Anvers*, **53**, 158–189.

Pelseener, P. (1934) La durée de la vie et l'âge de la maturité sexuelle chez certains mollusques. *Ann. Soc. Zool. Belg.*, **14**, 93.

Pemberton, R. T. (1970) Haemagglutinins from the slug *Limax flavus*. *Vox Sanguinis*, **18**, 74–76.

Perju, T., Rozalia, Z., Ghidra, V. and Cornea, O. (1981) Bioecology and control of *Solanum laciniatum* crop pests. *Bul. Inst. Agron. Clut-Napoca Seria Agric.*, **35**, 121–126.

Pernetta, J. C. (1976) Diets of the shrews *Sorex araneus* L. and *Sorex minutus* L. in Wytham grassland. *J. Animal Ecol.*, **45**, 899–912.

Personne, P. and Andre, J. (1964) Existence de glycogene mitochondrial dans le spermatozoide de la testacelle. *J. Microsc. (Paris)*, **3**, 643–650.

Pessah, I. N. and Sokolove, P. G. (1983a) Determination and characterization of cholinesterases in localized tissues of giant garden slugs *Limax maximus* (Linnaeus). *Comp. Biochem. Physiol. C. Comp. Pharmacol.*, **74**, 281–290.

Pessah, I. N. and Sokolove, P. G. (1983b) The interaction of organophosphate and carbamate insecticides with cholinesterases in the terrestrial pulmonate *Limax maximus* (Linnaeus). *Comp. Biochem. Physiol. C. Comp. Pharmacol.*, **74**, 291–298.

Petersen, H. and Luxton, M. (1982) A comparative analysis of soil fauna populations and their role in decomposition processes. *Oikos*, **39**, 288–354.

Petter, A. J. (1974) Le cycle evolutif de *Morerastrongylus andersoni* (Peter, 1972). *Ann. Parasitol. Humaine Comp.*, **49**, 69–82.

Phifer, C. B. and Prior, D. J. (1985) Body hydration and haemolymph osmolality affect feeding and its neural correlate in the terrestrial gastropod (*Limax maximus*). *J. Exp. Biol.*, **118**, 405–422.

Philip, E. G. (1987) *Tandonia rustica* (Millet), a slug new to the British Isles. *J. Conchol.*, **32**, 302.

Phillips, R. A. and Watson, H. (1930) *Milax gracilis* (Leydig) in the British Isles. *J. Conchol.*, **19**, 65–93.

Phillipson, J. (1983) Slug numbers, biomass and respiratory metabolism in a beech woodland: Wythan Woods, Oxford (England, UK). *Oecologia (Berlin)*, **60**, 38–45.

Pickett, J. A. and Stephenson, J. W. (1980) Plant volatiles and components influencing behaviour of the field slug *Deroceras reticulatum*. *J. Chem. Ecol.*, **6**, 435–444.

Piéron, H. (1928) Sensibilité à la pesanteur et réactions géotropiques chez les limaces. *Ann. Physiol. Phys. Biol.*, **4**, 44.

Pilsbry, H. A. (1948) Land Mollusca of North America north of Mexico. *Acad. Nat. Sci. Philadelphia*, Monograph 3, **2**, 521–1113.

Pinder, L. C. V. (1969) The biology and behaviour of some slugs of economic importance, *Agriolimax reticulatus*, *Arion hortensis* and *Milax budapestensis*. PhD thesis, Univ. of Newcastle-upon-Tyne.

Pinder, L. C. V. (1974) The ecology of slugs in potato crops, with special reference to the differential susceptibility of potato cultivars to slug damage. *J. Appl. Ecol.*, **11**, 439–451.

Pitcher, R. S. (1950) Slug damage to raspbery canes. *Rep. East Malling Res. Sta.*, *1949*, **37**, 131.

Pitchford, G. W. (1954) Blackbird feeding on *Arion subfuscus*. *J. Conchol.*, **24**, 24.

Pitchford, G. W. (1956) *Agriolimax reticulatus* (Müller) feeding on carrion. *J. Conchol.*, **24**, 96.

Platt, T. R. and Samuel, W. M. (1984) Mode of entry of 1st-stage larvae of *Parelaphostrongylus odocoilei* (Nematoda, Metastrongyloidea) into 4 species of terrestrial gastropods. *Proc. Helminthol. Soc. Washington*, **51**, 205–207.

Platts, E. A. and Speight, M. C. D. (1988) The taxonomy and distribution of the Kerry slug *Geomalacus maculosus* Allman 1843 (Mollusca Arionidae) with a discussion of its status as a threatened species. *Irish Nat. J.*, **22**, 417–430.

Plisetskaya, E. M. and Deyrup-Olsen, I. (1987) An insulin-like substance in the blood of the slug *Prophysaon foliolatum* Arionidae in the course of tail regeneration. *Comp. Biochem. Physiol. A. Comp. Physiol.*, **87**, 781–784.

Poivre, C. (1972) Observations sur le comportement predateur de l'orvet (*Anguis fragilis* L.): II. Capture de diverses proies. *Terre et la Vie*, **29**, 63–70.

Popham, J. D. and D'Auria, M. (1980) *Arion ater* (Mollusca Pulmonata) as an indicator of terrestrial environmental pollution. *Water, Air, Soil Poll.*, **14**, 115–124.

Port, C. M and Port, G. R. (1986) The biology and behaviour of slugs in relation to crop damage and control. *Agric. Zool. Rev.*, **1**, 255–299.

Potts, W. T. W. (1967) Excretion in the Molluscs. *Biol. Rev.*, **42**, 1–41.

Poulicek, M. and Jaspar-Versali, M. F. (1982) Essai d'interprétation d'un cycle saisonnier de la limacelle chez quelques pulmones limacidae. *Malacologia*, **22**, 241–244.

Poulicek, M. and Voss-Foucart, M. F. (1980) Seasonal variation of chemical composition of *Agriolimax reticulatus* (Gastropoda, Limacidae). *Arch. Zool. Exp. Generale*, **121**, 77–86.

Poulin, G. and O'Neil, L. C. (1969) Observations sur les prédateurs de la limace noire, *Arion ater* (L.) (Gastéropodes, Pulmonés, Arionidés). *Phyto-protection*, **50**, 1-6.

Prior, D. J. (1981) Hydration related behaviour and the effects of osmotic stress on motor function in the slugs *Limax maximus* and *Limax pseudoflavus*. In *Advances in Physiological Sciences*, Vol. 23, *Neurobiology of Invertebrates - Mechanisms of Integration* (ed. J. Salanki), Pergamon Press, Oxford, pp. 131-146.

Prior, D. J. (1982) Osmotic control of drinking behaviour in terrestrial slugs. *Amer. Zool.*, **22**, 978.

Prior, D. J. (1983a) Hydration-induced modulation of feeding responsiveness in terrestrial slugs. *J. Exp. Zool.*, **227**, 15-22.

Prior, D. J. (1983b) The relationship between age and body size of individuals in isolated clusters of the terrestrial slug, *Limax maximus*. *J. Exp. Zool.*, **225**, 321-324.

Prior, D. J. (1985) Water regulatory behaviour in terrestrial gastropods. *Biol. Rev.*, **60**, 403-424.

Prior, D. J. (1989) Contact-rehydration in slugs: a water regulatory behaviour. In *Slugs and Snails in World Agriculture* (ed. I. Henderson), British Crop Protection Council, Thornton Heath, pp. 217-223.

Prior, D. J. and Delaney, K. (1986) Activation of buccal neuron B1 in the edible slug, *Limax maximus*, mimics the action of exogenous SCP_B. *Amer. Zool.*, **26**, 126.

Prior, D. J. and Gelperin, A. (1974) Behavioural and physiological studies on locomotion in the giant garden slug, *Limax maximus*. *Malacol. Rev.*, **7**, 50-51.

Prior, D. J. and Grega, D. S. (1982) Effects of temperature on the endogenous activity and synaptic interactions of the salivary burster neurons in the terrestrial slug, *Limax maximus*. *J. Exp. Biol.*, **98**, 415-428.

Prior, D. J., Hume, M., Varga, D. and Hess, S. D. (1983) Physiological and behavioural aspects of water balance and respiratory function in the terrestrial slug, *Limax maximus*. *J. Exp. Biol.*, **104**, 111-128.

Prior, D. J. and Uglem, G. L. (1984a) Analysis of contact-rehydration in terrestrial gastropods: Osmotic control of drinking behaviour. *J. Exp. Zool.*, **111**, 63-73.

Prior, D. J. and Uglem, G. L. (1984b) Analysis of contact-rehydration in terrestrial gastropods: Absorption of ^{14}C-inulin through the epithelium of the foot. *J. Exp. Biol.*, **111**, 75-80.

Prior, D. J. and Watson, W. H. (1988) The molluscan neuropeptide SCP_B increases the responsiveness of the feeding motor program of *Limax maximus*. *J. Neurobiol.*, **19**, 87-106.

Prystupa, B. D., Holliday, N. J. and Webster, G. R. B. (1987) Molluscicide efficacy against the marsh slug *Deroceras laeve* (Stylommatophora, Limacidae) on strawberries in Mabitoba, Canada. *J. Econ. Entomol.*, **80**, 936-943.

Purdy, H. A. (1928) The improbability of tobacco mosaic transmission by slugs. *Amer. J. Bot.*, **15**, 100-101.

Pusswald, A. W. (1948) Beitrage zum Wasserhaushalt der Pulmonaten. *Z. Vergleich. Physiol.* **31**, 227-248.

Quattrini, D. (1962) La neurosecrezione nei Gasteropodi Polmonati. (Osservazioni in *Milax gagates*). *Monit. Zool. Italiano*, **70**, 1–04.

Quattrini, D. (1966a) Struttura e Ultrastruttura della Prostada dei Molluschi. I. Osservazioni in *Vaginulus borellianus*. *Monit. Zool. Italiano*, **74**, 3–29.

Quattrini, D. (1966b) Un altro reperto di fibrille endonucleari nelle cellule nervose dei molluschi gasteropodi. Osservazioni in *Vaginulus borellianus* (Colosi). *Caryologia*, **19**, 41–45.

Quattrini, D. (1967) Structure and ultrastructure of the molluscan prostate. Observations on *Milax gagates* (Draparnaud) (Gastropoda, Pulmonata, Stylommatophora). *Monit. Zool. Italiano*, **1**, 109–128.

Quattrini, D. (1970) La reproduzione di *Milax gagates* (Draparnaud) (Mollusca, Gastropoda, Pulmonata). *Boll. Soc. Italiana Biol. Sper.*, **46**, 802–804.

Quattrini, D. and Focardi, S. (1976) Observations sur les cristaux elaborés par la glande vestibulaire de *Milax nigricans* (Schultz) (Gastropoda, Pulmonata). *Haliotis*, **6**, 281–285.

Quattrini, D. and Lanza, B. (1965) Ricerche sulla biologia dei Veronicellidae (Gastropoda Soleolifera). II. Struttura della gonade orogeneri e spermatogenesi in *Vaginulus borellianus* (Colosi) e in *Laevicaulis alte* (Férussac). *Morit. Zool. Italiano*, **73**, 3–60.

Quattrini, D. and Lanza, B. (1968) Osservazioni sulla gonade di *Veronicella sloanei* (Cuvier) (Gastropoda, Soleolifera, Veronicellidae). *Boll. Soc. Italiana Biol. Sper.*, **44**, 2023–2026.

Quattrini, D. and Sacchi, T. B. (1971) La podocisti di *Milax gagates* (Draparnaud) (Mollusca, Gastropoda, Pulmonata): Ricerche al microscopio ottico ed elettronico. *Arch. Italiano Anat. Embriol.*, **76**, 39–52.

Quick, H. E. (1947) *Arion ater* (L.) and *A. rufus* (L.) in Britain and their specific differences. *J. Conchol.*, **22**, 249–261.

Quick, H. E. (1949) *Synopses of the British fauna. No. 8. Slugs (mollusca).* (Testacellidae; Arionidae, Limacidae). Linnean Society of London.

Quick, H. E. (1950) The spermatophore of *Milax sowerbii* (Férussac). *J. Conchol.*, **23**, 111–112.

Quick, H. E. (1951) *Agriolimax laevis* (Müller) feeding on mealy bugs *J. Conchol.*, **23**, 146.

Quick, H. E. (1952) Rediscovery of *Arion lusitanicus* Mabille in Britain. *Proc. Malacol. Soc. London*, **29**, 93–101.

Quick, H. E. (1960) British Slugs (Pulmonata: Testacellidae, Arionidae, Limacidae). *Bull. Brit. Mus. (Nat. Hist.) Zool. Ser.*, **6**, 106–226.

Rai, J. P. N. and Tripathi, R. S. (1985) Effect of herbivory by the slug, *Mariaella dussumieri*, and certain insects on growth and competitive success of two sympatric annual weeds. *Agric. Eco. Env.*, **13**, 125–138.

Rajagopal, A. S. (1973) A new species of slug (Stylommatophora, Arionidae) from Kumaun Himalayan Range. *Zool. Anzeiger*, **190**, 416–420.

Ramsbottom, J. (1953) *Mushrooms and Toadstools*, Collins, London.

Ramsell, J. and Paul, N. D. (1990) Preferential grouping by molluscs of plants infected by rust fungi. *Oikos*, **58**, 145–150.

Rathcke, B. (1985) Slugs as generalist herbivores: Tests of 3 hypotheses on plant choices. *Ecology*, **66**, 828–836.

Raut, S. K. and Mandal, R. N. (1986) Disease in the pestiferous slug *Laevicaulis alte* (Gastropoda, Veronicellidae). *Malacol. Rev.*, **19**, 106.

Raut, S. K. and Panigarhi, A. (1988a) Influence of temperature on hatching of eggs of the pestiferous slug *Laevicaulis alte* Férussac. *Boll. Malacol.*, **24**, 61–65.

Raut, S. K. and Panigarhi, A. (1988b) Egg-nesting in the garden slug, *Laevicaulis alte* (Férussac) (Gastropoda, Soleolifera). *Malacol. Rev.*, **21**, 101–107.

Raut, S. K. and Panigarhi, A. (1990). Feeding rhythm in the garden slug *Laevicaulis alte* (Soleolifera: Veronicellidae). *Malacol. Rev.*, **23**, 39–46.

Raven, C. L. (1966) *Morphogenesis: The Analysis of Molluscan Development*, 2nd edn, Pergamon Press, Oxford.

Raven, C. P. (1975) Development. In *Pulmonates*, Vol. 1 (eds V. Fretter and J. Peake), Academic Press, New York and London, pp. 367–400.

Raw, F. (1951) The ecology of the garden chafer, *Phyllopertha horticola* (L.) with preliminary observations on control measures. *Bull. Entomol. Res.*, **42**, 605.

Raw, F. (1959) Estimating earthworm populations by using formalin. *Nature (London)*, **184**, 1661.

Raw, F. (1966) The soil fauna as a food source for moles. *J. Zool.*, **149**, 50–54.

Raw, G. R. (1970) *CIPAC Handbook Volume 1. Analysis of Technical and Formulated Pesticides*, Collaborative International Pesticides Analytical Council Limited.

Rayner, J. M. (1975) Experiments on the control of slugs in potatoes by means of molluscicidal baits. *Plant Pathol.*, **24**, 167–171.

Rayner, J. M., Brock, A. M., French, H. J. and Lewis, S. (1978) Further experiments on the control of slugs in potatoes by means of molluscicidal baits. *Plant Pathol.*, **27**, 186–193.

Reader, P. M. and Southwood, T. R. E. (1981) The relationship between palatability to invertebrates and the successional status of a plant. *Oecologia (Berlin)*, **51**, 271–275.

Recio, A., Marigomez, J. A., Angulo, E. and Moya, J. (1988a) Zinc treatment of the digestive gland of the slug *Arion ater* L. 1. Cellular distribution of zinc and calcium. *Bull. Env. Contam. Toxicol.*, **41**, 858–864.

Recio, A., Marigomez, J. A., Angulo, E. and Moya, J. (1988b) Zinc treatment of the digestive gland of the slug *Arion ater* L. 2. Sublethal effects at the histological level. *Bull. Env. Contam. Toxicol.*, **41**, 865–871.

Reddy, V. V., Reddy, D. C. S., Jayaram, V., Sowjanya, K. and Naidu, B. P. (1981) Levels of inorganic and organic constituents in the blood of the slug, *Laevicaulis alte*, in response to starvation stress. *Proc. Indian Acad. Sci. Animal Sci.*, **90**, 571–576.

Reger, J. F. (1973) A fine structure study on haemocyanin formation in the slug *Limax* sp. *J. Ultrastruct. Res.*, **43**, 377–387.

Reger, J. F. and Fitzgerald, M. E. C. (1981) Membrane specializations in tentacular retractor muscle of the gastropod, *Limax* sp. *Tissue & Cell*, **13**, 535–540.

Reger, J. F. and Fitzgerald, M. E. C. (1983) Studies on the fine structure of the mitochondrial derivative in spermatozoa of a gastropod (*Limax* sp.). *Tissue & Cell*, **14**, 775–784.

Reidenbach, J. M., Vala, J. C. and Ghamizi, M. (1989) The slug-killing Sciomyzidae (Diptera): potential agents in the biological control of crop pest molluscs. In *Slugs and Snails in World Agriculture* (ed I. Henderson), British Crop Protection Council, Thornton Heath, pp. 273–280.

Reingold, S. C. and Gelperin, A. (1980) Feeding motor programme in *Limax* II.

Modulation by sensory inputs in intact animals and isolated central nervous systems. *J. Exp. Biol.*, **85**, 1–19.

Reingold, S. C., Sejnowski, T. J., Gelperin, A. and Kelly, D. B. (1981) Tritium-labelled 2-deoxyglucose autoradiography in a molluscan (*Limax maximus*) nervous system. *Brain Res.*, **208**, 416–420.

Renzoni, A. (1969) Observation on the tentacles of *Vaginulus borellianus* Colosi (Mollusca: Gastropoda). *Veliger*, **12**, 176–181.

Reuse, C. (1983) On the taxonomic significance of the internal shell in the identification of European slugs of the families Limacidae, and Milacidae (Gastropoda, Pulmonata). *Biol. Jb.*, **51**, 180–200.

Reygrobellet, D. (1970) Rôle de la sole pédieuse dans l'histogénèse régénératrice de l'extrémité posterior de quelques Limacidés. *C. R. Acad. Sci. (Paris)*, **270**, 2850–2852.

Reynolds, B. D. (1936). *Colpoda steini*: a facultative parasite of the land slug, *Agriolimax agrestis*. *J. Parasitol.*, **22**, 48–53.

Rhoades, D. F. and Cates, R. G. (1976) Towards a general theory of plant anti-herbivore chemistry. *Recent Adv. Phytochem.*, **10**, 168–213.

Rice, L. R., Lincoln, D. E. and Langenheim, J. H. (1978) Palatability of mono-terpenoid compositional types to a molluscan herbivore *Ariolimax dolicho-phallus*. *Biochem. Syst. Ecol.*, **7**, 289–298.

Richardson, B. and Whittaker, J. B. (1982) The effect of varying the reference material on ranking of acceptability indices of plant species to a polyphagous herbivore, *Agriolimax reticulatus*. *Oikos*, **39**, 237–240.

Richter, E. (1935) Der Bau der Zwitterdrüse und die Entstehung der Geschlechtszellen bei *Agriolimax agrestis*. *Z. Naturwiss. (Jena)*, **69**, 507–544.

Richter, K. O. (1976) A method for individually marking slugs. *J. Mollusc. Stud.*, **42**, 146–151.

Richter, K. O. (1979) Aspects of nutrient cycling by *Ariolimax columbianus* (Mollusca, Arionidae) in Pacific Northwest coniferous forests. *Pedobiologia*, **19**, 60–74.

Richter, K. O. (1980a) Movement, defense and nutrition as functions of the caudal mucus plug in *Ariolimax columbianus*. *Veliger*, **23**, 43–47.

Richter, K. O. (1980b) Evolutionary aspects of mycophagy in *Ariolimax columbianus* and other slugs. In *Soil Biology as Related to Land Use Practices (Proc. VIIth Internat. Coll. Soil Zoology)* (ed. D. L. Dindal), Office of Pesticides and Toxic Substances, EPA, Washington DC, pp. 616–636.

Ricou, G. (1961) Contribution à l'étude des Mollusques nuisibles en Normandie. *Rev. Soc. Savantes Haute-Normandie – Sciences*, No. 21, 30–44.

Ricou, G. (1964) Relations entre l'activité des limaces et la temperature. *Overdruk Med. Landbouw. Opzoekings. Staat Gent*, **29**, 1071–1079.

Ridgway, J. W. (1971) Studies on the nutrition of the slug, *Arion ater*. PhD thesis, Univ. of Bradford.

Ridgway, J. W. and Wright, A. A. (1975) The effect of deficiencies of B vitamins on the growth of *Arion ater* L. *Comp. Biochem. Physiol. A. Comp. Physiol.*, **51**, 727–732.

Riemschneider, R. and Hecker, A. (1979) Determination of the lethal dose of metaldehyde and evidence of a synergetic effect in snails. *Z. Pflanzenkrank. Pflanzenschutz*, **86**, 479–482.

Rising, T. L. and Armitage, K. B. (1969) Acclimation to temperature by the terrestrial gastropods, *Limax maximus and Philomycus carolinianus*: Oxygen consumption and temperature preference. *Comp. Biochem. Physiol.*, **30**, 1091–1114.

Roach, D. K. (1963) Analysis of the haemolymph of *Arion ater* L. (Gastropoda: Pulmonata). *J. Exp. Biol.*, **40**, 613–623.

Roach, D. K. (1966) Studies of some aspects of the physiology of *Arion ater* L. PhD thesis, Univ. of Wales.

Roach, D. K. (1968) Rhythmic muscular activity in the alimentary tract of *Arion ater* (L.) (Gastopoda: Pulmonata). *Comp. Biochem. Physiol.*, **24**, 865–878.

Robinson, W. H. (1965) Biology of a phorid feeding on slug eggs. *Bull. Entomol. Soc. Amer.*, **11**, 155.

Robinson, W. H. and Foote, B. A. (1968) Biology and immature stages of *Megaselia aequalis*, a phorid predator of slug eggs. *Ann. Entomol. Soc. Amer.*, **61**, 1587–1594.

Rogers, D. C. (1968) Fine structure of smooth muscle and neuromuscular junctions in the optic tentacles of *Helix aspersa* and *Limax flavus*. *Z. Zellforsch.*, **89**, 80–94.

Rogers, T. (1900) The eggs of the Kerry slug, *Geomalacus maculosus* Allman. *Irish Nat.*, **9**, 168–170.

Rogers, D. D., Chamblee, D. S., Mueller, J. P. and Campbell, W. V. (1985) Conventional and no-till establishment of ladino clover (*Trifolium repens* cultivar Tillman) as influenced by time of seeding and insect and grass suppression. *Agron. J.*, **77**, 531–538.

Rogers-Lewis, D. S. (1977a) Slug control in maincrop potatoes. *Arable Farming*, **4**, 59–62.

Rogers-Lewis, D. S. (1977b) A note on the importance of surface slug movement in relation to plot size for experiments with maincrop potatoes. *Exp. Husb.*, **32**, 68–69.

Rollo, C. D. (1982) The regulation of activity in populations of the terrestrial slug, *Limax maximus* (Gastropoda, Limacidae). *Res. Pop. Ecol. (Kyoto)*, **24**, 1–32.

Rollo, C. D. (1983a) Consequences of competition on the reproduction and mortality of three species of terrestrial slugs. *Res. Pop. Ecol. (Kyoto)*, **25**, 20–43.

Rollo, C. D. (1983b) Consequences of competition on the time budgets, growth and distribution of three species of terrestrial slugs. *Res. Pop. Ecol. (Kyoto)*, **25**, 44–68.

Rollo, C. D. (1983c) Alternative risk-taking styles: the case of time budgeting strategies. *Res. Pop. Ecol. (Kyoto)*, **25**, 321–335.

Rollo, C. D. (1988a) The feeding of terrestrial slugs in relation to food characteristics. Starvation, maturation and life history. *Malacologia*, **28**, 29–40.

Rollo, C. D. (1988b) A quantitative analysis of food consumption for the terrestrial mollusca. Allometry, food hydration and temperature. *Malacologia*, **28**, 41–52.

Rollo, C. D. (1989) Experimental and analytical methodologies for studying molluscan activity. In *Slugs and Snails in World Agriculture* (ed. I. Henderson), British Crop Protection Council, Thornton Heath, pp. 343–348.

Rollo, C. D. and Ellis, C, R. (1974) Sampling methods for the slugs, *Deroceras reticulatum* (Müller), *D. laeve* (Müller), and *Arion fasciatus* Nilsson in Ontario corn fields. *Proc. Entomol. Soc. Ontario*, **105**, 89–95.

Rollo, C. D., Vertinsky, I. B., Wellington, W. G. and Kanetkar, V. K. (1983a) Description and testing of a comprehensive simulation model of the ecology of terrestrial gastropods in unstable evironments. *Res. Pop. Ecol. (Kyoto)*, **25**, 150–179.

Rollo, C. D., Vertinsky, I. B., Wellington, W. G. and Kanetkar, V. K. (1983b) Alternative risk-taking styles: The case of time-budgeting strategies of terrestrial gastropods. *Res. Pop. Ecol. (Kyoto)*, **25**, 321–335.

Rollo, C. D. and Wellington, W. G. (1975) Terrestrial slugs in the vicinity of Vancouver, British Columbia. *Nautilus*, **89**, 107–115.

Rollo, C. D. and Wellington, W. G. (1977) Why slugs squabble. *Nat. Hist.*, **86**, 46–51.

Rollo, C. D. and Wellington, W. G. (1979) Intra- and inter-specific agonistic behaviour among terrestrial slugs (Pulmonata: Stylommatophora). *Canad. J. Zool.*, **57**, 846–855.

Rollo, C. D. and Wellington, W. G. (1981) Environmental orientation by terrestrial Mollusca with particular reference to homing behaviour. *Canad. J. Zool.*, **59**, 225–239.

Rose, J. H. (1960) The field slug *Agriolimax reticulatus* as a vector of the lungworm *Cystocaulus ocreatus*. *Nature (London)*, **185**, 180.

Rose, M. and Hamon, M. (1939) Sur l'influence des hormones sexualles de synthèse chez le Mollusque Gastéropode Pulmoné *Milax gagates* Draparnaud. *C. R. Séances Soc. Biol.*, **131**, 937–939.

Rosser, J. (1982) A study of peptidergic neurones in the pedal ganglion of *Deroceras reticulatum* (Pulmonata: Limacidae). *Malacologia*, **22**, 615–619.

Roth, B. (1986) Notes on three European land mollusks introduced to California (USA). *Bull. S. California Acad. Sci.*, **85**, 22–28.

Roth, B. and Lindberg, D. R. (1981) Terrestrial mollusks of Attu, Aleutian Islands, Alaska, USA. *Arctic*, **34**, 43–47.

Roth, B. and Pressley, P. H. (1983) New range information on two western American slugs (Gastropoda, Pulmonata, Arionidae). *Bull. S. California Acad. Sci.*, **82**, 71–78.

Roth, J., Lucocq, J. M. and Charest, P. M. (1984) Light and electron microscopic demonstration of sialic acid residues with the lectin from *Limax flavus*: A cytological affinity technique with use of fetuin-gold complexes. *J. Histochem. Cytochem.*, **32**, 1167–1176.

Rousset-Galangau, V. (1972) Presence de deux categories de spermies chez *Milax gagates* et *Agriolimax agrestis* (Mollusca, Gastropoda, Pulmonata, Limacidae): Étude comparée des ultrastructures au cours de la spermiogenese. *Ann. Sci. Nat. Zool. Biol., Animale*, **14**, 319–331.

Rowley, M. A., Loker, E. S., Pagels, J. F. and Montali, R. J. (1987) Terrestrial gastropod hosts of *Parelaphostrongylus tenuis* at the National Zoological Park's conservation and research center, Virginia, USA. *J. Parasitol.*, **73**, 1084–1089.

Roy, A. (1963) Étude de l'acclimation thermique chez la limace *Arion circumscriptus*. *Canad. J. Zool.*, **41**, 671–698.

Roy, A. (1969) Analyse des facteurs du taux de métabolisme chex la limace *Arion circumscriptus*. *Rev. Canad. Biol.*, **28**, 33–43.

Rózsa, S. K. (1962a) A reflex mechanism changing the activity in gastropods upon osmotic effects. *Proc. 5th Meeting Hungarian Biol. Soc.*, p. 44.

Rózsa, S. K. (1962b) Elektrofiziologiai adatok pulmonatak osmoregulaciojahoz. *Acta Biol. Deb.*, **8**, 69–75.

Rózsa, S. K. (1984) The pharmacology of molluscan neurons. *Prog. in Neurobiol.*, **23**, 79–150.

Rueda, A. (1989) Artificial diet for laboratory maintenance of the veronicellid slug, *Sarasinula plebeia*. In *Slugs and Snails in World Agriculture* (ed I. Henderson), British Crop Protection Council, Thornton Heath, pp. 361–366.

Runham, N. W. (1963) A study of the replacement mechanism of the pulmonate radula. *Qtly J. Microsc. Soc.*, **104**, 271–277.

Runham, N. W. (1969) The use of the scanning electron microscope in the study of the gastropod radula: The radulae of *Agriolimax reticulatus* and *Nucella lapillus*. *Malacologia*, **9**, 179–185.

Runham, N. W. (1975) Alimentary canal. In *Pulmonates*, Vol. 1 (eds V. Fretter and J. Peake), Academic Press, New York and London, pp. 53–104.

Runham, N. W. (1982) Hermaphroditism in the Stylommatophora. *Malacologia*, **22**, 121–123.

Runham, N. W. (1989) Snail farming in the United Kingdom. In *Slugs and Snails in World Agriculture* (ed I. Henderson), British Crop Protection Council, Thornton Heath, pp. 49–56.

Runham, N. W., Bailey, T. G. and Laryea, A. A. (1973) Studies of the endocrine control of the reproductive tract of the grey field slug *Agriolimax reticulatus*. *Malacologia*, **14**, 135–142.

Runham, N. W. and Hogg, N. (1979) The gonad and its development in *Deroceras reticulatum* (Pulmonata, Limacidae). *Malacologia*, **18**, 391–399.

Runham, N. W. and Hunter, P. J. (1970) *Terrestrial Slugs*, Hutchinson University Library, London.

Runham, N. W., Isarankura, K. and Smith, B. J. (1965) Methods for narcotizing and anaesthetizing gastropods. *Malacologia*, **2**, 231–238.

Runham, N. W. and Laryea, A. A. (1968) Studies on the maturation of the reproductive system of *Agriolimax reticulatus* (Pulmonata Limacidae). *Malacologia*, **7**, 93–108.

Runham, N. W. and Thornton, P. R. (1967) Mechanical wear of the gastropod radula: a scanning electron microscope study, *J. Zool. (London)*, **153**, 445–452.

Ruppel, R. F. (1959) Effectiveness of sevin against the grey garden slug. *J. Econ. Entomol.*, **52**, 360.

Ryder, T. A. and Bowen, I. D. (1977a) Endocytosis and aspects of autophagy in the foot epithelium of the slug *Agriolimax reticulatus*. *Cell Tissue Res.*, **181**, 129–142.

Ryder, T. A. and Bowen, I. D. (1977b) Studies on transmembrane and paracellular phenomena in the foot of the slug *Agriolimax reticulatus*. *Cell Tissue Res.*, **183**, 143–152.

Ryder, T. A. and Bowen, I. D. (1977c) The slug foot as a site of uptake of copper molluscicide. *J. Invert. Pathol.*, **30**, 381–386.

Ryder, T. A. and Bowen, I. D. (1977d) The use of X-ray microanalysis to demonstrate the uptake of the molluscicide copper sulphate by slug eggs. *Histochemistry*, **52**, 55–60.

Sabelli, B., Scanabissi, F. S. and Merloni, M. (1978) Distribution of germ cells in the gonadic acina of *Deroceras reticulatum* (Müller) (Gastropoda, Pulmonata, Stylommatophora). *Monit. Zool. Italiano*, **12**, 95–106.

Sahley, C. L. (1985) Dietary choline augments associative memory function in *Limax maximus*. *J. Neurobiol.*, **17**, 113–120.

Sahley, C. L., Gelperin, A. and Rudy, J. W. (1981a) 1-trial associative learning modifies food odor preferences of a terrestrial mollusk. *Proc. Nat. Acad. Sci. USA*, **78**, 640–642.

Sahley, C. L., Rudy, J. W. and Gelperin, A. (1981b) An analysis of associative learning in a terrestrial mollusk (*Limax maximus*): 1. Higher order conditioning, blocking and a transient unconditioned stimulus pre-exposure effect. *J. Comp. Physiol. A. Sensory, Neural Behav. Physiol.*, **144**, 1–8.

Sahley, C. L., Gelperin, A. and Rudy, J. W. (1981c) Food aversion learning in the terrestrial mollusc *Limax maximus*. A model system in which to study the neural basis of associative learning. In *Advances in Physiological Sciences*, Vol. 23, *Neurobiology of Invertebrates – Mechanisms of Integration* (ed. J. Salanki), Pergamon Press, Oxford, pp. 267–284.

Sahley, C. L., Martin, K. W. and Gelperin, A. (1990) Analysis of associative learning in the terrestrial molluscs, *Limax maximus*. II. Appetitive learning. *J. Comp. Physiol. A. Sensory, Neural Behav. Physiol.*, **167**, 339–346.

Salanki, J. and van Bay, T. (1975) Sensory input characteristics at the chemical stimulation of the lip in the snail *Helix pomatia*. *Ann. Inst. Biol. (Tihany) Hung. Acad. Sci.*, **42**, 115–128.

Salt, G. and Hollick, F. S. J. (1944) Studies of wireworm populations. I. A census of wireworms in pasture. *Ann. Appl. Biol.*, **31**, 52.

Samuel, W. M., Platt, T. R. and Knispel-Krause, S. M. (1985) Gastropod intermediate hosts and transmission of *Parelaphostrongylus odocoilei*, a muscle-inhabiting nematode of mule deer, *Odocoileus hemionus heminous* in Jasper National Park, Alberta (Canada). *Canad. J. Zool.*, **63**, 928–932.

Samways, M. J. (1981) *Biological Control of Pests and Weeds*, Edward Arnold, London.

Sano, T. (1977) Slug eating the white rust of *Chrysanthemum morifolium*. *Trans. Mycol. Soc. Japan*, **18**, 202.

Sathananthan, A. H. (1970) Studies on mitochondria in the early development of the slugs, *Arion ater rufus* L. *J. Embryol, Exp. Morphol.*, **24**, 555–582.

Sathananthan, A. H. (1972) Cytological and cytochemical studies of centrifuged eggs of the slug, *Arion ater rufus* L. *J. Embryol. Exp. Morphol.*, **27**, 1–13.

Sawyer, W., Deyrup-Olsen, I. and Martin, A. W. (1984) Immunological and biological characteristics of the vasotocin-like activity in the head ganglia of gastropod mollusks. *Gen. Comp. Endocrinol.*, **54**, 97–108.

Schagene, K. A., Welsford, I. G., Prior, D. J. and Banta, P. A. (1989) Behavioural effects of injection of small cardioactive peptide SCP_B on the slug, *Limax maximus*. *J. Exp. Biol.*, **143**, 553–557.

Schermer, A. (1958) Spreeuwen voeren naak te slakken aan hun jungen. *Levende Natuur*, **61**, 191–192.

Schmidt-Nielsen, C. R., Taylor, C. R. and Shkolnik, A. (1971) Desert snails: problems of heat, water and food. *J. Exp. Biol.*, **55**, 385–398.

Schoettli, G. and Seiler, H. G. (1970) Uptake and localization of radioactive zinc in the visceral complex of the land pulmonate, *Arion rufus. Experientia (Basel)*, **26**, 1212–1213.

Schrey, E. (1981) Ecological investigations on the feeding of starlings (*Sturnus vulgaris*) on Heligoland (West Germany). *Vogelwelt*, **102**, 219–232.

Schrim, M. and Byers, R. A. (1980) A method for sampling three slug species attacking sod-seeded legumes. *Melsheimer Entomol. Ser.*, **29**, 9–11.

Schuurmans-Stekhoven, J. H. (1920) Über die Atmung der Schnecken *Limax agrestis* und *Helix pomatia*. *Tijdschrift. Ned. Dierk. Vereen. Helder*, **18**, 1–43.

Schwalb, H. H. (1960) Beitrage zur Biologie der einheimschen Lampyriden *Lampyris noctiluca* (Geoffr.) und *Phausis splendidula* (Lec.) und experimentelle analyse ihres Beutefang und Sexualverhaltens. *Zool. Jb.*, **88**, 399–550.

Schwartzkopff, J. (1956) Herzfrequency und Körpergrosse bei Mollusken. *Zool. Jb.*, **20**, 463–469.

Scott, G. C., Griffiths, D. C. and Stephenson, J. W. (1977) A laboratory method for testing seed treatments for the control of slugs in cereals. *Proc. 1977 British Crop Protection Conf. - Pests and Diseases*, 129–134.

Scott, G. C., Pickett, J. A., Smith, M. C. and Woodcock, C. M. (1984) Seed treatments for controlling slugs in winter wheat. *Proc. 1984 British Crop Protection Conf. - Pests and Diseases*, 133–138.

Segal, E. (1960) Studies on annual rhythms of egg laying in slugs under constant conditions. *Anat. Rec.*, **134**, 636–637.

Segal, E. (1961) Acclimation in molluscs. *Amer. Zool.*, **1**, 235–244.

Segal, E. (1963) A temperature dependent abnormality in the slug *Limax flavus* L. I. Appearance and incidence. *J. Exp. Zool.*, **153**, 159–170.

Seixas, M. M. P. (1976) Terrestrial gastropods from the Portuguese fauna. *Boletim da Sociedade Portuguesa de Ciencias Naturais*, **16**, 21–46.

Sejnowski, T. J., Reingold, S. C., Kelley, D. B. and Gelperin, A. (1980) Localization of tritium labelled 2-deoxyglucose in single molluscan neurons. *Nature (London)*, **287**, 449–451.

Selim, S. I. (1979) The control of snail and slug pests in agriculture. I. Analysis for and dissipation of metaldehyde in vegetable crops II. Beer as a slug attractant. PhD thesis, Univ. of California.

Semlitsch, R. D. and Moran, G. B. (1984) Ecology of the redbelly snake (*Storeria occiptomaculata*) using mesic habitats in South Carolina (USA). *Amer. Mid. Nat.*, **111**, 33–40.

Senseman, D. M. (1977) Starch: a potent feeding stimulant for the terrestrial slug *Ariolimax californicus*. *J. Chem. Ecol.*, **3**, 707–716.

Senseman, D. M. (1978) Short-term control of food intake by the terrestrial slug, *Ariolimax*. *J. Comp. Physiol. A. Sensory, Neural Behav. Physiol.*, **124**, 37–48.

Senseman, D. M. and Gelperin, A. (1974) Comparative aspects of the morphology and physiology of a single identifiable neuron in *Helix aspersa*, *Limax maximus* and *Ariolimax californica*. *Malacol. Rev.*, **7**, 51–52.

Sergeeva, T. K. (1982) Methods and present state of studies of trophic relations between predatory invertebrates: A serological analysis of feeding. *Zool. Zh.*, **61**, 109–119.

Sergeeva, T. K. (1984) Change of trophic objects in the ontogenesis of the Ragionidae (Diptera). *Zh. Obsh. Biol.*, **45**, 124–131.

Shamsuddin, M. and Al-Barrak, N. S. H. (1988) Observations on *Monacha obstructa*, new record (Helicidae) and its larval trematodes (Brachylaemidae) from Iraq. *Bull. Iraq Nat. Hist. Mus. (Univ. of Baghdad)*, **8**, 67–88.

Shapiro, Y. S. (1981) Forecasting the number of *Deroceras* slugs (Stylommatophora, Agriolimacidae) in an agricultural district. *Biol. Nauk. (Moscow)*, **10**, 103–106.

Sharma, K. D., Renapurkar, D. M., Bhopale, M. K., Nathan, J., Boraskar, A. and Chotani, S. (1981) Study of a focus of *Angiostrongylus cantonensis*, new record in Geater Bombay, India. *Bull. Haffkine Inst.*, **9**, 38–46.

Shibata, D. M. and Rollo, C. D. (1988) Intraspecific variation in the growth rate of gastropods: Five hypotheses. *Entomol. Soc. Canada*, **0**(146), 199–213.

Shikov, E. V. (1979) Distribution of slugs of the genus *Deroceras* in flood plains of large rivers. *Soviet J. Ecol.*, **10**, 453–455.

Shileiko, A. A. (1967) Biology of reproduction and juvenile adaptations of the slug *Parmacella ibera*. *Zool. Zh.*, **46**, 946–948.

Shilova, S. A., Denisova, A. V., Dmitrieva, G. A., Voronova, L. D. and Bardier, M. N. (1971) Effect of some insecticides upon the common mole. *Zool. Zh.*, **50**, 886–892.

Shirbhate, R. (1987) Functional aspects of gastropod mucus. PhD thesis, Univ. of Durham.

Simpson, U. R. and Neal, C. (1982) *Angiostrongylus vasorum* infection in dogs and slugs. *The Veterinary Rec.*, **111**, 303–304.

Simroth, H. (1905) Über zwei seltene Missbildungen an Nacktschnecken. *Z. Wissen. Zool.*, **82**, 494–522.

Sivik, F. P. (1954) A technique for slug culture. *The Nautilis*, **67**, 129–130.

Skaren, U. (1978) Feeding behaviour, coprophagy and passage of foodstuffs in a captive least shrew. *Acta Theriol.*, **23**, 131–140.

Skelding, J. M. and Newell, P. F. (1974) On the functions of the pore cells in the connective tissue of terrestrial pulmonate molluscs. *Cell Tissue Res.*, **156**, 381–390.

Skorping, A. and Halvorsen, O. (1980) The susceptibility of terrestrial gastropods to experimental infection with *Elaphostrongylus rangiferi* (Nematoda: Metastrongyloidea). *Z. Parasitenkunde*, **62**, 7–14.

Smith, B. J. (1965) The secretions of the reproductive tract of the garden slug *Arion ater*. *Ann. N.Y. Acad. Sci.*, **118**, 997.

Smith, B. J. (1966a) Maturation of the reproductive tract of *Arion ater* (Pulmonata: Arionidae). *Malacologia*, **4**, 325–349.

Smith, B. J. (1966b) The structure of the central nervous system of slug *Arion ater* L., with notes on the cytoplasmic inclusions of the neurons. *J. Comp. Neurol.*, **126**, 437–452.

Smith, B. J. (1967) Correlation between neurosecretory changes and maturation of the reproductive tract of *Arion ater* (Stylommatophora: Arionidae). *Malacologia*, **5**, 285–298.

Smith, B. J. and Dartnall, A. J. (1976) Veronicellid slugs in the Northern Territory with notes on other land molluscs. *J. Malacol. Soc. Australia*, **3**, 186.

Smith, F. F. and Boswell, A. L. (1970) New baits and attractants for slugs. *J. Econ. Entomol.*, **63**, 1919–1922.

Sokolove, P. G., Beiswanger, C. M., Prior, D. J. and Gelperin, A. (1977) A circadian rhythm in the locomotor behaviour of the giant slug, *Limax maximus*. *J. Exp. Biol.*, **66**, 57–64.

Sokolove, P. G., Kirgan, J. and Tarr, R. (1981) Red light insensitivity of the extraocular pathway for photoperiodic stimulation of reproductive development in the slug *Limax maximus. J. Exp. Zool.*, **215**, 219–224.

Sokolove, P. G. and McCrone, E. J. (1978) Reproductive maturation in the slug, *Limax maximus*, and the effects of artificial photoperiod. *J. Comp. Physiol. A. Sensory, Neural Behav. Physiol.*, **125**, 317–326.

Sokolove, P. G., Melrose, G. R., Gordon, T. M. and O'Neil, M. C. (1983) Stimulation of spermatogonial DNA synthesis in slug (*Limax maxmus*) gonad by a factor released from cerebral ganglia under the influence of long days. *Gen. Comp. Endocrinol.*, **50**, 95–104.

Solem, A. (1978) Classification of Land Mollusca. In *Pulmonates*, Vol. 2A (eds V. Fretter and J. Peake), Academic Press, New York and London, pp. 49–97.

Solhøy, T. (1981) Terrestrial invertebrates of the Faroe Islands: 4. Slugs and snails (Gastropoda): Checklist, distribution and habitats. *Fauna Norvegica Ser. A*, **2**, 14–27.

South, A. (1964) Estimation of slug populations. *Ann. Appl. Biol.*, **53**, 251–258.

South, A. (1965) Biology and ecology of *Agriolimax reticulatus* (Müll.) and other slugs: Spatial distribution. *J. Animal Ecol.*, **34**, 403–417.

South, A. (1973a) Dégâts causés par les Limaces en Grande-Bretagne. *Haliotis*, **3**, 19–25.

South, A. (1973b) Estimation des Populations de Limaces. *Haliotis*, **3**, 89–95.

South, A. (1974) Changes in composition of the terrestrial mollusc fauna. In *The Changing Flora and Fauna of Britain* (ed. D. L. Hawksworth), Academic Press, New York and London, pp. 255–274.

South, A. (1980) A technique for the assessment of predation by birds and mammals on the slug *Deroceras reticulatum* (Müller) (Pulmonata: Limacidae). *J. Conchol.*, **30**, 229–234.

South, A. (1982) A comparison of the life cycles of *Deroceras reticulatum* (Müller) and *Arion intermedius* Normand (Pulmonata: Stylommatophora) at different temperatures under laboratory conditions. *J. Mollusc. Stud.*, **48**, 233–244.

South, A. (1989a) A comparison of the life cycles of the slugs *Deroceras reticulatum* (Müller) and *Arion intermedius* Normand on permanent pasture. *J. Mollusc. Stud.*, **55**, 9–22.

South, A. (1989b) The effect of weather and other factors on numbers of slugs on permanent pasture. In *Slugs and Snails in World Agriculture* (ed. I. Henderson), British Crop Protection Council, Thornton Health, pp. 355–360.

Southwood, T. R. E. (1978) *Ecological Methods*, 2nd edn, Chapman and Hall, London.

Sparks, A. K. (1972) *Invertebrate Pathology (Non-communicable diseases)*, Academic Press, New York and London.

Spaull, A. M. and Eldon, S. (1990) Is it possible to limit slug damage using choice of winter wheat cultivars? *1990 British Crop Protection Conf. – Pests and Diseases*, **2**, 703–708.

Spielman, A. and Lemma, A. (1973) Endod extract, a plant-derived molluscicide: Toxicity for mosquitos. *Amer. J. Trop. Med. Hygiene*, **22**, 802–804.

Spiridonov, S. E. (1985) *Angiostoma asamati*, new species (Angiostomatidae: Rhabditida) of nematodes from slugs (Mollusca). *Helminthologia*, **22**, 253–261.

Stadnichenko, A. P. (1969) Effect of larval forms of trematodes on the reproductive organs of their intermediate hosts. *Parazitologiya*, **3**, 53–57.

Stapley, J. H. (1949) *Pests of Farm Crops*, Spon, London.

Stephenson, J. W. (1959) Aldrin controlling slug and wireworm damage to potatoes. *Plant Pathol.*, **8**, 53–54.

Stephenson, J. W. (1962) A culture method for slugs. *Proc. Malacol. Soc. London*, **35**, 43–45.

Stephenson, J. W. (1965) The effect of irrigation on damage to potato tubers by slugs. *Euro. Potato J.*, **8**, 145–149.

Stephenson, J. W. (1966) Notes on the rearing and behaviour in soil of *Milax budapestensis* (Hazay). *J. Conchol.*, **26**, 141–145.

Stephenson, J. W. (1967) The distribution of slugs in a potato crop. *J. Appl. Ecol.*, **4**, 129–135.

Stephenson, J. W. (1972) Gelatin as a carrier for S^2-Cyanoethyl N-[(methylcarbamoyl)oxy] thioacetimidate, an experimental molluscicide. *Pesticide Sci.*, **3**, 1–7.

Stephenson, J. W. (1975a) Laboratory observations on the effect of soil compaction on slug damage to winter wheat. *Plant Pathol.*, **24**, 9–11.

Stephenson, J. W. (1975b) Laboratory observations on the distribution of *Agriolimax reticulatus* (Müller) in different aggregate fractions of garden loam. *Plant Pathol.*, **24**, 12–15.

Stephenson, J. W. (1978) A bioassay for comparing the acceptability to slugs of plant extracts or solutions of pure chemicals. *J. Mollusc. Stud.*, **44**, 340–343.

Stephenson, J. W. (1979) The functioning of the sense organs associated with feeding behaviour in *Deroceras reticulatum*. *J. Mollusc. Stud.*, **45**, 167–171.

Stephenson, J. W. and Bardner, R. (1976) Slugs in agriculture. In *Rothamsted Exp. Station Rep. 1976*, Part 2, pp. 169–187.

Stephenson, J. W. and Dibley, G. C. (1975) Electric fence for retaining slugs in outdoor enclosures. *Lab. Practice*, December 1975, p. 815.

Stephenson, J. W. and Knutson, L. V. (1966) A resumé of recent studies of invertebrates associated with slugs. *J. Econ. Entomol.*, **59**, 356–360.

Stern, G. (1970) Production et bilan énergetique chez la limac rouge. *Terre et la Vie*, **117**, 403–424.

Stern, G. (1975) Effect de la température sur la production et la consommation chez *Agriolimax reticulatus* (Müll.) en periode de croissance. *Bull. Écologie*, **6**, 501–509.

Stern, G. (1979) Weight and energy balance during growth and reproduction in *Agriolimax laevis* (Pulmonata: Limacidae). *Bull. Soc. Zool. France*, **104**, 147–160.

Stokes, B. M. (1958) The worm-eating slugs *Testacella scutulum* Sowerby and *T. haliotidea* Draparnaud in captivity. *Proc. Malacol. Soc. London*, **33**, 11–20.

Stone, B. A. and Morton, J. E. (1958) The distribution of cellulases and related enzymes in Mollusca. *Proc. Malacol. Soc. London*, **33**, 127.

Storey, M. A. (1985) The varietal susceptibility of potato crops to slug damage. PhD thesis, CNAA.

Strickland, A. H. (1965) Pest control and productivity in British agriculture. *J. Roy. Soc. Arts*, **113**, 62–81.

Strueve-Kusenberg, R. (1982) Succession and trophic structure of soil animal communities in different suburban fallow areas. In *Urban Ecology: The*

Second European Ecological Symposium (eds R. Burnkamm, J. A. Lee and M. R. D. Seaward), Blackwell Scientific Publications, Oxford.

Stubbs, A. G. (1934) *Testacella* eating *Milax*. *J. Conchol.*, **20,** 149.

Studdert, V. P. (1985) Epidemiological features of snail and slug bait poisoning in dogs and cats. *Australian Vet. J.*, **62,** 269–271.

Sumner, A. T. (1966) The fine structure of the digestive gland cells of *Helix, Succinea* and *Testacella*. *J. Roy. Microsc. Soc.*, **85,** 181–192.

Sumner, A. T. (1969) The distribution of some hydrolytic enzymes in the cells of the digestive gland of certain lamellibranchs and gastropods. *J. Zool. (London)*, **158,** 277–291.

Suzuki, H., Watanabe, M., Tsukahara, Y. and Tasaki, K. (1979) Duplex system in the simple retina of a gastropod mollusc. *J. Comp. Physiol. A. Sensory, Neural Behav. Physiol.*, **133,** 125–130.

Symonds, B. V. (1973) The early prediction of slug damage to potatoes. *Plant Pathol.*, **22,** 30–34.

Symonds, B. V. (1975) Evaluation of potential molluscicides for the control of the field slug, *Agriolimax reticulatus* (Müll.). *Plant Pathol.*, **24,** 1–9.

Symondson, W. O. C. (1989) Biological control of slugs by carabids. In *Slugs and Snails in World Agriculture* (ed. I. Henderson), British Crop Protection Council, Thornton Heath, pp. 295–300.

Szabó, M. (1935a) Pathologische veränderungen bei den Schnecken *Allanttani Közlemenyek (Budapest)*, **32,** 132–135.

Szabó, I. (1935b) Senescence and death in invertebrate animals. *Revista Biol.*, **19,** 377–436.

Szabó, I. and Szabó, M. (1929) Lebensdauer, Wachstum und Altern. Studiert bei der Nacktschnecken *Agriolimax agrestis* L. *Biologia Generalis*, **5,** 95–118.

Szabó, I. and Szabó, M. (1930a) Todesurachen und Pathologische Erscheinungen bei Pulmonaten. *Arch. Mollusk.*, **62,** 123–130.

Szabó, I. and Szabó, M. (1930b) Vorläubige mitteilung über die an der Nacktschnecke *Agriolimax agrestis* beobachteten Altersveränderungen. *Arbeiten Ungarisch Biol. Forsch.-Inst.*, **3,** 350–357.

Szabó, I. and Szabó, M. (1931a) Todesursachen und Pathologische erscheinungen bei Pulmonaten. II. Haut Karankheiten bei Nacktschnecken. *Arch. Mollusk.*, **63,** 156–160.

Szabó, I. and Szabó, M. (1931b) Histologische Studien über den Zussammenhang der verschiedenen Alterserscheinungen bei Schnecken. I. Bindegewebsvermehrung und Atrophie II. Alterspigment. *Z. Ver. Physiol.*, **15,** 329–351.

Szabó, I. and Szabó, M. (1931c) Altersveränderung und Pathologische Erscheinungen in der Koerperwand von *Limax flavus*. *Biol. Zentral.*, **51,** 695–701.

Szabó, I. and Szabó, M. (1933) Histologische Studien über den Zussammenhang der verschiedenen Alterserscheinungen bei Schnecken. III. *Arch. Mollusk.*, **65,** 11–15.

Szabó, I. and Szabó, M. (1934a) Alterserscheinungen und alterstod bei Nacktschnecken. *Biol. Zentral.*, **54,** 471–477.

Szabó, I. and Szabó, M. (1934b) Epitheliale Geschwulstbildung bei einem wirbellosen Tier *Limax flavus*. *Z. Krebsforsch.*, **40,** 540–545.

Szabó, I. and Szabó, M. (1934c) Histologische Untersuchungen über die

Genitalorgane der Ackerschnecke *Agriolimax agrestis* L. *Biol. Generalis (Wien)*, **10**, 425–456.

Szabó, I. and Szabó, M. (1936) Histologische Untersuchungen über den Zusammenhang zwischen Langlebigkeit und Fortpflanzung. *Zool. Anzeiger*, **113**, 143–153.

Takahashi, Y., Nishimura, Y. and Yamagishi, N. (1973) Electron microscopic studies on spermiogenesis of *Limax flavus* L. *J. Nara Med. Ass.*, **24**, 401–410.

Takeda, N. (1977) Stimulation of egg-laying by nerve extracts in slugs. *Nature (London)*, **267**, 513–514.

Takeda, N. (1979) Induction of egg-laying by steroid hormones in slugs. *Comp. Biochem. Physiol. A. Comp. Physiol.*, **62**, 273–278.

Takeda, N. (1982) Source of the tentacular hormone in terrestrial pulmonates. *Experientia (Basel)*, **38**, 1058–1060.

Takeda, N. and Sugiyama, K. (1984) Gonadal regeneration and sex steroid hormones in some terrestrial pulmonates. *Venus, Japan. J. Malacol.*, **43**, 72–85.

Takeda, N., Ohtake, S.-I. and Sugiyama, K. (1987) Evidence for neurosecretory control of the optic gland in terrestrial pulmonates. *Gen. Comp. Endocrinol.*, **65**, 306–316.

Tamamaki, N. (1989) The accessory photosensory organ of the terrestrial slug, *Limax flavus* L. (Gastropoda, Pulmonata). Morphological and electrophysiological study. *Zool. Sci. (Tokyo)*, **6**, 877–884.

Tan, B. S. (1987) Influence of some ecological factors on the first stages of fir regeneration in the mountain forests of the Jura, France. *Revue Écol. Biol. Sol*, **24**, 623–636.

Tattersfield, P. (1990) Terrestrial mollusc faunas from some south Pennine woodlands. *J. Conchol.*, **33**, 355–374.

Taylor, E. L. (1935) *Syngamus trachea*: the longevity of the infective larvae in the earthworm: slugs and snails as intermediate hosts. *J. Comp. Pathol. Therapeuti.*, **48**, 149–156.

Taylor, J. W. (1902–1907) *Monograph of the Land and Freshwater Mollusca of the British Isles (Testacellidae, Limacidae, Arionidae)*, Parts 8–13, Taylor Brothers, Leeds.

Ten Cate, J. (1923) Quelques recherches sur la locomotion des limaces. *Arch. Neerland. Physiol.*, **8**, 177–393.

Tervet, I. W. and Esslemont, J. M. (1938) A fungous parasite of the eggs of the gray field slug. *J. Quekett Microsc. Club (London)*, **1**, 1–3.

Theobald, F. V. (1895) Mollusca injurious to farmers and gardeners. *Zoologist*, **19**, 201–211.

Theolis, R., Weech, P. K., Marcel, Y. L. and Milne, R. W. (1984) Characterization of antigenic determinants of human apolipoprotein B: Distribution on tryptic fragments of low density lipoprotein. *Arteriosclerosis*, **4**, 498–509.

Thomas, D. C. (1944) Discussion on slugs. II. Field sampling for slugs. *Ann. Appl. Biol.*, **31**, 160–164.

Thomas, D. C. (1947) Some observations on damage to potatoes by slugs. *Ann. Appl. Biol.*, **34**, 246–251.

Thomas, D. C. (1948) The use of metaldehyde against slugs. *Ann. Appl. Biol.*, **35**, 207–227.

Thomas, K. (1988) Why slugs drink like a fish. *Whats Brewing*, November 1988, p. 7.

Thompson, G. A. (1966) The biosynthesis of ether-containing phospholipids in the slug, *Arion ater*. II. The role of the glyceryl ether lipids as plasmalogen precursors. *Biochemistry*, 5, 1290–1296.

Thompson, G. A. (1968) The biosynthesis of ether-containing phospholipids in the slug. *Arion ater*. III. Origin of the vinylic ether bond of plasmalogens. *Biochim. Biophys. Acta*, 152, 409–411.

Thompson, G. A. and Hanaham, D. J. (1963) Identification of α-glyceryl ether phospholipids as major lipid constituents in two species of terrestrial slug. *J. Biol. Chem.*, 238, 2628–2631.

Thompson, G. A. and Lees, P. (1965) Studies of the α-glyceryl ether lipids occurring in molluscan tissues. *Biochim. Biophys. Acta*, 98, 151–159.

Thompson, H. W. (1928) Further tests of poison baits in South Wales. *Welsh J. Agric.*, 4, 342–347.

Thompson, J. C. (1958) *Tetrahymena rostrata* as a facultative parasite in the grey garden slug. *Virginia J. Sci.*, 9, 315–318.

Tillier, S. (1981) South American and Juan Fernandez succineid slugs (Pulmonata). *J. Mollusc. Stud.*, 47, 125–146.

Tillier, S. (1983) Secondary structures in the lung and in the ureter of stylommatophoran slugs (Pulmonata). *Bull. Soc. Zool. France*, 108, 9–20.

Tillier, S. (1984) Patterns of digestive tract morphology in the limacisation of helicarionid, succineid and athoracophorid snails and slugs (Mollusca: Pulmonata). *Malacologia*, 25, 173–192.

Tillier, S. (1989) Comparative morphology, phylogeny and classification of land snails and slugs (Gastropoda, Pulmonata, Stylommatophora). *Malacologia*, 30, 1–304.

Tischler, W.-H. (1976) Studies of the animal colonization on carrion in different strata of woodland ecosystems. *Pedobiologia*, 16, 99–105.

Tod, M. E. (1970) The significance of predation by soil invertebrates on field populations of *Agriolimax reticulatus* (Gastropoda: Limacidae). PhD thesis, Univ. of Edinburgh.

Tod, M. E. (1973) Notes on beetle predation of molluscs. *The Entomologist*, 106, 196–201.

Tomlin, J. R. (1935) Slug and beetle. *J. Conchol.*, 20, 165.

Tompa, A. S. (1980) Studies on the reproductive biology of gastropods. III. Calcium provision and the evolution of terrestrial eggs among gastropods. *J. Conchol.*, 30, 145–154.

Trelka, D. G. (1972) The behaviour of predatory larvae of *Tetanocera plebeia* (Diptera: Sciomyzidae), and toxicological and neurological aspects of a toxic salivary secretion used to immobilize slugs. PhD thesis, Cornell Univ.

Trelka, D. C. and Berg, C. O. (1977) Behavioural studies of the slug-killing larva of two species of *Tetanocera* (Diptera: Sciomyzidae). *Proc. Entomol. Soc. Washington*, 79, 475–486.

Trelka, D. G. and Foote, B. A. (1970) Biology of slug-killing *Tetanocera* (Diptera: Sciomyzidae). *Ann. Entomol. Soc. America*, 63, 877–895.

Triebskorn, R. (1989) Ultrastructural changes in the digestive tract of *Deroceras reticulatum* (Mueller) induced by a carbamate molluscicide and by metaldehyde. *Malacologia*, 31, 141–156.

Triebskorn, R. and Ebert, D. (1989) The importance of mucus production in slug reaction to molluscicides and the impact of molluscicides on the mucus producing system. In *Slugs and Snails in World Agriculture* (ed. I. Henderson), British Crop Protection Council, Thornton Heath, pp. 373–378.

Triebskorn, R., Kuenast, C., Huber, R. and Brem, G. (1990) Tracing a carbon-14 labelled carbamate molluscicide through the digestive system of *Deroceras reticulatum* (Mueller). *Pest. Sci.*, **28**, 321–330.

Tsuyama, S. Siganuma, T., Ihida, K. and Murata, F. (1986) Ultracytochemistry of glycosylation sites of rat colonic mucus cells with labelled lectins. *Acta Histochem. Cyctochemi.*, **19**, 555–566.

Turchetti, T. and Chelazzi, G. (1984) Possible role of slugs as vectors of the chestnut blight fungus. *Euro. J. Forest Pathol.*, **14**, 125–127.

Turk, F. A. and Phillips, S. M. (1946) A monograph of the slug mite *Riccardoella limacum* (Schrank). *Proc. Zool. Soc. London*, **115**, 448–471.

Turner, R. S. (1966) Patterns of form and function in the central nervous system of *Ariolimax*. *J. Comp. Neurol.*, **128**, 51–62.

Turner, R. S. (1951) The organisation of the nervous system of *Ariolimax columbianus*. *J. Comp. Neurol.*, **94**, 239–256.

Tuzet, O. and Galangau, V. (1967) Ultrastructure de la 'membrane ondulante' d'*Agriolimax agrestis* L. *C.R. Séances 'Acad. Sci., Serie D. Sciences Naturelles (Paris)*, **264**, 337–339.

Uchikawa, R., Noda, S. Matayoshi, S. and Sato, A. (1987) Food preferences of *Rattus rattus* for some terrestrial molluscs and experimental infection of *Rattus rattus* and *Angiostrongylus cantonensis*. *Japan. J. Parasitol.*, **36**, 94–99.

Uglem, G. L., Prior, D. J. and Hess, S. D. (1985) Paracellular water uptake and molecular sieving by the foot epithelium of terrestrial slugs. *J. Comp. Physiol. B. Biochem. Systemic Env. Physiol.*, **156**, 285–290.

Uhlenbruck, G. (1969) Blutgruppenforschung Heute. *Internist (Berlin)*, **10**, 33–42.

Urban, E. (1980) Lung nematodes (Protostrongylidae, Dictyocaulidae) in sheep of the Podhale region, Tatra Highlands, Poland: 2. Intermediate hosts of Protostrongylidae. *Acta Parasitol. Polonica*, **27**, 63–74.

Urquart, B. (1952) Slugs and natural regeneration. *Qtly J. Forestry*, **46**, 110–112.

Ushadevi, S. V. and Krishnamoorthy, R. V. (1980) Do slugs (*Mariella dussumieri*) have silver-track pheromone?. *Indian J. Exp. Biol.*, **18**, 1502–1504.

Vágvölgyi, J. (1952) A new sorting method for snails, applicable also for quantitative researches. *Ann. Hist.-Nat. Mus. Nat. Hung.*, **3**, 101–103.

Valkounova, J. and Prokopic, J. (1978) Morphology of cysticercoid of the cestode *Rodentotaenia crassiscolex*. *Vestnik Ceskoslovenske Spol. Zool.*, **42**, 303–310.

Valkounova, J. and Prokopic, J. (1979) Histochemistry of the cysticercoid of *Rodentotaenia crassiscolex*. *Folia Parasitol.*, **26**, 325–335.

Valovirta, I. (1968) Land molluscs in relation to acidity on hyperite hills in central Finland. *Ann. Zool. Fennici*, **5**, 245–253.

Valovirta, I. (1969) First record of *Milax gagates* in Finland. *Ann. Zool. Fennici*, **6**, 345–347.

Valovirta, I. (1979) Primary succession of land molluscs in an uplift archipelago of the Baltic. *Malacologia*, **18**, 169–176.

Van den Bruel, W. E. and Moens, R. (1956a) A propos des propriétés hélicides de la cyanamide calcique. *C.R. Séances Soc. Biol.*, 150, 2281–2282.

Van den Bruel, W. E. and Moens, R. (1956b) Une méthode de lutte efficace, utilisable en plein champ contre les limaces. *Parasitica*, 12, 8–15.

Van den Bruel, W. E. and Moens, R. (1956c) Proeven op bestrijding van slakken (*Agriolimax reticulatus* Müller). *Med. Landbouw. Opzoekings. Staat Gent*, 21, 401–410.

Van den Bruel, W. E. and Moens, R. (1957) Les propriétés des hélicides et la protection des cultures. *C.R. IVième Congress International de Lutte Contre les Ennemies des Plantes, Hamburg 1957*, 2, 1255–1275.

Van den Bruel, W. E. and Moens, R. (1958a) Remarques sur les facteurs écologiques influençant l'efficacité de la lutte contre les Limaces. *Parasitica*, 14, 135–147.

Van den Bruel, W. E. and Moens, R. (1958b) Praktische bestrijding van *Agriolimax reticulatus* (Müller) op rogge. *Med. Landbouw. Opzoekings. Staat Gent*, 23, 695–703.

Van den Bruel, W. E. and Moens, R. (1958c) Nouvelles observations sur les propriétés des helicides. *Bull. Inst. Agron. Stations Recherches Gembloux*, 26, 281–304.

Van den Drift, J. (1951) Analysis of the animal community in a beech forest floor. *Med. Inst. Toege Biol. Onder. Natuur*, 9, 1–168.

Van Minnen, J. and Sokolove, P. G. (1981) Neurosecretory cells in the central nervous system of the giant garden slug, *Limax maximus*. *J. Neurobiol.*, 12, 297–302.

Van Minnen, J. and Sokolove, P. G. (1984) Galactogen synthesis-stimulating factor in the slug, *Limax maximus*: Cellular localization and partial purification. *Gen. Comp. Endocrinol.*, 54, 114–122.

Van Minnen, J., Wijdenes, J. and Sokolove, P. G. (1983) Endocrine control of galactogen synthesis in the albumen gland of the slug, *Limax maximus*. *Gen. Comp. Endocrinol.*, 49, 307–314.

Van Mol, J.-J. (1961) Étude histologique de la gland cephalique au cours de la croissance chez *Arion rufus* Linne. *Ann. Soc. Roy. Zool. Belg.*, 91, 45–56.

Van Mol, J.-J. (1962) Anatomie et physiologie de la limace rouge (*A. rufus*). *Nat. Belges*, 43, 1–17.

Van Mol, J.-J. (1967) Étude morphologique et phylogenetique du ganglion cerebroide des Gasteropodes Pulmones/Mollusques. *Mem. Acad. Roy. Belg.*, 37, 1–168.

Van Mol, J.-J. (1970) *Révision des Urocyclidae (Mollusca, Gastropoda, Pulmonata). Anatomie, Systématique, Zoogéographie*, Part 1, Musée Royal de l'Afrique Centrale, Tervuren, Belgium.

Van Mol, J.-J., Sheridan, R. and Bouillon, J. (1970) Contribution à l'étude de la glande caudale des Pulmonès Stylommatophores: I. *Arion rufus* (L.): Morphologie, Histologie, Histochimie. *Ann. Soc. Roy. Zool. Belg.*, 100, 61–83.

Vanderburgh, D. J. and Anderson, R. C. (1987a) The relationship between nematodes of the genus *Cosmocercoides* Wilkie 1930 (Nematoda, Cosmocercoidea) in toads *Bufo americanus* and slugs *Deroceras laeve*. *Canad. J. Zool.*, 65, 1650–1661.

Vanderburgh, D. J. and Anderson, R. C. (1987b) Seasonal changes in prevalence and intensity of *Cosmocercoides dukae* (Nematoda, Cosmocercoidea) in

Deroceras laeve (Mollusca). *Canad. J. Zool.*, **65**, 1662–1665.

Vaught, K. C. (1989) *A Classification of the Living Mollusca* (eds R. T. Abbott and K. J. Boss), American Malacologists Inc., Melbourne, Florida, USA.

Verdcourt, B. (1981) Veronicellid slug introduced into Britain. *The Conchologist's Newsheet*, No. 79, December 1981, p. 356.

Verdonk, N. H. and van den Biggelaar, J. A. M. (1983) Early development and the formation of the germ layers. In *The Mollusca*, Vol. 3, *Development* (eds N. H. Verdonk, J. A. M. van den Biggelaar and A. S. Tompa), Academic Press, New York and London, pp. 91–122.

Vianey-Liaud, M. (1975) La variation pondérale au cours de la croissance de la Limace *Milax gagates. Arch. Zool. Exp. Gén.*, **116**, 5–25.

Villella, J. B. (1953) *Retinella indentata* (Say), a first intermediate host of *Entosiphonus thompsoni* Sinitsin (1931) (Trematoda, Brachylaemilidae). *J. Parasitol.*, **39**, 667.

Von Proschwitz, T. (1988a) *Arion lusitanicus* (?) on the Falkland Islands. *J. Conchol.*, **33**, 49.

Von Proschwitz, T. (1988b) The land snail fauna in strongly man-influenced and man-made habitats in Goteborg (south west Sweden) with some remarks to the anthropochorous fauna elements: 1. Open air biotopes. *Malakol. Abha. (Dresden)*, **13**, 143–158.

Voogt, P. A. (1967) Biosynthesis of 3β-sterols in a snail, *Arion rufus* L. from 1-^{14}C-acetate. *Arch. Internat. Physiol. Biochim.*, **75**, 492–500.

Voogt, P. A. (1972) Lipid and steroid components and metabolism in Mollusca. In *Chemical Zoology* (eds M. Florkin and B. T. Scheer), Academic Press, New York and London, pp. 245–300.

Vorbrodt, A. W., Dobrogowska, D. H., Lossinsky, A. S. and Wisniewski, H. M. (1986) Ultrastructural localization of lectin receptors on the luminal and abluminal aspects of brain micro-blood vessels. *J. Histochem. Cytochem.*, **34**, 251–262.

Wadham, M. D. and Wynn Parry, D. (1981) [Damage inflicted by the slug *Agriolimax reticulatus* on rice plants]. *Ann. Bot.*, **48**, 399–402.

Wagner, P. and Roth, J. (1988) Occurrence and distribution of sialic acid residues in developing rate glomerulus: Investigations with the *Limax flavus* and the wheat germ agglutinin. *Euro. J. Cell Biol.*, **47**, 259–269.

Waite, T. A. (1987) Behavioural control of water loss in the terrestrial slug *Deroceras reticulatum* Müller. Body-size constraints. *Veliger*, **30**, 134–137.

Waite, T. A. (1988) Huddling and posture adjustment to desiccating conditions in *Deroceras reticulatum* (Müller). *J. Mollusc. Stud.*, **54**, 249–250.

Waldén, H. W. (1960) On two anthropochorous, terrestrial gastropods, *Limax valentianus* Férussac and *Deroceras caruanae* (Pollonera), found in Sweden, with notes on some further species, living under synanthropous conditions. *Goteborgs Kung. Vet. Vitter Hets-Sam. Hand. Sjatte Foljden Ser. B*, **8**, 5–48.

Waldén, H. W. (1976) A nomenclatural list of the Land Mollusca of the British Isles. *J. Conchol.*, **29**, 21–25.

Waldén, H. W. (1981) Communities and diversity of land molluscs in Scandinavian woodlands. 1. High diversity communities in taluses and boulder slopes in SW Sweden. *J. Conchol.*, **30**, 351–372.

Walker, G. (1969) Studies on digestion of the slug *Agriolimax reticulatus* (Müller) (Mollusca, Pulmonata, Limacidae). PhD thesis, Univ. of Wales.

Walker, G. (1970) Light and electron microscope investigations on the salivary glands of the slug, *Agriolimax reticulatus*. *Protoplasma*, **71**, 111–126.

Walker, G. (1971) The cytology, histochemistry, and ultrastructure of the cell types found in the digestive gland of the slug, *Agriolimax reticulatus*. *Protoplasma*, **71**, 91–109.

Walker, G. (1972) The digestive system of the slug, *Agriolimax reticulatus* (Müller): Experiments on phagocytosis and nutrient absorption. *Proc. Malacol. Soc. London*, **40**, 33–43.

Wallace, G. D. and Rosen, L. (1969) Studies on eosinophilic meningitis: V. Molluscan hosts of *Angiostrongylus cantonensis* on Pacific Islands. *Amer. J. Trop. Med. Hygiene*, **18**, 206–216.

Wallis, D. I. and Wright, B. R. (1971) The tactile sense of the tentacles of the common slug *Arion ater* L. *J. Physiol. (London)*, **213**, 8.

Wallwork, J. A. (1975) Calorimetric studies on soil invertebrates and their ecological significance. In *Progress in Soil Zoology* (ed. J. Vanek.), W. Junk, pp. 231–240.

Wäreborn, I. (1969) Land molluscs and their environments in an oligotrophic area in southern Sweden. *Oikos*, **20**, 461–479.

Wareing, D. R. (1982) Activity and potato food preferences in the slug *Deroceras reticulatum*. MSc thesis, Univ. of Manchester.

Wareing, D. R. (1986) Directional trail following in *Deroceras reticulatum*. *J. Mollusc. Stud.*, **52**, 256–258.

Wareing, D. R. and Bailey, S. E. R. (1985) The effects of steady and cycling temperatures on the activity of the slug *Deroceras reticulatum*. *J. Mollusc. Stud.*, **51**, 257–266.

Wareing, D. R. and Bailey, S. E. R. (1989) Factors affecting slug damage and its control in potato crops. In *Slugs and Snails in World Agriculture* (ed. I. Henderson), British Crop Protection Council, Thornton Heath, pp. 113–120.

Warley, A. P. (1970) Some aspects of the biology, ecology and control of slugs in S. E. Scotland with particular reference to the potato crop. PhD thesis, Univ. of Edinburgh.

Warren, E. (1932) On a ciliate protozoan inhabiting the liver of a slug. *Ann. Natal Museum*, **7**, 1–53.

Watkins, B. and Simkiss, K. (1990) Interactions between soil bacteria and the molluscan alimentary tract. *J. Mollusc. Stud.*, **56**, 275–288.

Wattez, C. (1973) Effet de l'ablation des tentacules oculaires sur la gonade en croissance et en cours de régénération chez *Arion subfuscus* Draparnaud. *Gen. Comp. Endocrinol.*, **21**, 1–8.

Wattez, C. (1975) Effect of repeated injections of ocular tentacle extracts on the growing and regenerating gonad in Arionidae whose tentacles have been removed: Study of *Arion subfuscus* Draparnaud. *Gen. Comp. Endocrinol.*, **27**, 479–487.

Wattez, C. (1976) Rôle des tentacules oculaires dans la différenciation sexuelle d'*Arion subfuscus* Drap. Étude *in vivo* et *in vitro* par la méthode des cultures organotypiques. *Bull. Soc. Zool. France*, **101** (Suppt 4), 96–102.

Wattez, C. (1978) Effect of the cephalic complex (ocular tentacles-brain) on the development, *in vitro* culture, of infantile and juvenile gonads in *Arion subfuscus* Drap. *Gen. Comp. Endocrinol.*, **35**, 360–374.

Wattez, C. (1979) The control of sexual differentiation by the cephalic complex in the slug *Arion subfuscus*. *Malacologia*, **18**, 407–411.

Wattez, C. (1982) The influence of the brain on oocyte growth in the slug *Arion subfuscus*. *Malacologia*, **22**, 125–129.

Wattez, C. and Durchon, M. (1972) Influence des tentacules oculaires dans la différenciation génitale chez *Arion subfuscus* Draparnaud (Mollusque, Gasteropode, Pulmoné). *C. R. Hebdom. Séances Acad. Sci.*, *Serie D. Sciences Naturelles*, **274**, 2328–2331.

Watts, A. H. G. (1952) Spermatogenesis in the slug *Arion subfuscus*. *J. Morphol.*, **91**, 53–78.

Webb, G. R. (1961) The phylogeny of American Land Snails with emphasis on the Polygyridae, Arionidae and Ammonitellidae. *Gastropodia*, **1**, 31–52.

Webb, G. R. (1965) The sexology of three species of limacid slugs. *Gastropodia*, **1**, 53–60.

Webb, J. (1988) Jellied meals. *The Mycologist*, **2**, 65.

Webbe, G. and Lambert, J. D. H. (1983) Plants that kill snails and prospects for disease control. *Nature (London)*, **302**, 754.

Weber, J. M. and Mermod, C. (1985) Quantitative aspects of the life cycle of *Skrjabingylus nasicola*, a parasitic nematode of the frontal sinuses of mustelids. *Z. Parasitenkunde*, **71**, 631–638.

Webley, D. (1962) Experiments with slug baits in South Wales. *Ann. Appl. Biol.*, **50**, 129–136.

Webley, D. (1963) Experiments with slug baits in 1959. *Plant Pathol.*, **12**, 19–20.

Webley, D. (1964) Slug activity in relation to weather. *Ann. Appl. Biol.*, **53**, 407–414.

Webley, D. (1965) Aspects of trapping slugs with metaldehyde and bran. *Ann. Appl. Biol.*, **56**, 37–54.

Webley, D. (1970) Observations on the effects of distribution and number of slug pellets on the catch of slugs. *Ann. Appl. Biol.*, **66**, 347–352.

Wedgwood, M. A. and Bailey, S. E. R. (1986) The analysis of single meals in slugs feeding on molluscicidal baits. *J. Mollusc. Stud.*, **52**, 259–260.

Wedgwood, M. A. and Bailey (1988) The inhibitory effects of the molluscicide metaldehyde on feeding, locomotion and faecal elimination of three pest species of terrestrial slug. *Ann. Appl. Biol.*, **112**, 439–458.

Weiss, J. C. (1973) La lutte anti-limaces: historique et évolution. *Haliotis*, **3**, 107–112.

Weiss, H. E. (1981) Prevalence and significance of salmonellae in the slug *Arion rufus*. *Der Prak. Tierarzt*, **62**, 256–258.

Weiss, M. (1968) Zur embryonales und postembryonalen Entwicklung des Mitteldarmes bei Limaciden und Arioniden (Gastropoda, Pulmonata). *Revue Swiss Zool.*, **75**, 157–225.

Wells, M. J. and Buckley, D. S. K. L. (1972) Snails and trails. *Animal Behav.*, **20**, 345–355.

Wells, R. M. G. (1980) *Invertebrate Respiration*, Edward Arnold, London.

Welsford, I. G., Banta, P. A. and Prior, D. J. (1990) Size dependent responses to dehydration in the terrestrial slug *Limax maximus* L. Locomotor activity and huddling behaviour. *J. Exp. Zool.*, **253**, 229–234.

Welsford, I. G. and Prior, D. J. (1987) The effect of SCP_B application and buccal neuron B1 stimulation on heart activity in the slug *Limax maximus*. *Amer. Zool.*, **27**, 138A.

Welsford, I. G. and Prior, D. J. (1988) Developmental responsiveness of the heart and feeding system of the terrestrial slug *Limax maximus* to SCP$_B$. *Soc. Neurosci. Abstr.*, **14**, 292.

Welsford, I. G. and Prior, D. J. (1989) Modulation of crop and oesophageal activity by the feeding motor programme SCP's and buccal neuron B1 in the terrestrial slug *Limax maximus*. *Soc. Neurosci. Abstr.*, **15**, 737.

Welsford, I. G. and Prior, D. J. (1991) Modulation of heart activity in the terrestrial slug, *Limax maximus*, by the feeding motor programme, small cardioactive peptides and stimulation of buccal neuron B1. *J. Exp. Biol.*, **155**, 1–19.

Welty, L. E., Anderson, R. L., Delaney, R. H. and Hensleigh, P. F. (1981) Glyphosate timing effects on establishment of sod-seeded legumes and grasses. *Agron. J.*, **73**, 813–817.

Wester, R. E., Goth, R. W. and Webb, R. E. (1964) Transmission of downy mildew (*Phytophthora phaseoli*) of lima beans by slugs. *Phytopathology*, **54**, 749 (Abstract).

Wharton, A. L. and Ensor, H. (1969) The slug problem in peas for processing. *Proc. 5th British Insecticide and Fungicide Conf.*, 442–445.

Wheater, C. P. (1989) Prey detection by some predatory Coleoptera (Carabidae and Staphilinidae). *J. Zool. (London)*, **218**, 171–186.

Whelan, R. J. (1982a) An artificial medium for feeding choice experiments with slugs. *J. Appl. Ecol.*, **19**, 89–94.

Whelan, R. J. (1982b) Response of slugs to unacceptable food items. *J. Appl. Ecol.*, **19**, 79–88.

Whitaker, J. O. and French, T. W. (1984) Foods of 6 species of sympatric shrews from New Brunswick (Canada). *Canad. J. Zool.*, **62**, 622–626.

Whitaker, J. O., Cross, S. P. and Maser, C. (1983) Food of vagrant shrews (*Sorex vagrans*) from Grant county, Oregon (USA) as related to livestock grazing pressures. *Northwest Sci.*, **57**, 107–111.

White, A. R. (1959a) Infestation of slug mite (*Riccardoella limacum*) on various species of slugs. *Entomol. Monthly Mag.*, **95**, 14.

White, A. R. (1959b) Observations on slug activity in a Northumberland garden. *Plant Pathol.*, **8**, 62–68.

Wieland, S. J. and Gelperin, A. (1983) Dopamine elicits feeding motor program in *Limax maximus*. *J. Neurosci.*, **3**, 1735–1745.

Wieland, S. J., Jahn, E. and Gelperin, A. (1987) Localization and synthesis of monoamines in regions of *Limax* CNS controlling feeding behaviour. *Comp. Biochem. Physiol. C. Comp. Pharmacol. Toxicol.*, **86**, 125–130.

Wieland, S. J., Jahn, E. and Gelperin, A. (1989) Release of dopamine and serotonin from *Limax* ganglia *in vitro*. *Comp. Biochem. Physiol. C. Comp. Pharmacol. Toxicol.*, **94**, 183–188.

Wijdenes, J., van Minen, J. and Boeer, H. H. (1980) A comparative study on neurosecretion demostrated by the alcian blue–alcian yellow technique in 3 terrestrial pulmonates (Stylommatophora). *Cell Tissue Res.*, **210**, 47–56.

Wijdenes, J. and Runham, N. W. (1976) Studies on the function of the dorsal bodies of *Agriolimax reticulatus* (Mollusca: Pulmonata). *Gen. Comp. Endocrinol.*, **29**, 545–551.

Wijdenes, J. and Runham, N. W. (1977) Studies on the control of growth in

Agriolimax reticulatus. Gen. Comp. Endocrinol., **31**, 154–156.

Wiktor, A. (1958) Biology of feeding in snails. *Przeglad Zool.*, **2**, 125–146.

Wiktor, A. and Likharev, I. M. (1979) Phylogenetische Probleme bei Nackschnecken aus den Familien Limacidae und Milacidae (Gastropoda: Pulmonata). *Malacologia*, **18**, 123–131.

Wiktor, A. and Likharev, I. M. (1980) The pallial complex of Holarctic terrestrial slugs (Pulmonata, Stylommatophora) and its importance for classification. *Zool. Poloniae*, **27**, 409–448.

Wiktor, A. and Norris, A. (1982) The synonomy of *Limax maculatus* (Kaleniczenko, 1851) with notes on its European distribution. *J. Conchol.*, **31**, 75–77.

Wild, S. V. and Lawson, A. K. (1937) Enemies of the land and freshwater Mollusca of the British Isles. *J. Conchol.*, **20**, 351–361.

Williamson, M. H. (1959a) The separation of molluscs from woodland leaf litter. *J. Animal Ecol.*, **28**, 153.

Williamson, M. H. (1959b) Studies on the colour and genetics of the black slug. *Proc. Roy. Soc. Edinb.*, **27**, 87–93.

Windsor, D. A. (1959) *Colpoda steini* and *Tetrahymena limacis* in several terrestrial pulmonate gastropods collected in Illinois. *J. Protozool.*, **6**, 135.

Winfield, A. L., Wardlow, L. R. and Smith, B. F. (1967) Further observations on the susceptibility of maincrop potato cultivars to slug damage. *Plant Pathol.*, **16**, 136–138.

Wolff, H. G. (1969) Einige Ergebnisse zur Ultrastruktur der Statocysten von *Limax maximus, Limax flavus* und *Arion empiricorum* (Pulmonata). *Z. Zellforsch. Mikrosk. Anat.*, **100**, 251–270.

Wolff, H. G. (1970a) Statocystenfunktion bei einigen Landpulmonaten (Gastropoda): Verhaltens- und elektrophysiologische Untersuchungen. *Z. Ver. Physiol.*, **69**, 326–366.

Wolff, H. G. (1970b) Efferente Aktivitaet in den Statenerven einiger Landpulmonaten (Gastropoda). *Z. Ver. Physiol.*, **70**, 401–409.

Wondrak, G. (1968) Elecktronenoptische Untersuchungen der Korperdecke von *Arion rufus* L. (Pulmonata). *Protoplasma*, **66**, 151–171.

Wondrak, G. (1969a) Elecktronenoptische Untersuchungen der Drüsen- und Pigmentyellen aus der Körperdecke von *Arion rufus* (L) (Pulmonata). *Z. Mikrosk. - Anatom. Forsch. (Leipzig)*, **80**, 17–40.

Wondrak, G. (1969b) Die Ultrastruktur der Zellen aus dem Interstitiellen Bindegewebe von *Arion rufus* (L) Pulmonata. *Z. Zellforsch. Mikrosk. Anat.*, **95**, 249–262.

Wondrak, G. (1969c) Die Ultrastruktur der Sohlendrüsenyellen von *Arion rufus* L. *Malacologia*, **9**, 303–305.

Wondrak, G. (1977) The inversion of the chemoreceptor area of the tentacles of *Helix pomatia* L. and *Arion rufus* (L.) (Gastropod, Pulmonata). *Zool. Anzeiger*, **199**, 301–313.

Woodward, B. B. (1913) *The Life of Mollusca*, London.

Worthing, C. R. (ed.) (1983) *The Pesticide Manual*, 7th edn, British Crop Protection Council, Thornton Heath.

Wright, B. R. (1972) Sensory structure and function of *Arion ater* L. tentacles. PhD thesis, Univ. of Wales.

Wright, A. A. (1973) Evaluation of a synthetic diet for the rearing of the slug *Arion ater* L. *Comp. Biochem. Physiol. A. Comp. Physiol.*, **46**, 593–603.

Wright, A. A. and Williams, R. (1980) Effect of molluscicides on the consumption of bait by slugs. *J. Mollusc. Stud.*, **46**, 265–281.

Wright, C. A. (1974) Biochemical and immunological taxonomy of the Mollusca. In *Biochemical and Immunological Taxonomy of Animals* (ed. S. A. Wright), Academic Press, New York and London, pp. 351–385.

Yamamoto, K. (1977) The amphinucleolus of *Limax* oocyte: An electron microscopic study. *J. Nara Medical Ass.*, **28**, 341–352.

Yamane, T., Oestreicher, A. B. and Gelperin, A. (1989) Serotonin stimulated biochemical events in the procerebrum of *Limax*. *Cell Mol. Neurobiol.*, **9**, 447–460.

Yamashita, Y., Jones, R. M. and Nicholson, C. H. L. (1979) Feeding of slugs (*Deroceras* sp. and *Lehmannia nyctelia*) on subtropical pasture species, particularly Kenya white clover (*Trifolium semipilosum*) cultivar Safari. *J. Appl. Ecol.*, **16**, 307–318.

Young, A. G. and Port, G. R. (1989) The effect of microclimate on slug activity in the field. In *Slugs and Snails in World Agriculture* (ed. I. Henderson), British Crop Protection Council, Thornton Heath, pp. 263–269.

Young, A. G. and Port, G. R. (1991) The influence of soil moisture content on the activity of *Deroceras reticulatum* (Müller). *J. Mollusc. Stud.*, **57**, 138–140.

Young, A. G. and Wilkins, R. M. (1989a) The response of invertebrate acetylcholinesterase to molluscicides. In *Slugs and Snails in World Agriculture* (ed. I. Henderson), British Crop Protection Council, Thornton Heath, pp. 121–128.

Young, A. G. and Wilkins, R. M. (1989b) A new technique for assessing the contact toxicity of molluscicides to slugs. *J. Mollusc. Stud.*, **55**, 533–536.

Zeissler, H. (1975) Snails in the Valley of Reiser near Muehlhausen in Thuringia. *Malakol. Abh. (Dresden)*, **4**, 237–244.

Zeissler, H. (1980) Detailed research on the snail faunas in the Eichholz forest near Zwenkau, Leipzig district, East Germany. *Malakol. Abh. (Dresden)*, **6**, 269–300.

Zs-Nagy, I. and Sakharov, D. A. (1969) Axo-somatic synapses in procerebrum of gastropodia. *Experientia*, **25**, 258–259.

Index

Italic page numbers refer to illustrations. Only the more important references in the text are included for common slug species. Entries for *Arion ater, A. hortensis agg., Deroceras reticulatum* and *Limax maximus* are necessarily restricted to a taxonomic reference.